Reading Human Geography

THE POETICS AND POLITICS OF INQUIRY

Edited by
Trevor Barnes
and
Derek Gregory

A member of the Hodder Headline Group
LONDON • NEW YORK • SYDNEY • AUCKLAND

First published in Great Britain in 1997 by
Arnold, a member of the Hodder Headline Group,
338 Euston Road, London NW1 3BH
175 Fifth Avenue, New York, NY10010

Copublished in the US, Central and South America by
John Wiley & Sons, Inc.,
605 Third Avenue
New York, NY10158–0012

British Library Cataloguing in Publication Data
A catalogue record for this book is available from the British Library

Library of Congress Cataloging-in-Publication Data
????

ISBN 0 340 63208 9 (Pb)
ISBN 0 470 23537 3 (Wiley)

ISBN 0 340 63209 7 (Hb)
ISBN 0 470 23538 1 (Wiley)

Composition by J&L Composition Ltd, Filey, North Yorkshire
Printed and bound in Great Britain by J. W. Arrowsmith, Bristol

CONTENTS

ACKNOWLEDGEMENTS

The authors and publishers would like to thank the following for permission to use copyright material in this book:

Basil Blackwell Inc. for Kim England, 'Getting personal: reflexivity, positionality and feminist research', *Professional Geographer* 46 (1994), pages 80–89, David Harvey, 'Between space and time: reflections on the geographical imagination', *Annals of the AAG* 80 (1990), pages 418–34, and Margaret Fitzsimmons, 'The matter of nature', *Antipode* 21 (1989), pages 106–20; Blackwell Publishers for Linda McDowell, 'Understanding diversity: the problem of/for "theory"' from Johnston, Taylor and Watts (eds), *Geographies of Global Change*, pages 280–94, and for the covers of *Thirdspace* by Edward Soja and *Feminism and Geography* by Gillian Rose; Carfax Publishing Company for Audrey Kobayashi and Linda Peake, 'Unnatural discourse: "race" and gender in geography', *Gender, Place and Culture* 1 (1994), pages 225–43; Guilford Press, Inc. for the cover of *Logics of Dislocation* by Trevor Barnes; Harcourt Brace and Company Ltd for David Demeritt, 'Ecology, objectivity and critique in writings on nature and human societies', *Journal of Historical Geography* 20 (1994), pages 22–37; The Johns Hopkins University Press and Macmillan Publishers Ltd for *The Betweenness of Place* by Nicholas Entrikin, pages 6–26 and for the cover; Pion Ltd for the excerpts from David Livingstone, 'The spaces of knowledge: contributions towards a historical geography of science', *Environment and Planning D: Society and Space* 13 (1995), pages 5, and 15–34, David Slater, 'On the borders of social theory: learning from other regions', *Environment and Planning D: Society and Space* 10 (1992), 307–27, Michael Brown, 'Ironies of distance: an ongoing critique of the geographies of AIDS', *Environment and Planning D: Society and Space* 13 (1995), pages 159–83, Nigel Thrift, 'On the determination of social action', *Environment and Planning D: Society and Space* 1 (1983), pages 23–58, Gerry Pratt, 'Spatial metaphors and speaking positions', *Environment and Planning D: Society and Space* 10 (1992), pages 241–44, Cathy Nesmith and Sarah Radcliffe, 'Remapping Mother Earth: a geographical perspective on environmental feminism', *Environment and Planning D: Society and Space* 11 (1993), pages 79–94, and for the cover of *Birds in Egg* by Gunnar Olsson; Polity Press & University of Minnesota Press for *Feminism and Geo-*

graphy by Gillian Rose, pages 86–198; Routledge for David Ley, 'Fragmentation, coherence and the limits to theory in human geography', from Kobayashi and Mackenzie (eds), *Remaking Geography* (Unwin Hyman, 1989), pages 227–44, Andrew Sayer, 'Realism and geography' and Stephen Daniels, 'Arguments for a humanistic geography', both from Johnston (ed.), *The future of geography* (Methuen, 1985), pages 159–73 and 143–58 respectively, Felix Driver, 'Bodies in space: Foucault's account of disciplinary power' from Jones & Porter (eds), *Reassessing Foucault: Power, Medicine and the Body*, pages 113–23, and for the cover of *Mapping the Subject* edited by Steve Pile and Nigel Thrift; Royal Geographical Society with the Institute of British Geographers for Denis Cosgrove, 'Prospect, perspective and the evolution of the landscape idea', *Transactions, Institute of British Geographers* 10 (1985), pages 45–62 and Stephen Pile, 'Human agency and human geography revisited: a critique of "new models" of the self', *Transactions, Institute of British Geographers* 18 (1993), pages 122–139; Scandinavian University Press for 'Mapping meaning, denoting difference, imagining identity' by Michael Watts from *Geografiska Annaler Series B*, 73, pages 7–16; School of Geography, University of Oxford for the excerpts from David Harvey and Allen Scott, 'The practice of human geography: theory and empirical specificity in the transition from Fordism to flexible accumulation' from Macmillan (ed.), *Remodelling Geography* (1989), pages 220–26; University of California Press for the cover of *Traces on the Rhodian Shore* by Clarence J. Glacken; University of Minnesota Press for *Lines of power/limits of language* by Gunnar Olsson (1991), pages 167–181; University of Toronto Press for Brian Harley, 'Deconstructing the Map', *Cartographica* 26 (1989), pages 1–20.

INTRODUCTION

Reading Human Geography: The Poetics and Politics of Enquiry

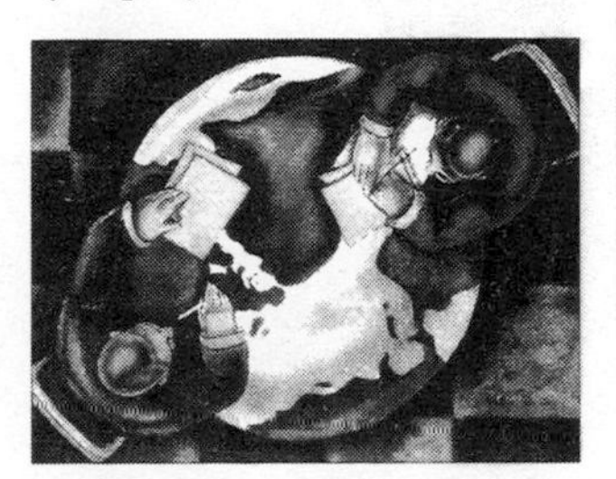

READING HUMAN GEOGRAPHY: THE POETICS AND POLITICS OF INQUIRY

Lost in the post? Human geography at the end of the 20th century

The geographical experiment has always been a contested and problematic enterprise, and its projects have varied so much over time and space that it is more accurate to speak in the plural. What most of us too readily treat as a universal discipline – a sort of 'Geography-with-a-capital-G' – is really only one sedimented and situated product of a series of intersecting historical geographies and colliding geographical experiments. But in the middle decades of the 20th century, particularly after the end of the Second World War, a local geography emerged in Europe (especially in Great Britain and Sweden) and in North America (especially in the USA and Canada) that not only commanded an unusual degree of internal agreement but also swept outwards to colonise other cultures of geographical inquiry. This self-styled 'new geography' was by no means unchallenged, and we do not mean to minimise the varied counter-reactions that it provoked: some were rearguard actions fought by traditionalists, but others were much more creative interventions that sought to establish a radically different philosophy for geographical inquiry. But we do think that the hegemony of spatial science, as this new geography came to be called, nonetheless involved an unusually high degree of consensus. Now consensus does not imply stability, and the shifting and accelerating development of spatial science continued to convulse both teaching and research for 20 years or more. But in general we suggest that its projects involved tacit agreement to reconstruct geography as a discipline modelled on a particular conception of *science* that in its turn required a distinctive constellation of *objectivity* and *truth*.

This constellation can be described in several different ways, but it had three strategic dimensions. First, it instilled a remarkable degree of certainty about the power of its empirical observations and its analytical methods: there was thus nothing problematic about its representations of the world, so that spatial science could free human geography from its burdensome baggage of artful subjectivities. Second, it conducted a systematic search for a hidden order underlying the endless differences of the world: coherences could be detected at the level of either pattern ('spatial form') or generative mechanism ('spatial process'), and spatial science could thus transcend the parochial limitations of traditional, descriptive geography and disclose a fundamental spatial organisation to the world. Third, it treated inquiry as inherently cumulative, not in the traditional geographical sense of compiling an exhaustive inventory of the globe but in the deeper sense of integrating individual discoveries into a single and systematic science of the spatial.

These dimensions can be summarised as the three C's: certainty, coherence, and cumulation. In their most general form they were by no means confined to geography, and the philosophy to which they appealed – positivism – had a considerable impact on many other disciplines too. But the construction of a positivist *geography* was particularly revolutionary for two main reasons. Traditional geographies relied on considerably less formal methodologies, whereas spatial science imposed new ideals of rigour and proof that introduced a much more explicit theorisation to geographical inquiry. And traditional geographies were bedevilled by the uneasy coexistence of physical geography and human geography, one drawn to the natural sciences and the other drawn to the humanities and social sciences, whereas spatial science was predicated on a common methodology ('the' scientific method) directed towards a common object (spatial organisation).

The readings that follow illustrate some of the most important ways in which this consensus has been dissolved by charting the emergence of a series of *post-positivist geographies*. There is no shortage of 'posts' in the late-20th-century academy, of course, and we certainly don't mean to erase the differences between them. There are substantial tensions within and between the various versions of postmodernism, post-structuralism, and post-colonialism that have commanded critical attention in recent years. But what they have in common is a vigorous rejection of the dominant constellation of objectivity and truth that we described as positivism (Gregory, 1994).

We suspect that most human geographers are now much less certain about their claims to truth than they were 10 years ago: they are much more willing to admit the partialities of their local knowledges, much more prepared to experiment with different strategies of representation and analysis, and much more attentive to the multiple voices of the people they seek to study. Most of them are also much more reluctant to advertise their accounts as closed and complete: instead of tying up all the loose ends and weaving a seamless and regular spatial pattern, they are much more likely to draw attention to the dangling threads and dropped stitches, to the tears and holes in their geographical fabrications. And many of them have probably abandoned a cumulative narrative of 'progress in geography': instead of fastening on a single and self-sufficient template for geographical inquiry, they are much more likely to inhabit the tense and creative spaces between different and dissonant theoretical traditions.

Unlike some commentators, however, we don't think that these changes have provoked a crisis, still less that geography is 'lost in the post(s)'. Not only are there what Gregory *et al.* (1994) identify as 'signals in the noise' – in other words, a number of common thematics have emerged in the contemporary refiguration of the Western academy – but there are also several continuities with the advances made under the sign of positivism. Ironically, the development of post-positivist geographies has reaffirmed the importance of theorisation – although 'theory' is now understood in a radically different way from the objectivist formulations of spatial science – and it has extended and even deepened the conversation between human and physical geography by reconfiguring 'science' as a fully social practice. But these are continuities in a different register, which find common ground in the realisation that the positivist constellation of objectivity and truth – of 'knowledge' – is better grasped as a constellation of *power* and knowledge. Although we have borrowed that phrase from the French philosopher Michel Foucault – who was in fact extraordinarily attentive to questions of space – we don't want to limit it to his own uses. But we do think that an important triangle is formed by *power, knowledge* and *geography*, and one that is symbolised by the painting that appears on the cover of our book. There is the power of two white men in business suits seemingly dividing up the world between them, the knowledge necessary for the task marked by the reading, note-taking and animated exchange of information, and the geography represented by the table in the shape of the globe. None of these are innocent terms and all of them can be challenged: what other modalities of power ought to be at work? What sort of knowledge is being produced, and what is excluded from its agenda? What kind of geography is being invoked, and what alternatives can be envisaged? Our basic argument is that to make sense of what is going on at this table – and at countless other, similar tables – it is necessary to examine the ways in which both poetics and politics are implicated in the construction of the triangle between power, knowledge and geography. We now want to explain what we mean by considering the two parts of our subtitle in more detail.

Poetics and geographical inquiry

It might seem odd that we foreground 'poetics'. Some geographers read and write poetry during their off-hours, and occasionally use it in their books and articles, but there is nothing conventionally poetic about most contemporary geographical writing. Some critics would say that post-positivist geographers are among the least poetic of all: that their prose is all too often stilted, leaden, inflated, and opaque. Although not wanting to defend poor writing, we think there can be good reasons for not always aspiring to graceful and lucid prose, and these revolve around the central concerns of poetics.

By poetics we have in mind the interpretation provided by the anthropologist and cultural critic James Clifford, and we took our subtitle from from his jointly edited book: *Writing culture: The poetics and politics of ethnography*. For Clifford, all ethnographic accounts – and, we would argue, geographical ones too – are rhetorical constructions, textual artefacts that seek to persuade us of their claims through an amalgam of 'academic' and 'literary' genres. But such accounts are not just texts; they also reach out to wider contexts of power and resistance, institutional constraint and innovation. For Clifford, and for us, it is this tension between text and context – or, if you prefer, between 'words' and 'worlds' – that calls for a careful appreciation of poetics.

Poetics is thus a critical practice that involves taking into account the force, exactness, and power of words themselves: 'Th' artillery of words', as Jonathan Swift put it in one of his own poems. Using words is a serious business and can have the most starkly material of consequences. The traditional children's rhyme is quite misleading. Sticks and stones are dangerous, of course, but so too are words: they can *all* break bones.

The first point about the poetics of inquiry, then, is the realisation that there is nothing 'mere' about a word. Words have the ability not only to represent but also to create worlds, to offer possibilities, to produce action. When Marx and Engels wrote in the *Communist Manifesto* 'Workers of the world, unite!', or when Martin Luther King said at the Washington Monument 'I have a dream . . . .', or when Margaret Thatcher declared 'There is no such thing as society', these were not 'mere words'. They had enormous practical consequences: they were intimately implicated in the Bolshevik Revolution in Russia, the political mobilisation of African-Americans and the struggle for civil rights in the USA, and the onward march of Thatcherism in the United Kingdom. Words, as poets and contributors to this volume know very well, are extraordinarily powerful. We need to use them with care, with sensitivity, and above all with a critical passion: a poetics of inquiry is thus also a politics of inquiry.

Words are also more than labels that get stuck to things – 'the mat', 'the cat' – but take on a force of their own that cannot be reduced to their representational function. When the poet Philip Larkin writes in *Dockery and Son* that 'Life is first boredom, then fear', or when W. B. Yeats writes in *The Second Coming*, 'Things fall apart; the centre cannot hold', we cannot reduce the individual words in those lines to literal, single meaings without undermining the power of the poems themselves. Words take on a non-representational role; they take on a meaning and a persuasive force even though they cannot be directly matched with objects in the world.

Such a non-representational view of language contradicts the clarity and certainty required by positivism, which typically conceives of language as transparent and unproblematic. From that perspective, language is always tethered to the real, and so the meaning of words is unambiguous and dependable. Words are supposed to be simply the verbal counterparts of the things they represent. But over the past 50 years philosophers have increasingly sided with the poets. Critics as different as Wittgenstein and Derrida have argued that the meanings of words are not self-evident and stable but often complex and mobile. It is extraordinarily difficult to police language, to confine words into single cells: so, for example, when Thomas Hobbes argued that metaphors ought to be removed from the language of politics because their meanings were capricious – and hence spurs to dissent – he made the glorious slip of advancing these claims in a text whose very title – *Leviathan* – made use of a metaphor of a monster to dramatise the state he was supposed to be defending. In short, we neither necessarily say what we mean, nor mean what we say.

We can see this in all kinds of ways. Words often have different levels of meaning that create overlap and ambiguity. For example, when Yeats wrote that 'the centre cannot hold' there is no single, compelling interpretation of his meaning. He could have meant that the moral centre of us as human beings is on the slide; or, then again, he could have been referring to the dislocation of Western Europe following the end of the First World War – he was writing in 1919 – and presciently anticipating the rise of European fascism; or, yet again, he could have had in mind the increasingly oppressive and violent measures carried out by Britain in order to maintain its hold over his native Ireland. Indeed, all of these interpretations have been plausibly argued. The significant issue is not which one is 'right', however, but that there are multiple interpretations. Even if Yeats had in mind a single meaning for the phrase, he could not determine its subsequent interpretations and reinscriptions. For words and their meanings quickly escape their authors. More than 60 years after Yeats wrote that line it was reused by the US geographer Edward Soja in what became a celebrated paper about postmodern Los Angeles. Yeats may have been prescient about some things, but he would have been hard-pressed to predict – let alone prevent – appropriations such as this!

Just as authors cannot direct the subsequent interpretation of their words, neither can they orchestrate the interpretations of their silences. For what is not said is often as meaningful – and sometimes as contentious – as what is said. Psychoanalytical theory makes us aware that much of what is most significant is never consciously articulated and that the gaps and erasures are often pregnant with meaning. In human geography the controversy surrounding David Harvey's critique of *The condition of postmodernity* provides a particularly vivid example. In this book Harvey offered an often brilliant geographical analysis of the Western articulations of late-20th-century economy and culture; but he was virtually silent about questions of gender and sexuality. Several critics claimed that this was, in large part, a consequence of the very language in which Harvey conducted his argument. So, for example, Rosalyn Deutsche argued that his use of visual metaphors produced a conceptual space in which Harvey appeared as an intellectual voyeur who gazes upon the world – itself conceived as a space rendered more or less transparent by the analytical power of the gaze – but on whom that world is never allowed to look in return: in short, Harvey's language both produces and reveals his own masculinist subject position. It is thus scarcely surprising, so Deutsche concluded, that Harvey should have failed to attend to the radical and insurgent critique of feminism.

The second point about the poetics of inquiry, then, is that we must be aware of the uncertainties and capriciousness of words, as well as the silences inbetween them. This is not to say that we must redouble our vigilance and so somehow make language transparent. Our awareness will always be partial and precarious. Neither is it to imply that we must struggle to straighten out those ambiguities and fill in those gaps. This too would be to revert to the positivist hankering after a vocabulary that is translucent. Rather, we must deal with language as it is – warts and all – and work *with* it. The result might not be poetry: but it will make us aware of poetics.

There is one final point about attending to the poetics of geographical inquiry: it takes time and effort. As we have said, writing cannot consist of automatically lining up words in the correct sequence to mimic the world, and for the same reason the practice of reading cannot involve mechanically matching ink marks to real-world objects. Both tasks are more complicated; they require dedicated practitioners: writers who try to be aware of what is they are *doing* when they write, and readers who are willing to make the effort to arrive at a critical understanding of the text. After all, none of the contributors to this volume are stupid, and they have good reasons for writing as they do. Dedicated readers, we suggest, are those who, when they encounter tough prose, continue chewing rather than spitting it out. The novelist Arthur Koestler once said that James Joyce's *Finnegan's Wake* was the perfect book for the perfect reader: Joyce gives you just enough words to make the connections necessary to understand, but it always remains a struggle. We are not suggesting that writers take the 20-odd years that it took Joyce to complete *Finnegan's Wake*, or readers the lifetime some have taken to interpret it. But this does show that, once you start to take poetics seriously, neither writing nor reading become any quicker or any easier: only much more revealing and much more rewarding.

What all this implies, we suggest, is that some of the most valuable lessons you can learn in human geography – as in the other humanities and social sciences – are how to read and how to write. Reading isn't about skimming through a text so that you can summarise its main points; it's about getting inside an argument and worrying away at its construction and consequences. And writing isn't about taking notes or dashing off an essay; it's about releasing and taking responsibility for the tremendous power of words.

Politics and geographical inquiry

This brings us directly to the politics of geographical inquiry. Positivist geography placed a premium on neutrality. Spatial science was supposed to be an 'innocent science' whose findings were neither contaminated by prejudice nor motivated by partisan interests. But some of the very first critiques of spatial science questioned these assumptions. From humanistic geography came the claim that *all* interpretation moves in a hermeneutic circle: that the very condition of our ability to understand one another is a framework of 'prejudgements' which are interrogated and reworked in the course of translation. From radical geography came the claim that *all* inquiries are 'interested': that spatial science was a form of knowledge that tacitly normalised the world of white, middle-class, men. These two counter-suggestions flowed into the development of an avowedly critical human geography that depends on a careful mapping of the circuits between knowledge and power.

Seen like this, politics is not something that – depending on one's inclinations and preferences – might be added on to human geography, a view which issues in 'applied geography' as an optional extra to purely academic inquiry. Human geography has always had practical implications, of course. It has played an important part in the logistics and planning of all sorts of military, state, and corporate adventures, and its more radical practitioners have been involved in (for example) agricultural reform, urban rehousing projects, and public health strategies in the Third World. The public and private institutions that frame geographical inquiry clearly have an impact on its research and teaching priorities too. The myth of the 'ivory tower' has long been discredited, and the turbulent politics of public policy and the powerful interests of academic institutions, commercial publishers, and corporate clients all have a profound and continuing affect on every discipline. It would thus be unreal to legislate about the 'nature' or 'spirit and purpose' of geography from a narrowly philosophical bench.

Human geographers, like many other scholars in the humanities and social sciences, have had an extraordinarily deferential attitude towards philosophers and their ability to adjudicate between competing claims and practices; but if those imperial claims have now been much reduced, as we think they have, this does not mean that we no longer have any need for philosophical reflection. On the contrary, what we need – desperately – is a conversation with (and not a course of instruction from) a different kind of philosophy. As we have already suggested, spatial science was codified by an appeal to positivism as a particular philosophy of science. But we now recognise that there are several other philosophies that ground the conduct of scientific inquiry much more directly in the practices of social life: as countless working scientists have argued, positivism is *not* synonymous with 'the' scientific method, which turns out to be much more variable, much more interpretative, and much more creative than such a rigid philosophy would allow. We know too that philosophy speaks to more than just 'Science', and that there are many other philosophies – political philosophies, moral philosophies, aesthetic philosophies – which can help to enlarge our geographical imaginations and sensitise us to the multiple ways in which claims to knowledge are always implicated in claims about power.

This matters so much because critical human geography *from the very start* insists that its inquiries are deeply political. This is so partly because its assumptions, concepts, and methods are inextricably interwined with grids of power – they are not innocent constructions, floating free – but also partly because the objectives of such a human geography are by no means confined to the *explanatory-diagnostic*: they also involve what critical theorists call the *anticipatory-utopian*. To be sure, the geographies that are described in these pages register a series of analytical claims about the world; many of these turn on the ways in which the relations of power that inhere within capitalism, colonialism, patriarchy, racism, homophobia, and other structures of discrimination and disadvantage are reproduced through the social production of nature and space, of places and landscapes. But it is not enough to leave matters there. A critical human geography must not only expose and elucidate these socio-spatial processes, it must also challenge their legitimacy. Whatever explanatory-diagnostic power it might possess, therefore, 'theory' is also an imaginative capacity to reconfigure the world and our place in it: to foreshadow a different human geography. We can hope that our inquiries will make a difference to the lives of other people – that they can enter into the making of a genuinely human geography that no longer trades on exploitation and erasure – but at the very least they should make a difference to our own being in the world. For thinking hard about questions of power, justice, and equity ought not to become an abstract or instrumental exercise: it ought to affect how we are in our dealings with others.

For these reasons we think that one of the most important responsibilities of a truly critical human geography is to expose what we might call the *taken-for-grantedness* of everyday life. This means showing how the worlds which we inhabit are the products of processes operating over varying time-scales *whose outcomes could have been different*: there is thus nothing inevitable about the situations in which we find ourselves. It also means showing how the worlds which we inhabit are the products of processes operating over varying geographical scales *which join our lives to those of countless others*: there is nothing isolated about the situations in which we find ourselves. To show all this, in detail and in concrete, is no easy task, and because such a project will call into question many established understandings and explanations it requires, as a necessary moment, the sort of critical reflections about geographical inquiry that occupy the following pages. We hope it will become clear that, even though many of the readings that we have chosen might *seem* unduly abstract – remote from mundane concerns and ordinary problems – they nonetheless enter very deeply into the production and transformation of everyday life.

It has been said that Marx's words are like bats, because if you read them attentively and seriously you can see both birds and mice in them. This is probably the most economical description of the connections we have tried to urge between power, knowledge, and geography, between the poetics and the politics of geographical inquiry. Like mice, our studies still keep us at ground level; but we can also hope that the power of our words might also help us to fly.

A user's guide

We want to emphasise that most of the essays in this book were not written primarily with students in mind: so if you find them difficult it is not because you are unusually dim!

The authors sometimes use specialised vocabularies (we have tried to help here: the first time we use a technical term in any of our own essays we mark it in **bold** to refer you to the Glossary at the back of the book). The essays also engage in an artful deployment of references and notes, they make reference to other debates, and sometimes they are written in what must seem like a set of secret codes. But it is learning those codes that is the basis of contemporary geographical inquiry. For whatever else human geographers do for a living, they also *write*: most importantly, we write lecture notes to deliver to our students, and books, articles, and essays that we also address to one another. All the writings that are found in this volume, then, are part of a wider public process within the academy. The sociologist of science Bruno Latour argued that success in academia – as anywhere else – requires allies, and that within the academy this is attained by writing in particular kinds of ways: by citing some people favourably and not others, by highlighting one debate and ignoring another, by inventing new terms and criticising old ones. All these strategies, as well as a host of others, are found in the readings below: they are the basis of the secret codes. But like all codes they take time to learn and decipher.

There is another reason why the articles might seem difficult. Writing is not only a public act but also a very private one. Everyone who writes – not just students! – struggles with the same problem of confronting a blank page (or screen). Finding the right words, ordering them correctly, and organising the argument are all demanding tasks. This is especially so when, as is often the case in the humanities and the social sciences, authors are still working out their argument: when they are thinking *through* the process of writing and not simply recording something they have already worked out in their heads. In any case, nobody knows in advance how things will work out: virtually all academic writing is 'refereed' – it is scrutinised by critical readers nominated by an editor or a publisher before it is accepted – so authors don't know whether their manuscript will pass that test. And – whatever they hope or fear – they have no idea whether their contribution will be widely cited or quickly forgotten, whether those citing it will be allies or critics, and whether it will end up in a reader in human geography. In other words, even though the essays in this collection might give the appearance of being neat and tidy – Latour uses the term 'black box' to describe this – at the time of their production, all of them were moments in an unfinished process. Unlike most textbook presentations that tidy away all the loose ends – and so present human geography as a special sort of black box – we think it more instructive to read authors while they are still at work: to read the original essays rather than the textbook summaries and, in doing so, to recognise that writing is a *process* – it doesn't come preformed.

So how did we choose the essays? Clearly, we were subject to various constraints. Some of them were imposed by the publisher (and here it is important to note the part played by the publishing industry – and especially its commissioning editors – in helping to form and transform the discipline: geography is not the product of geographers alone). The most demanding requirement was that, for economic reasons, we keep the book to around 200 000 words. We could have exceeded that figure many times over and still not included all the articles we wanted; we continued to agonise over our choices until the last possible moment, editing as much as we dared so as to include as many essays as we could.

In making those choices we used a number of criteria. First, we wanted all the selections to be by practicing geographers. This could be interpreted as either disciplinary chauvinism or insularity, but we intend neither of these inferences to be drawn: we know that much very good geography is done by those outside the formal confines of the discipline, and we know too that geography has been shaped by a large number of writers in other fields. These contributions are by no means absent from the essays that follow; but imposing this requirement on our *authors* made our task more manageable and also helped to ensure that the essays are more directly accessible to our would-be readers. We wanted students who read *Reading human geography* to do just that.

A second criterion was the need to be comprehensive in our coverage. In making our selction we wanted to identify some of the most important sites at which human geographers had done their work, which is why we have avoided structuring the book around a series of philosophical -isms and -ologies. We think geographical inquiry is demanding enough, without having to submit to the organising logic of other disciplines. Our most stubborn difficulty was then, the uneven distribution of geographical inquiry: in some areas it was reasonably straightforward to identify a cluster of key articles, but in others there are just too many, and we realise that there are probably as many ways of organising a book like this as there are human geographers This is *a* reader in human geography, not *the* reader in human geography.

Third, we decided to include no more than one single-authored article by the same author. One of the most emancipatory implications of post-positivism is its pluralism: even though our selections are limited to the English-speaking world (which is an extremely serious limitation) we have tried to include as many different authors as possible. This also explains why we have included none of our own past writings: apart from anything else, our editorial introductions probably already make our own voices too loud, and we realise that our selections also very much reflect who we are. So, in working with this selection, we would like you to think critically about not just the individual readings, but also about the structure in which we have placed them.

Finally, a word about the organisation of the book. It is divided into eight sections, each of which is prefaced by an editorial introduction. Each introduction is intended as an outline discussion of the section's major themes; it provides a thumbnail sketch of the larger intellectual context and the recent history of the debates, a summary of the selections that follow, and a guide to further reading.

The first three sections are concerned with various issues to do with theorising in human geography. 'Worlding Geography' is about how we enter into the theorising of geographical worlds; 'Grand Theory and Geographical Practice' illustrates a decade of changing styles of geographical theorisation; and 'Textuality and Human Geography' bears upon the difficulties of reading and writing under the sign of a theoretically informed human geography.

The three sections that follow consider some of the most important geographical concepts elaborated from a post-positivist position. 'The Politics of Nature' highlights the different senses in which 'nature' may be regarded as socially constructed; 'Space, Spatiality and Spatial Structure' charts the way in which 'space' may also be regarded as socially constructed; and 'Place and Landscape' uses those two ideas to reflect upon both the old and the new cultural geography.

The last two sections stress the complex relationships among these key concepts and the production of a genuinely human geography. 'Agency, Subjectivity and Human Geography' indicates some of the ways in which conceptions of agency and subjectivity have been transformed by the passage from a humanist to a post-structuralist geography; 'Geography and Difference' considers some of the ways in which geography is intimately involved in the variety of the human condition.

References

Clifford, James and Marcus, George E. (eds) 1986: *Writing culture: the poetics and politics of ethnography*. Berkeley: University of California Press.

Deutsche, Rosalyn 1991: Boy's town. *Environment and Planning D: Society and Space*, **9**; 5–30.

Gregory, Derek 1994: *Geographical imaginations*. Oxford: Blackwell.

Gregory, Derek, Smith, Graham and Martin, Ron (eds) 1994: *Human geography: society, space and social science*. London: Macmillan.

Harvey, David 1989: *The condition of postmodernity: an enquiry into the origins of cultural change*. Oxford: Blackwell.

Hobbes, Thomas [1652] 1962: *Leviathan*. London: Macmillan.

Koestler, Arthur 1964: *The act of creation*. London: Picador.

Larkin, Philip 1964: Dockerty and Son, *The Whitsun weddings*. London: Faber and Faber.

Latour, Bruno 1987: *Science in action: how to follow scientists and engineers through society*. Cambridge MA: Harvard University Press.

Soja, Edward W. 1986: Taking Los Angeles apart: some fragments of a critical human geography. *Environment and Planning D: Society and Space*, **4**; 455–72.

Yeats, William B. 1950: *Collected poems*. London: Macmillan.

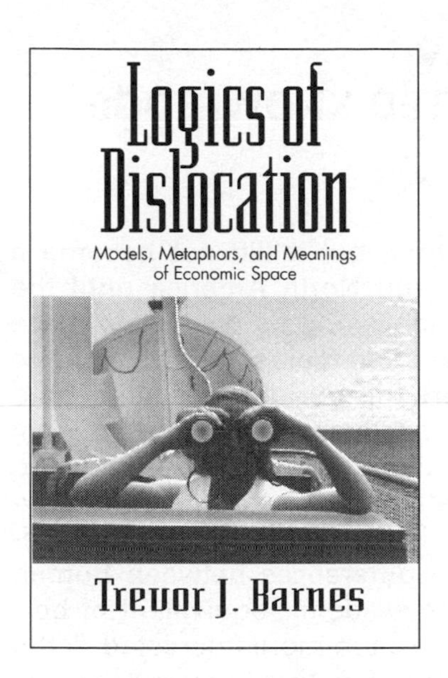

Worlding Geography: Geography as Situated Knowledge

1

WORLDING GEOGRAPHY: GEOGRAPHY AS SITUATED KNOWLEDGE

Crossing the bridge

Geography has a long intellectual history, but it did not become a formal university discipline in Europe and North America until the last decades of the 19th century. In establishing its distinctive place among other disciplines, geography was often represented as a bridge between the arts and the sciences. In part, this position was negotiated through the differentiations between distinctively human and physical geographies. The distance between the two geographies has changed many times: sometimes disappearing altogether, as human and physical geography have shared the same intellectual space, and sometimes increasing dramatically, with the differences between human and physical geography enforced by the academic equivalent of border fences and customs posts. But here we are more interested in the ways in which this bridging or 'mediating' function between the arts and the sciences has reappeared *within the trajectory of human geography itself.* In thinking about the poetics and politics of geographical inquiry (see the Introduction), we suggest that it is helpful to locate human geography within this wider academic division of labour for at least two reasons.

Geography in the ivory tower

First, as a result of their pivotal location between the arts and the sciences, human geographers have been able to draw on both those traditions in order to suggest the *disinterested* nature of their inquiries. Those human geographers who were most attracted to the arts could claim that their attempts to make sense of a place, a region or a landscape were not the impressionistic scribblings of tourists or travel writers: as disciplined scholars they were able to set aside superficial judgements, ignorant opinions, and ingrained prejudices to provide an informed and accurate description of the world. They might rely on a literary style and an artistic sensibility to do so, but neither of these stood in the way of claims to truth. Those human geographers who were most attracted to the sciences could insist that their geographies relied on 'hard facts' which had been established by a properly authenticated scientific method. Their geography was an objective science, in principle no different from physics or geology, and it too could describe and analyse the surface of the earth in a rigorous and systematic manner.

As we have set them out here, these two positions are only ideal types, and in practice human geographers traded on several different versions of them and, more importantly, combined them in different ways. The writings of Paul Vidal de la Blache at the turn of the 19th and

20th centuries provide one of the clearest examples of what we have in mind. One of the principal architects of the French school of regional geography, Vidal's best work has been compared to that of landscape painters and poets for the way in which he artfully evoked a sense of place – of *pays* (region) and *paysage* (landscape) – through the descriptive power of his prose, and the close connections between geography and history in his writings are well known; but Vidal was no stranger to physical geography or the methods of the physical sciences more generally, and he represented his version of regional geography as an unequivocally natural science.

However these two positions were worked out within human geography, common to both of them was an image of the scholar as a person – significantly, almost always a man – who had been elevated above the rest of the population, and who occupied a position from which he could survey the world with a detachment and a clarity that was denied to those closer to the ground (whose vision was supposed to be necessarily limited by their involvement in the mundane tasks of ordinary life). This idea of the scholar as somebody with privileged and panoramic vision has been called the intellectual as surveyor, as hero, and as sovereign. These three **metaphors** carry interesting implications about the connections between **power** and knowledge and their particular bearing on Western geographical inquiry. The *surveyor* speaks to the discipline's interests in mapping landscapes, making the world visible in specified ways, and allocating everything to its proper place; the *hero* speaks to geography's involvement in adventure, exploration, and discovery, travelling across oceans and penetrating to the heart of continents hitherto unknown to Western audiences and bringing back exotic tales redolent of danger and desire; and the *sovereign* speaks to geography's identification with the nation state, enlisting its services in the administration and defence of territory.

In thinking about these metaphors it is of course important to realise that they also have a more general meaning and that they are by no means confined to geography. On the contrary (and the second reason why all this matters), the image of privileged and panoramic vision was part and parcel of the idea of the modern university as an 'ivory tower', an elevated vantage point somehow sheltered from all the corruptions and distractions of politics and economy. Within its confines, artistic and scientific traditions were supposed to flourish without having constantly to establish their mundane significance, and a discipline that could mediate between those two traditions – between what the novelist and scientist C. P. Snow once famously identified as the 'two cultures', each with a 'curiously distorted image' of the other that made conversation between them virtually impossible – might be expected to inculcate a special sense of enduring intellectual value, of *progress* in geography, that would not be dependent on the transient contingencies of the market place.

This did not mean that academic inquiry had nothing to say about worldly questions. Indeed, some intellectual historians have argued that geography was admitted to the modern university precisely because it was useful. In effect, geography redeemed the promises of all three of the metaphors we identified: it helped to provide a logistical basis for modern warfare, to foster a sense of national and imperial identity, to administer colonies, and to compile resource inventories for commercial exploitation (Capel, 1981; Hudson, 1977). But these strategic functions were supposed to involve the application of an otherwise neutral and detached body of knowledge to practical questions which arose *outside* the academy, and to all intents and purposes the clinical separation between human geography and political practice was scrupulously maintained. To be sure, many prominent geographers were political activists: we think, for example, of men like Halford Mackinder in Great Britain, or Isaiah Bowman in the United States. But in most cases geography was supposed to be constructed quite independently of politics and *then* brought to bear on worldly questions. This sense of separation was often reinforced by appeals to Science-with-a-capital-S, in which case geography was underwritten by philosophies of science such as (logical) **positivism** which insisted on value-free inquiry: it was then simply a betrayal of scholarly integrity to stray from the path of the disinterested pursuit of knowledge and become partisan.

Seen like this, the image of the university as an 'ivory tower' was, for much of the 20th century, neither the criticism nor the complaint that it was to become by the end of the century. It was intended, rather, as a guarantee of Truth.

Towards a critical human geography

These ideas have had a long and tenacious hold on geographical inquiry. But they were challenged with a particular force in the late 1960s and 1970s, when there was a sustained call on both sides of the North Atlantic for geography to cast aside its supposed neutrality in order to become 'relevant', 'engaged', 'committed'. The argument for an explicitly critical human geography was not an introspective one; political struggles within the academy over the nature and conduct of university education – over the institutions, ideas, and ideals that shaped public discussion and public policy – were embedded in a much wider matrix of political movements. Many of these campaigns framed their questions about social justice in terms of the politics of class, 'race', and neocolonialism, and their primary concern was a political economy of exploitation and discrimination. The ivory tower was beginning to topple.

As the 1970s turned into the 1980s, however, the radical agenda was itself radically restructured. This was, in part, brought about by what came to be called the 'crisis of Marxism'. Much of the original inspiration for the development of a radical geography had been drawn from Marxism, and although this was never a single stream of unchanging ideas – far from it – the rethinking and reconstruction of **historical materialism** and, in close concert, the development of a **post-Marxism** were given a special impetus by the collapse of Communism in Eastern Europe and the Soviet Union. New forms of democratic politics rose to prominence, and the articulation of environmental politics, cultural politics, and the politics of gender and sexuality in particular made it impossible to contain geographical inquiry within the contours of a newly spatialised political economy. This progressive widening of the critical counter-project was also impelled by the aggressively expanded scope of neoconservatism in both Europe and North America. The rise of the New Right had direct implications for the politics of class, 'race' and ethnicity, gender and sexuality, but it also registered its own demands for the relevance of post-secondary education. Colleges and universities found that their attempts to provide more diverse courses and to foster more inclusive classrooms were resisted as intrusive forms of 'political correctness' by those who championed the privileges and practices of a conservative education; the same institutions were also obliged to respond to increasingly utilitarian demands for skills and training. These developments were bound in to an increased corporatisation of post-secondary education, affecting both the conduct of university affairs and the direction of advanced teaching and research, which further reinforced critical interrogation of the connections between power and knowledge.

This is not the place to provide a detailed account of these profoundly concrete and intricately connected concerns, but in general we suggest that throughout this period the argument for a critical account of the situatedness of human geography was double headed.

First, many scholars claimed that, as a matter of historical record, geography had *never* been neutral or detached. They argued that it had always produced highly selective, partisan views of the world. In one sense, this is hardly surprising: to return to the example of Vidal de la Blache, whether his regional geography is read as an artistic description, a scientific anatomy, or a combination of the two, many scholars now accept that any close reading must take into account the ways in which Vidal's writings were shaped by his commitment to French nationalism and colonialism. But the critics of geography's supposed detachment were making a much stronger claim about the connections between power and knowledge than this. They were suggesting that grids of power are inscribed within the *conceptual structure* of the discipline. Since this is peculiarly difficult to grasp in the abstract – indeed, it runs counter to the way in which most people think of theoretical work – we want to provide two examples to illustrate how these claims work out in practice.

Within the tradition of cultural geography established by the US geographer Carl Sauer in the 1920s and 1930s, for example, 'landscape' was conceptualised as a systematic conjunction of the scenic elements of material culture: seen thus, landscapes had an essential unity and coherence. But later critics objected that this view obscures the ways in which the production of **cultural landscapes** is typically a process of struggle – of competition, contestation, and conflict – so that, seen from a radically different position, the apparent unity and coherence of a landscape becomes an inscription of power and domination. They also complained that the traditional conception of landscape ignored its own visual ideology: that the representation of landscape as a pictorial image, constructed according to the mathematical logic of linear perspective, is not an unproblematically 'natural' way of seeing the world but a highly particular and culturally constructed one. It represents the world as an object-world to be defined, possessed and manipulated in ways that closely resonate with the interests and desires of the property-owning bourgeois male (see Chapters 19 and 20).

Our other example is derived from **spatial science**, which captured the geographical imagination most decisively in the 1960s. One of its central preoccupations was the spatial organisation of the modern capitalist economy, and the so-called '*space-economy*' was conceptualised as an equilibrium structure shaped by the frictions of distance and the disembodied calculus of 'rational economic man'. Here, too, there was a gendered subtext, though scarcely very far from the surface, and it reappeared in displaced form in many different registers of human geography. But this conceptual system also used its geometric order and its generative logic of location – in large measure derived from the physical sciences – to suggest the intrinsic rationality of capitalist production and, by extension, its naturalness. Critics charged that these models served to legitimise a particular economic geography, that they concealed the antagonistic social relations that were enclosed within their spatial templates, and that they obscured the ways in which crisis and contradiction are written into the constitution of capitalist space-economies (see our introduction to Section Five).

We hope that these two examples make it clear that this side of the argument about a critical human geography turned on the intellectual foundations of geographical inquiry, on the ways in which its claims to knowledge are also particular assertions of power, and hence on the *partialities* which are built into the conceptual architecture of geography's supposed objectivity. But the other side of the argument turned on the discipline's ethical foundations. Second, then, many geographers insisted that it was impossible to remain neutral over questions of social justice. They maintained that no philosophy of science could provide a sufficient ground for an academic discipline. They accepted that philosophies of science might be able to legislate about the *methodology* of (some forms of) geographical inquiry – about sampling strategies, hypothesis testing, or statistical modelling, for example – but they argued that moral and political philosophies were essential for the foundation of an *ethics* of geographical practice. This was more than a matter of scruple and discretion; many more conventional forms of intellectual inquiry had explicitly recognised the importance of 'normative' propositions – characterisations of how the world *ought* to be –

but these were usually sharply distinguished from so-called 'positive' propositions about how the world, as a matter of fact, *is*. It was precisely this distinction that was now being called into question. At one level, these arguments drew attention to what several commentators described as 'the myth of a value-free geography'; but at another level they offered a still more radical challenge to the conventional separation between intellectual work and political practice. In effect, they abandoned the model of geography as a body of autonomous knowledge constituted outside the social practices to which it might be applied, in favour of a post-positivist model that represented geography as a **discourse** whose assumptions, concepts, and ways of working were always and everywhere earthed in the material grids of power that structure the social world. Put like that, it was not enough to disclose the values that were tacitly incorporated into geographical inquiry. It was also necessary to explore the social constitution of geography and geographers and to recognise that the discipline was not only *about* the world: it was also *in* and *of* the world.

One of the earliest and most passionate statements of these twin arguments about objectivity and ethics was David Harvey's *Social justice and the city*, published in 1973. In the course of this remarkable book, Harvey moved from a 'liberal' to an explicitly 'socialist' formulation of critical geographical inquiry. One of his fundamental insights was that the production of mainstream geography – and its reproduction through teaching, research, and consultancy – was complicit with the production and reproduction of capitalist society. Spatial science was not scientific at all, Harvey charged, but ideological. Like many critical projects in the 1970s, however, Harvey's critique remained within an opposition between science and **ideology** (cf. Gregory, 1978). In consequence, Harvey and others were still committed to the construction of a recognisably scientific geography whose critical analyses would be capable of disclosing the objective sociospatial structures framing and limiting what Marx once called 'man in the whole wealth of his being.' This project raised at least two key questions: how was it possible to construct a geography that was at once critical *and* objective, and was the subject of this new geography (still) to be quite literally 'man'?

Situations and positions

We believe that the ideas of Donna Haraway (1991) are particularly helpful in thinking through these issues. She begins one of her most influential essays by acknowledging that modern scientific practice often turns on the primacy of sight. Think, for example, of the importance that is routinely attached to 'observation' and 'evidence' (from the Latin *videre*, to see) in establishing claims to knowledge. The discipline of geography is no exception. Indeed, its interest in cartography and graphical display implicates it more deeply in technologies of vision than many, perhaps most, other disciplines. Haraway wants to caution us against what she calls 'the god-trick of seeing everything from nowhere', by which she means advancing claims to knowledge (which are also assertions of power) that imply 'a leap out of the marked body and into a conquering gaze from nowhere.' This is more or less what those traditional conceptions of landscape and conventional models of spatial science achieved (above, pages 15–16): namely, a distanced and separated view into an object world whose 'hidden order' was supposed to be made fully visible (and hence completely knowable) through their optics of inquiry.

Many feminist critics – including Haraway – argue that this is a **masculinist** way of knowing the world, and that it is characteristic of mainstream and many critical formulations. Harvey's subsequent work has been subjected to precisely this objection (Deutsche, 1991; see also Harvey, 1992). Critics counterpose this 'god-trick' to other ways of knowing (and being in) the world that, in contrast, acknowledge the *incompleteness* and *partiality* of their insights. According to these counter-views, all knowledge is produced from somewhere by somebody, and there is no independent position from which one can freely and fully observe the world in all its complex particulars. The production of knowledge always plunges parts of the world into darkness as it illuminates others. This has become something of a commonplace in intellectual history – including the history of geography – where there has been a concerted attempt to identify the cables that anchor texts to contexts. Haraway calls these partial perspectives *situated knowledges*, but what makes her contribution so arresting, we think, is her belief that a recognition of the situatedness of knowledge, far from dissolving all claims to truth into so many local **relativisms**, *is instead the very condition of* ***objectivity***.

As Haraway uses the term, 'objectivity' is about responsibility and affiliation. In the first place, she insists on the importance of loosening our customary reliance on the perspectives of the powerful and prominent, and of learning to trust what she calls 'the vantage points of the subjugated'. Those on the margins, she suggests, are likely to have far more insight into the reach of (and resistance to) power in everyday life: a position with which radical geography has long been in sympathy. But situating our knowledge involves more than delinking ourselves from the uncritical recitation of dominant ideologies and making ourselves accountable to marginalised and disadvantaged populations. For Haraway also draws attention to the impossibility of establishing a grand synthesis in which different points of view are reconciled, and this is much more problematic for those radical traditions that claim to provide a **metanarrative** whose privileged position can subsume all others as secondary. In the second place, therefore, Haraway proposes a deliberate attempt to reach out from particular positions in order to construct webs of connection, lines of flight *whose arguments do not deny difference but which also recognise mutuality*. These strategic encounters in the politico-intellectual landscape correspond to what she calls 'shared conversations' in philosophical discussions about competing claims to knowledge (technically, **epistemology**) and, more prosaically, to solidarity in politics. The continuous construction of these webs of connection, of circuits of conversation and circles of solidarity, affects more than the ways in which we know the world: it has a profoundly moral or politico-ethical dimension because it also impacts on how we *are* in the world. Haraway argues that these attempts to 'reach out' enter into the construction of our own subjectivity:

> The knowing self is partial in all its guises, never finished, whole, simply there and original; it is always constructed and stitched together imperfectly, and *therefore* able to join with another, to see together without claiming to be another. Here is the promise of objectivity: a scientific knower seeks the subject position not of identity but of objectivity; that is, partial connection. There is no way to "be" simultaneously in all, or wholly in any, of the privileged (subjugated) positions structured by gender, race, nation and class . . . . The search for such a "full" and total position is the search for the fetishized perfect subject of oppositional history . . . .'

We think it important to underscore the *partiality* implied by Haraway's argument. It would be a major mistake to assume that acknowledging the situatedness of our knowledge is a prescription for 20:20 vision. To refuse to occupy a space of knowledge that ignores its own particular perspective is certainly an advance over claims to knowledge that dismiss others as the product of 'bias'; but, as Rose (1995a) argues, 'even acknowledging the specificity of a particular perspective' all too often assumes what she calls 'the transparent coherence of perspective itself.' In other words, it is always and everywhere a struggle to grasp our own partiality; our understanding of the situatedness of our knowledge is always itself partial, incomplete, ambiguous and vulnerable.

These suggestions seem particularly relevant to the practice of a geography that is concerned with understanding other people and other places. Seen like this, Haraway suggests, the variable ways in which we position ourselves – in space and in society – carry within them a responsibility for our practices; or, more directly, 'politics and ethics ground struggles for the contests over what may count as rational knowledge.' But these contests are marked by and mirrored within the tracks traced by our knowledges. As Haraway notes, 'The Western eye has fundamentally been a wandering eye, a travelling lens', and its movements have often been instruments of coercion and oppression. For this reason it is vitally important to disclose the bonds between modern geography and the practices of colonialism and imperialism, and to take very seriously indeed the continuing presence of these powers and privileges in our own present. Felix Driver (1992) has drawn upon the work of Edward Said to remind his geographical audience that 'the representation of the Other (places, people, races, gender) is intimately bound up with notions of the self.' But Said is important to Driver's project for other reasons too; his work cautions against assuming any simple, direct, and purely instrumental relationship between text and context. If our texts do not merely 'reflect' the imperatives of the material world – if they are, instead, situated cultural productions that provide a means by which that world is itself constituted – then we need to rethink many of our critical practices and modes of representation. Many critics have noted that **postmodernism**, for all its sensitivity to difference, remains a discourse of the metropolitan West, for example, while others have claimed that **post-colonialism**, for all its attempts to interupt and displace the continuing power of colonial discourse, remains caught in the embrace of high European theory (see Gregory, 1994; Sparke, 1994; Young, 1990; see also our introduction to Section Eight).

Worlding geography

These ideas frame most of the discussions in this book, but the essays in this section address three of the most important areas in which there have been attempts to 'world' (human) geography and to reflect on its situatedness. In thinking about the claims made in these essays, it is important to remember that they are not all cut from the same cloth – they are making different arguments, from different positions – so that the tensions between them are as important as the areas of congruence. But we want to signal some of the ways in which all three can be connected to Haraway's discussion of situated knowledge.

We have already noted that geographers have often appealed to 'science' as a guarantee of truth and objectivity. In the first essay, David Livingstone draws on the political philosophy of science, science studies, and the history of science to sketch out an approach to the situatedness of geographical practice that has several parallels with Haraway's project. In doing so, he effectively problematizes Snow's polemical distinction between 'two cultures' (above, page 13), and suggests – as have others, such as T. S. Kuhn and, for that matter, Haraway – that there are considerable insights to be gained by bringing the methods of the humanities to bear on the practices of the sciences: the two do not travel in hermetically sealed compartments (Bernstein, 1983). In essence, Livingstone suggests that science as it is conventionally understood and practised is a specialised and institutionalised form of what the anthropologist Clifford Geertz calls 'local knowledge'. But this does not make it a parochial practice. The capacity for scientific generalisation, for transcending the boundaries of particular sites of scientific inquiry, can be seen as a negotiated and conditional *achievement* that is affected through capillary networks of power–knowledge. Thus Livingstone (pages 26–7) draws attention to

> the role of the spatial setting in the production of experimental knowledge, the significance of the uneven distribution of scientific information, the diffusion tracks along which scientific ideas and their associated instrumental [apparatus] migrate, the management of laboratory space, the power rela-

> tions exhibited in the transmission of scientific lore from specialist space to public place, the political geography and social topography of scientific subcultures, and the institutionalisation and policing of the sites in which the reproduction of scientific cultures is effected.

This speaks to science in general but also to geography in particular, and it invites us to consider the ways in which 'local geographies' – the theories and explanations we offer from our particular situations and positions – are made to achieve a global purchase. What it does not do, however, at least in Livingstone's work to date, is examine the situatedness of its *own* account: Livingstone does not address the masculinism that looms so large in Haraway's critique, for example, and Gillian Rose (1995b, page 415) objects that, in effect, he offers 'a contextual history which refuses even to try to contextualise itself'.

Livingstone develops his argument in relation to historical geographies of Western science over more than three hundred years, and although he has much to say about the need to situate intellectual inquiry in space, he makes the important point that 'situating ourselves' is also about the need to interrogate the horizons of meaning within which our work takes its historical shape (see also Livingstone, 1993). Still, the question of 'whose history?' is an important one, as Driver (1992) has reminded us. And if, as he also suggests, it is a mistake to write history in such a way that the West is always privileged, always at the summit of its imperial History, then it is equally unacceptable to write geography in such a way that the West is always at the centre of its imperial Geography. In the second essay, David Slater identifies some of the ways in which, in the recent past, geographers in Europe and North America have constructed just such a position of power and privilege. What is unusual about Slater's argument is that it is not directed against the shibboleths of traditional regional geography or spatial science (though he would certainly be critical of their partialities too). Slater's critique is directed against some of the most prominent versions of critical human geography. Although he acknowledges that the conversations between human geography and social theory have socialised the discipline – removed it from the abstract, unchanging, and dehumanised world of spatial science – he worries that the concrete, turbulent, and peopled world that has been promoted in its place is still a highly determinate one. It is pervasively ethnocentric, so Slater suggests, and he identifies three main ways in which **ethnocentrism** – more specifically, 'Euro-Americanism' – has been smuggled into critical human geography. Classical Marxism, critical theory (or, more generally, **Western Marxism**), and postmodernism each provide different ways of privileging the West, but what they have in common, so Slater suggests, is an implicit view of the West as 'a self-contained entity' that can be understood 'of and by itself'. What is more, they tacitly assume that the record of the West is the standard yardstick against which other societies can be measured. These normative assumptions have been carried over into critical human geography, and it is through these three 'lineages of universalism', as Slater calls them, that the 'local geographies' of the West have been made to achieve global purchase. Setting himself against these arrogant assumptions, Slater draws attention to the importance of 'the borders of social theory' and castigates European and North American geographers for their reluctance to 'learn from other regions'. Echoing Haraway, he suggests that 'it is always the marginal or "peripheral" case which reveals that which does not appear immediately visible in what seem to be more "normal" cases.'

Perhaps one of the clearest examples of the worldliness and situatedness of geographical inquiry is fieldwork. According to Carl Sauer, the field was the arena in which the principal training of the geographer ought to take place. Fieldwork was assumed to impose its own disciplines: the rigorous and systematic cartographic survey of the land; the field sketch and the photograph recording the materiality of the landscape; the controlled transect triangulating, quartering, and dissecting the ground; and the meticulous recording of field observations. This disciplinary apparatus – which is itself a technology of vision – is embedded in that sense of geography as an empirical science which David Stoddart (1986) describes with such passion in *On geography and its history*. Given the importance traditionally ascribed to fieldwork, it is strange that there has been so little systematic reflection on its poetics and politics. In recent years, however, informed by parallel projects in anthropology and the insights of both feminist and post-colonial theory, there have been signs of a critical examination of fieldwork in geography. Kim England's essay was one of several contributions to a symposium on 'Women in the field' and in it she considers attempts to open geography 'to voices other than those of white, Western, middle-class, heterosexual men.' Her worry is that this might install a new, more subtle form of intellectual imperialism, in which geographers appropriate the voices of others and claim to speak *for* them as well as about them. Recalling Haraway's insights about our being in the world, England recognises the ways in which fieldwork involves reciprocity between researcher and researched ('the intersubjective nature of social life' and the 'dialogical' nature of inquiry) and insists that the grids of power which enter into those relationships cannot be set aside. Reflecting on her own experience of a research project she never carried out – an investigation of the ways in which lesbian identities are constructed in and through space – England admits that 'the complicated layering and interweaving of power relations between myself, my research assistant and the project became too much for me'. Hence she urges us (page 74) to think more deeply about the 'me' in order to think about the 'other':

> We do not parachute into the field with empty heads and a few pencils or a tape-recorder in our pockets ready to record "the facts" . . . We are differ-

ently positioned subjects with different biographies, we are not dematerialized, disembodied entities.

What this means, she suggests, is a careful consideration of the differences and distances between all those involved in fieldwork and of the ways in which those gaps are marked by power, privilege, and partiality. In short, England urges us to recognise that we conduct fieldwork not 'on the unmediated world of the researched, but on the world *between* ourselves and the researched' (cf. Katz, 1992). In worlding geography, then, we eventually have to acknowledge that in studying the world – in making it yield its secrets – we are also, inevitably, studying ourselves and giving up some of our own secrets.

References

Bernstein, R. 1983: *Beyond objectivism and relativism: Science, hermeneutics and praxis*. Oxford: Blackwell.

Capel, H. 1981: Institutionalization of geography and strategies of change. In Stoddart, D.R. (ed.), *Geography, ideology and social concern*. Oxford: Blackwell, 37–69.

Deutsche, R. 1991: Boys town. *Environment and Planning D: Society and Space* **9**; 5–30.

Driver, F. 1992: Geography's empire: histories of geographical knowledge. *Environment and Planning D: Society and Space* **10**; 23–40.

Gregory, D. 1978: *Ideology, science and human geography*. London: Hutchinson; New York: St Martin's Press.

Gregory, D. 1994: *Geographical imaginations*. Oxford, UK, and Cambridge, MA: Blackwell.

Haraway, D. 1991: Situated knowledges: the science question in feminism and the privilege of partial perspective, in *Simians, cyborgs and women: The reinvention of nature*. London: Routledge 183–201.

Harvey, D. 1973: *Social justice and the city*. London: Arnold.

Harvey, D. 1992: Postmodern morality plays. *Antipode* **24**; 300–26.

Hudson, B. 1977: The New Geography and the New Imperialism, 1870–1918. *Antipode* **9**; 12–19.

Katz, C. 1992: All the world is staged: intellectuals and the projects of ethnography. *Environment and Planning D: Society and Space* **10**; 495–510.

Livingstone, D. 1993: *The geographical tradition: Episodes in the history of a contested enterprise*. Oxford, UK, and Cambridge, MA: Blackwell.

Rose, G. 1995a: Distance, surface, elsewhere: a feminist critique of the space of phallocentric self/knowledge. *Environment and Planning D: Society and Space* **13**; 761-81.

Rose, G. 1995b: Tradition and paternity: same difference? *Transactions of the Institute of British Geographers* **20**; 414–16.

Sparke, M. 1994: White mythologies and anemic geographies: a review. *Environment and Planning D: Society and Space* **12**; 105–19.

Stoddart, D. R. 1986: *On geography and its history*. Oxford: Blackwell.

Young, R. 1990: *White mythologies: writing History and the West*. London: Routledge.

Suggested reading

On the 'ivory tower', see Zygmunt Bauman, *Legislators and interpreters: on modernity, postmodernity and intellectuals* (Cambridge, Polity Press, 1987). The original discussion of situated knowledge is in Haraway (1991). For a lively exchange on the implications of 'science studies' for the history of geography, see Trevor Barnes. 'Whatever happened to the philosophy of science?' *Environment and Planning A*, 25 (1993), pp. 301–4; the commentary by Keith Bassett '"Whatever happened to the philosophy of science?": some comments on Barnes', *Environment and Planning A*, 25 (1993) pp. 337–42; and Barnes's response, 'Five ways to leave your critic: a sociological scientific experiment in replying', *Environment and Planning A*, 26 (1994) pp. 1653–58.

The connections between geography, colonialism and imperialism are explored in Driver (1992), Livingstone (1993), and Anne Godlewska and Neil Smith (eds), *Geography and empire*. Oxford, UK, and Cambridge, MA, Blackwell, 1994). For discussions of postcolonialism and geography, see Derek Gregory and Daniel Clayton (eds), *Colonialism, postcolonialism and the production of space*. (Oxford, UK, and Cambridge, MA, Blackwell, in press).

For further reflections on fieldwork, see Andy Merrifield, 'Situated knowledge through exploration: reflections on Bunge's "Geographical Expeditions"', *Antipode*, 27 (1995), pp. 49–70; Katz (1992); Cindi Katz, 'Playing the field: questions of fieldwork in geography', *Professional Geographer*, 46 (1994), pp. 67–72.

1 David N. Livingstone

The Spaces of Knowledge: Contributions Towards a Historical Geography of Science

Excerpts from: *Environment and Planning D. Society and Space* 13, 5–34 (1995)

The recovery of geographical discourse

In recent years issues of space and place – motifs conventionally allocated to human geography – have come to occupy an increasingly prominent position within the humanities and social sciences. Thereby the discourse of geography, though hardly the discipline of geography, has begun to permeate spheres of intellectual endeavour hitherto seemingly immune, at least in any self-conscious way, to the claims of specificity and spatiality, territoriality and locality, site and situation. It is the concern with precisely issues of this sort that I refer to as geographical discourse – a discourse evidently of wider dimensions than the discipline of geography.[1] All this, of course, is in marked contrast to the classical tradition of social theory in the West whose leading exponents, as David Harvey (1985, page 141) has often reminded us, 'all have this in common: they prioritise time and history over space and geography and, where they treat of the latter at all, tend to view them unproblematically as the stable context or site for historical action'.

For geographers concerned with the history of their own tradition, there is something at once significant and disquieting about this recent twist of events. On the one hand, it is noteworthy that historians of science have begun to take account of the spatial. It is my impression, however, that historical geographers have remained oblivious to much of this work and accordingly I will try to review some of the ways in which spatial currents have begun to be registered within the history of science community. On the other hand, it is disquieting and, I think, ironic that historians of geography have taken so little account of the spatial in the histories they have produced. This is not to say that the 'geography of geography' has been entirely ignored; we have had numerous rehearsals of different national traditions, for example. But it is to suggest that geography's historians have too frequently been content to operate at the level of what Clifford Geertz (1973, page 21) in another context referred to as 'the wall-sized culturescapes of the nation, the epoch, the continent or the civilization'. Geographers, of all people, should surely be aware of the significance of scale in matters spatial.

A geography of science

Something of the character of the geography of science that I want to advocate can be seen from two recent events. First, the British Society for the History of Science organised a conference in March 1994 on the theme 'Making Space: Territorial Themes in the History of Science'. A simple listing of some of the topics will suffice to indicate just how 'spatialised' the history of science enterprise is becoming: the construction of boundaries around or between academic disciplines, laboratories, architectural formations, privileged sites, spatiality and representation (such as cartography, crystallography, stratigraphy), the specificity of knowledge and the local context, the politics of space, imperial and colonial science, the historiography of space, and cultural geographies of science. This list strikes me as presenting an agenda that cultural geographers have pursued in contexts *other than* that of scientific culture, and – I might add – that historians of geography have scarcely gestured towards . . .

Second, the promissory geographical note that this conference offers reflects some of the moves already being made within the history of science community. The 1991 thematic issue of *Science in Context*, devoted to 'The Place of Knowledge: The Spatial Setting and its Relations to the Production of Knowledge' illustrates the significance of this project *and* something of the forms an historical geography of science might assume. The editors of the collection – Adi Ophir and Steven Shapin – introduce their project with a robust rejection of the standard idealist accounts in which 'scientific ideas floated free in the air, as historians gazed up at them in wonder and admiration'. Traditionally, local factors were only accorded any explanatory role insofar as they might explain deviations from universal scientific objectivity. The significance of the spatial in scientific claims, in other words, was that of a sort of locational pathology. Now, as Ophir and Shapin make clear, there has begun to develop what they call 'an influential *localist* genre' derived from a wide range of empirical and theoretical studies – from Durkheimian social topology and Wittgensteinian linguistics to Kuhnian paradigmatic science and Goffman's symbolic interactionism (Ophir and Shapin, 1991, page 3). What it amounts to is a focusing on the places of knowledge and their role in claims about 'the ontological status of scientific objects and the epistemological standing of scientific statements' (page 5). In the hands of such practitioners of the history and sociology of science, this localist turn is given a wholly relativist twist because in large part their epistemological predilections derive from the relativist agenda defended by the so-called Edinburgh group of sociologists of science whose 'strong programme' has persistently analysed the grounds of the local credibility of scientific knowing.[2] But as I have already intimated I do not see that such historicism is an inevitable inference because there are no necessary links between the relativity of warranted credibility and relativism over substantive concepts of truth. Moreover, such philosophical issues are, I think, only indirectly relevant to the historian's task, for as Ophir and Shapin themselves put it, 'attention to place in the history of science emerged partly out of a willingness to ignore the problem of knowledge' (page 15).

Glimmerings of what a geography of scientific knowledge might amount to are thus indeed beginning to be glimpsed, as sociologists and historians of science have begun more explicitly to probe the role of the spatial setting in the production of experimental knowledge, the significance of the uneven distribution of scientific information, the diffusion tracks along which scientific ideas and their associated instrumental gadgetry migrate, the management of laboratory space, the power relations exhibited in the transmission of scientific lore from specialist space to public place, the political geography and social topography of scientific subcultures, and the institutionalisation and policing of the sites in which the reproduction of scientific cultures is effected. The cumulative effect of these investigations is to draw attention to local, regional, and national features of science – an enterprise hitherto regarded as prototypically universal.

I want now to try to earth one or two dimensions of this undertaking, and to begin to map out something of what a historical geography of science might look like, by surveying some recent contributions by historians and sociologists of science.[3] To date, these efforts remain fairly isolated ventures as yet unsystematically developed in terms of a coherent theory of place and space. The taxonomy that I have developed is thus, at best, rudimentary – a first approximation towards the cultivation of a spatialised historiography for science. What I present here, in other words, will be illustrative rather than exhaustive, suggestive rather than comprehensive.

Regionalisation of scientific style

Although it is only recently that the project of regionalising science has begun to bear the marks of a more serious engagement with social–spatial theory, the conviction that scientific styles are regionally differentiated has been long acknowledged by students of the history and sociology of science. In large part this is because of the diverse institutional arrangements within which science has been transacted in different national spaces. But it also reflects the variety of social, political, and religious commitments that characterise different regions. Accordingly it has seemed reasonable to investigate the comparative regional reception of key theories – such as Darwinism or Einsteinian relativity (see Glick, 1974; 1987). In his introductory comments to a collection of essays on the first of these projects, Thomas Glick (1988, page xi) expressed his interest in what he called the 'regional penetration of Darwinism', asking if it was more favourably received in some cities, regions, or disciplines than others. In a new preface to the 1988 reissue of the volume, Glick updated his historiographical reflections by surveying the broadening geographical scope of Darwin reception studies – notably to include China, Japan, and Norway – and noting how in these different places differing philosophical, religious, and social traditions conditioned the Darwinian encounter. In Norway, to just take one example, the prevailing philosophical ethos, religious opposition, and the lateness of the Norwegian translation of *The Origin of Species* (1889 compared with 1871 for Swedish and 1872 for Danish) were important factors influencing the diffusion of Darwinism.

Also at the continental scale of regional survey, the recent collection of essays on *The Rise of Scientific Europe* is described by one of its editors as contributing to the increasingly significant comparative approach among historians of science. The purpose is to compare 'scientific developments in different countries and at different times . . . to discern the effects of varying social, political and intellectual conditions' (Goodman, 1991, page 3). Accordingly, Iberian, Italian, Central European, French, English, and Scottish scientific traditions between the 15th and 17th centuries are the subjects of scrutiny. At the level of intercontinental comparison, what turns out to have been of crucial importance to the stimulus of European science was its level and character of urbanisation, the existence of a range of competing nation-states, and the oligarchic nature of town rule. Multiple centres of political authority and a myriad autonomous urban centres fostered the competition, capitalism, and commerce, that were intimately connected with the nascent European science. London's Gresham College, endowed by the wealthy merchant Sir Thomas Gresham and perhaps the main site of scientific endeavour in Elizabethan England, is one notable incidence of the key links between the rising science and trade. Besides, the *mentalité* of interstate rivalry, commercial competitiveness, and religious zealotry all contributed to the cultivation of a set of scientific cultures in Western Europe at odds with their counterparts in Balkan Europe.

Rehearsing the scientific import of institutional, political, religious, and social difference, however, does not exhaust the relevance of regionalism to an understanding of the scientific enterprise. The cognitive content of scientific claims may also be conditioned by national styles of scientific investigation. This position has been defended in Malcolm Nicolson's (1989) investigation of the divergent classification systems devised by French and American plant ecologists. Despite the internationalism of modern science, Nicolson maintains that the study of plant ecology has remained highly differentiated and characterised by distinctive regional styles. The taxonomic schema derived by Josias Braun-Blanquet . . . for vegetational classification stands in marked contrast to that evolved by Frederick E Clements, the American ecologist. Whence the difference, then, given that such radically different classificatory systems have each proven to be instrumentally successful? Initially, Nicolson acknowledges that the differing vegetational environments of the Alps and the American prairies may well have influenced the choice of classificatory criteria and the differing areal sizes of system units, producing in the case of Braun-Blanquet a system based solely on floristic composition and in the case of Clements one derived from plant physiology and morphology. But this 'ecology of ecologists' explanation fails to explain a wide range of differences, notably, 'why the Clementsian unit was held to be concrete and organismic and the Sigma unit [the name derived from the station in Montpellier where Braun-Blanquet worked] to be abstract' (Nicolson, 1989, page 145). Moreover, Nicolson is able to point to a range of counter instances from British and Russian ecology where the ecological constructivist model fails to provide an explanation. Instead, Nicolson looks to contextual factors 'to indicate how ideas of the plant community . . . are constructed within social contexts and structured by social purposes both internal and external to communities of trained scientists' (page 174).[4] In this particular case Clements's dislike of what he regarded as excessive taxonomic splitting reflected the Land-Grant-college context of Mid-West science where his research was geared at once to the problems of farmers, foresters, and grazers, *and* to the cultivation of a standardised terminology that relied on strongly deductive laws of vegetation. This context was markedly different from the 'pure' French science of the Sigmatists where the lay constituency exerted much less pressure on Braun-Blanquet; indeed the French scientific ideology discouraged practical orientation in the provincial faculties of science.[5] In these ways, different regional scientific styles insinuate their way into the cognitive content of scientific knowledge. Nicolson, to put it another way, has provided for us a social geography of plant geography.

Scientific regionalisation, of course, operates at the subnational scale too. In the 1970s, Yves Conry's study of *L'Introduction du Darwinisme en France au XIXe Siècle* (1974) exposed the incoherences in speaking about some unified abstraction called 'French science' or 'French biology', and, in his investigation of the cultural effects of municipal politics on science in Manchester, Arnold Thackray (1974) elaborated a research programme which took seriously local circumstance in shaping scientific inquiry – as did Shapin in his 1972 scrutiny of the cultural uses of provincial science as revealed in the Pottery Philosophical Society. Something of this different scale of regional interrogation can be illustrated by reference to the collection of essays published in 1983 under the editorship of Ian Inkster and Jack Morrell which grappled with ways of explaining what they called 'the cultural geography of science' (Inkster, 1983, page 28). Eschewing economic reductionism in their cultivation of a social history of British scientific culture between 1780 and 1850, in their volume they set out to address such fundamentally locational questions as 'Why was the scientific culture of Nottingham and Norwich during the period . . . seemingly so less diffused, institutionalized and vigorous than that of, say, York, Liverpool or Derby?' (page 13). This approach is in basic accord with Barry Barnes's concern to regard scientific ideas as tools employed by particular groups 'to achieve their purposes in particular situations' (page 14). So, besides charting the variations in metropolitan–provincial relationships – sometimes dysfunctional, sometimes optimal for the diffusing of scientific culture – they sought ways of accounting for the deceptive simplicity of the place-name adjectives in Bristol science, Edinburgh science, and Newcastle science, Yorkshire geology, the London scientific communities, and so on. The Edinburgh Philosophical Society, founded in 1832, for instance, is shown to have been a child of bourgeois needs and demands, and the *in*utility of the science cultivated at the Bristol Institution between 1820 and 1860 reflected, in a time of economic decline, the confidence of its local social elite (Shapin, 1983; Neave, 1983). The character of Newcastle's major scientific institution was intimately connected with the social networks of the city's dissenting substructure, and the Geological and Polytechnic Society of the West Riding was purposely intended to depart from the 'polite' scientific tradition in favour of practical utility (Orange, 1983; Morrell, 1983). In all of these, the regional features of scientific culture persistently reassert themselves.

Political topography of scientific commitment

If the scientific endeavour has displayed identifiably regional features during its history, it is entirely reasonable to ask whether commitment to science as a venture or the adoption of particular theories may not also follow the contours of political persuasion. That scientific research now reflects different patterns of government (and other) funding in different settings is only too clear; but in other more subtle ways, including the modulation of science's substantive claims, political factors are now known to have substantially conditioned the enterprise.

In their prosopographical account of the early years of the British Association for the Advancement of Science, Jack Morrell and Arnold Thackray (1981) directed the attention of historians of science to the ways in which that enterprise was conditioned by political and cultural affairs. In an era of political unrest in the wake of the Industrial Revolution, when active workers were turning variously to Chartism or Methodism and provincial capitalists routinely opting for the new progressive science, the British Association (BA) inaugurated its peripatetic circuit of industrial cities. Aided by the communications revolution of the day, turnpike roads, canals, passenger railways, and so on, the BA's inner core (Morrell and Thackray dub them the 'Gentlemen of Science') found it possible to hold up a moral vision under the guise of scientific neutrality. Given their shared orientation to Whig causes and Broad Church alignments, they represented a 'definite meliorist, centrist, reforming political attitude' designed, through the nurturing of 'geographical union', to eradicate the worrisome unrest generated by political fissiparism. In this way spaces of political resistance could be circumvented by 'uniting elements of the better classes in the varied towns and cities of the Empire' (Morrell and Thackray, 1981, pages 25, 12, and 11). At the BA the rising middle classes, the aristocracy, and the gentry, could meet in congenial union to pursue universal scientific truth. In this way BA science stood for all that was moderate and latitudinarian.

By contrast, links between science and political radicalism are discernible in certain geographical locales during the early decades of the 19th century, prior to the inauguration of the BA. Around the turn of the 19th century, a majority of members in the Lunar Society of Birmingham, and a number of influential figures in the Manchester Literary and Philosophical Society espoused a fervent radicalism. Certainly too much can be made of these particular associations, but there is evidence that the situation in provincial centres such as Manchester, Liverpool, and Derby, where local scientific leaders were frequently of Radical or Whig persuasion, was different from that in London where radical science was more typically practised in fringe societies and marginal institutions, not least because of the tight hold that Sir Joseph Banks kept on admission to the Royal Society (Russell, 1983, chapter 3). Just how closely the character of provincial and metropolitan science in general can be mapped onto political topography remains to be investigated.

Nevertheless we are now in possession of one quite remarkable account of the significance of political topography and social geography on the reception of pre-Darwinian evolution, namely, Adrian Desmond's *The Politics of Evolution* (1989). Indeed prior to the import of the new evolutionary morphology into London, Desmond speaks of the 'social geography of the debate' over Lamarckian evolution in Edinburgh where the new doctrine persistently turned up among 'working-class atheists and socialists, among medical demagogues and radical phrenologists'. That is to say, it 'was taken up by groups that flatly rejected aristocratic authority' (pages 59 and 60). This circumstance sets the scene for the complex and richly textured narrative that Desmond has to tell concerning the fate of evolution within the London medical circles. Yet the heart of the argument is readily recounted: the transformist, law-bound, deterministic science that was imported into Britain from Paris in the 1830s spread like wildfire among those young doctors who, marginalised within the medical establishment and outcasts from the gentlemanly science of the day, mobilised it in the cause of radical assaults on professional injustice, political expediency, and a hierarchical social order bolstered by priestcraft, providence, and Paleyan natural theology.

Desmond specifically focuses on what he refers to as the scientific 'low-life', namely those radicals hitherto ignored by students of the intellectual elite. In this underworld of medicine, serviced by secular anatomy schools and radical nonconformist colleges, Lamarckian evolution easily gained a foothold. It became a means of challenging the Anglican Tory stronghold of the Royal College of Physicians and the Royal College of Surgeons. The migration tracts of this revolutionary scientific and social philosophy from Paris to Edinburgh and on to London, which Desmond uncovers with great skill, together with his mapping of the social topography of the anatomical factions in Edinburgh and London, thus expose a political geography of science for too long hidden beneath the abstractions of a disengaged history of scientific ideas.

The ties connecting scientific commitment with urban political topography are also exposed in Iwan Morus, Simon Schaffer, and Jim Secord's (1992) portrayal of 'Scientific London'. During the early decades of the 19th century, they write, 'New maps of the sciences matched the capital's social geography' (page 129). As the metropolis consumed the new scientific offerings, struggles inevitably emerged over the management of science and its 'proper place . . . in popular culture' (page 130). Accordingly, imperial Tory London – centred on such spaces as the Royal Society and the Royal Institution – cultivated its brand of scientific endeavour; reformist Whig London mobilised scientific know-how for sanitary, engineering, and administrative purposes; while radical London wielded scientific performance in the cause of materialism – all under the guise of apolitical, universal scientific truth. These different constituencies, all with their different political ideals, thus disseminated very different scientific knowledges precisely because in all these spaces science *was* politics.

Social space of scientific sites

Scientific sites are designed to provide the locales for particular types of co-presence. These range from institutions, conferences, and scientific societies to field sites and laboratory spaces. Different kinds of social relation characterise these spaces, and these can have an important bearing on the products of scientific analysis. The microworld – or perhaps hyperspace – of the laboratory, for example, is radically different space from that of fieldwork, the one, as Dorinda Outram [1996] shows, privileging the sedentary distant gaze of the observer, the other 'the heroic quest of the naturalist–explorer' (Driver, 1994). Of crucial importance is the issue of control, namely, who are those sufficiently accredited to participate in the learned society, to share results, to read the dials, and so on. Here I will just focus on two particular arenas of investigation – the experimental laboratory and the scientific society.

There is, perhaps, no more appropriate place to begin than with the establishment of systematic experimentation as the sine qua non of the new science of the 17th century. This is the arena Shapin chooses for his investigation of experimentation, asking the following fundamentally spatial questions of the enterprise:

> I want to know where experimental science was done. In what physical and social settings? Who was in attendance at the scenes in which experimental knowledge was produced and evaluated? How were they arrayed in physical and social space? What were the conditions of access to these places, and how were transactions across their thresholds managed? . . . This essay offers reasons for systematically studying the venues of the knowledge. I want to display the network of connections between the physical and social setting of inquiry and the position of its products on the map of knowledge. I shall try to demonstrate how the siting of knowledge-making practices contributed toward a practical solution of epistemological problems. The physical and symbolic siting of experimental work was a way of bounding and disciplining the community of practitioners, it was a way of policing experimental discourse, and it was a way of publicly warranting that the knowledge produced in such places was reliable and authentic. That is to say, the place of experiment counted as a partial answer to the fundamental question, Why ought one to give one's assent to experimental knowledge claims? (1988, pages 373–374).

The geographical discourse that this extract embodies is, I think, quite remarkable, replete as it is with spatial categories, both material and metaphorical. This is perhaps not surprising because Shapin's history of science displays numerous signs of interaction with the sociology of Goffman, Sennett, and Giddens. Accordingly the boundary between, and circulation among, public and private spaces assumes considerable importance because conditions of knowing are crucially different on either side of the threshold. This means that the production of scientific knowledge, its character, conditions, and content, is an inherently spatially organised activity. In this scenario, warranted knowledge crucially depends on reliable testimony; because certain key observations can only be conducted, as Shapin puts it, 'by geographically privileged persons', the wider possession of the knowledge generated depends on testimonial attestation (page 375). [In 17th century England, we should recall, the geography of credibility followed, in general terms, 'the contours of English society' (page 376).] This is precisely what experimental sites are designed to give in a strenuously disciplined arena – warranted credibility.

These judgments, of course, resonate with Giddens's observations more generally on what he called the 'disembedding' mechanisms of modernity. By this he means 'the "lifting out" of social relations from local contexts of interaction and their restructuring across indefinite spans of time-space' (1990, page 21). These are specially evident in the 'expert systems' – both technical and professional – of the modern world; but they are no different in fundamental character from the 'disembedding' of 17th-century laboratory knowledges and their transmission beyond experimental space. For both rely upon what Giddens calls 'trust' (page 29) – the need to accept the testimony of those *absent* in time and space, the reliability of those who were, to reuse Shapin's words, 'geographically privileged'.

Laboratory space, at the same time, was a new kind of space, and one which 'had necessarily to be carved out of and rearranged from existing domains of accepted public and private activity and existing stipulations about the proper uses of spaces' (Shapin, 1988, page 386). By examining the experimental sites of Robert Boyle, the Royal Society, and Robert Hooke, Shapin reveals how the transactions across the thresholds were carefully, though generally informally, managed. Besides, in the actual experimental arena, a fundamental distinction functioned between those operators or assistants or servants who tended the apparatus and those who, by interpretation, manufactured knowledge. The former were invisible participants, the latter active knowledge producers. As Shapin puts it, 'technicians had skill but lacked the qualifications to make knowledge' (page 395). The social relations that pertained in this private space, where the 'trying' of an experiment was carried out, were different from those operating in the public space where the 'showing' of the experiment took place. Indeed it was only when the item of knowledge had – in a literal sense – 'come out into society' that the production process was regarded as complete (page 399). Thus the construction of experimental knowledge was intimately bound up with circulation, that is, with transmission from the sphere of private trial to the arena of public discourse. In large measure the shift from private to public, from the solitary to the communal, from 'trying' to 'showing', from delving to demonstrating, was at once a move from the context of scientific discovery to the context of justification, *and* an exercise in refinement for public consumption. The public arena – in contrast to private space – was no place for displaying either 'nature's recalcitrance' or 'the uncertainty of experimental integrity' (Shapin, 1990, page 207).[6]

The spatial diffusion of experimental science, in turn, required effecting social reproduction in different places. This issue is taken up by Shapin and Schaffer in their investigation of 17th-century experimental life entitled *Leviathan and the Air-pump* (1985). Let me concentrate on one particular aspect of their story, the replication of experimental equipment in the 1660s. The specific case that Shapin and Schaffer are concerned with is Boyle's air-pump – a machine designed to produce a vacuum. The specific scientific claims, however, need not detain us here; what is important is the general import of the study. What they show is that '*replication* is basic to fact-production in experimental science' (page 225). Accordingly it was necessary, during the 1660s, that the claims of Boyle should be tested by replication in different places. The problem was how to judge when proper replication had actually been achieved. How were experimenters elsewhere to be assured that they had successfully reproduced Boyle's machine, and thus confirmed his findings? As Shapin and Schaffer point out, 'The only way to do this was to use Boyle's phenomena as *calibrations* of their own machines. To be able to produce such phenomena would mean that a new machine could be counted as a good one. Thus, before any experimenter could judge whether his machine was working well, he would have to accept Boyle's phenomena as matters of fact' (page 226). The circularity is obvious; but what is equally obvious is that the testing of scientific claims required the replication of both the gadgetry and an accredited team of observers – namely the reproduction of experimental space. Significantly, the key detractors from Boyle's findings were outside the experimental community. Thus the dissemination of Boyle's science, including its migration tracks across Europe, was intimately bound up with the reduplication of Boyle's experimental space in different locales, for only in that way could the new physics travel.

The spatial analysis of laboratory life, of course, need not be restricted to the period of the Scientific Revolution. Drawing on [Claude] Lévi-Strauss's (1967, page 285) conviction that 'spatial configuration seems to be almost a projective representation of the social structure', Bill Hillier and Alan Penn (1991) focus their attention on the modern research laboratory. By comparing the open-space structure of two laboratories – one of a large public charity, the other of an academic institution – they found fundamental configurational differences in layout which had implications for either the reinforcing or the weakening of those boundaries that produce 'localism' or segregation, and thus in turn for the production of scientific knowledge. Boundaries effecting absence and presence thus assume considerable importance. And of course the Goffmanesque distinction between front and back regions, between public and private space, snakes its way through the experimental programme more generally. David Gooding (1985) has charted the passage from Michael Faraday's basement experiments at the Royal Institution to his public demonstrations in the lecture theatre. And H M Collins, writing on the modern public experiments of the nuclear industry, tellingly notes that the demonstrations are effective precisely because the smooth public display conceals, 'the untidy craft' of the scientist; demonstrations work – as he superbly catches the character of the circuit from private to public – by 'caging Nature's caprices in thick walls of faultless display' (1988, page 728). In the theatre of experiment, to change the figure, Nature speaks her lines, but her voice is never heard.

Given the site-dependent character of laboratory life, it is not surprising that its products should be described as a form of local knowledge. This is precisely how Joseph Rouse (1987), drawing on the influential critiques of Nancy Cartwright and Ian Hacking, depicts it because, he argues, 'Scientific knowledge is first and foremost knowing one's way about in the laboratory' (page 72). Thereby Rouse distances himself from Taylor's early argument that 'theoretical understanding aims as a disengaged perspective' and from Martin Heidegger's assertion that genuine 'theoretical representation is indifferent to local situations' (pages 69 and 77). To be sure, for Heidegger, objects have 'locations' – mathematised positionings in space – but, scientifically at least, they do not have 'places'. For Heidegger, it is precisely this delocalisation of scientific objects that discriminates between scientific theorising and practical dealings. Rouse eschews such claims. For him scientific theory is crucially connected with *local* laboratory culture, with particular 'know-how', and with the on-site availability of appropriate technology. Taking scientific *theory* seriously means taking laboratory *practice* seriously. As he puts it, 'scientific knowledge is fundamentally local knowledge, embodied in practices that are not fully abstractable into theories and context-free rules for their application

. . . The instruments, and the microworlds established by using them, are the proximate referents of scientific claims. Furthermore, scientists' knowledge depends upon their craft knowledge of the workings of these devices' (Rouse, 1987, page 108). All this means that what looks like the universalism of science – its transferability from one arena to another – turns out to be more to do with the standardisation or even customisation of local precedent. Migration beyond laboratory space is thus 'not to be understood in terms of the instantiation of universally valid knowledge claims . . . It must be understood in terms of the adaptation of one local knowledge to create another. We go from one local knowledge to another rather than from universal theories to their particular instantiations' (page 72). And of course this is not just a transfer between different material spaces; it is transacted across social spaces too for scientific claims are established within what Rouse calls 'rhetorical space', because 'scientific arguments settle for rational persuasion of peers instead of context-independent truth' (page 120). All this talk of space thus points to a redirection within scientific philosophy, away from theoretical prescription, towards practical description. So it is understandable that Rouse should find Foucault's historical interrogation of spaces and power fruitful for his own inquiries, for the manufacture and manipulation of laboratory phenomena are 'part of a network of power relations running throughout modern societies' (page 212). Grouping, distributing, separating, and partitioning in laboratory space are all intimately connected with the spatial organisation of power and knowledge – just as they are elsewhere. Bench spaces, sterile areas, storage spaces, radiation areas, and so on are typical spatial divisions within laboratories, and are 'the sine qua non of experimental science' (page 222). All in all, for Rouse, 'The local laboratory site turns out to be the place where the empirical character of science is constructed through the experimenter's local, practical know-how' (page 125).

(Much of this analysis, I suggest, is indicative of what might be called the spatialisation of David Hume's problem of induction. For just as Hume was only too aware of the logical problems of generalising from inductive particulars, so contemporary sociologists of science query the hitherto taken-for-granted universalisation of the 'disembedding' of locally produced knowledge.)

Of course laboratory space itself is not an unproblematic, undifferentiated arena. To the contrary. As Michael Lynch (1991) has shown, it is constituted through a series of what he calls 'topical contextures'. By this, Lynch refers to the suite of activities through which knowledge is produced. The rich texture of actions which typify laboratory life is not mappable onto 'an invariant spatial matrix', but is rather 'topically bound to particular constellations of details'. Drawing on Maurice Merleau-Ponty's phenomenology and Foucault's 'discursive formations', Lynch illustrates the multispatial character of laboratory worlds by contrasting the technological interventions of traditional observation – opticism, as he calls it – which incorporates lenses, prisms, microscopes, cameras, dioramas – with 'digitality' which involves image processing, 'pixelated' space, keyboards, and display monitors. The microspatial movements of the retina and the finger-tip, as the embodied sites of opticism and digitality, respectively, constitute the different zones of spatial engagement where 'the scientific action is'. The point of Lynch's diagnosis is to reach beyond, or beneath, the 'surface' morphology of laboratory space to the 'ecology of local spaces integrated with disciplinary practices' around which laboratory life revolves (pages 53 and 74).

The laboratory, however, is certainly not the only space in which scientific action takes place. In a different context, that of the diffusion of science in 1830s Edinburgh, Shapin explored the ways in which strategies to circumvent 'the boundaries of participation' operated among the city's different social groups. Here an increasingly less deferential mercantile group intruded itself into scientific culture through its creation of a new site of scientific inquiry – the Edinburgh Philosophical Association – and thereby outmanoeuvred the gatekeeping tactics of the hegemonic old order. By providing a variety of cheap lecture courses on diverse scientific subjects, the Association fulfilled the needs and desires of a commercial class for whom – as *The Scotsman* reported at the time – 'the gates of Colleges have not been wont to lift up their heads' (Shapin, 1983, page 155). Subsequently, as Shapin tellingly reveals, the Edinburgh Philosophical Association was itself implicated in gatekeeping connivance when it reacted in a hostile fashion to a proposed Edinburgh Society for Aiding the General Diffusion of Science, because it wanted to retain political control of cultural enterprises. Throughout, the desire to control the social space of scientific sites is plainly evident.

In a remarkably fine-grained investigation of the establishment of the Devonian system in the 1830s, Martin Rudwick (1985) centres much of his investigation on the Geological Society of London. Here a certain type of social space was produced in order to facilitate the kinds of social relation characterised by what Rudwick refers to as 'gentlemanly' codes of behaviour, codes which governed the organisation and conduct of proceedings. Of the available institutional models that the nascent Geological Society had available – the mineral resource centre, the scientific dining club, the learned society – most of the founders and early members initially favoured adding 'a gentlemen's geological dining club to other informal groups of London men of science' (Rudwick, 1985, page 20). This was at least partly strategic, for as a dining club it would pose no threat to the scientific hegemony exerted by the Royal Society. Moreover, the high price of the meal would ensure a restriction of membership to the gentlemen amateurs and thereby deter professional surveyors from participation. With this control on social relations, the new organisation soon moved in the direction of a learned society with a formal constitution, the acquisition of its own premises, and the publication of research in its *Philosophical Transactions*. The chamber's innovative spatial arrangements – set out like a small-scale House of Commons with two sets of seats facing each other – permitted gentlemanly confrontation and debate, something hitherto discouraged at the Royal Society. Such procedures seemed to undermine the public image of natural science as an objective, Baconian, fact-finding activity, for it evidently allowed for the possibility of dissension; yet by its control of the social space, the Geological Society could conduct its affairs in 'the private and gentlemanly milieu of its own meetings, while maintaining toward the wider public its earlier politic stance of corporate theoretical neutrality' (page 26).

Of course, the elite members of the Geological Society's gentlemanly clientele occupied other related spaces too: there was considerable overlap with the core membership of the Société Géologique – a space from which the old scriptural geologists were conspicuously absent. Moreover, many of them lived in central London within walking distance or a short cab ride of each other. Sir Roderick Murchison, George Scrope and Sir Philip Edgerton were in Belgravia, Charles Darwin was off Oxford Street, Charles Lyell in Bloomsbury, while Adam Sedgwick, William Phillips, and Henry De la Beche held lodgings along Whitehall and the Strand. Thereby locational proximity facilitated intellectual congeniality and institutional co-presence. And these topographies of relative competence and involvement were, as Rudwick so impressively illustrates, intimately bound up with their production of sanctioned geological knowledge.

In reviewing what I have referred to as the social space of scientific sites I have focused my attention on the laboratory and the scientific society as key exemplars of scientific locations. But it would be mistaken to treat these as the exclusive localised arenas of scientific knowledge production. Indeed it would be more correct to consider them as emblematic of the scientific enterprise in action. Nowhere, perhaps, is this more pointedly elucidated than in Bruno Latour's (1987; 1990) investigations. Let me illustrate. Crucial to his portrait of scientific practice are what Latour calls the 'centres of calculation' that facilitate the mobilisation of information, and thereby domination, at a distance.[7] Tellingly, he begins his analysis by recounting the 18th century expedition of the Comte de la Pérouse with its cartographic objectives. At first sight, of course, it looks as though the '*implicit* geography of the natives is made *explicit* by the geographers; the *local* knowledge of the savages becomes the *universal* knowledge of the cartographers' (Latour, 1987, page 216). But this is too precipitate. What is going on, rather, is the European geographers' capacity to 'bring home' unfamiliar events, places, peoples. This 'mobilisation of worlds' (page 223) – splendidly illustrated by the encoding of distant European cartographers – is symbolic of scientific enterprises more generally. Through natural history museums, botanical and zoological gardens, statistical offices, and so on, the domination of distant local ethno-knowledges is effected just as efficiently as the way 'the cartographer *dominates* the world that dominated Lapérouse' (page 224). Such claims, of course, can be made to *sound* as though botany did not exist until there were botanical gardens, zoology until there were natural history museums, demography until there were statistical offices. Whether or not this is so, and what it might *mean* in any case to make such an assertion, I remain unsure. But I have no doubt that there is something significant about the capacity that such locations have to reconstruct 'different spaces and different times . . . *inside the networks* built to mobilise, cumulate and recombine the world' (page 228). Whether in geological cross-sections, on photographic plates, in computer printout, on astronomical charts, on terminal screens, through scaled-down engineering models or scaled-up electron images of cells, these centres of calculation construct manageable *local* representational spaces – really local hyperspaces or microworlds – through which scientists move and inside which worlds can be mobilised.[8] So, as with 17th century cartography, it is not (according to Latour) that 'the local knowledge of the Chinese' is opposed 'to the universal knowledge of the European, but only two local knowledges, one of them having the shape of a network transporting back and forth immutable mobiles at a distance' (page 229).

Latour's emphasis on the mobility, circulation, and distant domination of scientific information, together with the fact that laboratory space may encompass every scale from the domestic to the world-regional, serves to recall our attention to the intimate connections between science and empire. On the one hand George Basalla's (1967) early model of the diffusion of Western science reminds us that the 'bringing home' of esoteric data is only the first move in the Westernisation of non-Western traditions which is itself succeeded by the struggle to establish an independent scientific culture. Of course this schema has undergone modification as a consequence of numerous empirical investigations which have succeeded in revealing the complexity and richness of these reciprocal processes (Leclerc, 1972; Brockway, 1979; MacLeod, 1982; Pyenson, 1982; 1984; Reingold and Rothenberg, 1987; MacKenzie, 1990; Pratt, 1992). Throughout, the fundamentally *geographical* character of this circuit is reaffirmed. And nowhere perhaps is this more clearly exposed than in the Victorian archive – that sociopsychological impulse instantiated in such spaces as the Royal Geographical Society and the British Museum. Indeed according to Thomas Richards (1993) (whose focus is particularly on Tibet) the 'Victorian archive appears as a prototype for a global system of domination through circulation, an apparatus for controlling territory by producing, distributing, and consuming information about it' (page 17). The necessity of the archive, of course, was obvious: the empire was far away and its management all too often centred on the shuffling of its archival records.

On the other hand, whereas we have already attended to the carving of laboratories out of 17th-century domestic space, other writers remind us that laboratories can be as extensive as the Pacific realm or the heart of Africa. In their collection of essays on *Western Science in the Pacific*, for example, Roy MacLeod and Philip Rehbock (1988) insist that the Pacific 'became a veritable school for science, and a vast classroom for educating the European mind' (page 1). To them it proved impossible to neglect 'the historical relationships that have long inspired motives for scientific discovery, geographical exploration, territorial conquest, colonial settlement, and trade' (page 4). Again, Sir Roderick Murchison used whole continents – notably Africa – to test his own Silurian theories and as the arenas over which his taxonomic control of geology could be extended (Secord, 1982; Stafford, 1988; 1989). Here again the constitutive role of spatiality is reasserted. Any comprehensive geography of science will therefore need to engage in elucidating the complex relationships between science and empire.

Coda

It is clear, I think, from even this thumbnail sketch that a geography of science will need to attend to spatial considerations at a variety of scales. Indeed, it will be one of the key methodological issues of such an undertaking to ascertain just what is the appropriate spatial scale at which to conduct any specific historical investigation, and then to determine how the various scales are to be related. Again, the spatiality of Western scientific experience is no less related than other dimensions of social life to the spatial transformations of advanced capitalism. An earlier need for spatial proximity or co-presence, bound up as it was with the available technology of information communication, has evidently been transformed with the development of electronically transmitted information. This 'collapsing' of space and time has inevitably produced a different geography of the scientific enterprise.[9] All this serves to remind us that space itself is anything but transparent (Gregory, 1991); to the contrary, because different relations have pertained between space and place at different times, the problems of representing these in a spatialised historiography of science are of considerable proportions.[10] Then there are issues to do with the spatial distribution of scientific information arising from such factors as differential access to what Nigel Thrift (1985) calls 'stocks of knowledge', its seizure by groups interested in science as 'cultural capital', and the patterns of communication networks conditioning the dissemination of knowledge. And of course there are the issues to do with the gendering of scientific space – at least half [of] humanity is absent from 'gentlemanly' science – and the implications that this circumstance has for warranted assertibility in science. I cannot begin to elaborate on the sorts of issue involved here, but they would need to be addressed in any fully developed historical geography of science.

An historical geography of geography: a rudimentary agenda

Ironic though it may seem, it is my impression that, although historians of geography have failed to attend to the spatial components of their tradition's history in one sense, they have been overzealous in inscribing it in another. What I mean by this is that the history of geography (as itself a science, or perhaps better, *Wissenschaft*) has frequently been written with little reference to the placing of geographical knowledge in its various settings. There has thus been, I would judge, a neglect of what I refer to as 'the spaces of geographical knowledge' and I suggest that a closer engagement with the recent writings of the social historians of science (not to mention social theory more generally) might foster a greater historiographical sophistication within the tradition. By the same token, historians of geography have all too frequently been concerned with the policing of geography's conceptual territory, with boundary making and discourse maintenance, in other words, with gatekeeping. They have, I fear, been overly preoccupied with the management or manufacture of what might be called geography's conceptual space. A few desultory comments on these topics will now suffice.

The spaces of geographical knowledge

Geography's history, it must be said, has frequently been rehearsed in terms of national traditions. We have heard of the German school, the French tradition, the British school, and so on;[11] indeed the textbook chronicles that still circulate frequently proceed on this basis.[12] Some of these surveys have certainly been helpful in contextualising the geographical tradition in different milieux. But they have, all too often, operated at the level of those 'wall-sized culturescapes of the nation' to which I referred earlier. Moreover, they all too frequently proceed by a Whiggish chronicling of the 'progress' of geography in different countries. Just how the spaces of geographical knowledge make a difference to the tradition's cognitive claims, subjects of scrutiny, and methods deployed, is left unexamined.

To adopt an historiography for geography that takes seriously the situatedness of knowing would mean, I suggest, abandoning normative history and looking to those contingent factors that, as we have seen, shape scientific inquiry. It will mean locating particular geographical theories, methodologies, representations, schools of inquiry, and so on, in their intellectual context, their social space, their physical setting. And it will mean resisting the tendency to privilege certain definitions of the subject's conceptual terrain over others. When Geertz (1973, page 22) observes that 'The notion that one can find the essence of national societies, civilizations, great religions, or whatever summed up and simplified in so-called 'typical' small towns and villages is palpable nonsense', we easily nod in consigned agreement. But the 'Jonesville-is-the-USA' model seems perilously parallel to thinking that Hettnerian chorology or locational analysis or *Kulturlandschaft* or the personality of place or – for that matter – the reciprocal constitution of society and space *is* geography. Taking seriously the historical geography of geography will enable us to escape the tyranny of such essentialist readings.

Surprisingly even Soja who has been so energetic in his efforts to inscribe spatiality into social theory ignores the geography of geographical knowledge in his analysis of the demise of human geography as a theoretical force in the middle decades of the 20th century. For having provided an essentially political reading of the intellectual history of the subordination or expurgation of space from social theory – precisely at a time when the spatial organisation of life was being reorganised at just about every scale – he accounts for the 'mid-century involution of Modern Geography' by reference to such grand themes as the intellectual bankruptcy of environmental determinism, the excesses of a generic *Geopolitik*, the arid chronicling of areal differentiation, the preponderance of a Kantian derived historicism, or the abstractions of locational analysis (Soja, 1989, pages 34–38).[13] Whatever the philosophical cogency of these dismissals, they favour macroscale repudiation over serious historical–geographical probing. And this is especially bothersome when the few examples of what Soja holds up as cases of good practice are so specifically earthed as the urban ecology of the Chicago school, the regional historiography of the *Annales* tradition, or the frontier theorists of Turnerian inspiration.

Geography's conceptual space

If historians of geography have left untended the spatial components of their own tradition's history, they have frequently been assiduous in prosecuting their role as adjudicators of geography's conceptual space, as gatekeepers to the disciplinary culture. Their mapping of geography's intellectual terrain is no more innocent than the spatial discourse of Carter's (1987) early Australian surveyors, their choice of labels no less emblematic than antipodean place-naming. For the naming of theories, schools, movements, is – to reuse Carter's words – 'to invent them, to bring them into cultural circulation'. Richard Hartshorne's project (1939) of determining the nature of geography from scrutinising its history is illustrative. By restricting his survey very largely to the geography of one particular space–time arena – that of early modern Germany – and then universalising his findings, he produced an orthodoxy to which allegiance was meant to be given. And by talking of deviations from the course of historical development, digressions from the high road of received wisdom, Hartshorne occupied himself in the project of boundary creation and surveillance. His history was fundamentally a gatekeeping exercise. But other historians have engaged in precisely the same operation, though culling from a different set of heroes. David Stoddart (1986, page ix), for example, affirms: 'Not surprisingly, my heroes are not the usual ones – the Ritters, Ratzels, Hettners, entombed by conventional wisdom. My geography springs from Forster, Darwin, Huxley: and it works'.

I do not cite these two very different cases to condemn. To the contrary. The fact of the matter is that, because all history is committed, because all history writing is apologetic, historians *are* gatekeepers. The question to be asked is what kind of gatekeeping exercise our historians are involved in. Gatekeepers can keep gates open, or keep them closed. As Said (1993, page 2), taking his cue from T S Eliot, insists, 'how we formulate or represent the past shapes our understanding and views of the present'. My claim is that taking seriously the spaces of geographical knowledge will enable us to be less defensive, because less essentialist, about geography's conceptual space.

If I am correct, the 'core symbols' – to use one of Geertz's expressions – of science in general and geography in particular are different at different points in time and in different spatial settings. This is the assumption, I believe, that should undergird the project of inscribing a historical geography of these knowledges. The significance of this task is of considerable proportions. As MacIntyre has eloquently argued, it is only when claims are earthed in the specificities of tradition that we can begin to construct an intelligible dramatic narrative that enables us to distinguish better and worse readings of how we have arrived at where we are, and more, to make any stab at understanding the ethical and epistemological crises that induce significant change. 'It is only when theories are located in history, when we view demands for justification in highly particular contexts of a historical kind, that we are freed from either dogmatism or capitulation to scepticism' (MacIntyre, 1980, page 74).

Notes

1 Some of these issues are taken up in Gregory (1994).
2 Foundational texts for this programme are Barnes (1974) and Bloor (1976).
3 Elsewhere I have outlined, and sought to exemplify, what a historical geography of

the encounter between science and religion might look like. See Livingstone (1994).

4 Nicolson, in my view rightly, insists that his reading should not be taken to mean that he denies that these two different 'forms of plant ecology were also the product of human interaction with the external reality of vegetation' (1989, page 174).

5 Ronald Tobey (1981) argues that Clements's organicism was derived, at least in part, from the social thinking of Lester Frank Ward. Earlier James Malin (1946; 1952) found Clements's organicism – as a form of collectivism – threatening to the American tradition of individualism and democracy.

6 Here Shapin also discusses the differences between Isaac Newton and Robert Boyle on the role of public demonstration.

7 The techniques of exercising long-distance control have been investigated by John Law (1986) who focuses his analysis on 16th-century Portuguese navigation and specifies 'documents, devices and drilled people' as the key resources. These facilitated 'western long-distance control'. This moves the debate about the triumph of Western science away from matters to do with cognitive capacity to issues of reproduction, circulation, communication, and well-drilled human agents.

8 On issues to do with representation in scientific practice see the essays in Fyfe and Law (1988) and Lynch and Woolgar (1990).

9 On the significance of the changing relations of time and space for understanding modernity, see Giddens (1990; 1991).

10 The ways in which we have learned to think about space and territory is the subject of Revel's 1991 essay.

11 See, for example, Church (1951); van Valkenberg (1951); Berdoulay (1981); Buttimer (1971); Schultz (1980); Freeman (1980a; 1980b).

12 So, for example, Dickinson (1969; 1976); James (1972); Fuson (1969). The historiography of works like these is discussed in Livingstone (1992).

13 Similar reservations are expressed in Matless (1992).

Selected references

Barnes B, 1974 *Scientific Knowledge and Sociological Theory* (Routledge and Kegan Paul, London)

Basalla G, 1967, 'The spread of Western science' *Science* **156** 611–622

Berdoulay V, 1981 *La Formation de l'École Française de Géographie (1870–1914)* (Bibliothèque Nationale, Paris)

Bloor D, 1976 *Knowledge and Social Imagery* (Routledge and Kegan Paul, London)

Brockway L, 1979 *Science and Colonial Expansion: The Role of the British Royal Botanical Gardens* (Academic Press, London)

Buttimer A, 1971 *Society and Milieu in the French Geographic Tradition* (Rand McNally, Chicago, IL)

Carter P, 1987 *The Road to Botany Bay: An Essay in Spatial History* (Faber and Faber, London)

Church R J H, 1951, 'The French school of geography', in *Geography in the Twentieth Century: A Study of Growth, Fields, Techniques, Aims and Trends* Ed. G Taylor (Methuen, London) pp 70–90

Collins H M, 1988, 'Public experiments and displays of virtuosity: the core-set revisited' *Social Studies of Science* **18** 725–748

Conry Y, 1974 *L'Introduction de Darwinisme en France au XIXe Siècle* (J Vrin, Paris)

Desmond A, 1989 *The Politics of Evolution: Morphology, Medicine, and Reform in Radical London* (University of Chicago Press, Chicago, IL)

Dickinson R E, 1969 *The Makers of Modern Geography* (Routledge and Kegan Paul, London)

Dickinson R E, 1976 *Regional Concept: The Anglo-American Leaders* (Routledge and Kegan Paul, London)

Driver F, 1994, 'Making space' *Ecumene* **1** 386–390

Freeman T W, 1980a *A History of Modern British Geography* (Longman, Harlow, Essex)

Freeman T W, 1980b, 'The British school of geography' *Organon* **14** 205–216

Fuson R, 1969 *A Geography of Geography: Origins and Developments of the Discipline* (Brown, Dubuque, IA)

Fyfe G, Law J (Eds), 1988 *Picturing Power: Visual Deception and Social Relations* (Routledge, London)

Geertz C, 1973, 'Thick description: an interpretive theory of culture', in *The Interpretation of Cultures: Selected Essays* (Basic Books, New York) pp 3–30

Giddens A, 1990 *The Consequences of Modernity* (Polity Press, Cambridge)

Giddens A, 1991 *Modernity and Self-identity: Self and Society in the Late Modern Age* (Polity Press, Cambridge)

Glick T F (Ed.), 1974 *The Comparative Reception of Darwinism* (University of Texas Press, Austin, TX)

Glick T F, 1987 *The Comparative Reception of Relativity* (D Reidel, Dordrecht)

Glick T F, 1988, 'Reception studies since 1974', in *The Comparative Reception of Darwinism* Ed. T F Glick (University of Chicago Press, Chicago, IL) pp xi–xxviii

Gooding D, 1985, 'In nature's school: Faraday as an experimentalist', in *Faraday Rediscovered: Essays on the Life and Work of Michael Faraday, 1797–1867* Eds D Gooding, F A L James (Macmillan, London) pp 106–135

Goodman D, 1991, 'Europe's awakening', in *The Rise of Scientific Europe, 1500–1800* Eds D Goodman, C A Russell (Eds) (Hodder and Stoughton, Sevenoaks, Kent) pp 1–30

Gregory D, 1991, 'Interventions in the historical geography of modernity: social theory, spatiality and the politics of representation' *Geografiska Annaler* **73B** 17–44

Gregory D, 1994 *Geographical Imaginations* (Blackwell, Oxford)

Hartshorne R, 1939 *The Nature of Geography: A Critical Survey of Current Thought in the Light of the Past* (Association of American Geographers, Lancaster, PA)

Harvey D, 1985, 'The geopolitics of capitalism', in *Social Relations and Spatial Structures* Eds D Gregory, J Urry (Macmillan, London) pp 128–163

Hillier B, Penn A, 1991, 'Visible colleges: structure and randomness in the place of discovery' *Science in Context* **4** 23–49

Inkster I, 1983, 'Introduction: aspects of the history of science and science culture in Britain, 1780–1850 and beyond', in *Metropolis and Province: Science in British Culture, 1780–1850* Eds I Inkster, J Morrell (University of Pennsylvania Press, Philadelphia, PA) pp 11–54

Inkster I, Morrell J (Eds), 1983 *Metropolis and Province: Science in British Culture, 1780–1850* (University of Pennsylvania Press, Philadelphia, PA)

James P E, 1972 *All Possible Worlds: A History of Geographical Ideas* (Bobbs-Merrill, Indianapolis, IN)

Latour B, 1987 *Science in Action: How to Follow Engineers and Scientists Through Society* (Harvard University Press, Cambridge, MA)

Latour B, 1990, 'Drawing things together', in *Representation in Scientific Practice* Eds M Lynch, S Woolgar (MIT Press, Cambridge, MA) pp 19–68

Law J, 1986, 'On the methods of long-distance control: vessels, navigation and the

Portuguese route to India', in *Power, Action and Belief: A New Sociology of Knowledge?* Ed. J Law (Routledge and Kegan Paul, London) pp 234–263

Leclerc G, 1972 *Anthropologie et Colonialisme: Essai sur l'Histoire de l'Africanisme* (Éditions du Seuil, Paris)

Lévi-Strauss C, 1967 *Structural Anthropology* (Anchor Books, New York)

Livingstone D N, 1992 *The Geographical Tradition: Episodes in the History of a Contested Enterprise* (Blackwell, Oxford)

Livingstone D N, 1994, 'Science and religion: foreword to the historical geography of an encounter' *Journal of Historical Geography* **20** 367–383

Lynch M, 1991, 'Laboratory space and the technological complex: an investigation of topical contextures' *Science in Context* **4** 51–78

Lynch M, Woolgar S (Eds), 1990 *Representation in Scientific Practice* (MIT Press, Cambridge, MA)

MacIntyre A, 1980, 'Epistemological crises, dramatic narrative, and the philosophy of science', in *Paradigms and Revolutions: Appraisals and Applications of Thomas Kuhn's Philosophy of Science* Ed. G Gutting (University of Notre Dame Press, Notre Dame, IN) pp 54–74

MacKenzie J M (Ed.), 1990 *Imperialism and the Natural World* (Manchester University Press, Manchester)

MacLeod R, 1982, 'On visiting the ''moving metropolis'': reflections on the architecture of imperial science' *Historical Records of Australian Science* **5** 1–15

MacLeod R, Rehbock P F, 1988 *Nature in Its Greatest Extent: Western Science in the Pacific* (University of Hawaii Press, Honolulu, HI)

Malin J C, 1946 *Essays on Historiography* (Malin, Lawrence, KA)

Malin J C, 1952, 'Man, the state of nature and climax: as illustrated by some problems of the North American grassland' *Scientific Monthly* **74** 29–37

Matless D, 1992, 'A modern stream: water, landscape, modernism, and geography' *Environment and Planning D: Society and Space* **10** 569–588

Morrell J, 1983, 'Economic and ornamental geology: the geological and polytechnic society of the East Riding of Yorkshire, 1837–53', in *Metropolis and Province: Science in British Culture, 1780–1850* Eds I Inkster, J Morrell (University of Pennsylvania Press, Philadelphia, PA) pp 231–256

Morrell J, Thackray A, 1981 *Gentlemen of Science: Early Years of the British Association for the Advancement of Science* (Clarendon Press, Oxford)

Morus I, Schaffer S, Secord J, 1992, 'Scientific London', in *London – World City 1800–1840* Ed. C Fox (Yale University Press, New Haven, CT) pp 129–142

Neave M, 1983, 'Science in a commercial city: Bristol 1820–60', in *Metropolis and Province: Science in British Culture, 1780–1850* Eds I Inkster, J Morrell (University of Pennsylvania Press, Philadelphia, PA) pp 179–204.

Nicolson M, 1989, 'National styles, divergent classifications: a comparative case study from the history of French and American plant ecology' *Knowledge and Society: Studies in the Sociology of Science Past and Present* **8** 139–186

Ophir A, Shapin S, 1991, 'The place of knowledge: a methodological survey' *Science in Context* **4** 3–21

Orange D, 1983, 'Rational dissent and provincial science: William Turner and the Newcastle Literary and Philosophical Society', in *Metropolis and Province: Science in British Culture, 1780–1850* Eds I Inkster, J Morrell (University of Pennsylvania Press, Philadelphia, PA) pp 205–230

Outram D, 1996 'Inner and outer spaces in natural history', in *Cultures of Natural History: From Curiosity to Crisis* Eds J Secord, E Spary, N Jardine (Cambridge University Press, Cambridge).

Pratt M L, 1992 *Imperial Eyes: Travel Writing and Transculturation* (Routledge, London)

Pyenson L, 1982, 'Cultural imperialism and exact sciences: German expansion overseas, 1900–1930' *History of Science* **20** 1–43

Pyenson L, 1984, 'Astronomy and imperialism: J. A. C. Oudemans, the topography of the East Indies, and the rise of the Utrecht Observatory, 1850–1900' *Historia Scientarum* **26** 39–81

Reingold N, Rothenberg M (Eds), 1987 *Scientific Colonialism: A Cross-cultural Comparison* (Smithsonian Institution Press, Washington, DC)

Revel J, 1991, 'Knowledge of the territory' *Science in Context* **4** 131–161

Richards T, 1993 *The Imperial Archive: Knowledge and the Fantasy of Empire* (Verso, London)

Rouse J, 1987 *Knowledge and Power Toward a Political Philosophy of Science* (Cornell University Press, Ithaca, NY)

Rudwick M J S, 1985 *The Great Devonian Controversy: The Shaping of Scientific Knowledge among Gentlemanly Specialists* (University of Chicago Press, Chicago, IL)

Russell C, 1983 *Science and Social Change, 1700–1900* (Macmillan, London)

Said E W, 1993 *Culture and Imperialism* (Chatto and Windus, London)

Schultz H-D, 1980 *Die Deutschsprachige Geographie von 1800 bis 1970: Ein Beitrag zur Geschichte ihrer Methodologie* (Geographische Institut der Freien Universität, Berlin)

Secord J A, 1982, 'King of Siluria: Roderick Murchison and the imperial theme in nineteenth-century British geology' *Victorian Studies* **25** 413–442

Shapin S, 1972, 'The Pottery Philosophical Society, 1819–1835: an examination of the cultural uses of provincial science' *Science Studies* **2** 311–336

Shapin S, 1983, ' "Nibbling at the teats of science": Edinburgh and the diffusion of science in the 1830s', in *Metropolis and Province: Science in British Culture, 1780–1850* Eds I Inkster, J Morrell (University of Pennsylvania Press, Philadelphia, PA) pp 151–178

Shapin S, 1988, 'The house of experiment in seventeenth-century England' *Isis* **79** 373–404

Shapin S, 1990, 'Science and solitude in seventeenth-century England' *Science in Context* **4** 191–218

Shapin S, Schaffer S, 1985 *Leviathan and the Air-pump: Hobbes, Boyle, and the Experimental Life* (Princeton University Press, Princeton, PA)

Soja E W, 1989 *Postmodern Geographies: The Reassertion of Space in Critical Social Theory* (Verso, London)

Stafford R A, 1988, 'Roderick Murchison and the structure of Africa: a geological prediction and its consequences for British expansion' *Annals of Science* **45** 1–40

Stafford R A, 1989 *Scientist of Empire: Sir Roderick Murchison, Scientific Exploration and Victorian Imperialism* (Cambridge University Press, Cambridge)

Stoddart D R, 1986 *On Geography and Its History* (Basil Blackwell, Oxford)

Thackray A, 1974, 'Natural knowledge in cultural context: the Manchester model' *American Historical Review* **79** 672–709

Thrift N, 1983, 'On the determination of social action in space and time' *Environment and Planning D: Society and Space* **1** 23–57

Thrift N, 1985, 'Flies and germs: a geography of knowledge', in *Social Relations and Spatial Structures* Eds D Gregory, J Urry (Macmillan, London) pp 366–403

Tobey R C, 1981 *Saving the Prairies: The Life Cycle of the founding School of American Plant Ecology, 1895–1955* (University of California Press, Berkeley, CA)

van Valkenburg S, 1951, 'The German school of geography', in *Geography in the*

Twentieth Century: A Study of Growth, Fields, Techniques, Aims and Trends Eds G Taylor (Methuen, London) pp 91–115

2 David Slater

On the Borders of Social Theory: Learning from Other Regions

Excerpts from: *Environment and Planning D: Society and Space* 10, 307–327 (1992)

Introduction

Although it may be reasonably argued that within critical geography . . . the tendency to centre theory on the economic, and the inclination to develop conceptual interpretations in an apparently gender-free world (androcentrism), have come under increasing attack, critiques of that *other* centrism – ethnocentrism – are somewhat thin on the ground. I want to suggest that the phenomenon of ethnocentric universalism, although present across a broad array of scientific discourses, has roots in two main radical origins which have helped to shape some of the main contours of critical geography, especially in its Anglo-American variant: Marxist political economy and critical theory, as associated with Habermas. Also, although far more ambiguous and polysemic, there is a third origin where elements of ethnocentrism and a masked universalism are present: namely within the postmodern intervention, or interruption, as Spivak (1988a) might call it. . . .

The interrogation of conventional urban and regional studies and the development of alternative theoretical approaches are not only rooted in the three origins of Marxism, critical theory, and the postmodern interruption; the various streams of feminist theory have also been crucial in the formulation of a good deal of critical geographical work. However, as it is in the first three origins that the need for a critique of ethnocentrism is so urgent, and as it is precisely within the women's movement and feminist analysis that the debate over universalism and Euro-Americanism has progressed so far, I shall refer more to this fourth possible origin as a source of the critique itself. This does not mean to imply that within the discussions of gender in geography there is no evidence of ethnocentrism, but rather that a consideration of such elements falls outside the scope of this particular intervention. It ought to be made quite clear at the outset that, although, in directly referring to the works of some of the more influential writers within 'critical geography', as well as in other more general fields of enquiry, I endeavour to identify and evaluate critically elements of an ethnocentric or universalist inscription, the intention is not to impugn these authors in any all-encompassing manner. Rather, through the consideration of a variety of texts, one of my main objectives is to illustrate the ways in which a Euro-Americanist discourse inhabits and mo(u)lds much of the fabric of theoretical analysis.

My argument will be organized as follows. First, I shall provide a few general remarks concerning three lineages of universalism, which will act as an introduction to a more detailed examination of what I shall here refer to as 'Euro-Americanism'. In this second section I shall take my examples from the work of critical geographers, and/or related enquiry which is directly concerned with urban and regional issues. In the final part, I shall present elements of a case for learning from the periphery. My position will be that not only is it crucial to interrogate all forms of Western ethnocentrism, but that by critically scrutinizing the historical constitution of the relations between the First World and the societies of the periphery the realities of the West can be better comprehended. In fact, I would argue that without such a connection First World geographers will not be able to grasp, certainly not in any effective way, the meanings and dispositions of the societies in which they live, and in this important sense will remain 'intellectual prisoners of the West'.

Lineages of universalism

Marxism

Although by now it is well established that the theoretical project of classical Marxism was gender-blind, and that within the contemporary currents of Marxist thought there has been a continual need to interrogate the continuation of this androcentric origin, the strong tendency towards a Marxist universalim has received less scrutiny, certainly at least within the sphere of critical geography. . . .

There are three points I want to make here: (1) some of the earlier highly ethnocentric positions expressed by Marx and Engels . . . in relation to development theory, have generally passed without critical comment in socialist urban and regional analysis (I shall return to some of the relevant passages in the second section of this paper); (2) the universalist inscription of Marx's theoretical work, with its implicit projection of the broader relevance of West European conditions (although later modified and further restricted . . .) has been widely written into the texts of critical geography; and (3) although it must be said that during Marx's time, knowledge, discussion, and information concerning peripheral societies were extremely limited, no such caveat applies to contemporary Marxists, who in some cases, and on this question, still tend to write today as if they were living with Marx. These observations, which I shall subsequently substantiate, with reference to the works of, for example, Lefebvre and Harvey, raise the question of the discursive specificity of critical geography, as exemplified in its widest context. In other words, in contrast to practically all other critical fields of social science enquiry, in the example of critical geography, there is a pervasive tendency to construct research agendas, formulate key issues for theoretical debate, and draw on bodies of literature as if the West were somehow a self-contained entity. I shall return to this peculiarity at the end of the discussion.

Critical social theory

A second origin of universalist thought can be traced to the Frankfurt school, and in particular to Habermas, with his development of critical social theory. In his analysis of the political and philosophical traditions that are seen as grounding Western liberal capitalism, and its possible socialist alternatives, one finds continuing evidence of a universalist and Eurocentric commitment. For example, in a recent study on the 'normative content of modernity', the reader is informed that in modern societies the discursive structures of the various public spheres 'owe a universalist tendency that is hardly concealed . . . the European Enlightenment elaborated this experience and took it up into its programmatic formulas' (Habermas, 1987, page 360). When Habermas talks of 'Occidental rationalism', he does so as if the West has been a self-contained entity, separated from the history of colonialism and imperialism; but not only is the history of colonialism the history of the West, but it is also, as Bhabha (1990, page 218) puts it, 'a *counter-history* to the normative, traditional history of the West'. Furthermore, by creating a world of universals, in order to imagine a certain representation of the West as universal for the rest of the world, what need is there, as Dallmayr (1989, page 9) asks, to engage in interaction and transcultural dialogue in order to learn politically and 'achieve a measure of political concord'? Under what has recently been referred to as the 'rationalist ''dictatorship'' of the Enlightenment' (Laclau, 1990, page 4), the use of any universal concept, such as 'class', 'race', or 'human being', can be highly limiting and detrimentally partial in trying to understand the ways in which systems of meaning and social organization are constructed within specific discourses; and this universality becomes pernicious when it is rooted in Western ethnocentrism.

Postmodernism

If we follow the idea that postmodernism is not simply an exit from modernity, but rather an internal rift or fissure, which looked at creatively can lead to all kinds of political rethinking, then it is relevant to recall one of Foucault's earlier texts. Over two decades ago he wrote that 'in attempting to uncover the deepest strata of Western culture, I am restoring to our silent and apparently immobile soil its rifts, its instability, its flaws; and it is the same ground that is once more stirring under our feet' (Foucault, 1970, page xxiv). But it is exactly in the context of that stated intention of uncovering the deepest strata of Western culture that it is possible to locate the contours of a Eurocentric focus. Spivak (1988b, page 291), for example, argues that Foucault's work as a whole tends to occlude a 'reading of the broader narratives of imperialism', and, as she expresses it, 'to buy a self-contained version of the West is to ignore its production by the imperialist project' (page 291). This kind of criticism goes together with an unequivocal recognition of Foucault as a brilliant thinker of 'power-in-spacing'; what Spivak is rightly pointing to is the need to chart all those diverse expressions of what she calls that 'sanctioned ignorance' of the imperialist project.

In contrast to Foucault, Derrida has given more time to investigating the often complex nature of ethnocentrism in the texts of influential European thinkers; for example, he effectively deconstructs the sentimental ethnocentrism of Lévi-Strauss, noting how the critique of ethnocentrism, a theme so dear to the author of *Tristes Tropiques*, 'has most often the sole function of constituting the other as a model of original and natural goodness'. Non-European peoples were to be studied, after Rousseau, as the 'index to a hidden good Nature, as a native soil recovered . . . with reference to which one could outline the structure, the growth, and above all the degradation of our society and our culture' (Derrida, 1976, pages 114–115). In Lévi-Strauss's own words, 'if the West has produced anthropologists, it is because it was so tormented by remorse' (quoted in Derrida, 1976, page 337). In this kind of vision, 'non-Western' peoples are essentialized around notions of the nobility and goodness of the primitive. Contradiction and difference are erased from their histories, and, as Fabian (1983) has effectively argued, in Lévi-Strauss's anthropological mission 'traditional' peoples are situated in another, previous time; their coevalness with the 'modern West' is denied. The critical importance attached by Derrida to these kinds of issues is supported by his stand on apartheid (Derrida, 1985), expressed in an essay on the 'last word in racism', which stands in marked contrast to the innovative orientations of much if not all of the recent work of postmodern thinkers such as Baudrillard and Lyotard.

With Baudrillard, there are a number of interesting ironies. In contrast to many other radical European intellectuals of the 1970s Baudrillard confronted both the universalist narrative of historical materialism and the ethnocentrism of Western Marxism. Of the former he writes that in Marxism, 'history is transhistoricized: it redoubles on itself and thus is universalized' (Baudrillard, 1975, pages 47–48). This is the case, as in historical materialism critical concepts such as labour power and surplus value are not seen as explosive and mortal but are constituted as universal, expressing an 'objective reality'; they thus cease to be analytical, and so the 'religion of meaning begins'. On the issue of ethnocentrism, Baudrillard first interrogates Western culture in general, writing that other cultures were entered into its museum 'as vestiges of its own images . . . it reinterpreted them on its own model, and thus precluded the radical interrogation these ''different'' cultures implied for it', and 'its reflection on itself leads only to the universalization of its own principles' (pages 88–89). It is then argued that the limits of the materialist interpretation of earlier societies are the same because in the last analysis historical materialism simply 'naturalizes' earlier societies 'under the sign of the mode of production'; these societies and those of the Third World are not comprehended – rather they become that other territory within which the analysis of the economic contradictions of Western societies is projected and implanted.

It is now a somewhat strange irony . . . that Baudrillard has recently been selected by Spivak (1988c, page 18) to serve as an example of the discourse of postmodernism with its sanctioned ignorance of the history of imperialism. Spivak quotes from Baudrillard's 1983 essay on the 'silent majorities', where there is indeed no recognition of the movements of resistance of Third World peoples, and wherein for Baudrillard in general 'the mass is only mass because its social energy has already frozen' (Baudrillard, 1983, page 26), and this during the phase of mass media culture, of the 'glaciation of meaning' (page 35). Even more ironic is the fact that in this same essay, some thirty pages later, Baudrillard does refer to the way in which colonization 'violently initiated . . . primitive societies into the expansive and centrifugal norm of Western systems' (page 59). Thus, I would prefer to suggest that Baudrillard has been well aware of the realities of the colonialist and imperialist project, with its conquests and violence, but that in his more recent work he is silent on the reality of resistance and the mobilizations and actions of the subaltern groups of the periphery; they are made to seem invisible, or, in shadowy references, the South becomes the object of a familiar Eurocentric condescension – the countries of the Third World will never internalize the values of democracy and technological progress' (Baudrillard, 1989, page 78).

In contrast to Baudrillard's earlier texts, there is a clear sense in which his more recent writing can be linked to Lyotard's considerations on the postmodern condition. In both cases, the Third World is largely present through its absence. As Lyotard (1986) states at the outset of his analysis of the postmodern condition, he is discussing the 'most highly developed societies'. At the same time the tendency to essentialize the 'traditional' and the 'developing', or in another more philosophical text (Lyotard, 1988, page 156) 'savage narratives', sits uneasily with the war on totality, and the critique of Western universalism. In fact, given some of the current realities of the 'most highly developed societies', it might be worthwhile subverting this implicit dichotomy, in postmodern style, so that we can consider the 'savage' inside the 'modern'.

From a rather different point of departure, Jameson, although generally regarded as a key figure of the postmodern turn in literary theory, would be difficult to include in the category of Lyotard's (1986, page 41) writers who have lost the nostalgia for a 'lost narrative'. In a number of articles, he has posited the need for a conception of the 'social totality' (Jameson, 1988, page 355), stressed the importance of the analyzing of global capital and class, and reaffirmed the need for 'systemic transformation', adding too a belief in the future reemergence of a new 'international proletariat' (Jameson, 1989a, page 44). Also, in contrast to Baudrillard and Lyotard, Jameson's interventions rarely marginalize the Third World; in fact he has underlined the crucial interlocking of First and Third World realities, writing that in terms of culture, awareness is central, and 'it would not be bad to generate the awareness that we in the superstate are at all times a presence in third world realities, that our affluence and power are in the process of doing something to them' (Jameson, 1989b, page 17). Equally brief but interesting observations are made on the significance of resistance in the periphery, where a connection is drawn with the idea of the emergence of collective subjects, beyond thc 'old bourgeois ego and schizophrenic subject of our organization society today' (page 21). It must be stressed here, however, that although Jameson does refer to the phenomenon of resistance, he does not develop any analysis of protest or social movements at the periphery; nor does he refer to any of the extensive literature on this theme. . . .

From this brief reading of certain representations of the postmodern genre, overall, the relationship with the Third World as 'other' would seem to be ambiguous, ambivalent, and certainly not free from ethnocentric traits, although as I have indicated there are important variations among the authors cited. Western ethnocentrism is certainly not explicitly defended, . . . and in the writings of Jameson and the earlier Baudrillard, the exploitative nature of the West's encounter with the periphery is dealt with or at least alluded to. . . . But, as I shall subsequently argue, the potential for renewal and reconstruction, present in the politics of the postmodern, will be nullified if the West is continually imagined as the transcendental pivot of analytical reflection.

Mapping Euro-Americanism

One of the most significant features of the contemporary debate on postmodernity is the fact that although phrases such as the end of social criticism or the end of history may well be prominently visible, the possibility of an 'end to ethnocentrism' has still to be voiced. Here, in fact, a common thread can be found with the critical theory of Habermas, and the foundational texts of classical Marxism. In this section, I want to trace out some of the recurrent elements of a universalism which is rooted in certain Western or Euro-American predilections concerning the elaboration of social theory and its contextual deployment; and, in this discussion, the context will be provided by critical geographical studies.

The persistence of absence

Theoretical development within a given domain of knowledge is very often propelled forward by the identification of important gaps or lacunae, the conceptualization of which can open up new pathways of thought. But if those 'forgotten zones' are structurally produced, the result of a deeper and occluded bias, their illumination will require far more effort and collective enterprise than otherwise would be the case. In much of the work of the contemporary theorists of spatiality, the complex territorial realities of peripheral societies and the changing nature of geopolitical imperatives, as they impinge on centre–periphery relations, rarely break the surface of Euro-American introspection. Those waves of conceptual innovation, which move out from the academic heartlands of the West, carry with them built-in assumptions and predetermined omissions which retard a genuine global expansion of meaning.

Influenced by Foucault, Hebdige (1990, pages vi–vii) has recently argued in a note on 'subjects in space' that spatial relations are no less complex and contradictory than historical processes. Space itself is refigured as inhabited and heterogeneous, 'as a moving cluster of points of intersection for manifold axes of power which can't be reduced to a unified plane or organized into a single narrative'; it is also suggested that some 'geo-political partitions', for example the North–South axis, have proved less permeable than others (for example, the Berlin Wall). The geopolitical divide of the North–South axis, with its antecedents in colonialism, and its contemporary expression in new forms of imperialism, rarely receives any concerted analysis in the work of geographers such as Gregory (1989), Harvey (1989), and Soja (1989).[1] In the domains of anthropology, literary theory, historical sociology, and feminist studies, the long-established discussions of colonial discourses, of the creative realities of postcolonial texts, and the impact of the contemporary imperial order do not seem to find consistent echoes in the thematic priorities of critical geographers. In addition to the androcentric drive of much of the work of the malestream of critical geography, a phenomenon that has been and continues to be interrogated, the Eurocentric or more broadly Euro-Americanist focus of critical geography creates a series of what Derrida (1981, page 3) has termed the 'blank spaces' of texts.

Thematic absences and the structured silences of texts also mean that within given provinces of knowledge and enquiry certain debates do not unfold. In a recent paper on locality Cooke (1989), who seems implicitly to endorse Rorty's view that ethnocentrism is unavoidable, makes a number of comments on imperialism and the world order which are in some ways reflective of an absence of debate. For Cooke there has been a failure of established historical modes of organization, for example, of corporatism, statism, and imperialism, to deal with real world crises (page 16). Moreover, it is suggested that there has been a flattening of the hierarchy of world and imperial power, and an 'undermining of its structure of internal social domination' (page 16). As a consequence, a new configuration of individual dependencies has emerged, in which 'negotiation and involvement are key features' (page 17). As a further illustration, Cooke adds that in contrast to thirty years ago when the British government high-handedly sent its troops into Egypt at the time of the Suez crisis, today it has to 'negotiate with its political equals' (page 17).

It ought to be indicated here that Cooke's article overall is stimulating and well argued, and his references to imperialism are not central to his argument. But it is exactly in the interstices of the text, on the margins of the narrative, that we encounter comments that are symptomatic of a much wider problem. A perspective which presumes that imperialism has failed, that its structure of internal social domination has been undermined, and that we are now living in an epoch of negotiation and involvement, where imperial powers are no longer able to send their troops into the recalcitrant periphery, is a perspective strangely out of touch with the geopolitical realities of world power. Prior to 1989, the invasion of Grenada in 1983, the 1982 Malvinas/Falklands War, the strategy of low-intensity warfare waged against Nicaragua, and the bombing of Libya were only some of the more overt and obvious examples of an imperialism which is hardly 'undermined' or moribund. What for some critical theorists of the urban and regional may be a blank space is in other domains of critical knowledge an area of the most central significance. I shall return to this point below.

There is another absence; the absence of Third World voices in the discussion of contemporary issues and theoretical strategies. In the representation of what are presumed to be the central axes of argument and debate the key authors who are cited and granted authority are invariably of Euro-American origin. This applies not only to texts which deal specifically with European or North American societies, but also to studies of peripheral societies, where the work of autochthonous intellectuals is not infrequently bypassed.[2] One of the negative consequences of this kind of absence is that there is no symbiosis of learning. Whereas the latest ideas of First World geographers are diffused to the periphery, and frequently adopted, the knowledge and theoretical ideas of Third World scholars do not appear on the agendas of their Western counterparts. The need here, therefore, is for a displacement; as Julien and Mercer (1988, page 6) express it, summarizing Stuart Hall, the displacement of the centred discourses of the West, 'entails putting into question its universalist character and its transcendental claims to speak for everyone, while being itself everywhere and nowhere'.

In general, I would argue that in so much of today's critical geography, the West is implicitly viewed as a self-contained entity, as if somehow it can be apprehended and comprehended of and by itself. There appears to be little realization that it is in the outskirts that the system often reveals its true face, or as Cocks (1989, page 4) defines it, in her book on the 'oppositional imagination', 'the political advantage in looking at peripheries and extremities is that power is exposed in what it drives from the center of life to the edges, and in what it incites as its own antitheses'.

Assumptions of universality

Concepts such as the spatial division of labour, the hypermobility of capital, territorial power, and localization are often used without any specification of their pertaining cultural and historical contexts. There is an underlying assumption of universal applicability, but how is it possible to refer to *the* spatial division of labour when the content of this division can change radically from one sociopolitical context to another, or from one historical juncture to another? Rather, this projected universality tends to conceal a particularity based, to a large extent, on the specific experiences of the USA and the United Kingdom. This projected universality, which exerts far more influence through being implicit and partially submerged, rests on the assumption that 'the West' (notwithstanding its complexities and contradictions) constitutes the primary referent in theory and practice. In relation to feminist theory, and 'making sense of the local', Probyn (1990, page 176) appropriately writes, 'in creating our own centers and our own locals, we tend to forget that our own centers displace others into the peripheries of our making'.

Some of the problems that surface as a result of the pervasiveness of universalist assumptions can be illustrated from Harvey's (1989) fascinating and thought-provoking analysis of the 'politics of space'. In his discussion of the Enlightenment project, he begins by admitting to a somewhat ethnocentric method of concentrating on the European case. The reader is then presented with a concise but one has to say largely uncritical survey of time and space in relation to the discourse of the Enlightenment. It is mentioned, somewhat en passant, that Enlightenment thought perceived the other as 'necessarily having . . . a specific *place* in a spatial order that was ethnocentrically conceived to have homogeneous and absolute qualities' (page 252), but this potential opening to a wider and more critical discussion is not developed.[3] At the same time, when Harvey goes on to outline a few ideas concerning 'the dilemmas of the politics of space in any kind of project to transform society' (page 254), we can see the continuation of a universalist current.

Leaning on Lefebvre, who suggested that one can find a permanent tension between the free appropriation of space for individual and social purposes and the domination of space through private property, the state, and other forms of class and social power, Harvey extracts a number of explicit dilemmas. From a concern with establishing the varying social principles by and around which space is 'pulverized' and fragmented, through an identification of the differing aspects of the 'production of space' in the France and USA of the eighteenth century, to the suggestion that a growing commodification began to render place subservient to transformations of space, Harvey ends with what he refers to as the most serious dilemma of all. This is the fact that space can only be conquered through the production of space, that is, through the development of transport and communications, systems of administration, and human settlement and occupancy; this production of space then forms a fixed frame, so that Harvey can argue that in the context of capital accumulation this 'fixity of spatial organization becomes heightened into an absolute contradiction' (page 258). The result is to 'unleash capitalism's powers of "creative destruction" upon the geographical landscape, sparking violent movements of opposition'. Harvey then seeks to generalize this point, writing that not only may crises of overaccumulation be connected to 'time–space compression', but also 'crises in cultural and political forms'.

Apart from the orthodox base–superstructure connection, to which I shall return below, the main problem with this discussion of the dilemmas involved in the politics of space is that although it is argued that these tensions or dilemmas are at the heart of 'any kind of project to transform society', the examples are taken from the Euro-American heartland of history, during the eighteenth century. Does this mean, therefore, that not only is there, in relation to central societies, a clear historical continuity from the Enlightenment era to the present, but that in addition one can posit a universal relevance for all societies? Or are some or all of these dilemmas specific or *particular* to the Euro-American core? Further, are there *other* dilemmas of the politics of space that escape Harvey's classification?

According to Harvey, Enlightenment thinkers 'sought a better society', and all the projects of this era had in common a 'relatively unified common-sense of what space and time were about and why their rational ordering was important' (page 258). Unfortunately, Enlightenment visions of this 'better-society' were frequently racist. The Scottish philosopher of the Enlightenment, Hume, writing half way through the eighteenth century, clearly conflated race with intelligence; he remarked, for instance, that, 'I am apt to suspect the negroes . . . to be naturally inferior to the whites . . . there never was a civilized nation of any other complexion than white, nor even any individual eminent in action or speculation' (quoted in Gates, 1985, page 10). Following Hume's lead, Kant writing in the 1760s declared that 'so fundamental is the difference between [the black and white] races of man . . . it appears to be as great in regard to mental capacities as in color'; and, moreover, for this leading philosopher of the Enlightenment age, to be 'black' was also to be 'stupid' (page 10).

The Enlightenment legitimated and paved the way for colonial conquests and the subordination of 'peoples without history'. Said (1985) documents this process in considerable detail. During the eighteenth century Europe moved itself outwards; and through, for instance, the various India Companies colonies were already created and ethnocentric perspectives installed. There was also historical confrontation which gave Europe the opportunity to deal more assertively with non-European cultures, its 'own previously unreachable temporal and cultural frontiers' (Said, 1985, page 120). Further, there was a steady dissolution of the self-contained visions of Europe – a broadening of horizons, and an appreciation of 'exotic locales'. Last, the classifications of humankind were extended to include other peoples and races; thus, when an Oriental was referred to, 'it was in terms of such genetic universals as his "primitive" state, his primary characteristics, his particular spiritual background' (page 120).

In relation to the critical tradition of historical materialism, the key source for both Harvey's and Lefebvre's projects, the central link with Enlightenment discourse comes through Hegel. The ethnocentrism embedded in Hegel's philosophical work has certainly not escaped critical scrutiny, but the depth of its impact on the Marxist tradition is not always emphasized. To establish the influence one may refer to Hegel's *Philosophy of Right* (1967), originally published in 1821. Here Hegel argues that the intrinsic value of courage as a mental disposition is located in the 'genuine, absolute, final end, the sovereignty of the state', and that the nation, that 'ethical substance', cannot become a nation-state without the due accompaniment of objective law and an explicitly established rational consumption (page 219). For Hegel the nation-state is 'mind in its substantive rationality and immediate actuality and is therefore the absolute power on earth' (page 212). It follows from this that every state is sovereign and autonomous against its neighbours, *unless* the referent is a people on a 'low level of civilization'. This means that the 'civilized nations' are justified in treating as barbarians those who 'lag behind them in institutions which are the essential moments of the state' (page 219). Thus, Hegel goes on, the civilized nation is conscious that the rights of barbarians are unequal to its own and 'treats their autonomy as only a formality'. Of the four world historical realms (the Oriental, the Greek, the Roman, and the Germanic) Hegel draws a fixed hierarchy with the Oriental as the least developed – 'its inner calm is merely the calm of non-political life and immersion in feebleness and exhaustion' – and the Germanic, 'the principle of the north', the most developed, where the mind 'receives in its inner life its truth and concrete essence, while in objectivity it is at home and reconciled with itself' (pages 219–222).

The posited principle of European superiority – 'thought and the universal' – is employed in the service of colonialist conquest. For Hegel the true courage of 'civilized nations is readiness for sacrifice in the service of the state'; this courage is not the personal mettle of the individual but the alignment of the many with 'the universal' (page 296). In India, Hegel writes, five hundred men conquered twenty thousand who were not cowards but only 'lacked this disposition to work in close cooperation with others.'[4]

In his discussion of civil society, Hegel makes it clear that the developed or mature civil society is 'driven' by its inner dialectic into colonizing activity, by which it supplies to a part of its population a return to life on the family basis in a new land and so also supplies itself with a new demand and field for its industry. At the same time, the public authority must provide for the interests that 'lead beyond the borders of its society', and its primary purpose is to 'actualize and maintain the universal contained within the particularity of civil society' (pages 151–152).

In terms of subsequent perspectives on colonialism, the universal and the imputed inferiority of other peoples, the writings of Marx and Engels bears witness to the Hegelian legacy. At the end of the 1850s, Engels, in a short article on Persia and China, referred to the notion of 'barbaric nations', and, in the context of the difficulties perceived in the extension of European military organization to the Persian region, the reader was reminded of the 'ignorance, impatience and prejudice of the Oriental' (Engels, 1968, page 122).[5] In a somewhat more analytical mode, Marx (1968, page 81), discussing the 'future results of British rule in India', declared that Indian society had no known history, and that, therefore, the question was not whether the English had a right to conquer India, 'but whether we are to prefer India conquered by the Turk, by the Persian, by the Russian, to India conquered by the Briton'.[6] As societies such as that of India are not granted any known history, recalling Hegel's 'peoples without history', it is only a small step to argue that their conquest by an already presupposed superior nation or culture must be in their long-term historical interest. To argue otherwise would be to indulge in obscurantist sentimentality.[7]

After discussing the Enlightenment, Harvey continues his analysis of 'time–space compression' by focusing on the rise of modernism as a cultural force; the reader is brought through the nineteenth century and introduced to the twentieth. In a chapter of just under twenty-five pages, imperialism is assigned one paragraph, in which it is observed that by World War 1 the 'world's spaces were deterritorialized, stripped of their preceding significations, and then reterritorialized according to the convenience of colonial and imperial administration' (1989, page 264). Could it be that one of the significant dilemmas facing colonial administrations concerned the politics of territorial power in those other peripheral worlds? The reference to deterritorialization and reterritorialization, which come directly from Deleuze and Guattari (1984), implicitly points to questions of culture, of community, and of conceptions of territory which Harvey hints at with his 'significations'.

Imperialist expansion, the will to spatial power, was not separated from previous cultural and economic developments; as Said (1985) so convincingly demonstrates, by the end of the nineteenth century the geographical space of the Orient had been penetrated, worked over, and taken hold of. For such an enterprise, a geographical imagination was needed. One French thinker of the early twentieth century averted that colonization represented the expansive, reproductive force of a mature people; 'it is its enlargement and its multiplication through space . . . it is the subjection of the universe or a vast part of it to that people's language, customs, ideas, and laws' (Said, 1985, page 219). But in reality that spatial project of subjection met with all kinds of resistances, with a complex of insurgent spatial practices. Within the territories of colonial rule, the 'production of space', to follow Lefebvre, reflected an imperialist imperative that was not only socioeconomic but cultural and political. The deployment of and resistance to territorial power under colonialism suggest another kind of spatial politics; a spatial politics that subverts the Euro-American centricity of critical geography.

Meaning, including the spatial, is socially constructed. Hence, the deployment of a particular universal concept such as 'class' or 'capital' or 'gender' outside a consideration of the discursive constructions which give these concepts different meanings in different societies, will disguise the particularity being given to the 'universal'. In this case, and in many others, the particularity and specificity of the Euro-American 'universal' remain invisible, submerged beneath the assumptions of generality. Harvey, in writing about modernization, asserts that money fuses the political and the economic into a genuine political economy of overwhelming power relations; 'the common material languages of money and commodities provide a universal basis within market capitalism for linking everyone into an identical system of market valuation and so procuring the reproduction of social life through an objectively grounded system of social bonding' (1989, page 102). For Harvey, money unifies because of its capacity to 'accommodate individualism, otherness, and extraordinary social fragmentation' (page 103). In this context, as with Harvey's use of other Marxist concepts, money is being used as a terminal abstraction to bestow meaning on differential social practice and subjectivity. It is social agents who give money a variety of meanings and not money which assigns meanings to agents. Conversely, for Harvey, the system of 'social bonding' and the 'universal basis' for the reproduction of social life have to be explained by the power of economic abstractions over the construction of social meaning by human agents.

Throughout Harvey's text, universalism and econocentrism reinforce each other. In referring to nationalism, at the end of his chapter on modernism as a cultural force, reference is made to the trauma of World War 2 and to the idea that geopolitical and aesthetic interventions always seem to imply nationalist, 'and hence unavoidably reactionary, politics' (page 283). In an earlier discussion of Heidegger and National Socialism, Harvey includes a reference to the Sandinistas, noting that there can also be a left-wing version of the 'aestheticization of politics', and that the 'rhetoric of national liberation movements' has a close link to the appeal to the 'mythology of place and person' (page 209). Leaving aside the dissonance between the two passages, I want to make two points.

First, it is clear that Harvey's generally negative view of nationalism has been influenced by the repugnant discourse of Nazi geopolitics, and on that basis there is a strong tendency, as Harvey remains within the orbit of the Euro-American world, to view nationalism *tout court* with a good deal of suspicion. The 'rhetoric' of national liberation will then do for all the complexities and divergences of the national question in the Third World.

Second, nationalism is seen as a deviation from the centrality of the class struggle; it is a force which is suspect because it transgresses the boundaries of class meaning. After Lefebvre, it is argued that the whole history of territorial organization, colonization and imperialism, uneven geographical development, urban and rural contradictions, and geopolitical conflict all testify to the 'importance of such struggles within the history of capitalism'

(Harvey, 1989, page 237). Even when it is concluded that race, gender, and religion should be recuperated, it is within the overall frame of class politics, with its emphasis on the 'unity of the emancipatory struggle', and the power of money and capital circulation (page 355). However, for the periphery at least, the symbols of the 'national-popular' have been and remain of crucial political significance; they cannot be slotted into the straitjacket of class; they are open and indeterminate, calling for the *construction* of meaning and not for the application of an already constituted class formula.[8]

Running through Harvey's approach, and he is certainly not alone, there is an implicitly drawn political frontier. On one side of the divide one has forces and tendencies which belong to the conservative other: nationalism, democracy, the individual, aesthetics, myths, symbols, and desire, and on the opposed side one has class, socialism, collectivity, the workers, and struggle and toil. The task is to counteridentify, not to subvert and rearticulate; the meanings are given, the divide is drawn, the abstractions have power. And those abstractions are not concretely situated, nor are the views which are expressed; as Schick (1990), puts it, in her critical comment on First World feminism, the key issue is that views must be articulated as 'situated' opinions and not as universal truths spoken from the vantage point of a more 'advanced' society.

The problem of 'worlding'

It is possible to talk of the double bind of ethnocentric universalism. Thus, there is not only the supposition that the West acts as the primary referent for theory and philosophical reflection, but also a frequent inclination to express an interest in the periphery in the context of information retrieval, the incorporation of token, often stereotyped, themes such as the 'culture of tradition' or 'violence', and above all the simplification of the heterogeneity of the periphery. Further, not only is the West the pivot of theoretical reflection, but it is constituted, in opposition to the 'rest', as complex, differentiated, heterogeneous, and innovative.

The dark side of modernity, within the societies of the West, is not infrequently bypassed. Interestingly, and somewhat symptomatically, the radical critique of modernization theory, which found a home in geography as well as in other domains, tended to concentrate its fire on the inadequate and slanted portrayal of the Third World together, of course, with the needed interrogation of the thesis that relations between the First World and the Third World were beneficial for the latter. What was missing was a sustained critique of the discursive construction of the 'First World', and in particular of the bases on which 'worlding' unfolded. As a clear example of this tendency, within modernization theory, the stunning idealization of Western society, and in particular of the USA, was not given any critical priority.

The principles on which 'worlding' has taken place and continues to take place have received insufficient treatment in critical geography. This has a number of dimensions; it connects with the relative absence of a sociocultural and political critique of the foundation of Western societies; it manifests itself in the context of the paucity of research on the colonial and Third World other, and it finds expression in the design of theoretical debate and the construction of analytical agendas. As has been observed elsewhere, it is not uncommon for the authority of the centre to be based on its absence – that is, its specificity remains unacknowledged.[9] The centre is invisible, so that one of the primary functions of critique is to make this invisibility *visible*, just as the feminist critique has made the gender-blind nature of much social and economic theory visible.

The purpose, however, is not, to paraphrase Hall, to engage in the politics of simple reversal; the idea ought not to be to replace 'the essential, bad, white (First World) social subject' by the 'essential, good, black (Third World) social subject' but to recognize diversity, openness, and the plurality of agency and subjectivity.[10] To communicate essentialist notions of the Third World as 'positive' is reminiscent of what West (1990, page 28) refers to as the strategy of many Third World 'authoritarian bureaucratic elites who deploy essentialist rhetorics about "homogeneous national communities" and "positive images"' in order to 'repress and regiment their diverse and heterogeneous populations'. Similarly, although, within a certain postmodern current, it may be trendy to be marginal, 'in' to be different and peripheral, there is at the same time a tendency here to fail to specify the types of 'marginality', an inclination to romanticize poverty, and an orientation which translates the enforced 'creativity' of everyday survival into aesthetic (and anaesthetized) images of 'self-help', primitive ingenuity, and learning to become 'squatter-wise'.

I have previously argued that critical geography, as well as to a considerable extent the broader (that is, in a disciplinary sense) domain of urban and regional research, is characterized by a high degree of Anglo-American centricity.[11] It has been noted above, and certainly not as a new observation, that Marxism carries a Eurocentric imprint; Soja (1989) refers to the same phenomenon, but strangely fails to anchor the insight into critical–postmodern geography. Gregory (1989) persists in his attempt to convince a geographical audience of the crucial significance of Habermas's project, and along the way seems quite unaware of and/or unconcerned about the profound Eurocentrism of Habermas's work.[12]

Into the 1990s it would seem increasingly evident, certainly in Soja's (1989) recent text, as well as in Harvey's work, that many of the ideas (especially on space) of the French philosopher and Marxist Lefebvre are being incorporated into one of the more influential currents of critical geography. Outside the Anglo-American frame, and expressive of an inimitable French intellectual cachet, Lefebvre's contribution can be seen as highly relevant for a rethinking and reassertion of the significance of space. Lefebvre's project is also instructive with respect to the general issue of Eurocentrism, and the specific problem of situating different 'worlds'.

In a specific reference to the possible meanings of 'periphery', Lefebvre (1976a, pages 115–116) distinguishes the following: (a) the 'so-called underdeveloped countries, particularly the ex-colonial ones'; (b) the regions which are distant from the centres within the capitalist countries themselves (for example, Sicily and the South in Italy); (c) the urban peripheries – 'the inhabitants of the suburbs, immigrant workers in the *bidonvilles*, etc'; and last, (d) the social and political peripheries – 'particularly youth and women, homosexuals, the desperate, the ''mad'' and the drugtakers'. Lefebvre goes on to suggest, with a thinly disguised reference to Foucault, that the 'leftism' which concentrates on the fourth kind of sociopolitical periphery is not wrong, for it directly or indirectly prepares, 'even puts into practice, a critique of power which is more radical than the critique that is addressed solely to the economic' (page 116). But he reaffirms, this 'tactic, which concentrates on the peripheries and only on the peripheries, simply ends up with a lot of pin-prick operations which are separated from each other in time and space . . . it neglects the centres and centrality; it neglects the global' (page 116). I shall return to this point below.

Apart from the fleeting reference to underdeveloped countries in this 1976 text, other considerations on the Third World rarely surface in Lefebvre's work. In his well-known text on the production of space, however, when the focus is specifically on the social, Lefebvre does introduce a few comments on the colonial city, especially in relation to Latin America. He connects the production of space within the colony with the imperatives of conquest: discover, people, pacify. Moreover, he explicitly highlights the links that bind the organization of space in colonial Latin America with accumulation in Western Europe (Lefebvre, 1974, pages 176–178).

Given these sorts of passages can it be concluded that Lefebvre avoids the all too customary assumptions of Western universality? Not entirely, as there are references in his work to the presupposed equivalence of social needs with the urban. For example, in the text on the survival of capitalism, it is posited that 'social needs today are, above all, urban needs' (Lefebvre, 1976a, page 37). Relevant for most of the First World, how far can this idea be geographically stretched? More telling, perhaps, is the way Lefebvre has recently discussed culture and the 'everyday'. For Lefebvre a discussion of culture in relation to Marxist theory appears to be unquestionably rooted in European experience. The concept of *la vie quotidienne* (daily life) is distinguished from *le quotidien* (the everyday), and for Lefebvre 'the everyday designates the entry of this daily life into modernity: the everyday as an object of a programming . . . whose unfolding is imposed by the market, by the system of equivalences, by marketing and advertisements' (Lefebvre, 1988, page 87). How might this set of designations relate to the Third World? For Lefebvre, there are hundreds of millions of poor people in the world aspiring to everyday life. These are the people who are outside modernity but aspiring to be on the inside, under the 'system of equivalences' that is the market. In calling for the elaboration of a much needed cultural project, Lefebvre moves in the direction of the social and political; 'more enhanced forms of democracy, such as direct democracy in cities; definition of a new citizenship; decentralization; participatory self-management and so on' (page 86). However, he excludes from this project the thousands of millions of poor people of the Third World, whose 'incomes are well below the social average' (page 88). These people, for Lefebvre, are still outside modernity, still not subjects of the 'everyday'. It has been continually shown that poor people in the exemplifying regions that Lefebvre mentions – Northeast Brazil, Upper Volta and the 'Mexican *campamientos*' – survive within a market system, and nor are they somehow outside the cultural boundaries of capitalist modernity, even though there are many other indigenous cultural traits which merge with or remain on the fringes of this modernity. The way Lefebvre constructs the main lines of his cultural project leads to an inevitable marginalization of Third World culture. Moreover, the complexity of these societies is flattened into an image of enduring poverty coupled with the aspiration to make it into the West's advancing project of modernity.

Unfortunately, Lefebvre's position here is all too symptomatic of a wider perspective in which it is assumed that the Third World is a cultural desert. Not only is the Third World caricatured within a Western-oriented frame, but simultaneously there is no awareness that the intellectual in the West can *learn* from the Third World other. The idea that these societies have deeply rooted cultures and experiences which connect with the debate on decentralization, direct democracy, notions of citizenship, movements of resistance, the place of culture in political development, and the contemporary analysis of globalization has not been grasped. Lefebvre is not alone.

Above I drew attention to Lefebvre's emphasis on the global and on the central. This was in the context of the discussion of the peripheral and Lefebvre's idea, contra Foucault, that the radical critique must not lose sight of the centres of power, of capitalism, and of state power. In a spatial sense Lefebvre (1974) stresses the imbrication of the global and the local; spatial levels do not cancel each other out – rather they are interpenetrated, imbricated. Space is also deeply political (Lefebvre, 1976b). But if space is political, how do we characterize the political spatiality of the Third World? Also, from Lefebvre's point on the global and the local, how might this imbrication, taken with the national and the regional, acquire changing meanings in peripheral contexts?[13]

To begin to approach these kinds of questions requires, in the first place, an appreciation of the fact that the coevalness of the global and the local applies to the Third World also. Let us for a moment remember one of Foucault's observations in his paper on other spaces. He wrote:

> The present epoch will perhaps be above all the epoch of space . . . we are in the epoch of simultaneity: we are in the epoch of juxtaposition, the epoch of the near and far, of the side-by-side, of the dispersed. We are at a moment, I believe, when our experience of the world is less that of a long life developing through time than that of a network that connects points and intersects with its own skein (Foucault, 1986, page 22).

But this epoch of space, of intersection, and of simultaneity is also home to visions of global development that assign the peoples of the periphery another time, and another space. As Fabian (1983) has convincingly shown, in anthropology, the 'other' was kept in another time – behind, backwards, primitive, traditional, underdeveloped. Coevalness was denied; that is, the problematic but crucial simultaneity of different, conflicting, and contradictory forms of consciousness was removed from the agenda of international relations. For Fabian, geopolitics has been based on chronopolitics. The spatial expansion of the West – conquest, plunder, colonialism, and imperialism – are rooted in a conception of time that establishes a binary division between the advanced centre and the backward peripheral other, that is not in the same time but in another previous backward time.

Returning to Lefebvre, I would argue that his oeuvre is also emblematic of this chronopolitical division. Equally, Lefebvre's narrative creates the notion of other spaces that belong in a different world, a world that is not coeval, not of the same time–space nexus. As a consequence, it is not possible to penetrate the full meaning of the 'global', nor of the 'local', nor to begin to try to comprehend their simultaneous intersection – intermeaning. I would argue that this point is relevant right into the heartland of critical geography.

The place of the South

Apart from one or two exceptions (Folch-Serra, 1989), the possible pathways of communication across the terrains of critical geography, postmodernism, and Third World studies still remain predominantly unexplored. The feminist critique of critical geography's androcentrism raises the crucial issue of alterity, but, unfortunately, tends to do so only within a Western orbit. Similarly, the postmodern interruption has done little to subvert the ethnocentric currents of First World discourse; and yet what could be more relevant than seeing the North as 'other'; of concentrating on the peculiarities of the West, of the persistence of the 'savage', and the 'traditional' within the 'civilized' and the 'modern'?[14]

At the same time, it seems clear that within many expressions of postmodern writing there is, as Poppi (1991, page 86) suggests, a 'celebration of difference'. But if that celebration carries with it the belief that synchronicity is the only legitimate site of historical evaluation, if the symbols, products, and meanings of other cultures are decontextualized, and then juxtaposed or combined with similarly extracted elements, the historical and political vitality of the distinct and of the distinguishable will be erased. It is the analysis of the distinct of what Laclau (1991) refers to as particularism – religion, ethnicity, nationalism – and, we can add, regionalism and the local, that is becoming increasingly vital across so many territories. In this sense, it is now necessary to return to and conclude with the question of learning from other regions.

First, as I argued previously in discussing the persistence of absence, learning from the realities of the South must include listening to Third World voices. The analytical and reflective contribution of non-Western intellectuals, working in both the North and the South needs to be taken far more seriously within the domain of critical geography. The penetrating analyses of Said and Spivak, which are characterized by creative approaches to subjectivity, space, and the other, provide perhaps one of the most immediately obvious reflections of such a contribution.[15]

Second, the critique of colonialism and imperialism, and of the dynamic nature of the varying modes of Western penetration of the periphery, including all the complexities and heterogeneities of the societies falling under that rubric, is essential to the learning process. As was noted previously, the history of colonialism is a counterhistory to the normative, traditional history of the West. In our context, and in an identical frame, it could be suggested that the geography of colonialism can be a countergeography. Within the West's critical geography implicit notions of the West as a self-contained entity must be continually confronted. By the development of critiques of colonialism and imperialism, we are not only helping to contribute to the understanding of Third World reality, but also of the reality of the West. As Fabian (1991, page 198) expresses it, in his comments on critical anthropology, 'we need critique . . . to *help ourselves* . . . the catch is, of course, that "ourselves" ought to be them as well as us'. In the context of the current resurgence of racism in the West, the imbrication of the South in the North and the North in the South has never been a more urgent and fundamental issue.

Last, in a geopolitical sense, I would argue that we can learn from other regions by realizing that it is always the 'marginal' or 'peripheral' case which reveals that which does not appear immediately visible in what seem to be more 'normal' cases. For example, the struggle against the exclusionary logic of racism in South Africa reveals the hidden forms of the same logic in our societies, and US aggression against the Sandinista regime exposed the ultimate limit of liberal regimes, just as tacit acceptance of the Israeli appropriation and illegal occupation of Arab land has revealed the duplicitous attitude of the West to international law. In this way, Managua, the 'occupied territories', and Soweto become the names of the frontiers through which our own political identities are constructed (Laclau, 1990).

The above three suggestions are meant to represent possible elements in a learning process. There are, of course, other features which, for example, relate to the deployment of spatial concepts (Slater, 1989), or touch on the highly complex field of nationalist struggles and popular culture. All I am indicating in this short conclusion is the outline of a possible opening and a reemphasis of previously discussed points.

On the outer rim, on the borders of critical geography, in its Western tradition, there is another starting point. That starting point is not just the expression of curiosity in the Third World other; the incorporation of stereotyped thematics. It requires both a looking within, an interrogation of the history and geography of the West and its encounter with its others, *and* a listening to other voices, literatures, representations. Fanon (1969, page 193) once talked about a 'literature of combat', combat has many terrains, including the fight for new ideas and new territories of enquiry and debate. Critical geography needs to be reconstructed not only through the feminist critique but also through an interrogation of its ethnocentric universalism. In this intervention I have tried to show why that task needs to be put on the agenda . . . *and now*.

Notes

1 It is important to indicate here that I am referring to these three more recent texts. In earlier work Harvey has dealt with aspects of colonialism and imperialism, and attempted to develop some theoretical ideas on their spatial dimensions. In contrast, Gregory tends to treat the West as a self-contained entity, whereas Soja does still maintain an interest in 'core–periphery' issues (Soja, 1989), even though these have

now been overshadowed by a notion that the postmodern world comes together (and falls apart) in the creative sprawl of Los Angeles.

2 I have made reference to this problem in an earlier article (Slater, 1989), but in that instance I was referring to the literature of mainstream development geographers who, although in some cases working for many years on Third World societies, somehow find it possible to ignore the work of their social science counterparts born and based in these societies.

3 For example, when Harvey (1989, page 249) talks about the development of accurate maps and chronometers during the Enlightenment period, and the fact that abstract maps were being used to define property rights in land, territorial boundaries, domains of administration, and social control and communication routes, no clear link is made to the spatial expansion of European power, and to the onset of colonialism. These connections can be found in Said's (1985) *Orientalism*, where, for instance, he reminds us of the statement of a Swiss–Prussian authority on international law who in 1758 invited European states to take possession of territory inhabited only by mere wandering tribes (page 216).

4 The reference is to Clive in the India of 1751. Hegel's text is marked by a series of highly pejorative comments on India – 'its complete degeneration and inner corruption' (1967, page 133) – 'stagnant and sunk in the most frightful and scandalous superstition' (page 151).

5 In other similar passages, Engels talks of the 'stupidity' and 'pedantic barbarism' of the Chinese (1968, page 124), or of that 'timid race', the Moors, with their 'very low moral character' (page 157).

6 In other passages, Marx expresses some feeling for the plight of the poor Indians, but as Said (1985, page 155) shows, in one of his typically eloquent and incisive commentaries, 'the vocabulary of emotion dissipated as it submitted to the lexicographical police action of Orientalist science and even Orientalist art' . . . 'an experience was dislodged by a dictionary definition: one can almost see that happen in Marx's Indian essays, where what finally occurs is that something forces him to scurry back to Goethe, there to stand in his protective Orientalized Orient.'

7 And here we have an immediate link with Warren's (1980) thoughts on colonialism, and the ideas, sometimes only thinly veiled, of other Anglo-American development specialists. For a critique see my earlier paper on economism and development theory (Slater, 1987).

8 I shall deal, although only briefly as this is not one of the major arguments of the paper, with the problem of the centrality of class analysis towards the end of the discussion; I have also examined this issue elsewhere (see Slater, 1992).

9 In discussing marginalization and contemporary culture, Ferguson (1990, page 9) suggests that too often the alternatives to dominant cultural power have been successfully segregated, 'so that many different bodies of marginalized creative production exist in uneasy isolation . . . Such isolation can only contribute to the security of a *political* power which implicitly defines itself as a representative of a stable center around which everyone else must be arranged'. Such a phenomenon has a wide variety of potential meanings, including the politics of knowledge within the academy.

10 This passage is referred to in Julien and Mercer (1988, page 5).

11 When writing this phrase, it occurred to me that one could also think of 'Anglo-American eccentricity', in the sense that certainly in the case of the United Kingdom many of the historical tendencies of its capitalist development and its state–society relations possess a certain 'peculiarity', in relation to other European countries, as well as to peripheral societies, that casts a surreal shadow over all

those universalist conceptualizations that emanate from this country. Of course, the term 'Anglo' has its problems as it conflates different ethnic and linguistic identities present within that United Kingdom.

12 In an interview in the 1980s, Habermas was asked if the conception of socialism developed in the course of the anti-imperialist struggles in the Third World had any lessons for the tasks of democratic socialism in the advanced capitalist world and if his own critical theory of the advanced capitalist world had any lessons for socialist forces in the Third World. He replied: 'I am tempted to say "no" in both cases. I am aware of the fact that this is a eurocentrically limited view. I would rather pass the question' (quoted in Pantham, 1988, page 187). In the same vein as Harvey, he admits to being eurocentric and then proceeds with his analysis. For an interesting critique of Habermas, set in the context of Indian political theory, see Pantham (1988).

13 For an analysis of a parallel debate with reference to feminist studies and Western ethnocentrism, see Mohanty's (1988) important contribution.

14 For example, from the advent of slavery, through the 'modernizing innovation' of colonialism, to the barbarism of Fascism, and to the machine-like violence of today's serial murderer, the dark side of the Occident can be brought into focus. But in much, although as I showed not in all, postmodern reflection, this darkness is not deployed to expose the 'light' of the modern.

15 In related papers, I have referred to the work of a variety of Third World social scientists (see Slater, 1989; 1992).

Selected references

Baudrillard J, 1975 *The Mirror of Production* (Telos, St Louis, MO)

Baudrillard J, 1983 *In the Shadow of the Silent Majorities* (Semiotext(e), New York)

Baudrillard J, 1989 *America* (Verso, London)

Bhabha H, 1990, 'The third space', in *Identity: Community, Culture, Difference* Ed. J Rutherford (Lawrence and Wishart, London) pp 207–221

Cocks J, 1989 *The Oppositional Imagination: Feminism, Critique and Political Theory* (Routledge, Chapman and Hall, Andover, Hants)

Cooke P, 1989, 'The contested terrain of locality studies' *Tijdschrift voor Economische en Sociale Geografie* **80**(1) 14–29

Dallmayr F, 1989 *Margins of Political Discourse* (State University of New York Press, Albany, NY)

Deleuze G, Guattari F, 1984 *Anti-Oedipus: Capitalism and Schizophrenia* (Athlone Press, London)

Derrida J, 1976 *Of Grammatology* (Johns Hopkins University Press, Baltimore, MD)

Derrida J, 1981 *Positions* (University of Chicago Press, Chicago, IL)

Derrida J, 1985, 'Racism's last word', in "*Race*", *Writing, and Difference* Ed. H L Gates Jr (University of Chicago Press, Chicago, IL) pp 329–338

Engels F, 1968, 'Persia and China', in *On Colonialism* K Marx, F Engels (Progress Publishers, Moscow) pp 120–125

Fabian J, 1983 *Time and the Other* (Columbia University Press, New York)

Fabian J, 1991, 'Dilemmas of critical anthropology', in *Constructing Knowledge: Authority and Critique in Social Science* Eds L Nencel, P Pels (Sage, Newbury Park, CA) pp 180–202

Fanon F, 1969 *The Wretched of the Earth* (Penguin Books, Harmondsworth, Middlesex)

Ferguson R, 1990, 'The invisible center', in *Out There: Marginalization and Contemporary Cultures* Eds R Ferguson, M Gever, T T Minha-ha, C West (The New

Museum of Contemporary Art and Massachusetts Institute of Technology, Cambridge, MA) pp 9–14

Folch-Serra M, 1989, 'Geography and post-modernism: linking humanism and development studies' *The Canadian Geographer* **33**(1) 66–75

Foucault M, 1970 *The Order of Things: The Archaeology of the Human Sciences* (Random House, New York)

Foucault M, 1986, 'Of other spaces' *Diacritics* (Spring) 22–27

Gates H L Jr, 1985, 'Introduction: writing "race" and the difference it makes', in *"Race", Writing and Difference* Ed. H L Gates Jr (University of Chicago Press, Chicago, IL) pp 1–20

Gregory D, 1989, 'The crisis of modernity? Human geography and critical social theory', in *New Models in Geography, Volume 2* Eds R Peet, N Thrift (Unwin Hyman, London) pp 348–385

Habermas J, 1987 *The Philosophical Discourse of Modernity* (Polity Press, Cambridge)

Harvey D, 1989 *The Condition of Postmodernity* (Basil Blackwell, Oxford)

Hebdige D, 1990, 'Subjects in space' *New Formations* number 11 (Summer) v–x

Hegel G W F, 1967 *Philosophy of Right* translated by T M Knox (Oxford University Press, Oxford)

Jameson F, 1988, 'Cognitive mapping', in *Marxism and the Interpretation of Culture* Eds C Nelson, L Grossberg (University of Illinois Press, Champaign, IL) pp 347–357

Jameson F, 1989a, 'Marxism and postmodernism' *New Left Review* **176** (July/August) 31–45

Jameson F, 1989b, 'Regarding postmodernism: a conversation with A Stephanson', in *Universal Abandon? The Politics of Postmodernism* Ed. A Ross (Edinburgh University Press, Edinburgh) pp 3–30.

Julien I, Mercer K, 1988, 'Introduction: de margin and de centre' *Screen* **29**(4) 2–10

Laclau E, 1990 *New Reflections on the Revolution of our Time* (Verso, London)

Laclau E, 1991, 'Interview: what comes after 1991?' *Marxism Today* (October) 16–19

Lefebvre H, 1969 *Dialectical Materialism* (Jonathan Cape, London)

Lefebvre H, 1974 *La Production de l'Espace* (Editions Anthropos, Paris)

Lefebvre H, 1976a *The Survival of Capitalism: Reproduction of the Relations of Production* (Allison and Busby, London)

Lefebvre H, 1976b, 'Reflections on the politics of space' *Antipode* **8**(2) 30–37

Lefebvre H, 1988, 'Toward a leftist cultural politics: remarks occasioned by the centenary of Marx's death', in *Marxism and the Interpretation of Culture* Eds C Nelson, L Grossberg (University of Illinois Press, Champaign, IL) pp 75–88

Lyotard J-F, 1986 *The Postmodern Condition: A Report on Knowledge* (Manchester University Press, Manchester)

Lyotard J-F, 1988 *The Differend: Phrases in Dispute* (Manchester University Press, Manchester)

Marx, K, 1968, 'The future results of the British rule in India', in *On Colonialism* Eds K Marx, F Engels (Progress Publishers, Moscow) pp 81–87

Mohanty C, 1988, 'Under Western eyes: feminist scholarship and colonial discourses' *Feminist Review* **30** (Autumn) 61–88

Pantham T, 1988, 'On modernity, rationality and morality: Habermas and Gandhi' *Indian Journal of Social Science* **1** 187–208

Poppi C, 1991, 'From the suburbs of the global village' *Third Text* **14** (Spring) 85–96

Probyn E, 1990, 'Travels in the postmodern: making sense of the local', in *Feminism/Postmodernism* Ed. L Nicholson (Routledge, Chapman and Hall, Andover, Hants) pp 176–189

Rorty R, 1985, 'Habermas and Lyotard on postmodernity', in *Habermas and Modernity* Ed. R J Bernstein (Polity Press, Cambridge) pp 161–175

Said E, 1985 *Orientalism* (Penguin Books, Harmondsworth, Middx)

Schick I C, 1990, 'Representing Middle Eastern women: feminism and colonial discourse' *Feminist Studies* **16** 345–380

Slater D, 1987, 'On development theory and the Warren thesis: arguments against the predominance of economism' *Environment and Planning D: Society and Space* **5** 263–282

Slater D, 1989, 'Peripheral capitalism and the regional problematic', in *New Models in Geography, Volume 2* Eds R Peet, N Thrift (Unwin Hyman, London) pp 267–294

Slater D, 1992, 'Theories of development and politics of the post-modern: exploring a border zone' *Development and Change* **23** 283–319

Soja E, 1989 *Postmodern Geographies* (Verso, London)

Spivak G, 1988a, 'Practical politics of the open end: an interview with Sarah Harasym' *Canadian Journal of Political and Social Theory* **XII**(1–2) 51–69

Spivak G, 1988b, 'Can the subaltern speak?', in *Marxism and the Interpretation of Culture* Eds C Nelson, L Grossberg (University of Illinois Press, Champaign, IL) pp 271–313

Spivak G, 1988c, 'Subaltern studies: deconstructing historiography', in *Selected Subaltern Studies* Eds R Guha, G Spivak (Oxford University Press, Oxford) pp 3–32

Warren B, 1980 *Imperialism – The Pioneer of Capitalism* (Verso, London)

West C, 1990, 'The new cultural politics of difference', in *Out There: Marginalization and Contemporary Cultures* Eds R Ferguson, M Gever, T T Minha-ha, C West (The New Museum of Contemporary Art and Massachusetts Institute of Technology, Cambridge, MA) pp 19–36

---

3 Kim V. L. England

Getting Personal: Reflexivity, Positionality, and Feminist Research

Reprinted in full from: *Professional Geographer* 46(1), 80–89 (1994)

> Think we must. Let us think in offices; in omnibuses; while we are standing in the crowd watching Coronations and Lord Mayor's Shows; let us think as we pass the Cenotaph; and in Whitehall; in the gallery of the House of Commons; in the Law Courts; let us think at baptisms and marriages and funerals. Let us never cease from thinking – what is this 'civilization' in which we find ourselves? What are these professions and why should we make money out of them? Where in short is it leading us, the procession of the sons of educated men? (Woolf 1938, 62–63).

Virginia Woolf's words speak to the process of making geography. She urges us to think about and to reflect on the spatial fabric of everyday life. She asks us to consider the structure of our social relations and how we are accountable for them and how our actions perpetuate those relations. She wants us to consider how things could be different.

In this paper, I discuss the process of making geography at a time when social scientists are increasingly suspicious of the possibility of 'objectivity' and value-free research, and when the acceptance of the socially constructed and situated nature of knowledge is increasingly commonplace. In particular, I focus on and problematize fieldwork, a term that I use as shorthand for those research methods where the researcher directly confronts those who are researched.[1] I approach this task as a feminist, but recognize that many of the issues that I am struggling with exist for researchers of other philosophical-political-methodological stripes.

Troubling questions, professional armor, and the threat of the personal

Feminism and poststructuralism have opened up geography to voices other than those of white, Western, middle-class, heterosexual men. This allows for a geography which, as Lowe and Short put it, 'neither dismisses nor denies structural factors, but allows a range of voices to speak' (1990, 8). While this makes for a more complete analysis of the complexities of the social world, it also raises new ethical issues. In our rush to be more inclusive and conceptualize difference and diversity, might we be guilty of appropriating the voices of 'others'? How do we deal with this when planning and conducting our research? And can we incorporate the voices of 'others' without colonizing them in a manner that reinforces patterns of domination? Can these types of dilemmas be resolved, and if so, how? Geographers have had relatively little to say about these troubling questions (important exceptions include Miles and Crush 1993; Moss, et al. 1993; Pile 1991; Sidaway 1992; S. J. Smith 1988). Instead, anthropologists and, to a lesser extent, sociologists have been leading the discussion on the ethics of fieldwork.[2]

Feminism and the so-called postmodern turn in the social sciences represent a serious challenge to the methodological hegemony of neopositivist empiricism. One of the main attractions of 'traditional' neopositivist methods is that they provide a firmly anchored epistemological security from which to venture out and conduct research. Neopositivist empiricism specifies a strict dichotomy between object and subject as a prerequisite for objectivity. Such an epistemology is supported by methods that position the researcher as an omnipotent expert in control of both passive research subjects and the research process. Years of positivist-inspired training have taught us that impersonal, neutral detachment is an important criterion for good research. In these discussions of detachment, distance, and impartiality, the personal is reduced to a mere nuisance or a possible threat to objectivity. This threat is easily dealt with. The neopositivist's professional armor includes a carefully constructed public self as a mysterious, impartial outsider, an observer freed of personality and bias.

Perhaps Stanley and Wise put it best when they said the 'western industrial scientific approach values the orderly, rational, quantifiable, predictable, abstract and theoretical: feminism spat in its eye' (1993, 66). The openness and culturally constructed nature of the social world, peppered with contradictions and complexities, needs to be embraced not dismissed. This means that 'the field' is constantly changing and that researchers may find that they have to maneuver around unexpected circumstances. The result is research where the only inevitability seems to be unreliability and unpredictability. This, in turn, ignites the need for a broader, less rigid conception of the 'appropriate' method that allows the researcher the flexibility to be more open to the challenges of fieldwork (Hondagneu-Sotelo 1988; Opie 1992).

For me, part of the feminist project has been to dismantle the smokescreen surrounding the canons of neopositivist research – impartiality and objectivist neutrality – which supposedly prevent the researcher from contaminating the data (and, presumably, vice versa). As well as being our object of inquiry, the world is an intersubjective creation and, as such, we cannot put out common-sense knowledge of social structures to one side. This immediately problematizes the observational distance of neopositivism because, as Stanley and Wise tell us, 'treating people like objects – sex objects or research objects – is morally unjustifiable' (1993, 168). Their point is that those who are researched should be treated like people and not as mere mines of information to be exploited by the researcher as the neutral collector of 'facts'.

In general, relationships with the researched may be reciprocal, asymmetrical, or potentially exploitative; and the researcher can adopt a stance of intimidation, ingratiation, self-promotion, or supplication (S. J. Smith 1988). Most feminists usually favor the role of supplicant, seeking reciprocal relationships based on empathy and mutual respect, and often sharing their knowledge with those they research. Supplication involves exposing and exploiting weaknesses regarding dependence on whoever is being researched for information and guidance. Thus the researcher explicitly acknowledges her/his reliance on the research subject to provide insight into the subtle nuances of meaning that structure and shape everyday lives. Fieldwork for the researcher-as-supplicant is predicated upon an unequivocal acceptance that the knowledge of the person being researched (at least regarding the particular questions being asked) is greater than that of the researcher. Essentially, the appeal of supplication lies in its potential for dealing with asymmetrical and potentially exploitative power relations by shifting a lot of power over to the research*ed*.

The intersubjective nature of social life means that the researcher and the people being researched have shared meanings and we should seek methods that develop this advantage. We can attempt to achieve an understanding of how social life is constituted by engaging in real or constructed dialogues in order to understand the people studied in their own terms (sometimes described as the insiders' view); hence the recent efforts to retrieve qualitative methods from the margins of social science. These methods offer the opportunity 'to convey the inner life and texture of the diverse social enclaves and personal circumstances of societies' (Jackson 1985, 157).

In essence I am arguing for a geography in which intersubjectivity and reflexivity play a central role. Reflexivity is often misunderstood as 'a confession to salacious indiscretions,' 'mere navel gazing', and even 'narcissistic and egoistic', the implication being that the researcher let the veil of objectivist neutrality slip (Okely 1992). Rather, reflexivity is self-critical sympathetic introspection and the self-conscious *analytical* scrutiny of the self as researcher. Indeed reflexivity is critical to the conduct of fieldwork; it induces self-discovery and can lead to insights and new hypotheses about the research questions. A more reflexive and flexible approach to fieldwork allows the researcher to be more open to any challenges to their theoretical position that fieldwork almost inevitably raises. Certainly a more reflexive geography must require careful consideration of the consequences of the interactions with those being investigated. And the reflexive 'I' of the researcher dismisses the observational distance of neopositivism and subverts the idea of the observer as an impersonal machine (Hondagneu-Sotelo 1988; Okely 1992; Opie 1992).

Failed research?

> In the social sciences the lore of objectivity relies on the separation of the intellectual project from its process of production. The false paths, the endless labors, the turns now this way and now that, the theories abandoned, and the data collected but never presented – all lie concealed behind the finished product. The article, the book, the text is evaluated on its own merits, independent of how it emerged. We are taught not to confound the process of discovery with the process of justification (Burawoy 1991, 8).

A further characteristic of neopositivist empiricism, as Burawoy indicates, is to ignore the actual *making* of geography. The concerns associated with *doing* research are usually ignored and accounts are produced from which the personal is banished. However, research is a *process* not just a product. Part of this process involves reflecting on, and learning from past research experiences, being able to re-evaluate our research critically, and, perhaps deciding, for various reasons, to abandon a research project. In short, I see research as an ongoing, intersubjective (or more broadly, a dialogic[3]) activity, and it is in this spirit that I want to discuss my dilemmas about 'doing' a recent research project about lesbians in Toronto.

Questions relating to sexualities have been placed firmly on the research agenda of cultural and feminist studies (Crimp 1992; de Lauretis 1991; D'Emilio 1992; Douglas 1990; Gamson 1991; Grosz 1989; Ross 1990) and geography (Bell 1991; Geltmaker 1992; Jackson 1989, 1991; Knopp 1987, 1990, 1992; Valentine 1993a, 1993b). In the last few years I have read this work with great interest, but have been disappointed that geographers have paid very little attention to lesbians (but see, Adler and Brenner 1992; Peake 1993; Valentine 1993a, 1993b; and Winchester and White 1988). Living in a city with a large, gay male and lesbian population, I began to consider developing a research project about the extensive lesbian communities of Toronto.

Most previous geographical work on sexual identities has focussed on the residential geography of gay men, especially their role in inner city revival. Inspired by Adler and Brenner's (1992) work on locating and characterizing the lesbian neighborhoods of a United States city, I used publicly available information (for example, 'The Pink Pages: Toronto's Gay and Lesbian Directory') to compile and map the postal codes of lesbian-positive and lesbian-owned services and amenities in Toronto. However, I wanted to move beyond merely uncovering spatial patterns to explore the sociospatial implications and political consequences of this particular form of urban restructuring. Moreover, given that the most recent work in geography has advanced the notion that sexualities and space are mutually constructed (Geltmaker 1992; Knopp 1992; Valentine 1993b), I felt it was important to explore how lesbian identities are constructed in and through space.

Recently there has been a surge of interest in urban-based marginalized groups (see, for example, Laws 1993; Rowe and Wolch 1990; Ruddick 1994; N. Smith 1993). This interest broadens the horizons of geography, promises new research directions, and asks new questions. Generally, marginalized groups seem better able to exist autonomously, or even anonymously, in central cities than elsewhere. Certainly lesbian (and gay) territories and spaces are relatively insulated 'safe places' away from heterosexism and homophobic prejudice.[4] They help provide a collective affirmation of identity, and allow for self-definition and self-exploration. However, the territorial claims of marginalized people are almost always contested more vigorously than those of more privileged groups. Despite gains made regarding prejudice and discrimination against numerous social groups, North America is still very heterosexist and homophobic. A chilling example of this was the extensive support of Amendment 9 in Oregon and Initiative 2 in Colorado (measures to overturn existing municipal laws protecting lesbians and gay men from discrimination in housing and employment) during the 1992 United States elections. So, for the minority of gay men and lesbians who live in self-identified neighborhoods, such self-exposition is not without its dangers. The more lesbian and gay communities imprint and reinvent their identities in space, the more vulnerable they become to surveillance and containment. The most obvious and pernicious outcome of this is lesbian/gay baiting, bashing, and, as recent Montreal cases sadly illustrate, murder.[5]

Sexual identities are negotiated, contested, and, quite literally, defended in and through space. Toronto's gay men and lesbians have been actively struggling against heterosexism and homophobia, and space has been a crucial component of this struggle. This is particularly apparent in lesbian and gay protests in 'public' spaces: 'homo kiss-ins' in shopping centers and straight bars; the annual lesbian and gay Pride Week Parade[6]; the frequent demonstrations around efforts to increase federal and provincial funding for AIDS prevention and research; and, the recent, very loud demonstrations about the Canada Customs seizure of lesbian-explicit erotica. The cheers of 'We're queer, we're here, get used to it' and 'We're fags and dykes and we're here to stay' are noisy expressions of anger *and* affirmations of identity. Derogatory, 'deviant' labels are turned on their head. T-shirts printed with 'DYKE' or 'I'm so queer I can't even think straight' reclaim meanings, disrupting and challenging the very process of categorizing and labelling. I read these events as lesbians (and gay men) occupying spaces that have been coded heterosexual. Spaces that are, supposedly, public are actually 'heterosexed' spaces that are not intended to be spaces for lesbians or gay men. In short, these protests and resistances of heterosexism and homophobia are inherently territorial and capture the link between identity, resistance, and space.

Clearly, I think that the intersection of gender, sexual identities, and space is a very fruitful area for geographic research, but I have not really progressed much beyond merely thinking about doing this research. Initially I had three major concerns. First, is it ethical to identify the place of the study? Other research did not reveal the location of the community studied (Adler and Brenner 1992; Lockard 1985; Valentine 1993a, 1993b). The reason was that some lesbians and gay men might not have wanted their communities 'outed', and there was the real fear of reprisals, including physical attack. Second, I had concerns regarding my research assistant. I had employed her mainly because of her intellectual abilities, but also because she is a lesbian and, as such, provided me with another means by which I could gain entry into the lesbian world. The complicated layering and interweaving of power relations between myself, my research assistant, and the project became too much for me. I began to engage in what I can only describe as the mental hand-wringing of a straight, white (my research assistant is an Afro-Caribbean Canadian), feminist academic. Finally, I made a few preliminary phone calls to, for example, the organizing committee of the Pride Week Parade. My calls were not returned. It is probable that my timing was not very good; I made my calls a few weeks before the parade took place. Then, I began to think about Gerda Wekerle choosing to exclude a nonprofit housing project for native women from her study of Canadian women's housing cooperatives because the women 'felt that they had already been overstudied' (1988, 103). I began to wonder whether, in an era of postmodernity marked by the celebration of 'otherness' in which, as Suzanne Westenhoefer (a lesbian stand-up comic) wryly put it, 'everyone wants to know a lesbian or to be with a lesbian or just to dress like one' (quoted in Salholz et al. 1993), we are engaged in the process of fetishizing 'the other' (Probyn 1993). Some of my discomfort about these three problems is captured by Liz Bondi:

> the post-modern venture is a 'new kind of gender tourism, whereby male theorists are able to take package trips into the world of femininity,' in which they 'get a bit of the other' in the knowledge that they have return tickets to the safe, familiar and, above all, empowering terrain of masculinity (Bondi 1990, 163).

I had to ask myself if I am guilty of something similar? Could I be accused of academic voyeurism? Am I trying to get on some cheap package tour of lesbianism in the hopes of gaining some fleeting understanding of, perhaps, the ultimate 'other', given that lesbians are not male, heterosexual, not always middle-class, and often not white? In the midst of academic discourse on the problems of appropriating the voices of marginalized people and the perils of postcolonialism, I worried that I might be, albeit unintentionally, colonizing lesbians in some kind of academic neoimperialism.

Appropriating the voices of 'others'; or when reflexivity is not enough

The questions prompted by my 'failed research' raise two sets of problems. The first revolves around the role of the researcher in the research encounter, the second around the nature of power relations in research about marginalized groups. I see fieldwork as a dialogical process in which the research situation is structured by both the researcher and the person being researched. Two issues flow from this point. The first is that the dialogical nature of research increases the probability that the research may be transformed by the input of the researched. The second is that dialogism means that the researcher is a visible and integral part of the research setting. Indeed, research is never complete 'until it includes an understanding of the active role of the analyst's *self* which is exercised throughout the research process' (S. J. Smith 1988, 18; also see Evans 1988; Pile 1991). We do not parachute into the field with empty heads and a few pencils or tape-recorder in our pockets ready to record the 'facts'. As Stanley and Wise point out:

> Whether we like it or not, researchers remain human beings complete with all the usual assembly of feelings, failings, and moods. And all of those things influence how we feel and understand what is going on. Our consciousness is always the medium through which the research occurs; there is no method or technique of doing research other than through the medium of the researcher (Stanley and Wise 1993, p. 157).

In short, the researcher is an instrument in her/his research and despite some commonalities (our education, and in many instances, our 'race' and class), geographers are not part of some universal monolith. We are differently positioned subjects with different biographies, we are not dematerialized, disembodied entities. This subjectivity does influence our research as is illustrated by, for example, the extensive literature on how the gender of the researcher and those being researched influences the nature of fieldwork (Geiger 1990; Herod 1993; Oakley 1981; Warren 1988). Moreover, we have different personal histories and lived experiences, and so, as Carol Warren (1988, 7) makes clear, the researcher as 'any person, without gender, personality, or historical location, who would objectively produce the same findings as any other person', is completely mythical.

The biography of the researcher directly affects fieldwork in two ways. First, different personal characteristics (be it that I am a white, straight English woman living in Canada or that I don't have a flair for quantitative methods) allow for certain insights, and as a consequence some researchers grasp some phenomena more easily and better than others. Indeed fieldwork 'requires imagination and creativity and, as such, is not for everyone' (Mills and Withers 1992, 163). At the same time, the everyday lives of the researched are doubly mediated by our presence and their response to our presence. I will illustrate this point with an example from my fieldwork experience. A couple of my previous projects have involved interviewing managers, almost all of whom are white men who are older than me. Occasionally they volunteered information that indicated that their firm had been engaged in practices that were, at best, marginally legal. My questions were not intended to elicit these responses (an example of the people being investigated shaping the nature of the research), and I have often wondered whether this information would have been so readily revealed to an older, more established male academic, especially one who did not supplicate, but instead intimidated the managers or was motivated by self-promotion. This experience reflects Linda McDowell's assertion that because women may be perceived by men that they interview as 'unthreatening or not "official", confidential documents [are] often made accessible, or difficult issues broached relatively freely' (McDowell 1988, 167; also see McDowell 1992a, 1992b). Certainly I think that a combination of my biography and my tendency towards supplication gained me access to information that might not be given so willingly to a differently positioned academic. The researcher cannot conveniently tuck away the personal behind the professional, because fieldwork *is* personal. As Okely notes 'those who protect the self from scrutiny could as well be labelled as self-satisfied and arrogant in presuming their presence and relations with others to be unproblematic' (1992, 2). A researcher is positioned by her/his gender, age, 'race'/ethnicity, sexual identity, and so on, as well as by her/his biography, all of which may inhibit or enable certain research method insights in the field (Hastrup 1992).

The second set of problems raised by my 'failed research' derives from the nature of power relations in the research encounter. My 'failed research' taught me that recognizing or even being sensitive to these power relations does not remove them. I would even argue that adopting the role of a supplicant may make it too easy for the researcher to 'submerge the instrumental and exploitative elements of participant observation beneath a wave of altruistic intent' (S. J. Smith 1988, 22). Fieldwork is inherently confrontational in that it is the purposeful disruption of other people's lives. Indeed, anthropologists even speak of the 'violence' of fieldwork, even if the violence is symbolic (Crapanzano 1977; Hastrup 1992; Rabinow 1977). In fact, exploitation and possibly betrayal are endemic to fieldwork. This is not to say that the research experience is always a negative one for the researched. Many of the women whom I have interviewed told me that they found the exercise quite cathartic and that it enabled them to reflect on and re-evaluate their life experiences. Despite this I think that fieldwork might actually expose the researched to greater risk and might be more intrusive and potentially more exploitative than more traditional methods (Finch 1984; Oakley 1981; Okely 1992; Stacey 1988, 1991). Judith Stacey summarizes these worries:

> Precisely because [these methods rely] upon human relationship, engagement and attachment, it places research subjects at grave risk of manipulation and betrayal by the [researcher] . . . For no matter how welcome, even enjoyable the fieldworker's presence may appear to 'natives', fieldwork represents an intrusion and intervention into a system of relationships, a system of relationships that the researcher is far freer than the researched to leave (Stacey 1988, 22–23).

Indeed I am concerned that appropriation (even if it is 'only' textual appropriation) is an inevitable consequence of fieldwork. This possibility is uncomfortable for those of us who want to engage in truly critical social science by translating our academic endeavors into political action. Yet, as researchers we cannot escape the contradictory position in which we find ourselves, in that the 'lives, loves, and tragedies that fieldwork informants share with a researcher are ultimately data, grist for the ethnographic mill, a mill that has a truly grinding power' (Stacey 1988, 23). Like Stacey, I have to admit there have been interviews when I have listened sympathetically to women telling me about the details of their lives (my role as participant) while also thinking how their words will make a great quote for my paper (my role as observer).

At the same time I am not convinced of the viability of some of the popular solutions for dealing with this (textual) appropriation. These include sharing the prepublication text with the researched for feedback and writing 'multi-vocal' texts that 'give voice' to the researched by, for example, including lengthy quotes from their interviews. Indeed, some feminists argue that these practices are vital parts of the research process. The intent is to minimize appropriation by avoiding misrepresentation and extending the idea of a reciprocal research alliance between the researcher and the researched. While we can revise our work in response to the reactions of the researched, surely the published text is the final construct and responsibility of the researcher? For example, it is the researcher who ultimately chooses which quotes (and, therefore, whose 'voices') to include. Also, is weaving lengthy quotes from interviews into the text a sufficient means of including 'others', especially when those quotes are actually responses to unsolicited questions that came about through the researcher's disruption of someone else's life (Okely 1992; Opie 1992; Stacey 1988, 1991)?

So where does all of this leave those who wish to conduct research with integrity about marginalized people? I am, quite frankly, still unsure about the answer to this question. However, at this point my position is this. The first step is to accept responsibility for the research, as Rachel Wasserfall remarks, researchers 'cannot pretend to present fully their informants' voices and have to take responsibility for their intrusions both in their informants' lives and the representations of those lives' (1993, 28). There also needs to be recognition that the research relationship is inherently hierarchical; this is simply part and parcel of the (conflictual) role of the researcher. I am not saying that we should not adopt strategies to counterbalance this inevitability, but reflexivity alone cannot dissolve this tension. Reflexivity *can* make us more aware of asymmetrical or exploitative relationships, but it cannot remove them.

Perhaps the more thorny question is whether, given the inevitability of unequal power relations in fieldwork, we should even be doing this research at all. I think any answer must be equivocal. What I have argued thus far is that the research encounter is structured by both the researcher and the research participants, and that the research, researched, *and* researcher might be transformed by the fieldwork experience. I want to take this argument a step further. I suggest that we approach the unequal power relations in the research encounter by exposing the partiality of our perspective. I am a straight woman who is sympathetic to the argument that lesbian geographers should do lesbian geography. However, I agree with Linda Peake who has argued that 'in their efforts to wrest control of developments in feminist theory [certain Black feminists] are delivering a potent rhetoric of political correctness that can strike panic in feminists who are sympathetic to their concerns' (1993, 419). Of course, all the sympathy in the world is not going to enable me to truly understand what it is like for another woman to live her life as a lesbian. However, researchers are part of the world that they study; as Dorothy Smith puts it, 'Like Jonah, she is inside the whale. Only of course she is one among the multiplicity of subjects whose coordered activity constitute the whale . . . she is of and inside the cosmos she seeks to understand' (1987, 142). There exists a continuum between the researcher and the researched. We do not conduct fieldwork on the unmediated world of the researched, but on the world *between* ourselves and the researched. At the same time this 'betweenness' is shaped by the researcher's biography, which filters the 'data' and our perceptions and interpretations of the fieldwork experience (Hastrup 1992; Hondagneu-Sotelo 1988; McDowell 1992b). So, should I decide to pursue my research project on the lesbian community, it will be in the full knowledge that I cannot speak for them and not myself. What I will be studying is a world that is already interpreted by people who are living their lives in it and my research would be an account of the 'betweenness' of their world and mine.

In short, I believe that we need to integrate ourselves into the research process, which admittedly is anxiety provoking in that it increases feelings of vulnerability. However, as Geraldine Pratt remarks 'establishing the grounds for taking a position and the right to speak – for oneself and certainly *about* others – is by no means unproblematic' (1992, 241, emphasis in the original). I believe it is important to be more open and honest about research and the limitations and partial nature of that research. We need to locate ourselves in our work and to reflect on how our location influences the questions we ask, how we conduct our research, and how we write our research.

Conclusion

I have discussed the process of making geographies that are sensitive to feminist and poststructuralist challenges to objectivist social science. I explored ethical questions that exist in most research, but are thrown into stark relief when there is an immediate relationship between the researcher and the people being investigated. I began with a critical discussion of neopositivist and feminist/critical methodology. I noted that the latter does not provide a clear set of rules to follow, but a series of 'maps' to guide research. I argued that greater reflection on the part of the researcher might produce more inclusive, more flexible, yet philosophically informed methodologies sensitive to the power relations inherent in fieldwork. Hence, I engaged in a reflexive inquiry into a 'failed' research project about gender, sexual identities, and space. That process raised further insight into the ethical nature of my research question, especially with regard to the dialogical relationship between the researcher and the researched. Of course, ethical problems, by their very nature, are not easily resolved and the solution that I offered illustrates the situated and partial nature of our understanding of 'others'. I argued that fieldwork is intensely personal, in that the positionality and biography of the researcher plays a central role in the research process, in the field as well as in the final text.

Notes

1 This includes those methods that are variously described as feminist, qualitative, interpretive, intensive, ethnographic, and critical. I recognize that each of these has its own unique contribution.
2 Of course, the primacy of anthropology here is partly related to classical anthropology's colonial heritage when anthropologists were often members of the colonial regime that dominated the country they studied (Driver 1992).
3 Dialogism is Mikhail Bakhtin's (1986) theory about encountering 'otherness' through the potential of dialogue between people (or with oneself). It involves the continual interaction between meanings, each of which has the potential of conditioning the others. Dialogism turns on the notion that people's responses are conditional and human circumstances are contingent (Folch-Serra 1990).
4 Homophobia is the irrational fear and hatred of lesbians, gay men, and bisexuals. Heterosexism refers to the privileging of heterosexuality over other sexual identities, and the assumption that heterosexuality describes the world.
5 Bashing appears to be on the increase in Toronto's most visible lesbian/gay neighborhood. Although bashing occurs throughout the year, it increases during the summer months when then main perpetrators – male youth – are out of school and come to this part of the city specifically to beat up gay men and lesbians. In 1990 the neighborhood community center established a 'bashing hotline' so that victims can call for support, but also to log the details of the attack. This information has been used to prompt better police response and sensitivity. In the summer of 1993 the Toronto Metropolitan Police (in cooperation with the City of Toronto's committee on lesbian and gay issues) began a campaign against bashing. This campaign includes public service announcements and bus shelter advertising that announce that 'Lesbian and gay bashing is a hate-motivated crime' and 'Being lesbian or gay is not a crime. Bashing is.'
6 Pride Week is in June and marks the anniversary of the 1969 Stonewall riots in Greenwich Village. These riots, a reaction to continued police raids on gay bars, are generally considered to have been the beginning of the lesbian and gay rights movement in the United States. It is celebrated in a number of cities around the world. The parade has a 13-year history in Toronto, but it was not until 1993 that the police designated it a community event, exempting it from policing costs.

References

Adler, S., and J. Brenner. 1992. Gender and space: Lesbians and gay men in the city. *International Journal of Urban and Regional Research* 16:24–34.

Bakhtin, M. 1986. *The Dialogical Imagination.* Austin: University of Texas Press.

Bell, D. J. 1991. Insignificant others: Lesbian and gay geographics. *Area* 23:323–29.

Bondi, L. 1990. Progress in geography and gender: Feminism and difference. *Progress in Human Geography* 14:438–45.

Burawoy, M. 1991. Reconstructing social theories. In *Ethnography Unbound: Power and Resistance in the Modern Metropolis*, ed. M. Burawoy, A. Burton, A. Arnett Ferguson, K. J. Fox, J. Gamson, N. Gartrell, L. Hurst, C. Kurzman, L. Salzinger, J. Schiffman, and S. Ui, 8–27. Berkeley: University of California Press.

Crapanzano, V. 1977. On the writing of ethnography. *Dialectial Anthropology* 2:69–73.

Crimp, D. 1992. Portraits of people with AIDS. In *Cultural Studies*, ed. L. Grossberg, C. Nelson, and P. Treichler, 117–33. London and New York: Routledge.

de Lauretis, T. 1991. Queer theory: Lesbian and gay sexualities. *Differences: A Journal of Feminist Cultural Studies* 3:iii–xviii.

D'Emilio, J. 1992. *Making Trouble: Essays on Gay History, Politics and the University*. London and New York: Routledge.

Douglas, C. 1990. *Love and Politics: Radical Feminist and Lesbian Theories*. San Francisco: ISM Press.

Driver, F. 1992. Geography's empire: Histories of geographical knowledge. *Environment and Planning D: Society and Space* 10:23–40.

Evans, D. 1988 Social interaction and conflict over residential growth: A structuration perspective. In *Qualitative Methods in Human Geography*, ed. J. Eyles and D. M. Smith, 118–35. Oxford: Polity Press.

Finch, J. 1984. 'It's great to have someone to talk to': The ethics and politics of interviewing women. In *Social Researching: Politics, Problems and Practice*, ed. C. Bell and H. Roberts, 70–87. London: Routledge.

Folch-Serra, M. 1990. Place, voice, space: Mikhail Bakhtin's dialogical landscape. *Environment and Planning D: Society and Space* 8:255–74.

Gamson, J. 1991. Silence, death and the invisible enemy: AIDS activism and social movement newness. In *Ethnography Unbound: Power and Resistance in the Modern Metropolis*, ed. M. Burawoy, A. Burton, A. Arnett Ferguson, K. J. Fox, J. Gamson, N. Gartrell, L. Hurst, C. Kurzman, L. Salzinger, J. Schiffman, and S. Ui, 35–57. Berkeley: University of California Press.

Geiger, S. 1990. What's so feminist about doing women's oral history. *Journal of Women's History* 2:169–82.

Geltmaker, T. 1992. The queer nation acts up: Health care, politics, and sexual diversity in the County of Angels. *Environment and Planning D: Society and Space* 10:609–50.

Grosz, E. A. 1989. *Sexual Subversions: Three French Feminists*. Sydney: Allen and Unwin.

Hastrup, K. 1992. Writing ethnography: State of the art. In *Anthropology and Autobiography*, ed. J. Okely and H. Callaway. 116–33. London and New York: Routledge.

Herod, A. 1993. Gender issues in the use of interviewing as a research method. *The Professional Geographer* 45:305–17.

Hondagneu-Sotelo, P. 1988. Gender and fieldwork. *Women's Studies International Forum* 11:611–18.

Jackson, P. 1985. Urban ethnography. *Progress in Human Geography* 9:157–76.

Jackson, P. 1989. Gender and sexuality. In *Maps of Meaning: An Introduction to Cultural Geography*, P. Jackson, 104–31. Boston: Unwin Hyman.

Jackson, P. 1991. The cultural politics of masculinity: Towards a social geography. *Transactions of the Institute of British Geographers* 16:199–213.

Knopp, L. 1987. Social theory, social movements and public policy: Recent accomplishments of the gay and lesbian movement in Minneapolis. *International Journal of Urban and Regional Research* 11:243–61.

Knopp, L. 1990. Some theoretical implications of gay involvement in an urban land market. *Political Geography Quarterly* 9:337–52.

Knopp, L. 1992. Sexuality and the spatial dynamics of capitalism. *Environment and Planning D: Society and Space* 10:651–69.

Laws, G. 1993. The land of old age: Society's changing attitudes toward built environments for elderly people. *Annals of the Association of American Geographers* 83:672–93.

Lockard, D. 1985. The lesbian community: An anthropological approach. *Journal of Homosexuality* 11:83–95.

Lowe, M., and J. Short. 1990. Progressive human geography. *Progress in Human Geography* 14:1–11.

McDowell, L. 1988. Coming in from the dark: Feminist research in geography. In *Research in Human Geography*, ed. J. Eyles, 154–73. Oxford: Blackwell.

McDowell, L. 1992a. Valid games? A response to Erica Schoenberger. *The Professional Geographer* 44:212–15.

McDowell, L. 1992b. Doing gender: Feminism, feminists and research methods in human geography. *Transactions of the Institute of British Geographers* 17:399–416.

Miles, M., and J. Crush. 1993. Personal narratives as interactive texts: Collecting and interpreting migrant life-histories. *The Professional Geographer* 45:84–94.

Mills, C. A., and C. W. J. Withers. 1992. Teaching qualitative geography as interpretative discourse. *Journal of Geography in Higher Education* 16:159–65.

Moss, P., J. Eyles, I. Dyck, and D. Rose. 1993. Focus: Feminism as method. *The Canadian Geographer* 37:48–61.

Oakley, A. 1981. Interviewing women: A contradiction in terms. In *Doing Feminist Research*, ed. H. Roberts, 30–61. London: Routledge and Kegan Paul.

Okely, J. 1992. Anthropology and autobiography: Participatory experience and embodied knowledge. In *Anthropology and Autobiography*, ed. J. Okely and H. Callaway, 1–28. London and New York: Routledge.

Opie, A. 1992. Qualitative research: Appropriation of the 'other' and empowerment. *Feminist Review* 40:52–69.

Peake, L. 1993. Challenging the patriarchal structuring of urban social space? *Environment and Planning D: Society and Space* 11:415–32.

Pile, S. 1991. Practising interpretative geography. *Transactions of the Institute of British Geographers* 16:458–69.

Pratt, G. 1992. Spatial metaphors and speaking positions. *Environment and Planning D: Society and Space* 10:241–44.

Probyn, E. 1993. *Sexing the Self: Gendered Positions in Cultural Studies*. London and New York: Routledge.

Rabinow, P. 1977. *Reflections on Fieldwork in Morocco*. Berkeley: University of California Press.

Ross, B. 1990. The house that Jill built: Lesbian feminist organizing in Toronto, 1976–1980. *Feminist Review* 35:75–91.

Rowe, S., and J. Wolch. 1990. Social networks in time and space: Homeless women in skid row, Los Angeles. *Annals of the Association of American Geographers* 80:184–204.

Ruddick, S. 1994. *Homeless in Hollywood: Mapping the Social Imaginary*. London and New York: Routledge.

Salholz, E., D. Glick, L. Beachy, C. Monserrate, P. King, J. Gordon, and T. Barrett. 1993. Pride and prejudice: Lesbians come out strong. *Newsweek* June 21:54–60.

Sidaway, J. D. 1992. In other worlds: On the politics of research by "First World" geographers in the "Third World." *Area* 24:403–08.

Smith, D. 1987. *The Everyday World as Problematic: A Feminist Sociology*. Boston: Northeastern University Press.

Smith, N. 1993. Homeless/global: Scaling places. In *Mapping the Futures: Local Cultures, Global Change*, ed. J. Bird, B. Curtis, T. Putnam, G. Robertson, and L. Tickner, 87–119. London and New York: Routledge.

Smith, S. J. 1988. Constructing local knowledge: The analysis of self in everyday life. In *Qualitative Methods in Human Geography*, J. Eyles and D. Smith, 17–38. Cambridge: Polity Press.

Stacey, J. 1988. Can there be a feminist ethnography? *Women's Studies International Forum* 11:21–27.

Stacey, J. 1991. Can there be a feminist ethnography? In *Women's Words: The Feminist Practice of Oral History*, ed. S. Berger Gluck and D. Patai, 111–19. New York and London: Routledge.

Stanley, L., and S. Wise. 1993. *Breaking Out Again: Feminist Ontology and Epistemology*. London and New York: Routledge.

Valentine, G. 1993a. Desperately seeking Susan: A geography of lesbian friendships. *Area* 25:109–16.

Valentine, G. 1993b. Negotiating and managing multiple sexual identities: Lesbian time–space strategies. *Transactions of the Institute of British Geographers* 18:237–48.

Warren, C. A. B. 1988. *Gender Issues in Field Research*. Newbury Park, CA: Sage.

Wasserfall, R. 1993. Reflexivity, feminism and difference. *Qualitative Sociology* 16:23–41.

Wekerle, G. 1988. Canadian women's housing cooperatives: Case studies in physical and social innovation. In *Life Spaces: Gender, Household, Employment*, ed. C. Andrew and B. Moore Milroy, 102–40. Vancouver: University of British Columbia Press.

Winchester, H., and P. White. 1988. The location of marginalised groups in the inner city. *Environment and Planning D: Society and Space* 6:37–54.

Woolf, V. 1938. *Three Guineas*. New York: Harcourt Brace Jovanovich.

Grand Theory and Geographical Practice

GRAND THEORY AND GEOGRAPHICAL PRACTICE

The grandness of grand theory

Much of the history of Anglo-American human geography in the second half of the 20th century has involved the search for a single, or a tightly-bounded, set of methodological principles that, once found, would provide unity and intelligibility to the disparate material studied. When located, such principles would function as a kind of philosopher's stone, transmuting the scattered base facts of the world into the pure gold of coherent explanation. No matter the kind of phenomenon investigated, it could always be slotted into a wider theoretical scheme. Nothing would be left out; everything would be explained.

The US sociologist C. Wright Mills (1959) labelled this quest the search for **grand theory**, defined as 'a systematic theory of the "nature of man [*sic*] and society"' (page 23). Mills's purpose in coining the term was critical, and formed part of his attack on the work of another US sociologist, Talcott Parsons (1951), who was possibly the grandest grand theorist of the 20th century. Parsons, a **structural functionalist**, argued that the existence and characteristics of any element within a social system should be accounted for by its functional role in the maintenance and reproduction of that system. To that end, Parsons deduced four types of needs that were necessary for system survival: adaption, goal attainment, integration, and latency. With this grid, the facts of the world were sorted and explained accordingly: things economic fitted in the adaption box, things political in the goal attainment box, and so on. As a grand theorist, then, Parson's achievement was to organise the world as if it were a giant filing cabinet; everything had its place, and every place had its thing. The social order was neatly catalogued, divided into separate components determined by an overarching organising principle.

From this example we can begin to delineate some of the general features of grand theory. There is the identification of a monolithic explanatory principle that gives meaning and direction to the other parts of the framework (here, functionalism). There is the mobilisation of that core principle to derive a series of subsidiary categories that are then related and manipulated at high levels of abstraction (in this case, the four types of system needs). Finally, there is a specific kind of empirical practice, one that is conceived as an exercise in finding independent, real-world examples to fit within the preconceived conceptual boxes (economic facts going under the adaption category, political facts under goal attainment, and so on).

Geography and grand theory

Whereas Parson's structural functionalism per se had little influence in geography, other grand theories have had an influence. Indeed, the history of human geography since the 1960s is one of a continual importation of grand theories devised elsewhere, a result partly of pressures to conform to other social sciences, and partly a lack of indigenous grand theory on which geographers could draw. One of the most pervasive and durable of those imported theories has been **rational choice theory**, itself a critical element of **neoclassical economics** which first entered the discipline during the 1960s, and formed part of human geography's **quantitative and theoretical revolution**. Like all grand theories, rational choice theory was sustained by the belief that every facet of life – literally from birth to death – could be understood by referring to a single principle: getting the most for the least. From this core postulate a set of conceptual corollaries were mathematically derived, and, once in place, geographers strove to find confirming empirical instances: the shopping habits of Nebraskan consumers (Berry, 1967), the planting practices of Kentish hop farmers (Harvey, 1963), and the location decisions of Mexican iron and steel magnates (Kennelly, 1954). The curious thing, though, was that, once found, the empirical examples were forgotten, or at least archived within a textbook. Theory always came first.

With the waning of rational choice theory in human geography from the early 1970s other grand theories jostled to take its place. For the following 20 years or so geographers engaged in a contest of competing 'isms', many of which bore the hallmarks of grand theory. Harvey's (1973) version of **structural Marxism** made capitalist accumulation its centre piece, with all other social relations consequent upon its vicissitudes; Guelke's (1974) **idealism** placed the rational actor as the explanatory mainspring, reducing all actions and events to the consequences of human reason; and Giddens's (1976) **structuration** theory took as pivotal the dualism of agency and structure, and from which a plethora of conceptual niceties were deduced and real-world examples found.

Common to each of these different grand theories was the idea that by applying the logic of some central principle(s), accurate representation and explanatory success was guaranteed, providing both certainty and clarity. The world when viewed from the lens of grand theory was not William James's 'blooming, buzzing, confusion' but took on a shape, a **totality**, an order, that allowed for understanding and, if necessary, intervention. One knew with assurance what was right and real and true.

Critiques of grand theory

As already noted, grand theory as a term originated as part of a critique; the adjective was derogatory not complimentary. For C. Wright Mills grand theory was equivalent to empty theory, one resolutely ignoring the teeming variety of everyday social life that should have been the object of its inquiry. Against Parsons, Mills (1959, pages

56–57) argued that 'there is no "grand theory", no one universal scheme in terms of which we can understand the unity of social structure, no one answer to the tired old problem question of social order'. Since the early 1980s human geographers have also increasingly questioned the use of grand theory in their own discipline. Criticism initially came from at least two directions. The first was from those associated with **humanistic geography** whose explicit purpose was to describe the varied **lifeworlds** of ordinary people; worlds squashed flat, according to them, by grand theory (one of the foci of Ley's reading below). The second, and more complicated, was from critical **realism**, a philosophy that among other things attempted to create an opening within theory for concreteness and contingency (set out in Sayer's piece below). In particular, critical realism lay behind the call in human geography during the mid-1980s for new place-based, local inquiries (for example, **locality studies**) that took geographical specificity and uniqueness seriously.

These critiques of grand theory, in turn, produced a counter-response particularly from the political left (see Harvey and Scott, this volume) that was couched in terms of the political necessity of a **totalising** theory that kept all the parts together. By failing to view capitalism as a whole, the argument went, theory missed the essential characteristic of the central object of its inquiry.

Since that response of the late 1980s, however, things have begun to change yet again with the increasing influence of **post-structuralism** (discussed by McDowell, this volume). Perhaps the most insistent claim here is the need to practice a 'modest' (Law, 1994, ch. 1) rather than a grand theory which, among other things, eschews imperial and universal principles, recognises the impossibility of constructing abstract schemes that mirror the world, and within its structure reflects upon the very act of its own theoretical and empirical practice, especially its own situatedness within an often Western and **masculinist discourse** (**reflexivity**). Putting these tenets into practice has proven difficult, though. Given the long socialisation of human geographers into the tradition of grand theorising, trying to do things in a different way, a post-structural way, goes against deeply ingrained habits and inclinations.

But those habits and inclinations must be changed if geographers are to practice post-structuralism, because as an intellectual movement it offers a counterpoint to each of the three defining characteristics of grand theory discussed above. First, there is the rejection by post-structualism of the monolithic explanatory principles of grand theory. Such principles create a totalised and seamless view of the world that is free of rifts and wrinkles. But, as post-structural writers such as Foucault and Derrida argue, social (and geographical) life is never that all-encompassing and uniform. Rather, it is permeated with cracks and creases, disjunctures and discontinuities, and theory must work with the fragments and contradictions, and not smooth them over (see also Section One).

Second, post-structuralism also shuns the idea that theory consists of a set of purely abstract concepts, derived from some superordinate logic of explanation, that are then held up against a concrete real world. As post-structuralists have again suggested, the abstract purity of theory is not always pure, the superordinate logic of explanation is not always logical, and the concreteness of the real world is not always concrete. Rather, it is better to think of theory as a means to make something that is less intelligible into something that is more so. This is a much more open-ended interpretation, admitting that almost anything can be the basis of theory, that there is no ineluctable force holding the elements of theory together, and that the thing explained need not take on some special form, such as quiddian bits of the world. In short, theory making should be seen as an active social practice, rather than a completed final product; as something that is pieced together, contingent and unfinished, and reflecting the context of its manufacture.

Finally, post-structuralism abandons the idea that empirical practice entails only the mechanical checking off of real-world examples against abstractly derived concepts. Facts do not exist autonomously in the world, waiting to be found, but, along with much else, they are socially produced and are not independent of the theory of which they are supposedly examples. More generally, facts possess complex social biographies that are enmeshed in complicated webs of **power**, vested interests, and apparatuses of persuasion. There is nothing 'mere' about a fact, and nothing mechanical about empirical practice.

In sum, post-structual theory promises something entirely different from grand theory. Ruptures and incongruities are highlighted rather than hidden; the piecemeal nature of theorising is embraced rather than eschewed; and the social nature of empirical practice is included as a vital part of the question, rather than excluded because it produces unwarranted answers.

Principled positions

Although David Harvey and Allen Scott's article was one of the last to be published in this section of readings, it is a good starting point to this topic because of its clarity in setting out both a particular kind of grand theory, Marxism, and its wider justification. Their review begins in the 1970s, a bleak period for Marxism. On the one hand, it was then that Marxism's structural functionalist version associated with Althusser (1969) suffered harsh criticism from humanists such as E. P. Thompson (1978), who argued that it denied human agency, making individuals dupes of only economic imperatives. On the other hand, it was also then that the empirical world no longer seemed explicable by traditional Marxist precepts: rampant inflation, high unemployment, falling productivity, deindustrialisation, and the beginnings of political neoconservatism shouldn't have been happening but they were. Against this perceived failure, some human geographers turned to other approaches such as critical realism, or embryonic versions of various post-prefixed theory. But the result, according to Harvey and Scott, was not better theory, but no theory at all; there was only a fixation on things empirical and local. Moreover, these surficial empirical descriptions and fragmented accounts missed the central point about capitalism which was that to be comprehended it must be represented as a totality. For Harvey and Scott there is nothing wrong with the project of grand theory; in fact, given the nature of capitalism it is indispensible. Admittedly, grand theories, such as Marxism, require occasional refurbishment, but their wider mandate, the thing that makes them grand, must remain both for intellectual and for political reasons.

In other of their works, Harvey and Scott's purpose has been to update and modify Marxism so that it has something specific to say about the late-20th-century industrial scene; in this case, theorised as a transition between an older system of **Fordism** and a newer one of **flexible accumulation**. Remaining invariant, though, is the imperative 'to build theory of sufficient power to keep the totalizing behaviour of [capitalism] . . . clearly in view' (page 94).

Because of their commitment to representing that totality through theoretical synthesis, all three basic characteristics of grand theory are seen in Harvey and Scott's account. First, there is the concern with searching for some larger explanatory principle, in their case, **historical materialism**. Second, there is an attempt to use that explanatory or 'generative' principle to derive a set of analytical abstract categories or, as Harvey and Scott call them, 'concrete abstractions'. Finally, there is the search for real-world examples to exemplify their prederived theoretical formulations – the American mid-West as a paragon case of Fordism, the third Italy as a paradigm of flexible accumulation.

David Ley's article that follows offers the diametrically opposite view. Rather than too little theory, the problem for Ley is that there is too much. For one thing it has created a war of 'isms' resulting in disciplinary confusion, uncertainty, and disarray. Like Harvey and Scott, Ley bemoans fragmentation, but he sees it in entirely different terms. For him the problem is not too many facts, but too many theoretical perspectives. For another thing, and this is perhaps where Ley comes closest to C. Wright Mills's original critique, users of theory are increasingly uncaring about its purchase on the world. There is emerging, as Ley says, '. . . a great divorce between [theory's] abstractions and the empirical lifeworld' (page 103). Rather than a practical pursuit, theory has become a form of artistic enterprise 'admired less for its ability to illustrate reality, than for itself, as an intellectual product with an elegance and coherent logic, with a *beauty* of its own' (page 102, emphasis in original). This gives an ironic twist to the etymological origins of theory which are with the Greek word *teoria*, meaning distance: in human geography it now appears that the more distant theory is from the 'real world' the more it is revered; it is theory for theory's sake. The geographical world, as a consequence, is treated shabbily. It is either tossed off only as an illustrative example of some already-worked-out abstract conception, or it is ignored altogether, thereby severing theory from the world it supposedly explains.

There is one other point about theory that Ley makes which is that it possesses a sociology and a concomitant ideology. Theory does not come out of the sky like bolts of lightning but is produced in particular contexts by people with particular interests (see the Introduction). In this light, Ley asks, what does the proliferation of so much abstract theory tell us about those who currently profess it? His argument is that it reflects a wider set of cultural and political mores that he labels '**modernism**' and which over much of this century has influenced a spectrum of activities both intellectual and practical. But nothing lasts forever, and a modernist sensibility is giving way to a **postmodernist** one, which for Ley, at least, holds out the prospects of a different theoretical practice which is 'orientated' but not 'fixated' on theory.

Perhaps the most complex of the responses to grand theory has been Andrew Sayer's critical realism. Based upon the writings of two philosophers, Roy Bhasker and Rom Harré, critical realism represents a complete philosophical system. Its primary virtue for Sayer is in providing an alternative to the traditional philosophy of science – **positivism** – that had been introduced into human geography during the 1960s as part of the quantitative and theoretical revolution. Positivism, as Sayer argues, is concerned with identifying empirical regularities or constant conjunctions of events that are then expressed in the form of law-like statements, 'if A, then B'. In turn, recognition of such laws are the basis of making causal statements. In contrast, critical realism argues against this positivist conception. The mere presence or absence of an empirical association between events, particularly in the 'open system' of the social world, can never amount to much because of potential intervening factors that can distort or create spurious relationships, or even block them altogether. Instead, it is better to rethink the very nature of cause. That is the purpose of realism, presenting causes as a **necessary** power that inheres within a person or thing. Such 'causal liabilities', as they are called, are realised only if the right mechanism is present to activate them. For example, gunpowder has the necessary causal power to explode, but it only does so provided that the mechanism of throwing in a lighted match occurs. By expressing causes in these terms, Sayer allows a space for **contingent relations**. For events such as gunpowder exploding only occur if the right set of favourable circumstances are present to activate them. The upshot is that critical realists are loathe to make general statements about causation, preferring instead to work through limited sets of causal relations within particular contexts using case-study material.

In the sense that Sayer shuns overarching generalisations, avoids totalising representations, and eschews ironclad determinations, he is also seemingly against grand theory. This explains some of the invective implicitly directed at his work by Harvey and Scott. For them, a method that privileges limited case studies over grand theoretical synthesis has no hope in dealing with capitalism. Sayer's approach, though, is not equivalent to Ley's either. Sayer would be aghast at Ley's seeming praise for the virtues of generalisation. In this sense, Sayer's objective is to find a middle way, one that while rejecting the all-encompassing nature of grand theory at the same time upholds many of its attributes such as abstraction, rigorous conceptual distinction, and a general method.

Finally, Linda McDowell explicitly introduces a post-structural perspective into the debate. Her argument is that the very nature of theory must be rethought. It can no longer be conceived as the 'view from nowhere', but must be critically scrutinised for its hidden assumptions, vested interests, and implicit relations of power. This is the importance for human geography of the recent importation into the discipline of feminist, **post-colonial**, and post-structural theories. In different ways, each of these three perspectives challenge traditional conceptions by forcefully illustrating whose voices get left out from the theories from nowhere, and why. The problem in moving to this post-prefixed world, though, is an implied **relativism**. Once all theory is conceived as the view from somewhere, truth becomes relative to the particular context in which its claim originated. But how, then, are we to choose among different theories? Where are we to find our moral centre?

Coping with both this **epistemological** and moral relativism is the central problem of McDowell's paper. She does so by invoking the idea of 'principled positions'. Her argument is that although relativism is inescapable it should not paralyse us in making judgements. Although we may never know the Truth, we must still make theoretical choices. But rather than thinking that such choices are somehow given objectively by the facts of the world, which is the view from nowhere, the implication of post-prefixed theory is that the burden of responsibility for choosing falls back on us. Adopting a post-prefixed perspective, then, throws into stark relief the moral and political choices that we must necessarily make. They cannot be fudged or ignored; we are forced to take principled positions.

References

Althusser, L. 1969: *For Marx*. London: Verso.

Berry, B. J. L. 1967: *Geography of market centers and retail distribution*. Englewood Cliffs, NJ: Prentice-Hall.

Giddens, A. 1976: *New rules of sociological method*. London: Hutchinson.

Guelke, L. 1974: An idealist alternative in human geography. *Annals of the Association of American Geographers*, **64**; 193–202.

Harvey, D. 1963: Locational change in the Kentish hop industry and the analysis of land use patterns. *Transactions of the Institute of British Geographers*, **33**; 123–44.
Harvey, D. 1973: *Social justice and the city*. London: Arnold.
Kennelly, R. A. 1954: The location of the Mexican steel industry, In Smith, R. H. T. Taffe, E. J. and King. L. J. (eds), *Readings in economic geography*. Stokie, IL: Rand McNally, 126–57.
Law, J. 1994: *Organizing modernity*. Oxford: Blackwell.
Mills, C. W. 1959: *The sociological imagination*. Oxford: Oxford University Press.
Parsons, T. 1951: *The social system*. London: Routledge & Kegan Paul.
Thompson, E. P. 1978: *The poverty of theory and other essays*. London: Merlin.

Suggested reading

In philosophy the issue of grand theory is often bound up with the issue of rationality; rationality is a 'foolproof' method for organising enquiry and empirical practice. A classic collection dealing with rationality is edited by Brian Wilson, *Rationality* (Oxford, Blackwell, 1970). Michael Curry explores some of these same issues in a geographical context in his article 'On rationality: contemporary geography and the search for a foolproof method', *Geoforum*, 16 (1985), pp. 109–18. Quentin Skinner's collection, *The return of grand theory in the human sciences* (Cambridge, Cambridge University Press, 1985), contains a number of relevant articles, including a discussion of some writers not normally considered grand theorists, such as Foucault and Derrida. David Harvey develops the stance he takes in the reading in this volume in 'Class relations, social justice and the politics of difference', in Judith Squires (ed.) *Principled positions: Postmodernity and the rediscovery of value* (Brighton, Wheatsheaf, 1993), pp. 85–120, and Andrew Sayer's full exposition of critical realism is found in his *Method as social science* (Oxford, Blackwell, 2nd ed., 1991). For more on post-structualist theory, including elaborations on David Ley's position, see the collection in James Duncan and David Ley (eds) *Place/* ***culture****/representation* (London, Routledge, 1993).

4 David Harvey and Allen J. Scott

The Practice of Human Geography: Theory and Empirical Specificity in the Transition from Fordism to Flexible Accumulation

Excerpt from: B. MacMillan (ed.), *Remodelling geography*, chapter 16. Oxford: Blackwell (1989)

From the mid-1970s an increasing disparity arose between prevailing theory and the actual evolution of most of the advanced capitalist societies. These discrepancies began, slowly at first, and then at an accelerating rate, to undermine confidence in the general value of Marxian theory as both scientific synthesis and as a programme of political action. In particular, by the time that Thatcherism and Reaganism began their rise, it had become apparent that the state was now in full retreat from many of its earlier social commitments and, astonishingly, the result was not so much a resurgence of the political and legitimation predicaments that its intervention was supposed to have resolved in the first place, but a general acquiescence in the politics of neoconservatism.

The massive re-structuring of the capitalist economy that had been going on for some ten or fifteen years was now approaching a major watershed. By the end of the 1970s, and certainly by the early 1980s, the Fordist regime of accumulation was being displaced by flexible accumulation as the dominant way of doing capitalist business. The resurgence of economic competition and entrepreneurial activity at all levels in the economy undermined theories of monopoly or state–monopoly capitalism and certainly made any implied teleology in earlier transitions moot. The rise of high-technology, artisanal, and service sectors of production, the fragmentation of institutionalized working-class power (the unions in particular), the increasing dualism in labour markets and the appearance of thriving new industrial regions and spaces in areas formerly shunned by industrial capital, suggested that the traditional views of the ability of working-class power to shape the trajectories of capitalist development were open to question. There were, as it turned out, very few arenas of Marxian enquiry that were not called seriously into question by the crisis of Fordism and the rise of flexible accumulation and by its associated politics of neoconservatism.

The movement of events provoked deep cleavages within the field of Marxian theory. Just as the bloom was fading from structuralist interpretations, for example, E. P. Thompson launched a trenchant attack upon its conceptual underpinnings. Thompson isolated for especial critique its rigid sense of social relations, its totalizing closure, and its depreciation of the role of consciousness in historical eventuation. Many of his criticisms were well-founded and his attack certainly opened the way for a new respect for history among Marxian social scientists. Structuralism, particularly in its Althusserian version, was depicted by Thompson as a pathological aberration on a par with Stalinism, and indeed conducive to the latter. At the same time, those who focused rather exclusively on the role of working-class power in the supposed teleology of capitalist history also had to face up to the fact that the working class as a whole (in spite of continuing struggles on the part of some segments such as the air controllers in the United States or the miners in Britain), seemed not only powerless but even partially to accept a role of complicity in the neoconservative thrust.

In view of the increasing mismatch between much of grand theory as it was then conceived and these emerging realities, and in view also of the damaging cross-fire among the major figures of Marxism as to how to react to these new circumstances, it is hardly surprising that we have since experienced a decisive retreat from theoretical work and an increasing fragmentation of research concerns. Along the way, strong counter-movements such as deconstructionism, post-modernism and post-structuralism, pragmatism, and a sort of 'new naturalism' (reinforced by realist philosophy) in the social sciences have also taken their toll and helped intensify the withdrawal from attempts at theoretical synthesis.

In human geography we have, as a consequence, seen a re-assertion of the primacy of empirical research and a fixation on the specificity of the local, as opposed to a continuing concern for elucidating the generality of capitalism in its totality. Here again, we do not wish to deny the important and fruitful lines of research that have been opened up by these developments, and to some degree this retreat into the dense empirical details of specific historical and geographical situations can be seen as a sensible and necessary step in the struggle to come to terms with the puzzling realities of the current (transitional) conjuncture. At the same time, as at least some of this work has decisively turned its back on any theoretical engagement with the nature of capitalism as a whole, so has it also become increasingly sterile in both scientific and political terms.

Recent developments in the area of conceptual elaboration have also raised a number of problems. By the early 1980s much theoretical work on the left in human geography had been reduced to a variety of pallid, depoliticized and ultimately vacuous versions of the theory of structure and agency. We say vacuous precisely because by this time, in many accounts, the very notions of structure and agency had been shorn of any very definite content in terms of political economy or the laws of motion of capitalism as a whole. The idea of human agency itself came to be seen as an implicit suggestion that there might not even be such laws of motion, and there was a tendency to ascribe unplanned, macroscopic social outcomes to the workings of 'unintended consequences'. This is, of course, in one sense correct since such outcomes (like flexible accumulation itself) are the overall result of myriad individual choices, decisions and behaviours. But the problem here is that the very language of 'unintended consequences' suggests that broad social change is somehow only a dissolved side-effect of the inchoate swirl of human agency. This swirl, real as it may be, is, we insist, bounded and shaped by the stubborn logic of capitalism as a whole, and rationally explicable as such. This, surely, is the only reasonable interpretation that can be put upon structuration theory once it is grounded in concrete historical referents. At its best (as in some of the more polished realist accounts) the tendencies we are criticizing have given birth to research that evinces a sensitive understanding of the palimpsests of local events, and strong insights into the interplay between the fragments and contingencies of capitalist development. At its worst it dissolves into descriptive recitations of regional characteristics, with occasional Marxist atmospherics as remote stage scenery. In any case, how all of this related to capitalism in its entirety (or, indeed, if capitalism in its entirety could even now be said to be a sustainable concept) remained far from clear.

We maintain that something decisive has been lost in much recent Marxist work in human geography, and that the disengagement from explicit theoretical work rooted in political economy has been an understandable but self-defeating retreat. In particular, we take a stand against the notion that the current phase of capitalism is marked by steadily disorganizing tendencies in which the coherence and unity of the capitalist system is dissolving away. We are also opposed to the slippage that begins with the notion that history and geography are constructed by human agents in unique places with an open-ended future, and ends with the notion that anything goes, especially at the micro-scale. Indeed, we want to suggest that the issue of the local is considerably less potent than some of its protagonists seem to imply. Above all, we argue that a re-assessment of current analytical priorities dominated by empiricist lines of inquiry (infused with notions of contingency, the open-endedness of outcomes, and idealist views of agency) is long overdue.

We recognize the potential for misunderstanding in these remarks, and we hasten to add two provisos: (1) that the detailed realities of historical and geographical transformation are not immediately deducible from general theorizations, and (2) that we in no sense wish to deny the significance of freedom of action, of imagination, and of will in the construction of alternative realities. But these provisos also imply a third point, viz.: in order to comprehend the complexity of real historical and geographical processes in capitalism we must be armed with, and be prepared to operationalize, theoretical ideas about the workings of capitalism as a total system. Only in that way can we get behind the fragmentary, the contingent and the ephemeral characteristics of the modern (post-modern?) world, and tackle its underlying systematicities.

The theoretical imperative

We still live in a world dominated by capitalism. There are, therefore, scientific and political imperatives to build theory of sufficient power to keep the totalizing behaviour of this mode of production clearly in view, particularly when the surface confusions of contingency and ephemeral change mask underlying dynamics. We use 'theory' here in its usual Marxist sense, to mean the creation of the intellectual preconditions for self-consciousness of the structures of capitalist domination coupled with the construction of coherent representations and analytical tools to facilitate the struggle for human emancipation. Our ability to know the world, and to represent it truthfully, is essential to this emancipatory enterprise.

All representations of the world, including those put forward under the aegis of pure empiricism, carry implicit theoretical presuppositions and codings. It is therefore particularly important to continue the task of clarifying the presuppositions underlying research in human geography at a time when the increasing flexibility of contemporary practices of accumulation obscures underlying realities, when there is little agreement as to what constitutes a valid conceptual framework, when meta-theory has in any case been largely reduced to background atmospherics, and when the analytical foreground to research is occupied by a multiplicity of competing and fragmented discourses focused on the local and the empirical.

What we are, in effect, calling for here, is a major theoretical effort that transcends the disconnected plethora of approaches and findings that have been generated these past few years in the course of trying to come to grips with the surface appearances of flexible accumulation. These approaches and findings, we submit, must either be synthesized and integrated into some more general theory of the spatial dynamics of capitalism, or be rejected as mere surface gestures and representations that mystify rather than clarify underlying meanings. This task must be pursued in the full recognition that regimes of accumulation do change over time, and that the arenas of accumulation shift around in geographical space. There are, nevertheless, certain durable aspects of capitalism that we wish to insist upon in striving to build a holistic theory able to capture its totalizing behaviour. And that means dealing with basic concepts of class relations, capital accumulation, commodity exchange, money forms, finance capital, state formation and the various manifestations of oppression endemic to capitalism. We need, then, to keep the theory of the totality of capitalism very much in the foreground of all analysis.

We should be explicit, however, as to what we mean by 'totality' and the way in which a holistic theory might be produced. This is a crucial matter since there has been much criticism these past few years of attempts to build any kind of totalizing discourse or holistic meta-theory. We wish first to acknowledge the force of some of this criticism. For example, we reject the idea, derived mainly from Lukács, that an understanding of the social totality has some ontological priority over an understanding of internal relations, subsystems, individuals and the like. This kind of argument has had particularly nefarious effects when researchers fall into the lazy habit of hypostatizing current theories of the totality. It also subsumes away the different logics and relationships that might properly be identified at meso- and micro-levels of analysis. We do not accept, either, that kind of totalizing discourse, often found among the leftist avant-garde in the inter-war years (and which was revived in some Marxist circles in the 1960s) that presumes either some dialectical necessity (teleology) in the transition to socialism or some total dominion over the future through the powers of reason mobilized in conjunction with the productive force of modern technology. Such discourses are quite antithetical to our conception of the role of individuals and other agents in processes of historical–geographical transformation.

We think of the production of holistic theory as a *project* of understanding the totalizing behaviours of capitalism. By the totalizing behaviours of capitalism we mean the way in which, e.g. (in addition to such familiar questions as the logic of commodity production or the operation of labour markets), the production of information, the marketing of that information through the media, the organization of pleasure and entertainment, the production of new knowledge, the division of labour within the household, and so on, are now mediated by capitalist social relations. At the same time, a reading of the financial press indicates immediately the far-flung global extension of capitalism and its imbrication in the social and political life of peoples scattered across the entire surface of the earth. It is this totalizing activity that holistic theory seeks to grasp; and plainly it will take major efforts of theoretical and empirical analysis to do so. How, then, are we to come to grips with this totality in the midst of all manner of fragmentations and empirical specificities?

Detailed investigation of the particularities of historical geography provides one gateway to comprehension of a totality that is always differentiated and multi-layered, always the product of human action, no matter how much individuals are held captive in their own material and ideological constructions under capitalist relations of production. Passage through these particularities can lead us to an understanding of the real universal qualities of capitalism. It is in terms of these universal qualities that we think of the totality, rather than in terms of a 'thing' that can be understood abstractly. Our task is to identify these universals within the fleeting, the ephemeral, the contingent and the fragmented aspects of daily life under conditions of flexible accumulation.

Recent inquiries provide us with abundant raw materials for such a theoretical project. But there seems to have been little effort devoted so far to the tasks of bringing these hard-won insights – sometimes cast in pure empirical form and in other instances incompletely theorized as part of some realist endeavour – back into the fold of a general theory of the political economy of capitalism. Which brings us to the problem of how that might be done.

Abstractions, we insist, should always in the first instance be rooted in the analysis of daily life. Their formation depends, therefore, upon the detailed appropriation of historical–geographical materials in a manner that respects the integrity and variety of human experience. The materialist method, however, entails a search for those 'concrete abstractions' through which the capitalist mode of production (or any other mode of production for that matter) is bound together into a working whole. Theory construction means the conceptual representation of such concrete abstractions and their linkage into a coherent analysis through a careful reconstruction of the necessary relationships that connect them together and ensure the reproduction of capitalism as a viable social system. Money is a prime example of a concrete abstraction. We deal with it daily, spend much of our life either getting it or disposing of it, and in many respects find ourselves bound to it, even ruled by concern for it. Analysis reveals how some form of money is inherent in commodity exchange and we can also see how money becomes the way in which the value of social labour is represented under capitalism to the degree that production for exchange becomes generalized. Careful study of actual processes permits us to identify other concrete abstractions such as 'commodity', 'division of labour', 'profit', 'labour power' and 'capital', at the same time that we gain information on the relationships between them. The very identification of such concrete abstractions opens up all kinds of possibilities for speculative theorizing. To the degree that money becomes a vital source of social power, for example, so the monetization of daily life, including those aspects that humanists are wont to treat under the rubric of the life world, becomes a real possibility (with many compelling examples that lie readily to hand), thus suggesting the totalizing hold that the sociality of money, and by extension the circulation of capital, can have over even intimate aspects of experience. These are the kinds of speculative engagements that can arise out of an imaginative consideration of a single concrete abstraction like money.

The task of theorizing here is three-fold. First, we strive to show how the various concrete abstractions that we can identify through historical materialist inquiry are *necessarily* linked. Analysis of relationships allows us to show, for example, how the concrete abstraction 'commodity' with its two faces of use and exchange value produces of necessity a money form that is itself divided between its roles as a measure of the value of social labour and a pure facilitator of circulation. It may then be possible to show how the tensions latent within the money form foreshadow the circulation of capital and the buying and selling of labour power. Theory construction here means the representation of the binding relationships that give the capitalist mode of production its contradictory coherence. Theoretical argument of this sort always has the capacity to produce new insights and findings, and this provides a second gateway to the creation of new understanding.

Secondly, then, we can seek out underlying concepts that have the power to synthesize and explain the links between such concrete abstractions but which are not themselves directly identified through the appropriation of historical and geographical materials. Such notions as the hidden hand of the market, the coercive laws of competition, class relations, equalization of the rate of profit, the necessity of accumulation for accumulation's sake, the annihilation of space by time, and the like, can be imputed as essential to the dynamics of a capitalist mode of production. What we are looking for here are, as it were, certain generative principles that help to explain the linkage between concrete abstractions just as they explain the dynamics of any capitalist mode of production.

The third task is to set this whole apparatus to work, now built into a synthetic but incomplete statement of the necessary laws of motion of capitalism, so as to interpret the historical geography of capitalism. There is, here, a moment in the research process of trying to interpret the actual dynamic theoretically. This can either be done retroactively (in order to explain past events) or through political practice (where the role of theory is to provide a better-informed basis for fighting for emancipation from oppression). In either case, the gap between the theoretical representations and historical geographical events, depicted in terms of concrete abstractions, forms a space out of which new concrete abstractions can grow. Here lies a third gateway to the creation of new knowledge. The reintegration of these into the theory allows the project of holistic theory construction to move forward to the point where we might hope to reflect the totalizing dynamics of capitalism, as in a mirror.

The building of such a theoretical structure through historical materialist research is, of course, a collective endeavour and there is much room for debate over the status and form of linkage between different concrete abstractions as well as over the interpretation of underlying terms and their projection as laws of motion of capitalism. Out of this there always arises a danger of arcane theoretical debate about theoretical issues (a kind of closure around 'theory for theory's sake' or some notion of 'pure theoretical practice'). That danger can be avoided if we scrupulously adhere to a basic methodological precept: namely that (1) the unity of theorizing and (2) the constant revision of theory by historical materialist inquiry, are both fundamental to the project. This means, then, that our current task is to appropriate as much historical–geographical material as we can, to search this material for new concrete abstractions and to match that process by an extension of the framework of theoretical argument which we already possess concerning the basic laws of motion of capitalism. Only through such procedures can we hope to build a general theory of the space-economy of capitalism under condition where flexible accumulation is becoming more and more the dominant form of economic organization.

5 David Ley

Fragmentation, Coherence, and Limits to Theory in Human Geography

Excerpt from: A. Kobayashi and S. Mackenzie (eds), *Remaking human geography*, chapter 12. London: Unwin Hyman (1989)

Pervasive fragmentation

There is more uncertainty about the direction of the discipline today than at any time in the last 20 years. The confidence of, first, the quantitative and, later, the structuralist paradigms has faltered, as a critical and sometimes adversarial scholarship has demonstrated how the excessive claims of an avant-garde have not been attained, and most likely never could be. What are we left with? Few would identify a new consensus, when opinions are so polarized and conflict seems to be so self-evident. Indeed for many practitioners, possibly a majority, confusion would be a more appropriate description of contemporary human geography as a discipline. One recent writer has characterized the field as forming 'a non-relativistic philosophical pluralism'

(Paterson 1983, page 175). Another sees the discipline as 'branching toward anarchy' (Johnston 1983, page 189), with 'no consensus, no paradigm-dominance, only a series of mutual accommodations' (page 188). Brian Berry (1980), a former champion of quantitative spatial analysis, is even more apprehensive: 'Pluralism in contemporary geography has gone beyond the creative liberty of variety and degenerated into license that threatens the future of the profession'. The implication seems to be that anything goes, that the doors of a huge theoretical and philosophical emporium have been thrown open, and an uncritical ransacking of the stock is in progress, where the measure of a successful fit is that the product is either suitably new enough or suitably old enough to gain respect. But in this excess human geographers, it seems, are not alone. Daniel Bell (1976, page 13) sees such intellectual libertarianism as a broader trait of modernity: 'Modern culture is defined by this extraordinary freedom to ransack the world storehouse and to engorge any and every style it comes upon'.

It is not difficult to find evidence to support these assessments of the discipline. In 1983 *Geographical Analysis* published a series of invited short essays on the topic of theory in the social sciences. The stimulus to the dozen participants was a brief paper 'Some thoughts about theory in the social sciences' which had appeared in an earlier issue (Papageorgiou 1982), where the author had pondered his subject as he wandered reflectively through a mystical stone garden. The participants (including myself) were given the disarming task of responding to his reverie. Rather like the interpretation of a Rorschach test, the outcome of 'Papa's TAT' was a range of replies among the 12 respondents, which provided a perfect illustration of Gunnar Olsson's claim that theoretical discourse reveals as much or more about the mental categories of the writer as it does about the real world about which he or she is writing. Thus in response to Papageorgiou, Britton Harris (1983) proferred 'I believe that a concern with human health has been one of the driving forces leading to the expansion of biology'; Bob Bennett (1983) saw 'the dilemma between individual and territorial equity' as the central issue, requiring the accounting framework of a negative income tax; Olsson (1983) lisped his way via King Oedipus through 'Expressed impressions of impressed expression'; Anne Buttimer (1983) gently and classically invited Wittgenstein to accompany us through 'Teoria, Ryoanji, and the Place Pompidou'; rather grandly, and defensively, I rallied around 'Metaphysics and the place of values in science' (Ley & Pratt 1983); while smarting at real or imagined injuries Peter Gould (1983) took a walk through a vineyard meditating on theory and perversity in the social sciences. The point in all of this was that, given a common stimulus, we took off in a dozen different and non-intersecting directions. What Habermas (1983) saw as the autonomy of the segments has clearly provided further separations within the realm of science itself.

A more serious example of theoretical fragmentation occurs in the edited volume, *A search for common ground* (Gould & Olsson 1982), where, as the title suggests, a self-conscious search for convergence was undertaken. Again there were 12 contributors – one wonders if there is something symbolic about this figure – whose papers were originally presented to a small symposium at a lake retreat in Italy. Those invited represented a cross-section of the discipline, Europeans and North Americans, men and women, graduate students and senior scholars, and representatives of different perspectives in the discipline. Each author was specifically requested to direct his or her thoughts to the search for common ground and in his epilogue Olsson (1982) notes that among the varied contributors 'the closeness of the individual pieces is amazing'. Certainly, the identification of issues shows considerable agreement: the sense, if not for all authors the actuality, of an era emphasizing integration and synthesis, hearkening to the humanist critique for enquiry to pass beyond behaviour to the subjectivity and intersubjectivity of experience, yet conscious of the institutional argument to place experience in a broader explanatory context, a context which may not always be readily visible but is none the less real. There is, then, as Olsson suggests, some ground for optimism for a new disciplinary convergence. But the maturing of this vision into substantial scholarship will depend on the ability of geography's avant-garde to slacken their grasp of the theoretical and methodological vehicles which have brought them along their diverse paths to common ground. Reflecting the state of the discipline, there is a cacophony of theoretical voices in this volume: time geography, structuration, phenomenology, Marxism, Q-analysis, linguistic philosophy, semiotics, each of them one senses holding in part an aesthetic appeal to contributors.

Fragmentation! The same fragmentation as we shall see in other cultural spheres of modernism such as art and architecture. Is the medium the message? Has a form of theoretical elegance, a form of art for art's sake, led some of the authors to their preferred theory and methodology? It is striking how often theory is admired for its beauty, as an aesthetic object. In a presidential address, Leontief (1971) described econometric theory as 'both impressive and in many ways beautiful'. Wittgenstein's biographer detects the same aesthetic in the abstractions of both his philosophy and his house design: '(the house's) beauty is of the same simple and static kind that belongs to the sentences of the *Tractatus*' (Von Wright 1958, page 11). And Charles Jencks (1985) has crisply chastised modern architecture in general for having mistaken pristine beauty for substantive meaning. Is there, perhaps, a symbiotic relationship between the elegance of theory, of a distinctive style, and the self-realization of the intellectual creator? And might the individualism of 'artistic expression' be the most formidable obstacle in the search for common ground?

Theory and meta-theory

Before taking up this question, we should comment briefly on present contenders for paradigm dominance. Several recent writers have identified three: logical positivism, humanism, and a critical position, often but not necessarily a radical political economy (Jackson & Smith 1984, Johnston 1983). While each of the three has its nuances, for present purposes the trichotomy is satisfactory. Logical positivism arrived in geography in the 1960s, and offered valuable standards of rigour and an objective procedure of verifiability through hypothesis testing. But positivism has always been more vulnerable in explanation beyond observable phenomena: humanists identified its weakness in dealing with intangibles, the subjective and intersubjective worlds, and in particular its incapacity to deal with questions of meaning. Structuralists in turn criticized its empiricism, its confinement to an observation language which rendered it unable to account for hidden and underlying causes. Associations might be identified, but not necessarily causal relations, particularly in complex social situations, the very essence of most social science problems. But humanism and structuralism also have their emphases and their blind spots (Daniels 1985, Duncan & Ley 1982).

There have been several attempts to synthesize these three candidates for paradigmatic status into an overarching macrotheory which recognizes the utility, but also the partial insights, of each of them. We might mention three such attempts. The first, Berger & Luckmann's *The social construction of reality* (1966), is an unusually concise and readable volume that, beginning from the starting point of the social individual in everyday life, builds up a more complex portrait of a structured and pre-existing social reality which we inherit, sustain, and in turn modify. A second candidate is the somewhat awesome output of the Frankfurt School, especially Jurgen Habermas (Bernstein 1985, McCarthy 1978), and his identification of the technical, the hermeneutic, and the critical-emancipatory approaches to knowledge, a tripartite set which coincides to some degree with the three contenders for paradigmatic status in human geography. While Habermas' range and scholarship are seminal, it may be fair to state that to the present his argument is both largely abstract and, also, has not achieved a formal integration of the three realms of discourse that he identifies. The third attempt at synthesis is the structuration approach of Anthony Giddens (1979). At present this probably has the greatest currency in theoretical human geography (Gregson 1987, Moos & Dear 1986), and again it contains an awesome breadth of scholarship to which it brings an ambitious attempt at integration. But Giddens' outline is, perhaps, less a thesis than a framework, an inventory of important issues, but issues which are aggregated rather than synthesized. In unsophisticated hands, it can become little more than a checklist, proffering the advice that all things are dialectically related, but without specifying lines of determination.

Each of these attempts at overarching theory has not fully achieved its purposes. But there is a further important point to note here. Each of them represents a second degree abstraction, a theorization of theories, and each is therefore doubly removed from the empirical world. Each encounters substantial problems in integrating, for example, economy and culture, or individual and structure, the so-called problem of human agency, a problem identified by humanists which has become a central preoccupation for theoretical human geography.

But here we must pause and make a striking observation. While *theoretical* integration of such dualisms as agency and structure are demanding, a *practical* resolution of them is made routinely by heavily empirical traditions in the humanities and social sciences and by real human beings every day. The high school history question asking students to contrast immediate and underlying causes of the First World War, for example, directly crosses a chasm which would make theorists tremble. In other historical and cultural disciplines and in the arts, the integration of agency and structure routinely occurs. A novel like *War and peace*, for example, is successful in large measure for its effective interweaving of individuals with their social and historical context. Such endeavours are concerned with telling a story, rather than erecting a theory. These empirical traditions are oriented to issues of understanding and interpretation, and are notable in their disregard of, or even open scepticism toward, questions of theory. One thinks for example of a scholar of the stature of E. P. Thompson and his monumental historiographic work, *The making of the English working class* (1968). But Thompson is also the author of *The poverty of theory* (1978), a skilful assault on certain highly theoretical forms of Marxism. Yet at the same time as he challenges the pursuit of theory, Thompson himself does not reject generalization; it is not the unique *per se* that Thompson or even Tolstoy are interested in.

What is the difference between generalization and theory, and why is it important? Generalization emerges commonly from inductive study and lies closer to the data in its construction of conceptual heuristics; theory in contrast may represent a priori deduction, the imposition of more formal intellectual abstractions upon the data. Francis Bacon characterized this dualism of inductive versus deductive method in a striking metaphor several hundred years ago; 'Empiricists are like ants, they collect and put to use; but rationalists, like spiders, spin threads out of themselves'. Idealism, in this sense, may not be far removed from deductive, theory-building methods. In his critique of structuralism, Thompson observes how the ideas, the a priori categories of the researcher, may be granted a privileged status over empirical circumstances: 'capitalism has become Idea, which unfolds itself in history' (Thompson 1978, page 253). Theory building has become increasingly insulated from the empirical world, in phase with its detachment 'from the hermeneutics of everyday communication'. Increasingly, we see suggestions such as that made regarding central place theory that the theory ultimately may not be testable in that its basic postulates are nowhere met in the real world. As such the theory is admired less for its ability to illuminate reality, than for itself, as an intellectual product with an elegance and coherent logic, with a *beauty* of its own. This is what the autonomy of the segment may imply, a methodological aesthetic, an internalization of scholarship where the intellectual product, most notably theory, is admired for itself. Theory is become not a medium but the message.

How has this come about? And why?

Critique: rethinking the place of theory

'By our theories you shall know us', the rousing finale of *Explanation in geography* (Harvey 1969, page 486), both reflected and inspired a powerful current of thought in human geography over the past 20 years. The avant-garde in both logical positivism and structuralism, as innovative and vigorous intellectual movements, placed theoretical development at the top of their research agenda. Ian Burton ushered in the quantitative revolution and theoretical geography in 1963 with his remark that 'any branch of geography claiming to be scientific has need for the development of theory' (Burton 1963). This argument was advanced unambiguously by three of the seminal volumes of the 1960s: Bunge's *Theoretical geography* (1966), Chorley & Haggett's *Models in geography* (1967) and Harvey's *Explanation in geography* (1969). By the early 1970s the message had become institutionalized and had diffused rapidly into the more sophisticated introductory texts, where students read statements like 'Theory is the matrix of all science, and the science with an underdeveloped theoretical arm is very much like a ship without a rudder: it drifts rather aimlessly and gets nowhere except by sheer good fortune' (Abler *et al.* 1971, page 45). More recently while positivists have ventured in new directions such as systems theory and information theory, with forays into catastrophe theory and polyhedral dynamics, structuralists have placed theory in at least as privileged a position, and in general have held a sceptical view of empirical research (often denigrated as 'empiricism'), a scepticism which has varied from a criticism of empirical study as preoccupied with superficial appearances, to more severe political charges of its bourgeois and ideological nature. The imperative, in Paul Hirst's words, is 'the necessity of theory' (Hirst 1979). For much innovative work in human geography (and elsewhere), Anne Buttimer is surely right in her suggestion that theory 'became that chameleon-like goddess, who claimed reverence and unquestioning faith' (Buttimer 1983).

There is no need to repeat here the compelling case in favour of a theoretical orientation in social science. The arguments raised in the 1960s are substantial enough: the purely factual nature of idiographic research, its non-cumulative character, the limited search for explanation, all pointed to an incomplete intellectual agenda, which was by no means restricted to human geography. But with hindsight the triumph of theory has been too complete and too non-reflexive, for critical debate on the relative merits of theory and empirical study was essentially stillborn, so that the pursuit of theory has become a given, unreflected and taken for granted. The argument which follows should be clearly understood; while supporting an *orientation* to theory it is critical of a *fixation* upon theory.

Why should this development be seen as problematic? Our word theory comes from the Greek *teoria* or distancing, gaining a separation from the situation at hand. But how much distancing? At what point has a representation of reality been lost? In distancing itself from the idiographic tradition, theoretical endeavour has often achieved a great divorce between its abstractions and the empirical lifeworld, a rupture which we will shortly see finds its affinities in other disciplines and other intellectual fields besides the social sciences. There is an important distinction to be drawn here between the abstractions of theoretical work over the past 20 years, and conceptual heuristics where separation from everyday geographies is more modest.

The problematic distancing of *teoria* from the world we know might be illustrated from David Harvey's work. Following the call to theory in 1969, came a less confident message in 1972: 'There is a clear disparity between the sophisticated theoretical and methodological framework which we are using and our ability to say anything really meaningful about events as they unfold around us' (Harvey 1972). There began a new engagement with Marxist theory, but ten years later little seems to have changed, as we read in the forthright preface to *The limits to capital*:

> I think it is possible to . . . transcend the seeming boundaries between theory, abstractly formulated, and history, concretely recorded; between the conceptual clarity of theory and the seemingly endless muddles of political practice. But time and space force me to write down the theory as an abstract conception, without reference to history. In this sense the present work is, I fear, but a pale apology for a magnificent conception. And a violation of the ideals of historical materialism to boot. In self-defence I have to say that no one else seems to have found a way to integrate theory and history (Harvey 1982, page xiv).

What are some of the limitations of a privileged status for theory?

First we find a tendency to instrumentalism, where the empirical record is introduced into the argument only as an adornment or an illustrative device, rather than as a means for generating or formally confirming theoretical claims. The historian Spencer has challenged this use of the historic record, where the theoretician 'loses his openness to the historical-phenomenon-in-itself, and he uses the historical reality to illustrate his theory, thus reversing the desirable relationship of these terms' (Spencer 1977). Raymond Williams is equally critical of a 'theoretistic tendency (which) extends its presumptive interpretations and categories in what is always, essentially, a search for illustrative instances' (Williams 1980, page 34). A cavalier approach to historical conditions is encouraged in a structuralist account with its inbuilt biases against 'empirical appearances' where 'theory is about a theoretical object, and this latter is already several removes from empirical reality' (Glucksmann 1974, page 115). Representation of the empirical world was often treated with disdain by Lévi-Strauss, who upheld the structuralist manifesto that 'the ultimate goal of the human sciences is not to constitute but to dissolve man' (cited in Poster 1978, page 319).

The same tendency is found in econometrics and other highly theoretical areas of economics. In a critique of the loss of empirical representation in economics, Leontief (1971) observed how 'The weak and all too slowly growing empirical foundation cannot support the proliferating superstructure of pure, or should I say, speculative economic theory'. More recently, in an analysis of articles in the *American Economic Review* over almost a five-year period, he noted that a large majority used no data at all, while less than 1% employed data specifically collected by the author to test theoretical propositions (Kuttner 1985). In another guise, some of the humanistic literature is equally incomplete, as eclectic sources are selectively culled to illustrate rather than to test propositions (Daniels 1985).

This raises a related difficulty in the task of verification. If the empirical record is disqualified as an appropriate judge, then how is theory to be verified? Ultimately, as we noted earlier, it was suggested that central place theory may not be testable in as much as its basic postulates were nowhere met in the real world. A similar conundrum faces structuralism and perhaps realism, for in devaluing the empirical world no alternative procedure can convincingly be employed to test the accuracy of theoretical claims. When explanation is referred to deep levels beyond the reach of empirical verification it is possible for self-referential assertions to be presented as self-evident truths. Verification resides in a spiralling internalization of discourse with an ever more elaborate detailing of the purely internal logic of the theory. The result is an ever-increasing abstraction and formalism of discourse until the divorce with the empirical world is complete and reality is made over in a purely formal language. Mystification has occurred, and through the creation of an internalized speech community, theory is sealed against the distractions of what Barbara Kennedy (1979) has called 'a naughty world'.

Access to the community is limited to those who speak the language and belong to an intersubjective social world of shared assumptions. The result can be the elaboration of language games, the pursuit of what Francis Bacon rejected as 'the idols of the tribe'. This anthropological metaphor has been borrowed in a fable by the economist Leijonhufvud (1973) entitled, 'Life among the Econ'. The methodological fetish of the profession can best be understood, he suggests, as tribal rituals: 'Among the Econ . . . status is tied to the manufacture of certain types of implements, called "modls" . . . most of these "modls" seem to be of little or no practical use, [which] probably accounts for the backwardness and abject cultural poverty of the tribe . . . ' (cited in Kuttner 1985). As Kuttner demonstrates persuasively, this 'autonomy of the segments' and their 'separation from the hermeneutics of everyday communication' is expressed particularly keenly in economics, a situation of some concern in light of the profession's privileged status in advising public policy.

These trends, then, are by no means limited to human geography, but are present throughout the social sciences and, as we shall see, beyond. In a distinguished lecture to the American Anthropological Association, for example, Flannery (1982) identified a malaise in anthropology associated with the loss of culture as a central paradigm, a divisive fragmentation of the field, and an exalted status for method and theory. There is a striking parallel between the social sciences and the evolution of other intellectual realms in the 20th century including the arts and architecture. Abstraction and the elaboration of intricate codes and protocols have similarly ruptured the link between intellectual constructions in the arts and architecture and representation of the everyday lifeworld. In architectural modernism the built form is essentially the projection of a theory, the imposition of a rationalist abstraction upon historical reality. Local places, emerging slowly and organically, rooted in vernacular traditions and regional cultures, have been systematically erased in favour of a superior architectural logic. Folk knowledge, with its local intricacies, has given way to a universal logic held by the specialist which frequently derides its local constituency. Walter Gropius, an early master of modern architecture, did not consult the workers whose housing he designed, for he regarded them as 'intellectually undeveloped'; his colleague Mies van der Rohe never gave his clients the possibility of choosing among alternative designs: 'How can he choose? He hasn't the capacity to choose' (cited in Knox 1987). Only the specialist possesses the theory, and it is a means of separation from the logics of everyday life.

In a broader sociology of contemporary culture, Bernice Martin has identified in the arts a similar modernist theme 'which focuses on form rather than substantive content', so that 'The pictures (or music) get simpler and barer while the legitimacy theory gets more complex and rococo'. The art exists simply to illustrate the theory while 'the theory itself becomes the real art work' (Martin 1981, page 88). This underlies Tom Wolfe's complaint that 'these days, without a theory to go with it, I can't see a painting' (Wolfe 1975, page 6), and his invective against modern architecture as the product of academic compounds where theory is exalted over practice (Wolfe 1981). No less than in the social sciences, much of the art of the 20th century has separated itself from the hermeneutics of everyday life, developing its own inaccessible codes and protocols (Rookmaaker 1973). We are reminded of Olsson's remark that abstraction encapsulates the empirical but also the mythical: 'to read urban theory as a text about cities is less penetrating than to approach it as a text of modernity . . . through which both the particular of external reality and the general of internal myth are kept alive' (Olsson 1983). We can see, then, a more general trend emerging where theory as it embraces the mythology of a speech community becomes self-referential and self-confirming.

To the extent that this characterization is accurate it has two further implications that can only be outlined here. First, the acceptance of the notion of the value basis of knowledge means that theoretical positions are both objective and subjective – or, rather, intersubjective, for they indicate a shared meaning system. Like all meaning systems, theories evoke a social world of like-minded others with distinctive social interests. In both senses of the word, theory is ideological. It implicates value systems, even if these are unrecognized and unstated. It is 'pure theory, wanting to derive everything from itself, that succumbs to unacknowledged external conditions and becomes ideological' (Habermas 1971, page 314).

What are the social interests represented by theoretical knowledge? Writers as diverse as Habermas and Bell, as well as a number of geographers, have pointed to the centrality of information and theoretical knowledge in advanced societies. Access to theoretical knowledge is a characteristic of the highly educated, the occupational groups loosely defined as the quaternary sector or the new class. The guild advantage of these groups according to Gouldner (1979), their cultural capital, resides precisely in their membership of a specialized speech community, securing access to specialized knowledge. Thus theoretical knowledge must be seen not only as idea but also as ideology, supportive of professionalism, of new class interests.

Second, new class ideology (like all ideologies) has political as well as social connotations. To evoke the expert is to evoke a power relationship which may take several forms, such as the conferring of prestige, of influence, or of political authority, each of them implying unequal social relations. Derek Gregory (1980) has noted how systems theory is an ideology of control, and perhaps the best introductory text of a positivist geography concludes with a call to theory in moulding future geographies: 'The various theories of human spatial behaviour are as much alternative versions of the world of tomorrow as they are mirrors of the way the world is today . . . our theories are our future' (Abler *et al.* 1971, page 575). This is the deepest dilemma, for if our theories incorporate the reductionism and fragmenation exemplified in theoretical human geography over the past 20 years, then so too will our futures. Theoretical reductionism might well support an existential reductionism, as people are compressed into the matchstick figures that a corporate society models and its social engineers plan for. The suppression of human nature in theory first reflects and then justifies the suppression of humans in practice. We become what we model. And as 'intellectually undeveloped' we are not given the possibility to choose.

Possibilities for integration

It is helpful, in short, to see the fragmentation of contemporary human geography as part of a broader expression, not only in social science, but also in the arts, of the cultural codes of modernism. The 'dissolution of man', in Lévi-Strauss' phrase, of human proportion and sensibilities, was accomplished in the arts and architecture as in the social sciences in the search for theoretical universals beyond history and beyond culture, in solutions deeply impregnated by technique, the removal of a regional context and of a recognizable representation of everyday life. The result has been abstraction and the development of private and self-referential codes, inaccessible to a broader public. This is the process Habermas evokes in his commentary on 'the autonomy of the segments treated by the specialist and their separation from the hermeneutics of everyday communication'. Such abstraction and fragmentation have been criticized before; Leszek Kolakowski (1968) saw them as *The alienation of reason*, and earlier Max Weber, in a related context, also commented critically on 'this fragmentation of the soul' which accompanied the political domination of instrumental reason. In art their argument was anticipated in Goya's etching 'The dream of reason produces monsters', depicting the irrationality of a seemingly rational world.

While remaining cautious to idealistic references to *geist*, there was a cultural movement, a spirit of the age of modernism, with reciprocal lines of contact between the arts, architecture and social science (Ley 1986). There are numerous crossovers: Le Corbusier, the father of modern architecture, was himself a painter and in close contact with avant-garde artists in Europe; he also drew extensively upon the geographer Walter Christaller in his development of urban and regional plans of the built environment in the 1930s (Frampton 1980, page 182). Christaller's view of space coincided with that of the modern architect for whom 'space is seen as isotropic, homogeneous in every direction . . . abstract, limited by boundaries and edges, and rational or logically inferable from part to whole, or whole to part' (Jencks 1981, page 118). Such spaces reveal a broader cultural character of modernism: its theoretical and technical codes were internally logical but self-referential, in part technique for technique's sake, a code shared by a small speech community, but divorced from everyday empirical contexts. There are significant linkages here with the fragmentation and methodological preoccupation we noted earlier as characterizing human geography's *Search for common ground.* At this point, too, we must broaden our definition of modernism to include not only excessive rationalism, the decentring of recognizably human content, but also the oppositional protest of excessive subjectivity expressed by such 20th-century movements as surrealism and some of the post-structuralists. When Gunnar Olsson (1979), who skilfully brings these writers together, concludes one of his papers, 'There are only two of us in the cell; I and my mind', then an absolute threshold has surely been crossed in the self-referential lexicon of the individual self. Might not this be the scientific equivalent of art for art's sake? For abstraction and fragmentation, in art following from the exaltation of technique, of style, in social science contained in the privileged status of theory and method, are related faces of the culture of modernism, whether they represent rational geometries or a highly personal subjectivity.

So we arrive at the fragmentation of self-conscious styles, of intolerant mutual accommodations, of theses and antitheses; but where in this 'autonomy of the segments' can points of synthesis be found? Kenneth Clark, looking at modern art, detected a similar impasse between the abstractions of the competing avant-gardes of rationalism and subjectivity: both 'cubism and superrealism [surrealism], far from being the dawn of a style, are the end of a period of self-consciousness, inbreeding and exhaustion . . . no new style will grow out of a preoccupation with art for its own sake. It can only arise from a new interest in subject matter' (cited in Stern 1980).

If the argument set out here has any merit concerning the interrelatedness of social science with other expressions of modern culture, perhaps contemporary developments in the arts and architecture may contain some clues to a striving toward a new integration in human geography. Post-modernism is a broad-based cultural reaction within the arts and architecture to the abstraction and power of professional elites in modern culture. It is fragile and far from unitary, but the varied strands do have some important points of convergence. First, as intimated in Clark's final sentence there is 'a new interest in subject matter' in arts and architecture, in 'representational as opposed to abstract or conceptual modes' (Jencks 1986, Stern 1980). The attention to specific social and regional contexts, human proportions, and historic traditions in post-modern architecture, for example, all denied by the modern movement, have led in Europe to a revival of traditional vernacular forms (*Architectural Design* 1987, Klotz 1985). In North America, where post-modern architecture is less rooted historically, many design cues are being drawn more eclectically from classical forms. Representation as opposed to abstraction communicates not only to the mind but also to the senses. In their attention to decoration and symbolism, post-modern landscapes aim to be familiar, pleasing, picturesque, even meaningful, speaking unlike modern design to 'the imperatives of the spirit' (Jencks & Chaitkin 1982, page 217). If we might, simplistically, summarize the post-modern project, it is far more fully and self-consciously, to 'make man the measure'. The double coding of post-modern architecture to both the specialist and the user is, in Stern's words, 'an act of communication' to a broader public (in geography, cf. Hasson 1983). Writing of post-modern art more generally, Hutcheon (1986–87) observes how it 'directly engages audiences in the processes of signification. It therefore denies the alienation and transcendence of social milieu that characterized modernism.' In this manner, an attempt is made to restore contact with 'the hermeneutics of everyday communication'.

But this is not to say that the post-modern project has succeeded. Indeed critics abound, though one group of them in particular is primarily defending the earlier separation between the specialist and everyday culture. The challenge to the geographer too is how tenaciously we cling to our 'isms', to a set of abstractions, partial insights and specialized methodological and theoretical vehicles. How easily will we allow the contributions of humanistic geography or historical materialism (or indeed positivism) to become part of an enriched synthesis in geography, and to what extent will we defend our differences, and make a fetish of them, so that like the Econ in Leijonhufvud's fable we proceed toward 'more ornate, ceremonial modls' ? At the same time as signs of fragmentation abound, there is also the groping toward a convergence which Gould & Olsson (1982) correctly perceived amid the confusion of tongues. There are tantalizing possibilities for a renewed human geography, one which is more representational (or empirical) than the recent past, more contextual in its regional, cultural and historical specificity, which seeks to integrate facts and meanings, both the objective and the subjective. There are implications, too, for synthesis as opposed to analysis, for the place of theory which now lies much closer to the ground in conceptual heuristics and lower-order generalizations, and for the style of discourse which reaches beyond the private lexicons of speech communities. Abstraction and fragmentation in human geography are related facets of a broader cultural condition, and as such they should not be confused with a law of nature.

References

Abler, R., Adams and J., and Gould, P. (eds) 1971: *Spatial Organization*. Englewood Cliffs, NJ: Prentice-Hall.

Architectural Design, 1987: Post-modernism and discontinuity, **57**(1–2), whole issue.

Bell, D. 1976: *The cultural contradictions of capitalism*. New York: Basic Books.

Bennett, R. J. 1983: Individual and territorial equity. *Geographical Analysis*, **15**; 50–7.

Berger, P. and Luckmann, T. 1966: *The social construction of reality*. Garden City, NY: Doubleday.

Bernstein, R. 1985: *Habermas and modernity*. Cambridge: Polity Press.

Berry, B. 1980: Creating future geographies. *Annals of the Association of American Geographers*, **70**; 449–58.

Bunge, W. 1966: *Theoretical geography*. Lund: Gleerup.

Burton, I. 1963: The quantitative revolution and theoretical geography. *Canadian Geographer*, **7**; 151–62.

Buttimer, A. 1983: Teoria, Ryoanji, and the place of Pompidou. *Geographical Analysis*, **15**; 42–6.

Chorley, R. and Hagget, P. (eds) 1967: *Models in geography*. London: Methuen.
Daniels, S. 1985: Arguments for a humanistic geography. In Johnson, R. (ed.), *The future of geography*. London: Methuen.
Duncan, J. and Ley, D. 1982: Structural marxism and human geography: a critical assessment. *Annals of the Association of American Geographers*, **72**; 30–59.
Flannery, K. 1982: The golden Marshalltown: a parable for the archeology of the 1980s. *American Anthropologist*, **84**; 265–78.
Frampton, K. 1980: *Modern architecture: A critical history*. New York: Oxford University Press.
Giddens, A. 1979: *Central problems in social theory*. London: Macmillan.
Glucksmann, M. 1974: *Structuralist analysis in contemporary social thought*. London: Routledge and Kegan Paul.
Gould, P. 1983: On the road to Colonus: or theory and perversity in the social sciences. *Geographial Analysis*, **15**; 35–40.
Gould, P. and Olsson, G. 1982: *A search for common ground*. London: Pion.
Gouldner, A. 1979: *The future of intellectuals and the rise of the new class*. New York: Seabury.
Gregory, D. 1980: The ideology of control: systems theory and geography. *Tijdschrift voor Economische en Social Geografie*, **71**; 327–42.
Gregson, N. 1987: Structuration theory: some thoughts on the possibility of empirical research. *Environment and Planning D: Society and Space*, **5**; 73–91.
Habermas, J. 1971: *Knowledge and human interests*. Boston, MA: Beacon Press.
Habermas, J. 1983: Modernity: an incomplete project. In Foster, H. (ed.), *The anti-aesthetic: Essays on postmodern culture*. Port Townsend, WA: Bay Press.
Harris, B. 1983: Interdisciplinary disciplines. *Geographical Analysis*, **15**; 47–50.
Harvey, D. 1969: *Explanation in geography*. London: Arnold.
Harvey, D. 1972: Revolutionary and counter-revolutionary theory in geography and the problems of ghetto formation. In Rose, H. (ed.), *Geography of the ghetto*. DeKalb, IL: Northern Illinois University Press.
Harvey, D. 1982: *The limits to capital*. Oxford: Blackwell.
Hasson, S. 1983: *The neighbourhood organization as a pedagogical project*. Institute of Regional and Urban Studies, Hebrew University of Jerusalem, Jerusalem.
Hirst, P. 1979: The necessity of theory. *Economy and Society*, **8**; 417–45.
Hutcheon, L. 1986–87: The politics of postmodernism: parody and history. *Cultural Critique*, **5**; 179–207.
Jackson, P. and Smith, S. 1984: *Exploring social geography*. London: Allen and Unwin.
Jencks, C. 1981: *The language of postmodern architecture*. New York: Rizzoli.
Jencks, C. 1985: Ornamentation and symbolism in postmodern architecture. Paper presented to the Emily Carr College of Art and Design, Vancouver, BC.
Jencks, C. 1986: *What is post-modernism?* New York: St. Martin's Press.
Jencks, C. and Chaitkin, W. 1982: *Architecture today*. New York: Harry Abrams.
Johnston, R. J. 1983: *Philosophy and human geography*. London: Arnold.
Kennedy, B. 1979: A naughty world. *Transactions of the Institute of British Geographers*, **4**; 550–8.
Klotz, H. 1985: *Postmodern visions*. New York: Abbeville Press.
Knox, P. 1987: The social production of the built environment: architects, architecture and the postmodern city. *Progress in Human Geography*, **11**; 354–77.
Kolakowski, L. 1968: *The alienation of reason*. New York: Doubleday.
Kuttner, R. 1985: The poverty of economics. *Atlantic Monthly*, **255**(2); 74–84.
Leijonhufvud, A. 1973: Life amond the Econ. *Western Economic Journal*, **11**; 327–37.

Leontief, W. 1971: Theoretical assumptions and nonobserved facts. *American Economic Review*, **61**; 1–7.
Ley, D. 1986: Modernism, post-modernism, and the struggle for place. Paper presented to the seminar series, *The Power of Place*, Department of Geography, Syracuse University, Syracuse, NY.
Ley, D. and Pratt, G. 1983: Is philosophy necessary? *Geographical Analysis*, **15**; 40–56.
McCarthy, T. 1978: *The critical theory of Jurgen Habermas*. London: Macmillan.
Martin, B. 1981: *A sociology of contemporary cultural change*. Oxford: Blackwell.
Moos, A. and Dear, M. 1986: Structuration theory in urban analysis: 1. Theoretical exegesis. *Environment and Planning A*, **18**; 1–52.
Olsson, G. 1979: Social science and human action or on hitting your head against the ceiling of language. In Gale, S. and Olsson, G. (eds), *Philosophy in Geography*. Dordrecht, Holland: Reidel.
Olsson, G. 1982: Epilogue: a ground for common search. In Gould, P. and Olson, G. (eds), *A search for common ground*. London: Pion.
Olsson, G. 1983: Expressed impressions of impressed expressions. *Geographical Analysis*, **15**; 60–4.
Papageorgiou, G. 1982: Some thoughts about theory in the social sciences. *Geographical Analysis*, **14**; 340–6.
Paterson, J. 1983: *David Harvey's geography*. London: Croom Helm.
Poster, M. 1978: *Existential Marxism in postwar France*. Princeton, NJ: Princeton University Press.
Rookmaaker, H. R. 1973: *Modern art and the death of culture*. Downers Grove, IL: Inter-Varsity Press.
Spencer, M. 1977: History and sociology: an analysis of Weber's 'The City'. *Sociology*, **11**; 507–25.
Stern, R. 1980: The doubles of post-modern. *Harvard Architecture Review*, **1**; 75–87.
Thompson, E. P. 1968: *The making of the English working class*. Harmondsworth, Middx: Penguin.
Thompson, E. P. 1978: *The poverty of theory and other essays*. London: Merlin Press.
Von Wright, G. 1958: Biographical Sketch. In Malcolm, N. (ed.), *Ludwig Wittgenstein, A memoir*. London: Oxford University Press.
Williams, R. 1980: *Culture*. London: Fontana Books.
Wolfe, T. 1975: *The painted word*. New York: Bantam Books.
Wolfe, T. 1981: *From bauhaus to our house*. New York: Farrar, Straus, Giroux.

6 Andrew Sayer
Realism and Geography

Reprinted in full from: R. Johnson (ed.), *The future of geography*, chapter 8. London: Methuen (1985)

In the last three decades, geographers have repeatedly cast around in the literature on the philosophy and methodology of science for ideas which could give their research greater purchase on the world. In so doing they have shifted position unusually frequently. Recently, some geographers have begun to consider realist philosophy as a possible source of guidance and in this chapter I shall try to explain why.[1]

It's easiest to understand this interest in realism if we first recall some of the methodological positions that went before. The traditional regional geography which preceded the quantitative revolution combined a distaste for theory and conceptual analysis with a treatment of geographical phenomena as *unique* and as not susceptible to explanation by reference to general laws. In other words it was both *atheoretical and idiographic.* In retrospect, geography at this stage seems to have been in an academic backwater, blissfully ignorant of the methodological and philosophical issues which enlivened other disciplines. The elders of the discipline were given to anodyne talk of the 'craft' of geography, which always seemed to elude definition, though whatever it was, we were given to believe that the profs had it. Consequently – this is relevant for the discussion of realism – they gave idiographic studies a bad name.

Probably the most refreshing and lasting contribution of the quantitative revolution was in breaking out of this complacent isolationism and mediocrity. In place of the study of the unique (idiographic method), it pushed the search for order, or regularity, i.e. a nomothetic method, which it held to be characteristic of 'science'. And 'science' easily won the battle for the minds of the younger geographers against the vapid claims of 'craft'.

It also celebrated *theory* as the means and end of geographical enquiry. But whatever was said about theory in the methodological manifestos of the 'new geography' (e.g. Chorley and Haggett 1967; Harvey 1969), in practice a rather narrow conception of theories as *ordering frameworks* came through. 'Theories' shaded into 'models', understood as formalizations of enduring regularities between variables. Once these invariances had been discovered and formalized, data could be 'plugged' into the model and used for prediction, or so it was hoped. Sometimes the ordering frameworks were derived deductively from first principles (e.g. central place models from principles of supply and demand); sometimes they were discovered inductively by fitting them to data (e.g. diffusion curves from data on rates of adoption of innovations).

There are two important points to note about this approach. First, while it was possible, by the deductive method, to generate idealized landscapes that were characterized by regularities, these turned out to bear little resemblance to actual ones. This then raised the problem of deciding whether the differences indicated errors in the model or the presence of interfering processes, or both. Where regularities were discovered inductively they turned out to be only very approximate, as well as variable across space and time, so that the models had to be refitted anew to each and every case. The goal of universally applicable models which could be used for the prediction of events in a variety of situations was not realized: ironically, the regularities and the values of the parameters that represented them turned out to be unique to each application. Second, in *practice*, the theories used by quantitative geographers said remarkably little about how we should conceptualize objects of interest and hence about the nature of the latter. The term 'data' was understood literally as 'given things' not as something problematic. Consequently, by default, this 'scientific' geography rested upon the unacknowledged and unexamined concepts of common sense.

Post-quantitative geography has slowly drifted away from these methods: the search for regularities has gradually weakened, and concern with conceptualization has grown. Contrast, for example, quantitative geography's modernization studies with radical geography's analyses of Third World development, in particular the former's astonishing neglect of the problem of the meaning of 'development' with the latter's near obsession with this question. Other interesting changes are a (possibly rather tentative) re-emergence of apparently idiographic regional studies, albeit of a relatively theoretical kind. Also history is coming back into geographical explanations after having been banished by quantitative geography as unscientific. Once again it is becoming respectable to be a specialist on a particular region whereas the nomothetic geographers would have found this incompatible with the search for general regularities in regional systems.

In my view, although few recognize them as such, all the above post-quantitative trends are moves towards a realist approach. The changes are clearest in radical and Marxist research in geography (sometimes rather unfortunately labelled 'structuralist'). However, this association of realism and radical geography is not a necessary one: some radical work has been done using a nomothetic deductive method (e.g. Scott 1980) and acceptance of realist philosophy does not entail acceptance of a radical theory of society – the latter must be justified by other means. Some aspects of humanistic geography might also be taken to be convergent with realism and occasionally even mainstream geographers stumble across realist methods, if only temporarily and unknowingly.

Having set the context in this possibly tendentious way, let us switch to a closer and more sober analysis of some of the issues at stake. I will begin by comparing and evaluating two basic methodological principles – one characteristic of the 'new', nomothetic or quantitative geography, the other of realism. I will then try to demonstrate the significance of these issues by reference to an extended example from geographical research. This will be followed by a discussion of the nature of theory. In conclusion, I will outline some of the methodological problems peculiar to geography and discuss whether realist geography is idiographic or nomothetic.

Two contrasting methodological principles

When stated baldly, methodological prescriptions have a habit of appearing bland, innocuous, or too obvious for words, but when we follow through their implications, their particularity and critical significance becomes clear. Consider the following pair of injunctions, the first broadly 'positivist'[2] (and nomothetic), the second realist.

(1) If we are to explain processes we must discover the regularities of universal laws governing their behaviour. Hence the thrust of research must be towards the discovery of order.
(2) If we are to explain why things behave as they do we must understand their structure and the properties which enable them to produce or suffer particular kinds of change.

Both of these probably sound unexceptionable, but there are two problems with 1. First, although this strategy appears to work in some natural sciences – particularly physics – it doesn't or hasn't in human geography; regularities in the latter tend to be approximate and temporally and spatially specific, or, as we have already noted, unique. There is a convenient excuse for this state of affairs, however, namely that nomothetic human geography and other social sciences are as yet immature and that if we only persist with this allegedly 'scientific' approach, order will be discovered. But even if we accept that social science is more youthful than natural science (not an unquestionable assumption) there is still the danger that such an argument is merely a protective device – rather like saying that if prayer doesn't work it's because you haven't been praying hard enough. At the very least, it seems unreasonable to refuse to consider the possibility that the search for regularities may not be *appropriate* given the different nature of objects studied by disciplines such as human geography. The second problem with 1 is that even where regularities do exist they don't provide explanations. That two or more variables behave in a regular, predictable fashion doesn't tell us why this is so, or, to put it another way, it tells us nothing about what *makes* it happen, what produces it. (This is why the conclusions of empirical research into regularities are so inconclusive!) Note that this is not simply the old problem of 'spurious correlation', familiar though that may be. Even if correlations or other indices of association and regularity concern variables which might reasonably be expected to be causally related, the mere existence of the (non-spurious) regularity says nothing about what produces it. Nor, for that matter, does the possession of a hypothesis or theory that they are related provide evidence for the regularity being causal. For example, we may succeed in fitting some sort of spatial interaction model to some migration data, or more particularly to certain regularities between variables hypothesized to be causally related to migration, such as the ratio of job vacancies in the regions of origin and destination. It's a safe bet that the parameters of the model will have to be adjusted for other data sets. Even if we feel able to call the relationship a 'regularity' and even if it fits with our expectations we have not shown what actually produces it. Note that this does not mean that regularities can never be causal, but rather that whether they are can only be determined by a different kind of analysis.

Before saying what this different kind of analysis would be like, let us stop to ask a crucial question: why is the search for regularities seen as so central if it does not provide explanation? The answer is that 1 (above) is misguided for the simple reason that *what causes something to happen has nothing to do with the matter of the number of times it has happened or been observed to happen and hence with whether it constitutes a regularity.* Unique events are caused no less than repeated ones, irregularities no less than regularities. Even showing that an individual event is an instance of a universal regularity or law fails to explain what produces it. In failing to grasp this simple point the whole debate of the 1950s and 1960s between the advocates of idiographic and nomothetic conceptions of geography was ill-founded.

We will return to the issue of regularities later, but we can now turn to what is in my view a more satisfactory route to explanation – the realist route expressed in 2. Discovering causes involves finding out what *produces* change, what *makes* things happen, what *allows* or *forces* change. The italicized verbs are all causal terms and they all allude to mechanisms or ways of acting of objects. These mechanisms exist in virtue of the structure of the objects that possess them. For example, people have the power to build houses or program computers in virtue of their anatomical and mental structures, as modified by experience and education. This is not a tautology, for there is a difference between such powers and the structures which 'ground' them. Sometimes powers which we might initially attribute to individuals (whether people or other objects) turn out to be grounded in structures of which those individuals are just a part. For example, the powers of the Prime Minister (*as* Prime Minister and not merely as Margaret Thatcher or whoever) derive not from the individual occupying that position but from the wider political structures on which existence of that position is dependent. Some mechanisms are thoroughly mundane and already familiar in everyday discourse; for example, when we talk about the *cultivation* of land we are referring to a familiar causal mechanism. Many other mechanisms, and especially social structures, are more elusive and their identification may be a central concern of research, both theoretical and empirical; for example the debates over whether the mechanisms which produce a segregation of low income households into poor quality housing are simply grounded in the actions of autonomous individuals such as housing officers or in the wider financial structures which support the (re)production of the built environment in capitalist society.

In explaining processes it is not enough merely to posit such mechanisms and structures: careful observation and description are needed to check whether the objects to which the properties are attributed do in fact have them and to eliminate other possible contenders which might equally be responsible for the effects we want to explain. Much of the business of research is concerned with precisely such questions. Now it might be thought that the standard methods of mainstream geography, including, centrally, the search for regularities, might be appropriate for this task; if certain conditions B are causally necessary for the existence of A, then whenever A occurs, so must B, thereby constituting a regularity. The latter is true, except for the qualification that obviously if the number of As is very small then the relationship will hardly constitute a 'regularity'. But the converse is not true: that is that B always accompanies A does not establish whether the connection is necessary. However, in many cases, the regularities produced by necessary conditions or relationship are not very interesting, either because they are already known (e.g. the relationship between brains and the power to think) or more generally because the regularity itself does not disclose its own status.

It might also be thought that causal mechanisms would produce regular effects, but this depends on circumstances. If the object possessing the mechanism and/or the conditions in which it operates change, then we shouldn't expect a regular outcome. For example, if we were studying a case of land use conflict, we wouldn't expect the actions of a particular pressure group to produce regular effects if either (a) its internal structure was changing (e.g. through internal disunity) or (b) the external political and legal environment were changing. So we cannot 'read off' effects simply from a knowledge of the existence of certain mechanisms; we must find out what conditions the mechanisms are operating in as well as discover the state of the mechanisms.

We can now answer the question posed earlier concerning why some disciplines, such as physics, seem to be able to progress by searching for regularities, while others, such as geography, do not. We have just seen that the production of regularities by causal mechanisms depends on two conditions – that the internal structure of the mechanisms and their external conditions must be constant. A system which satisfies these criteria is called a *closed* system; those which don't are termed *open.* Physicists find that many of the objects they study exist spontaneously in closed systems and that where they don't they can often be produced artificially, through the scientific experiment. Geographers – and also other social scientists, and some natural scientists (e.g. geologists) – deal with open systems. Indeed the distinctive quality of human beings that they can *learn* to understand and respond to their circumstances in novel ways guarantees that social systems are open. So the success of physics has little to do with its maturity and plenty to do with the nature of what it studies, with obvious implications for the lack of 'success' of the social sciences – particularly those that follow methodological prescription 1, above. Yet, even in physics, although the availability of closed systems and hence regularities facilitates prediction, physicists still have to switch to realist methods for *explaining* regularities. (In some cases, however, they rest content with non-explanatory prediction.)

This argument may at first sight encourage a certain gloom and despondency about the possibility of progress in open system subjects, but there are compensations. In the case of *social* systems, we have an advantage over our counterparts in natural science in that actions and many other social phenomena have an intrinsic meaning which we can *understand*, and for which no equivalent exists in natural science. Atoms don't act towards one another on the basis of any understanding: people usually do, even if it isn't a correct understanding. More particularly, people's *reasons* for doing things can often be taken as the *cause* of their actions (not always because they may not always know their reasons). So through already knowing or being able to ask people about these matters we have an 'internal access' to our objects of study which, although fallible, is simply not open to natural scientists. And this is also a reason why realists can and must accept a crucial element of the manifesto of humanistic geography: viz. we must understand what people understand, mean or intend by their actions and not merely rely on measurements of their overt physical behaviour.

The example of research on 'manufacturing shift'

At this point, having come through some rather compressed and abstract methodological argument, it may help to go through a more extended substantive example drawn from geographical work, and one which can demonstrate the consequences of following either of our two methodological principles. It concerns some recent research on 'manufacturing shift' in Britain; that is the tendency for metropolitan areas and large towns to suffer major losses of manufacturing employment while small towns and rural areas enjoy gains.

In this case there is a striking, if approximate and transient 'regularity': the larger the settlement the greater the job loss, with the smallest actually showing an increase, albeit small in absolute terms. The 'messy' and transitory nature of the 'regularity' reminds us that it is hardly a closed system – indeed what else could one expect of something like employment change?! Most researchers have sought explanations through strategy 1 (searching for regularities). Accordingly various hypotheses were tested, such as the possibility that the pattern reflected (i.e. was determined by) whether places had Assisted or Non-Assisted Area status. In this particular case, the hypothesis was rejected, for Non-Assisted Areas such as East Anglia and the South Coast, as well as parts of Assisted Areas (e.g. Northumberland), experienced employment growth. Conversely, the larger settlements within Assisted Areas seemed to suffer decline no less than those in Non-Assisted Areas. But let's just reconsider whether this really was a refutation of the hypothesis. Now most researchers cast the hypothesis in the form of an expectation of an *empirical regularity*: that employment growth or decline would vary according to whether the area was Assisted or Non-Assisted. But there is another way of casting such a hypothesis – one which refers to a mechanism and which is therefore more in keeping with a realist approach. We could specify how we thought regional development grants and other incentives might influence firms – not just by positing a simple regularity between incentives and employment change but by thinking through what *kinds* of firms would find them an attraction and why (e.g. in relation to financial resources, capital intensity and to competing influences). This recasting of the hypothesis would require a different (and possibly more time-consuming) style of research to check its empirical adequacy. Interviews with firms, taking into account their differing circumstances, might be required. And it would probably produce complex answers, hedged round with qualifications about intervening, mediating mechanisms – but then this is what decision-making is usually like. We might find that even though, at an aggregate level, there is no firm correlation between Assisted Area status and employment decline, there were nevertheless cases where location and *in situ* investment decisions were influenced by regional incentives. However, this need not always translate into employment increases, for grants can be used to replace workers by machines! This latter kind of mediating factor is very common in this field of study, but a realist approach is more likely to pick it up than one which concentrates on regularities. So rather than reject the hypothesis because the expected regularity did not occur it would be more reasonable to say that it is appropriate (albeit with certain qualifications) for some cases but not for the majority.

Now this may seem a rather pedestrian conclusion, but behind such issues lies a confusion which is endemic in mainstream methodology. If we ask what determines the location of investment the answer will probably be that several different mechanisms either do or could do. We might then try to find out which ones do, and *how they work* and produce certain effects when they do. These might be called 'intensive' research questions and answering them would require looking at some instances to see how the mechanism works. But discovering how a mechanism works does not tell us how widespread that mechanism is; for example in how many instances were regional incentives influential? The latter can be called an extensive research question: instead of asking how a mechanism works it enquires into the *extent* of certain phenomena, be they events, mechanisms or (as is usually the case) regularities. The basic confusion of so much mainstream methodology is that this second kind of question regarding distribution and extent is expected to answer the first, regarding causation. And in an open system, where mechanisms generally do not stand in a fixed relation to their effects, this expectation is especially disastrous. Yet it is virtually built into what are widely taught as 'scientific methods' – particularly in the cavalier transfer of methods of statistical inference often originally developed for the study of largely controlled and closed systems, such as experiments on plants.

Research on manufacturing shift has replicated these problems; most of it has been based on hypotheses which predict regularities and it has repeatedly confused the intensive and extensive research questions by expecting the latter to do the job of the former. One exception was research on the spatial consequences of industrial restructuring, the best-known example being that of Massey and Meegan on the electrical engineering industry (1979). This research did not put forward hypotheses, at least not in the form of expectations of empirical regularities among sequences and patterns of events. Rather it sought to research how a certain kind of mechanism – restructuring – worked and how its effects were mediated by various contingent conditions. Only subsequently, and through covering most of the instances in a small population of firms, was it able to offer answers to the extensive research question regarding the *extent* of the effects of restructuring. Conducting the research in this order was especially necessary given that often the same mechanism could produce different effects (e.g. the introduction of new technology can lead to additional or reduced employment, depending on conditions) and conversely, the same effect could result from different causes (e.g. job loss due to loss of markets or automation).

One response to the pressure to restructure was found to be the search for cheaper and more productive sources of labour, and so many firms relocated to take advantage of female labour. Later, another researcher, David Keeble, translated the restructuring thesis into a hypothesis about a possible regularity, an association between the availability of female labour and manufacturing shift (Keeble 1980). The evidence for such an association turned out to be weaker than some of the competing hypotheses concerning manufacturing shift. But this was hardly fair for (a) it failed to note that the original analysis did not claim that a shift to locations with abundant female labour was the only possible outcome of restructuring; and (b), and related to this, it misidentified what was primarily an intensive research project as one concerned with and assessable by extensive questions (Sayer 1982).

Keeble also noted a particularly strong association between manufacturing employment change and what he termed 'rurality', which led him to conclude that 'rurality' was a 'causal factor'. But in what sense could 'rurality' be said to be a causal factor, or in other words how does 'rurality' 'produce effects on' or 'operate or work upon' manufacturing? The obvious question (and a typically realist one) is; what is it *about* rurality that could produce a change in manufacturing? Clearly the concept of 'rurality' needs 'unpacking' in order to see if it covers any relevant constituent mechanisms and conditions, for otherwise its statistical association with manufacturing shift could be interpreted as an accidental one.

But while many, perhaps most, mainstream researchers would ask our realist question, their standard methodology of seeking further regularities between more specific variables would not provide an answer, for it would merely re-pose the underlying methodological problem: that regularities do not explain themselves.

So to summarize the key points of the discussion so far: although 'regularities' are often interesting, causation is not a matter of regularities in events but rather of the mechanisms which produce them. Likewise what is interesting about regularities is (a) simply their pervasiveness and (b) their determinants. Mechanisms will only produce the precise, enduring regularities that are the goal of quantitative geography's 'search for order' in closed systems and these are not available in human geography's object of study (except perhaps in so far as machines are included in the latter). In social systems, which are open, the activation and effects of causal mechanisms will depend on the nature of the objects which have them and the conditions in which they happen to exist. Given this openness, while we may find that the search for regularities and generalizations may contribute answers to questions regarding the quantities and distributions of phenomena (extensive questions) it will throw little light on what produces them (intensive questions).

A closer look at 'theory'

Having battled through the complexities of causal explanation we can now return to the question of the changing nature of *theory* in geography; as we shall see the two main competing conceptions of theory complement the two alternative routes to explanation explored above. This will also help to clarify how the existence of less familiar causal mechanisms and structures is established.

The ordering-framework conception of theory clearly fits neatly with the search for regularities as a route to explanation. On this view description and conceptualization of data is an unimportant preliminary to the 'real' business of 'science', the attempt to discover a framework into which data can be 'plugged' with a view to predicting certain unknowns. Accordingly, the bulk of what passes as theorizing is concerned with the development and refinement of such frameworks, most prestigiously in the form of mathematical models. If the categorizations in which the data are given are ever questioned it's usually in terms of their accuracy (e.g. susceptibility to measurement error) and only rarely in terms of their *conceptual* adequacy. This can readily be checked by scanning the pages of a few mainstream journals, for example *Environment and Planning A*, *Geographical Analysis*, particularly for quantitative deductive approaches, *Regional Studies*, *Urban Studies*, particularly for inductive approaches. (*Annals of the Association of American Geographers* and *Transactions of the Institute of British Geographers* tend to have a more eclectic mix.)

These priorities are inverted in a realist view of 'theory' where, as we noted earlier, conceptualization takes priority over the construction of ordering-frameworks; indeed if the latter are proposed at all then they are not intended as a predictive device. The connection between this emphasis and the realist view of causation lies in the fact that causation is seen as deriving from the properties of particular objects and that if these are to be grasped adequately then considerable care is needed in conceptualizing those objects. Without such care we are chronically vulnerable to the problem of attributing powers to the wrong objects or aspects of objects, in fact this is a central, perhaps *the* central, substantive problem which any enquiry into the nature of the world must face. To what should underdevelopment be attributed? We have now learned that earlier geographers repeatedly attributed to nature what was in fact due to the organization of society in the explanation of underdevelopment. But then we must ask, as we did for 'rurality' before, what is it *about* the structure of certain societies that makes them vulnerable to stagnation and decline? Is it a 'lack of capital' or a blockage of the masses' access to the means of production or what? And again in the case of industrial re-location to exploit female labour, we might ask what it is about women and society which makes them more productive (often) and cheaper to employ. Is it something inherent in their biology or a consequence of their socialization in a patriarchal society? And if the latter what aspects of socialization in particular?

The point to be underlined here is that little progress can be made in answering such questions without exhaustive scrutiny of our concepts of such objects. At first sight many mainstream geographers may be tempted to react dismissively to this by claiming that they already do it. But we should judge them by their actual research and compare it with that of radical researchers following a broadly realist approach. The contrast between the treatment of the question of the meaning of development in modernization studies and radical research is salutary. So a typical realist theoretical question would be, 'What do we mean by development/gentrification/urbanism/collective consumption/class/rent/community, or whatever?'

And in empirical work, whereas mainstream researchers tend to quickly 'collect' data and do some sort of quantitative analysis on it, realists would emphasize the scrutiny of the categories in which the data are set, *possibly* following up with some quantitative analysis. Empirical research is not just a question of determining magnitudes and distributions once the categories have been finalized; it is also important for checking the adequacy of our categorizations. Indeed, to a realist, description should never warrant the pejorative prefix 'mere', for the adequacy of our descriptions – our applications of concepts – dictates the adequacy of our explanations. Terms like 'service employment' or 'interest group' may look quite innocuous, and in many contexts they are, but as soon as we put some 'explanatory weight' upon them they may lead us astray. The danger of 'services' is that it covers such a heterogeneous scatter of activities that it's absurd to attribute to services certain unitary powers (e.g. static productivity). The danger of the term 'interest group' when applied to groups such as trade unions, tenants' groups, pressure groups, employers' and professional associations is that it tends to imply that each group stands in an equivalent position within the power structures of society so that when we use the term to explain, say, a land use conflict, we overlook crucial imbalances and asymmetries and differences in constraints between such groups.

It is for these reasons that 'abstraction' and 'unpacking' are 'buzzwords' in realist analyses. We try to reduce a complex entity into its component parts, abstracting them out one by one in order to consider their properties. To a limited extent we do this intuitively, but the favoured procedures of mainstream methodology divert attention away from this healthy habit towards the search for empirical regularities among whatever data – however categorized – are available, in other words among phenomena whose conceptualization is seen as unproblematic. By contrast, realism attempts to formalize and develop abstraction.

One way of doing this is by distinguishing between those properties of objects which they possess independently of their relationships to other objects and those which only exist through them. For example part of my behaviour as an owner-occupier is dependent on my relationship to the building society which has granted me a mortgage. Purely individualistic explanations of my actions would therefore be inadequate; my insertion into a specific structure of social relations (in this case concerning indebtedness) would have to be acknowledged as a necessary condition of my behaviour. By contrast, my recreational activity may be independent of my being indebted to a building society. Simply by asking what are the necessary conditions for the existence of objects of interest, we can make considerable progress in working out the (causal) structure of the systems in which they exist. And where necessary relations are discovered, we can make strong theoretical claims about them, for example the existence of A *presupposes* B. Empirical work will then be needed both to check such claims and to discover the configuration (including the spatial form) of the many relationships which are contingent (i.e. neither necessary nor impossible). For example, it is one thing to note that the survival of capital presupposes the reproduction of a labour force (among other things) but the nationality of that workforce is a contingent matter only discoverable through empirical research.

Problems of geography as synthesis

I now want to mention some realist ideas concerning the difficulty of doing geography. Prior to the quantitative revolution it was widely thought, following Kant, that geography and history were special subjects in that they were concerned with *synthesis*, in space and time respectively. The quantitative geographers put paid to that idea with their assertions that method was not determined or constrained by subject matter and that any discipline could embrace a nomothetic method: geography was not an exception. This was a bold thesis and it led to some interesting methodological experiments, but it has failed. It is not *only* that geographers deal with open systems but that they place more emphasis than many other disciplines on synthesizing the interactions of diverse systems. *Regions*, for example, are enormously complex and heterogeneous aggregates of diverse processes, some interdependent, others quite independent, and yet we frequently try to grasp the whole. Apart from history, other social sciences tend to limit themselves to particular types of object or system, be they political groupings, markets or interpersonal relations. Now I don't want to exaggerate the contrast, for it is one of degree, and in many cases other social scientists find themselves driven towards broad levels of synthesis; for example, sociologists studying working-class culture face a dauntingly large and complex set of issues. ('Political economy' is perhaps in a different category, at least in its empirical work attempting a very high level of synthesis.) At the very least, geography is far removed in character from those natural sciences which can objectify their abstractions in closed systems, isolating their objects of interest from the contexts in which they spontaneously occur. Physicists may protest that their objects are enormously complex too, but I doubt if they are *in the same sense.*

Consequently, given the heterogeneity and complexity of objects like regions, we cannot reasonably expect works of geographical or historical synthesis to provide exhaustive intensive studies on all aspects of a region. The best we can normally manage is an incomplete picture consisting of a combination of descriptive generalizations at an aggregate level (e.g. on changes in population and standards of living), some abstract theory concerning the nature of basic structures and mechanisms (e.g. concerning modes of production), and a handful of case studies involving intensive research showing how in a few, probably not very representative, cases these structures and mechanisms combine to produce concrete events. Look at any historical geography of the industrial revolution of twentieth-century Britain and you will find such combinations, although many will simply ignore abstract theories and rely on common-sense ideas. Certainly, *any* research is bound to be incomplete, because there are always more things to discover and better ways of conceptualizing what we think we already know; but works of synthesis such as these, which are particularly prominent in geography and history, are also incomplete in a different sense, which has more to do with the practicalities of this kind of research.

There are some kinds of research problems which we can't expect new kinds of philosophy and methodology to solve, indeed it is better that the latter disclose rather than conceal their existence as quantitative nomothetic geography tended to do.

Conclusion

A realist geography must be deeply concerned with theory, even in the conduct of empirical research, though in a different sense of the term 'theory' from that understood by mainstream geographers. In so far as it succeeds in discovering necessary connections in the world, its theoretical claims will have some generality, although a realist approach won't expect to find enduring regularities at the level of empirical events. It is not entirely clear which kind of generality a 'nomothetic' approach is concerned with, although most geographers seem to have assumed it is the second. If so, then realist geography might more properly be called idiographic. Given its concern with open systems and its preoccupation with synthesis an attempted nomothetic approach in the second sense can only lead to failure. Finally, if we are to call it idiographic, a realist geography would, in contrast to traditional regional geography, have every right to claim to be theoretical, explanatory and, if the word means anything, 'scientific'.

Notes

1 For fuller, and I hope more rigorous accounts of realism see my (1984) *Method in Social Science: A Realist Approach* (London, Hutchinson); and in relation to geography, my (1982) 'Explanation in economic geography', *Progress in Human Geography*, 6, 68–88 (and the references therein).

2 The meaning of the term 'positivist' is notoriously contested and those who are attacked under this heading frequently capitalize on the confusion by arguing about its meaning, thereby evading the thrust of the criticisms directed at them. I only use the label here because it may be familiar to some readers. What matters is not its 'correct meaning' but whether the actual practices – *whatever they may be called* – which I attack are defensible.

References

Chorley, R.J. and Haggett, P. (1967) *Models in Geography*, London, Methuen.
Harvey, D. (1969) *Explanation in Geography*, London, Edward Arnold.
Keeble, D.E. (1980) 'Industrial decline, regional policy and the urban–rural manufacturing shift in the United Kingdom', *Environment and Planning A*, 12, 945–62.
Massey, D. and Meegan, R.A. (1979) 'The geography of industrial reorganization', *Progress in Planning*, 10 (3), 155–237.
Sayer, A. (1982) 'Explaining manufacturing shift: a reply to Keeble', *Environment and Planning A*, 14, 119–25.
Scott, A. (1980) *The Urban Land Nexus*, London, Pion.

7 Linda McDowell

Understanding Diversity: the Problem of/for "Theory"

Reprinted in full from: R. Johnston, P. Taylor and M. Watts (eds), *Geographies of global change: remapping the world in the late twentieth century*, chapter 17. Oxford: Blackwell (1995)

One day, the philosopher of science Paul Feyerabend, then teaching at the University of California Berkeley, confessed to a female African-American student his reservations about teaching imperial Western philosophy to representatives from different cultures, who had their own ways of knowing and their own values. She replied: 'You teach. We choose'. (Interview with Feyerabend, *The Higher*, 10 Dec. 1993).

Introduction

This story is the key to this chapter. A set of challenges to the theoretical basis of disciplinary knowledge currently poses a major challenge to a discipline like geography whose *raison d'être* is the explanation of difference and diversity. Recent critiques of Western science, exposing its purportedly universal and objective truths as ethnocentric, masculinist, and historically specific assumptions, seem to have left us marooned in a quagmire of relativism in which different ways of knowing have apparently equal claims to validity. An angry debate, ranging across disciplinary boundaries, within and between the social sciences and the humanities, and reaching into the sciences, has erupted in which adherents of different perspectives exhort each other variously to celebrate difference or to hold on to old verities. The bad-tempered arguments within a related discipline, anthropology, give a flavor of the debate and the passions with which each side holds to its arguments for or against notions of rational science and cultural relativism (see Gellner 1992). Our own discipline, too, has seen fierce exchanges about the nature of theory between, for example, Marxists, feminists, and a number of adherents of various versions of postmodernism.[1]

I want to argue here, however, that abandoning old certainties does not necessarily entail the whole-hearted embrace of relativism. Recognizing different ways of knowing does not mean abrogating responsibility for distinguishing between them. For teachers, as well as students, one of the main purposes of theory is to permit and to justify such choices. As geographers, we need theoretical perspectives that not only permit the elucidation of the main outlines of difference and diversity, of the contradictory patterns of spatial differentiation in an increasingly complex world, where ever-tighter global interconnections coexist with extreme differences between localities; but perspectives that also allow us to say something about the *significance* of these differences. Thus, I argue here that geographers must develop *Principled Positions* (Squires 1993) as a basis for constructing a *politics* of difference.

In this chapter, I shall try to summarize the main features of three sets of theorists of diversity – feminists, poststructuralists, and postcolonial theorists. This is an almost impossible task as the literatures in each area are now huge, both within and without geography. It is an important task, however, as without a flavor of these debates it is impossible to address the question of relativism that arises if we argue that discourses are geographically and socially specific, or the possibility of distinguishing between different ways of knowing.

What is theory/What is theory for?

In debates about the status of knowledge, the term theory has a privileged position. A major consequence of the debates is that its definition has changed. A theoretical analysis of a geographic issue consists of a rule-governed activity. These rules establish 'what counts as evidence, what constitutes an argument, what appeals are to be made to empirical truth, explanatory adequacy, rhetorical effectiveness, or any other standard of judgement' (Landry and Maclean 1993, page 136). Conventionally, this has meant a scientific, or objective, analysis: this is usually what is meant by 'theory' in the technical sense and is the definition that informs geographic approaches such as spatial science or Marxist analyses. But increasingly, geographers recognize that theories themselves are 'texts' or narratives that tell a single and particular story, and not others that might have been told. Thus a second kind of theoretical analysis is becoming common in geography: analyses of theories themselves, revealing the ideological assumptions in the work; the contradictions, unvoiced assumptions, the limits to the texts' logic (see Barnes 1992). Scholars from a range of positions – feminist, poststructural, postmodern, and postcolonial – have raised new questions about whose point of view is represented in geographic writing. These scholars point out the absence of certain voices in geographic narratives – in the main, those of the powerless and dispossessed – arguing against conventional scientific methods that assume a correspondence between 'reality' and its representation by geographers, the idea that maps and texts make transparent a singular meaning for the reader. Rather, they argue that knowledge reflects and maintains power relations, that it is partial, contextual, and situated in particular times, places, and circumstances. Representations of these partial truths are produced by authors who are 'raced', gendered, and classed beings with a particular way of seeing the world. Further, these socially-produced texts have no necessary or fixed meaning; the reader is implicated in the construction of meaning. A reader's own set of assumptions and sociogeographic location affects how texts are read and interpreted.

These arguments are tremendously powerful and potentially liberating, raising new questions about power and knowledge and, it seems, particularly pertinent questions for geographers about the relationship between place and representations, about the social construction of geographic knowledge and our research methods. They are also unsettling, posing difficult questions about the significance of different views of the world. If there is no singular Truth any more, but diverse truths, how, we might ask Feyerabend's student, does she go about choosing *between* them?

To understand the extent of the challenge posed by what we might term the new theorists of diversity, it is helpful to reflect on the history of theoretical endeavor in the discipline. Surveys of the history of geographical explanation tend to give the impression that the replacement of one approach by another is orderly progression in which increasingly sophisticated theories replace earlier and, by implication, discredited perspectives. But this is a false impression. With the possible exception of spatial science – and even this approach never achieved complete dominance – a more accurate picture is one of competing knowledge claims, of interpretative communities with different views of the world and different methodological approaches. Geography, like most social sciences, has a history of multiple perspectives, competing paradigms, and theoretical diversity. However, these diverse ways of seeing the world were regarded as *alternatives.* The adherents of each approach argued that their view was the correct one. Each held to a notion of a single truth embodied in their own view of the world and uniquely revealed by their approach to research – beliefs that are crucial in arguments that geography is 'scientific' but are disrupted by the claims by the multiple 'others' who have recently appeared in geographic texts. The voices of women, working-class subjects, 'Third World' peoples, of the underclass and transgressors, make different claims – that the singular geographies we have been accustomed to are equally partial; they reflect the view of the world of the powerful rather than a rational, scientific, or objective view of reality as claimed. The theoretical challenge of representing the partial perspectives of multiple 'others' rather than what Haraway (1991) has termed 'the view from nowhere' is subsumed in part of my title: the problem *of* theory. The current problem *for* theory is how to respond to it.

The problem of theory: the status of 'Truth/truth'

Science, ideology, knowledge, discourse, narrative: these five words indicate alternative positions in the intellectual landscape of contemporary geography. The first two reflect englightenment views of Truth that underpin the liberal and Marxist perspectives in geography, whereas the latter three are common terms in new theoretical positions. Whereas Marxist analysts in geography are concerned to demonstrate what Marx himself termed 'the economics of untruth' in liberal theories of the world, more recently other geographers, especially those influenced by the French philosopher Foucault, have challenged the scientific basis of Marxist theory in their focus on 'the politics of truth' (the term is Foucault's). I shall briefly spell out these differences, but interested readers will find a more thorough discussion in Michele Barrett's *The Politics of Truth* (1991).

The first challenge to the scientific basis of positivist approaches in geography came from the work of Marxist scholars and from socialist feminists, who argued that the work of too many theorists bore little relationship to the reality that it claimed to portray. Positivist spatial science, for example, portrayed a landscape without power, poverty, or political struggles: areas to which Marxist and socialist feminist analysts drew particular attention. Although adherents of Marxian perspectives continue to hold to notions of scientific knowledge and Truth, they argue that positivist science is ideological, justification of the privileges of a class society dressed up as truth. Here, we see the idea that knowledge is positional or embedded, a version of the world that embodies a particular justification of the distribution of rewards and privileges. However, it is within that set of knowledges variously labeled postmodernist, poststructuralist, and postcolonialist, and in recent developments in feminist scholarship (McDowell 1992a and 1992b), that resistance to truth claims has been most strenuous. The rest of the chapter discusses these arguments.

Introducing the 'Others': feminism, poststructuralism and postcolonialism

What was it that happened to disrupt theoretical confidence in a singular notion of Truth? What were the challenges to geographers' ways of seeing the world? How were spaces opened up in which ideas about plurality, diversity, multiplicity, difference, disruption, and contradiction challenged the singular emancipatory vision of Western science, that central notion of development and progress that sustained both the liberal/scientific and the Marxist/socialist discourses? The answer lies partly in the political movements developing from the late 1960s onwards as well as in theoretical shifts in the academy. Within geography, a group of committed and radical academics were active not only in class struggles but also in feminist, green, and ecosocialist struggles. (The founding of the journal *Antipode* in 1969 gave these radical geographers a disciplinary standpoint.) Their combination of academic and political work raised questions about the assumptions behind the notions of social justice and progress that informed both liberal-humanist notions of individual equality and socialist versions of the future. A sustained attack was launched from several quarters on the supposedly universal 'we' of both these discourses. A choir of voices – women's, gay men, colonial subjects, prisoners, and other 'deviants' – was raised to protest that the supposed universal 'we' of humanist and socialist discourses in fact excluded their specific experiences, based as they were on a particular view of the world: a view through the eyes of the Western, male, bourgeois subject.

Theoretical developments in the critical social sciences coincided with these political movements to necessitate the re-evaluation of the privileged status of existing theoretical frameworks, with their commitment to notions of a single universal truth and a unitary, universal human subject (Barrett 1991). Instead of a search for Truth and universal laws, we have seen a number of geographers, especially those working in the social, historical, and cultural areas of the discipline, turning toward the theorization, and in some cases the celebration, of difference and diversity, in opposition to the totalizing tendencies in what have become known as the 'grand narratives' of liberal humanism and Marxism. Instead of searching for general laws these geographers insist on the particularity and plurality of knowledge, on analyses of the specificity of its social construction by certain social groups located, in particular, in space–time frameworks. While debates about the situatedness of knowledge have permeated the social sciences and humanities more generally, for geographers they strike a particular chord as they coincide with our long-held interest in difference, diversity, and place-related specificity.

The key players in the displacement of old certainties have, appropriately, been many and various and their arguments overlap. However, I shall give a flavor of their claims under three headings: feminism, poststructuralism and postcolonialism, indicating the implications for geographical work, although many of the key theoretical texts are outside our discipline.

Feminism

Despite the insistent claims of a parade of grand (and mainly Euopean) male theorists – Michel Foucault, Jacques Derrida, Jacques Lacan, Jean-François Lyotard, Richard Rorty et al. – some of the earliest attempts to 'deconstruct' the humanist subject were made by feminist theorists, 'as long ago as 1792 by Mary Wollstonecraft in her demand for women to be included in the entitlements claimed by the "Rights of Man"' (Soper 1990, pages 229–30). As innumerable feminists have demonstrated since, Western humanism succeeded in writing women out of the theoretical agenda by constructing the category 'Woman' in opposition to 'Man', in such a way that women were defined as the 'Other', as an absence or a lack, compared with the 'One', that is an idealized rational, disembodied subject or individual, that, in fact, embodied masculine (and Western) attributes. Through the construction of a set of binaries that equated masculinity with science, rationality, objectivity, and the public and political arenas of life, mapping femininity onto opposite characteristics – emotion, irrationality, subjectivity, the private and domestic sphere – women were relegated to an arena that apparently needed no theoretical investigation but might be taken for granted as 'natural'. By naming femininity in opposition to the category 'Man', which was assumed to be human, woman was situated in a very different category from that of man (Pateman and Gross 1986). Feminism has convincingly demonstrated how 'women's silence and exclusion from struggles over representation have been the condition of possibility for humanist thought: the position of woman has indeed been that of an internal exclusion within Western culture' (Martin 1988, page 13).

The early achievements of feminist scholarship involved establishing women's right to inclusion, based on an appeal to liberal notions of social justice. As women, too, were individuals, to deny them the rights and benefits available to men was patently unjust. By the mid-twentieth century, however, it became clear to feminists that a claim for equality within humanist thought was impossible. As the humanist subject was masculine, women would always be excluded. So, it is argued, humanism as such was inherently flawed and must be replaced by an alternative set of knowledge claims (although based on what claims and in what form is contested and uncertain). It is here, in the *deconstruction (the challenging and taking apart) of knowledge claims*, that recent feminist, poststructuralist, and postmodern theories unite to post a common challenge to the theoretical basis of conceptions of knowledge, truth, and science: a challenge that is resulting in the 'displacement of the problematics of science and ideology, in favour of an analysis of the fundamental *implications* [original emphasis] of power–knowledge and their historical transformations' (Morris 1988, page 33). By this, Morris means that we must look at whose interests theories support, how they vary across space and how they change over time. It is this emphasis on change that perhaps distinguishes poststructuralism from Marxism but there is also a significant difference in the way in which power relations are theorized, as I shall demonstrate below.

Poststructuralism

The term power/knowledge is a reference to the work of Foucault, who, although resistant to the labeling of his work, is generally regarded as a poststructuralist. What is important here is his rejection of the notion of ideology, and a reconceptualization of power relations as complex and unstable, part of the social construction of self rather than an external system 'imposed' on people from above. Let us take these claims in turn.

Foucault rejects the concept of ideology on several grounds: first, that it necessitates an opposite notion, that of truth; secondly, that it stands in a secondary position to the material or economic determination of society; and thirdly, that it rests on the humanist notion of an individual subject. Poststructuralists have developed instead the concept of 'discourse', a concept that is dependent upon Foucault's redefinition of power. Rather than seeing power as a social and political system in which a repressed group is dominated for the benefit of the oppressors, Foucault (1970, 1977, 1980) argued that language, texts, discourse, as well as social practices, are part of the maintenance of particular power relations in modern societies. He provided what he called a different 'grid' for deciphering history – a new way of looking at and analysing power which focused on the ways in which discourses, and the pleasures and powers they produce, have been deployed in the service of creating and maintaining hierarchical relations in Western societies (Foucault 1978). Thus power relations are maintained not through the false consciousness of the repressed, who will eventually act as a class for themselves, but rather as part of the very construction of identity. Thus an individual subject is constructed through a grid of discourse and practice. In poststructuralist theory there is no notion of a prior or essential moral self, as in liberal and Marxist theory, and so traditional ideas of 'truth' and 'self' are rejected. As Foucault argued, 'power seeps into the very grain of individuals, reaches right into their bodies, permeates their gestures, their posture, what they say, how they learn to live and work with other people' (quoted in Sheridan 1980, page 217). Our bodies, identity, and sense of self have no prior existence but are brought into play through language, discursive strategies, and representational practices, and so conventional distinctions between thought and action, between language and practice, are collapsed as all social actions have a discursive aspect.

While all discourses are changeable, open to contestation because they are social practices, Foucault termed the establishment of an interlocking set of successful or dominant discursive formations a *regime of truth.* The term truth, however, is used here in a different sense from conventional notions. It is impossible to distinguish between what is true or false by appealing to the 'facts' of the matter, as the 'facts' themselves are part of a particular discursive formation. The 'truth' of a situation is determined by the outcome of struggles between competing discourse.

Postcolonialism

It is within recent work under this heading, linked to Third World feminists' critiques of the ethnocentric biases of Western feminism, that perhaps the most powerful arguments about contested and incompatible knowledges, as well as multiple subject positions, have been advanced. Within Western theory, 'women of color' have criticized the racially exclusionary theory and practices of white, middle-class feminism, problematizing assumed commonalities as women. At the international level, the theoretical focus is on imperialism rather than racism, and here the most influential work has been that of Said in his arguments about the construction of a colonial 'Other' by the West, and the work of the Subaltern Studies group, who are rewriting the history of colonial India from the point of view of peasant insurgents. In both cases, despite his profound Eurocentrism, Foucault is a key inspiration, as he has been for a number of feminists who argue that women's exclusion from struggles over representation within Western culture are another demonstration of Foucault's concept of power (Martin 1988).

Said's *Orientalism* (1978) has probably been the most influential text in the development of new ways of thinking about Western imperialism. Drawing on Foucault's notion of discourse, Said argued that 'without examining Orientalism as a discourse one cannot possibly understand the enormously systematic discipline by which European culture was able to manage – and even produce – the Orient politically, sociologically, militarily, ideologically, scientifically, and imaginatively during the post-Enlightenment period' (page 3). Through a range of discursive strategies and social practice, people of the Orient were constructed as the 'Other' to the Western 'One', in ways such that 'European culture gained in strength and identity by setting itself off against the Orient as a sort of surrogate and even underground self' (Said 1978, page 3).

Whereas Said focuses on the discursive construction of the Orient in Western texts, the Subaltern Studies group is interested in the impact of colonialism on native cultures. Spivak (1988) argues that the imposition of colonial rule in India effectively ruptured and silenced indigenous cultures, a fracturing that she terms, after Foucault, 'epistemic violence'. The colonial subject is unable to answer back in a language that is untouched by imperial contact, and which may only be glimpsed in the interstices of colonial texts. The truly marginalized, the 'subaltern' of the group's title and Spivak's most well-known paper, is unable to speak at all.

These arguments have interesting implications for how geographers undertaking ethnographic and field-based work in postcolonial societies theorize and analyze the narratives collected from 'native informants'. The implications of Spivak's arguments are that no one falls outside the field of imperialist discourse. Whether engaged in peasant insurgency or postcolonial critique, the native is forced to speak the language of imperialism. However, rather than resulting in capitulation, Bhabha (1990) has argued that what he terms 'hybridization' is the result, a sort of double displacement whereby native mimicry of imperialist discourses embodies cultural and political resistance, disrupting authoritative representations of imperialist power. Thus rather than either a nostalgic search for lost origins or a desire to re-establish authentic traditions, the postcolonial critic analyses the ways in which resistance lies in hybridization. In the context of the West, where the huge migration flows related to slavery, imperialism, and global capitalism have resulted in peoples who are of neither First nor Third World, Bhabha's notion of hybridization also seems relevant. Indeed, the sociologists Gilroy (1993) and Hall (1991) have developed interesting parallel arguments about cultural translation. Gilroy, for example, in his work on African-American peoples, argues that a hybrid Black Atlantic culture has developed. Similarly, 'women of color' in the US are exploring the position of women who are the product of international and cultural borders – women of mixed origins, whose identity is constructed out of differences (Anzaldua 1987; Sandoval 1991).

In combination, the work summarized under the above three headings has altered the subject of geographic research, refocused theoretical and methodological assumptions, and challenged the political practice of 'radical' geographers. Many have realized that, in the words of Landry and Maclean (1993, pages 125–6), 'the white male heterosexual worker of the First World industrial nations just won't do any longer as the alternative or insurgent subject of history, the locus of resistance to ruling class dominance'. Drawing in particular on qualitative and ethnographic methods, with increasing attention to everyday social relations and to discursive strategies, a number of geographers have turned to detailed analyses of the 'other' – women in different positions (see my reviews – McDowell 1992a and b), female travelers (Domosh 1991), native peoples and ethnic minorities (K. Anderson 1992), the mad (Philo 1989), and the perverse (Valentine 1993), and of the transgressive landscapes of desire of 'alternative' sexualities (Bell 1994; Bell, Binney, Cream, and Valentine 1994). There has also been a 'discursive turn' in which the methods of discursive and textual analysis have become influential in the interpretation of landscapes of power and oppression, as well as social relations in localities. As Peet and Watts (1993a) have suggested, 'one of the great merits of the turn to discourse, broadly understood, . . . is the demands it makes for nuanced, richly textured empirical work' (page 248): a type of work admirably illustrated in Watts' book with Pred (Pred and Watts 1992).

Recently, attention also has turned to the discursive strategies of geographers, to textual analyses of our own writing. Geographic language is recognized as being not simply a medium expressing reality 'out there' but as playing an active role in creating social and cultural worlds. Interpretative schemes 'in part . . . create the reality that they seek to interpret' (Barnes 1992, page 118). Theories 'have the imaginative capacity to represent and reconfigure the world like *this* rather than like *that*' (original emphasis: Gregory 1994, page 182). In new histories of geography, theoretical perspectives are seen not as alternatives in a struggle for hegemonic status but as alternative, and incompatible, interpretative communities. The purpose of theoretical endeavor in our discipline is thus disrupted. Rather than a struggle to establish universalizing claims, the grand and certain aim of grand theory, we (not this time united by theoretical consensus, but a motley collection of geographers with different views of the world) have in common what Gregory terms a 'strategy of supplementarity – *dis*-closing what theory closes off' so that 'the certainties of theory can be capsized, its confidence interrupted and its conditional nature reasserted' (Gregory 1994, pages 181–2). The singular geographical imagination so eloquently defined by David Harvey in 1990 has become the partial and plural geographical imaginations of Derek Gregory's (1994) latest survey.

Problems *for* theory

Claims that geographies are multiple, that knowledge is contested and that space is discontinuous and fluid, a discursively constructed set of relationships that discipline, control, or privilege (Lefebvre 1991), a 'plane of contest' (Ashley 1987) rather than a set of fixed regions defined by Cartesian coordinates, have led to an exciting set of debates in the 'new' geography but also an unease. Fears of relativism, enunciated not only by those critical of claims that knowledge is embedded and situated but also by many who generally welcome the challenge to Enlightenment notions, are perhaps the most serious, although it is important to note Barrett's claim that poststructuralist critiques of truth and knowledge are 'not so much relativist as highly politicised' (Barrett 1991, page 161). Theoretical acceptance of plural and fragmented identities, of multiple and local sites of power, has been enormously empowering for those 'others' condemned as handmaidens, helpmeets, 'perverse', or slaves of Western 'civilisation', whether marginalized by the universalist pretensions of humanism, or left at home/outside by Marxists while the 'real' struggles take place elsewhere. But questions remain. How, for example, might we judge between competing knowledge claims? On what may we base our political opposition to the oppression and exclusions that are revealed in these new discourses? Are all claims to knowledge and power equally valid?

Many, geographers included (Driver 1992; Gregory 1994), have pointed to a paradox at the heart of the three theoretical positions discussed above: their critique of the exclusion of 'others' surely depends for its critical force on acceptance of modern ideals of autonomy, human rights, and dignity. Thus, as Soper (1990) asks, 'why concern ourselves with the exclusions from the "humanist" universal of women, or blacks or gay people, or the insane, or any other oppressed or marginal group, except on the conventional ethical grounds that all human beings are equally entitled to dignity and respect?' (Soper 1990, page 149).

It seems to me that here we must establish ways of criticizing universalistic claims without completely surrendering to particularism. And surely this is a more general example of the problem that geographers, in their focus on the particularity of place, have been grappling with for many years. Geographers have long accepted a form of relativist argument: that the specificities of local socioeconomic structures, differences in cultural attitudes and in ways of living, are part of the explanation for uneven development (McDowell and Massey 1984; Massey 1984). Drawing on this work, in which universal and particular explanations are neither counterposed nor seen as alternatives, we need to *recast our general notion of relativism* and no longer see it as counterposed to universalistic explanations. Rather, we must see relativism, in the sense of difference, as inescapable. Acceptance that there are different ways of knowing which are historically and geographically specific does not mean that we cannot make judgments about their value. Foucault drew our attention to the ways in which discourses, or speaking positions, are constituted by power relations. Thus oppositional knowledges of the subjugated are inherently and inescapably political, demanding judgment about the validity of different discourses on the basis of political considerations. We might, therefore, accept the multiple claims to power of some groups and reject those of others, in attempts to construct a multiple and inclusive notion of social justice that includes the claims of the non-white, non-masculine subject (Flax 1990; J. E. Young 1992). But as Barrett points out, this second move, one that she terms 'transformation', is much more difficult than the first, the transgressive deconstruction of dominant binary oppositions where feminism, poststructuralism and postcolonialism have been so successful. Once the categories of 'woman', or 'postcolonial subject', have been dismantled, it becomes more difficult to speak on their behalf. There is a tension between challenging the oppositional construction of others and continuing to make political demands on their behalf. A possible way forward is suggested by 'black' theorist Du Bois (1989, and see Gilroy 1993) in his theorization of 'double consciousness' that simultaneously allows for the deconstruction of the category black and for the need to assert the truth and authenticity of the black identity. Biddy Martin (1988) similarly has suggested a 'doubled strategy' for feminist struggles.

Conclusions: principled positions

The theoretical and moral implications of a rejection of both universalism and relativism as conventionally defined are complex. It is clear that the exercise of moral responsibility is culturally dependent and temporally specific. But the fact that political choices and everyday decisions are made in relation to a contestable code or set of conventions as to what is 'good' or 'bad' does not in itself detract from the element of judgment and the assertion of values in our choice. As Haraway (1991), a profoundly thoughtful theorist, has suggested, we need both to accept irreducible differences and yet hold on to what she terms a 'successor science project' in which 'projects of finite freedom, adequate material abundance, modest meaning in suffering and limited happiness' (page 187) are the focus of action. As Gregory suggests in his recent book: 'If we are to free ourselves from universalising our own parochialisms, we need to learn how to reach beyond particularities, to speak to larger questions without diminishing the significance of the places and the people to which they are accountable' (Gregory 1994, page 205).

Theorizing diversity does not mean abandoning these larger questions and projects. The challenge now for geographers is to find ways to respond to them.

Note

1 I avoid the use of the term postmodern in the rest of this chapter, preferring instead to use poststructural to refer to theoretical claims, and reserving postmodern for social and cultural shifts. However, like Rosenau (1992), I recognize that 'the terms overlap considerably and are sometimes considered synonymous. Few efforts have been made to distinguish between the two, probably because the difference appears of little consequence' (page 3).

References

Anderson, K. 1992: *Vancouver's Chinatown: Racial discourse in Canada 1875–1980*. Montreal: McGill-Queen's University Press.

Anzaldua, G. 1987: *Borderlands/La Frontera: The New Mestiza*. San Francisco, CA: Spinster/Aunt Lute.

Ashley, R. 1987: The geopolitics of geopolitical space: towards a critical social theory of international relations. *Alternatives*, **12**; 403–34.

Barnes, T. 1992: Reading the texts of theoretical economic geography. In Barnes, T. and Duncan, J. (eds), *Writing worlds: Text and metaphor in the representation of landscape*. London: Routledge.

Barrett, M. 1991: *The politics of truth: From Marx to Foucault*. Cambridge: Polity Press.

Bell, D. 1994: Erotic topographies. *Antipode*, **26**; 96–100.

Bell, D., Binney, J., Cream, J. and Valentine, G. 1994: *Landscapes of desire*. London: Routledge.

Bhabha, H. K. (ed.) 1990: *Nation and narration*. London: Routledge.

Domosh, M. 1991: Towards a feminist historiography of geography. *Transactions of the Institute of British Geographers: New Series*, **16**; 95–104.

Driver, F. 1992: Geography's empire: histories of geographical knowledge. *Environment and Planning D: Society and Space*, **10**; 23–40.

Du Bois, W. E. B. 1989: *The souls of black folk*. New York: Bantam.

Flax, J. 1990: *Thinking fragments: Psychoanalysis, feminism and postmodernism in the contemporary West*. Berkeley and Los Angeles, CA: University of California Press.

Foucault, M. 1970: *The order of things: An archeology of the human sciences*. London: Tavistock.

Foucault, M. 1977: *Discipline and punish: The birth of the prison*. London: Tavistock.

Foucault, M. 1978: *The history of sexuality 1: An introduction*. New York: Random House.

Foucault, M. 1980: *Power/Knowledge*. Brighton: Harvester Press.

Gellner, E. 1992: *Postmodernism, reason and religion*. London: Routledge.

Gilroy, P. 1993: *The black Atlantic. Modernity and double consciousness*. London: Verso.

Gregory, D. 1994: *Geographical imaginations*. Oxford: Basil Blackwell.

Hall, S. 1991: Ethnicity: identity and difference. *Radical America*, **23**; 9–20.

Haraway, D. 1991: *Simians, cyborgs and the rediscovery of nature*. London: Free Association Press.

Landry, D. and Maclean, G. 1993: *Materialist feminisms*. Oxford: Basil Blackwell.

Lefebvre, H. 1991: *The production of space*. Oxford: Basil Blackwell.

McDowell, L. 1992a: Space, place and gender relations, part I: feminist empiricism and the geography of gender relations. *Progress in Human Geography*, **17**; 157–79.

McDowell, L. 1992b: Space, place and gender relations, part II, difference, feminist geographies and geometries. *Progress in Human Geography*, **17**; 305–18.

McDowell, L. and Massey, D. 1984: A woman's place?. In Massey, D. and Allen, J. (eds), *Geography matters! A reader*. Cambridge: Cambridge University Press.

Martin, B. 1988: Feminism, criticism and Foucault. In Diamond, I. and Quinby, L. (eds), *Feminism and Foucault: Reflections on resistance*. Boston, MA: Northeastern University Press.

Massey, D. 1984: *Spatial division of labour*. London: Macmillan; New York: Methuen.

Morris, M. 1988: The Pirate's fiancee: feminists and philosophers, or maybe tonight it'll happen. In Diamond, I. and Quinby, L. (eds), *Feminism and Foucault: Reflection on resistance.* Boston, MA: Northeastern University Press.
Pateman, C. and Gross, E. (eds) 1986: *Feminist challenges.* Boston, MA: Northeastern University Press.
Peet, R. and Watts, M. J. 1993a: Development theory and environment in an age of market triumphalism. *Economic Geography*, **69**(3); 227–53.
Philo, C. 1989: 'Enough to drive one mad': the organization of space in nineteenth century lunatic asylums. In Wolch, J. and Dear, M. (eds), *The power of geography.* London: Macmillan, 258–90.
Pred, A. and Watts, M. J. 1992: *Reworking modernity: Capitalisms and symbolic dissent.* Brunswick, NJ: Rutgers University Press.
Rosenau, P. M. 1992: *Postmodernism and the social sciences: Insights, inroads and intrusions.* Princeton, NJ: Princeton.
Said, E. 1978: *Orientalism.* London: Routledge; New York: Pantheon.
Sandoval, C. 1991: *The global city.* Princeton, NJ: Princeton University Press.
Sheridan, A. 1980: *Michel Foucault: The will to truth.* New York: Tavistock.
Soper, K. 1990: Postmodernism, subjectivity and the question of value. In Squires, J. (ed.), *Principled positions: Postmodernism and the rediscovery of value.* London: Lawrence and Wishart.
Spivak, G. C. 1988: Can the subaltern speak? In Nelson, C. and Grossberg, L. (eds), *Marxism and the interpretation of culture.* Urbana, IL: University of Illinois Press, 271–313.
Squires, J. (ed.) 1993: *Principled positions: Postmodernism and the rediscovery of value.* London: Lawrence and Wishart.
Valentine, G. 1993: Negotiating and managing multiple sexual identities. Lesbian time–space strategies. *Transactions of the Institute of British Geographers: New Series*, **18**; 237–48.
Young, I. M. 1992: *Justice and the politics of difference.* Princeton, NJ: Princeton University Press.

Textuality and Human Geography

TEXTUALITY AND HUMAN GEOGRAPHY

Words and worlds

The philosopher Richard Rorty is fond of saying that language 'goes all the way down'. By that he means that language cannot be separated from the things, events, and activities that it names. For the world is as much produced by the words we use as it is reflected by them. There can be no preinterpreted brute reality because even to think of such an entity requires the use of language. As Gunnar Olsson (this volume, page 145) says below, without language 'nothing would be, not even Nothing'.

The 'linguistic turn' that Rorty and Olsson represent is in contrast to an earlier view of language which, following Eagleton (1983), is called naive **realism**. Here 'words are felt to link up with their thoughts or objects in essentially right and incontrovertible ways' (Eagleton, 1983, page 134). In this conception, language is unproblematic in that words are taken as the naturally given correlates of objects or ideas: a rose is a rose is a rose. The task of writing is then simply the mechanical one of lining up words in the right order. Anyone can do it provided that there is a dictionary at hand.

Of course, not everyone believed in this mirror view of language. As far back as the fifth century BC the Greek sophists argued that words could have a persuasive power of their own, a rhetorical force, that was independent of the thing that they represented. Although their view was later dismissed by Plato and Socrates as morally corrupt because it ostensibly denied any final truth, the idea that language makes worlds as much as representing them has persisted historically in a variety of forms. During the 20th century we can recognise at least three approaches that have sought to problematise the relationship between words and things, thus undermining naive realism.

- Ordinary language philosophers emerged after the end of the second world war and drew heavily upon Wittgenstein's later philosophy. Wittgenstein's early philosophy had proposed a so-called picture theory of meaning in which the most basic 'atomic facts' of the world were reflected by the most basic 'elementary propositions' within language. After teaching reading and writing in primary school during the inter-war years back in his native Austria, Wittgenstein came to reject his earlier thesis. He recognised that the meaning of language was not tied to objects in the world, but emerged in all kinds of different ways that could be established only by examining the case at hand. Once this was understood, the task of philosophers was then one of clarifying the wider context – the **language game** in Wittgenstein's terms – in which words took their meaning. More broadly, for ordinary language philosophers meaning of language came from the context of use, rather than reference to an object.

- Critical **hermeneutics** is associated with the continental philosophers Gadamer and Ricoeur, who in turn drew upon the late-19th-century German hermeneutician, Dilthey. Defined simply as the study of interpretation and meaning, hermeneutics originated with the exegesis of biblical texts. Over time, however, hermeneuticians progressively widened the scope of their method and their definition of what counted as a **text**. Originally only philological methods were employed, but increasingly a range of interpretive techniques were admitted. Remaining key, though, was the so-called hermeneutical circle first set out by Dilthey, and involving a tacking back and forth between individual parts of text and the wider whole including the presuppositions of the reader. Likewise, although originally only God's word was hermeneutically interpreted, it was gradually recognised that all kinds of texts, both written and non-written, could be analysed. Ricoeur (1971), for example, suggests that virtually anything within the social world could be treated as an interpretable text. As with the ordinary language philosophy, critical hermeneutics suggests that texts are not mere referential duplications of something outside of themselves, but take on a meaning that is produced through a historically specific and complex negotiation between, in this case, the interpreter and the interpreted.
- **Semiology** (or semiotics as Charles Peirce called it) emerged at the beginning of the 20th century with the work of the Swiss linguist Ferdinand de Saussure. Prosecuting a **structuralist** approach to language, Saussure distinguished between the **signifier**, the sound or image, and the **signified**, the meaning. For Saussure there was no necessary relationship between the two. Rather, language was defined only by the underlying structural relationship among the system of signifiers. Useful here is Saussure's distinction between *la langue* and *la parole. La parole* represents the everyday usage of language, whereas *la langue* is the deeper, unconscious set of rules that govern it. Given his structuralist position Saussure was primarily interested in *la langue,* and for that reason was not concerned with the evolving history of language (a diachronic analysis) but only with its contemporary internal relations (a synchronic analysis). As with the two other approaches, meaning for Saussure is produced within, rather than reflected through, language; language constitutes rather than represents social reality.

Geography and language

Despite the intellectual pedigree of these three 20th century approaches to language, all of which problematise the relationship between 'texts and lumps' (Rorty, 1985), human geographers resolutely ignored them until the mid-1980s. This is not surprising. Geography was the paragon descriptive science, and, in many ways, naive realism was made for the discipline. Admittedly, during the 1970s there was some discussion about writing, but it was couched in terms of literary style rather than a theory of language as such. Even Marxist discussions of **ideology** around the same period did nothing to dislodge naive realism. Marxists seemed to believe that once the fog of ideology was blown away, it was writing as usual.

A remarkable exception to this general nonchalance towards language was the work of Gunnar Olsson. An acolyte of geography's **quantitative and theoretical revolution**, from the late 1960s Olsson increasingly had doubts about **spatial science**, posing a series of awkward technical and later philosophical questions about explanation, representation and ultimately language itself. The result was *Birds in egg* (first published in 1975), 'a brilliantly flawed presentation' (Gregory, 1978, page 173), and one that drew particularly upon Wittgenstein's later philosophy and to a lesser extent hermeneutics. The nub of its thesis was that human geography's models say far more about the language that they are written in than the things that are written about. Although 'hitting our heads against the ceiling of language' (Olsson, 1980) is unavoidable, Olsson's important political point is that the kind of language against which we collide makes a difference. For example, the use of some vocabularies – Olsson singles out the Galilean tradition – can produce quite unsavoury political consequences. Olsson's task, then, is to experiment with words, to be the jester and the poet, to pull and stretch language in new ways so as to reveal and partially overcome its rigidities, biases, and silences, and in doing so head off unsavoury consequences. Not everyone has been happy with the result, but if for nothing else Olsson needs crediting for putting language on the agenda of human geography, and in doing so also preparing the discipline for **post-structuralism**, a movement defined as much as anything by its attitude towards language.

On discourse and deconstruction

The post-structural sensibility that first entered geography in the mid-1980s brought with it two central concepts, **discourse** and **deconstruction**. There are many definitions of discourse, but we focus on Foucault's which, as Philip (1985, page 69) writes, 'is best understood as a system of possibility for knowledge'. A discourse in this interpretation is defined by a set of rules that provide the preconditions for the existence of a field of knowledge. Those rules determine, for example, which statements can be made, their order, and their degree of validity (Philip, 1985, page 69). Such rules, however, do not form a methodological template to be consciously followed, but act, as it were, behind the backs of those who operate within them, constructing both the subject of research and the subjects undertaking the research. Whereas in his earlier writings Foucault presents discourses as almost free-floating, abstract conceptions, in his later writings he increasingly situates them on the ground within intricate social networks of power.

With respect to language, a discourse represents the limits of what can be said about the world; or, more precisely, a discourse is the world because there is nothing outside of it, no extradiscursive reality. In this sense, discourses often take on a naturalising function, universalising particular features of social life and presenting them as the norm. Though presenting themselves as fixed and eternal, discourses are inherently unstable and fragile. Because of their embeddedness within constellations of power, they are continually challenged and contested producing discontinuity and rupture. Acquisition of knowledge under different discourse regimes is thus neither rational nor progressive, but at best contingent and halting. Although there are truths, there is no Truth.

Whereas Foucault developed the idea of discourse in order to understand particular historical social and material practices, Derrida's notion of deconstruction was primarily fashioned for critique. That critique is aimed at **logocentricism**, defined as a belief in an ordered world. In attacking logocentricism Derrida radicalises both Saussure's and critical hermeneutic's respective views of language. Saussure was right in averring that there was no necessary relationship between signified and signifier, but he was wrong in suggesting that meaning is fully determined by the structural relationship among signifiers. Instead, for Derrida, meaning is always underdetermined because there is no fixed relationships within that system of signifiers to anchor it. Every signifier is related to every other signifier, making it undecidable. Similarly, although Derrida accepts the view of critical hermeneutics that texts require interpretation, he criticises them for suggesting that there is some final interpretation graspable through the use of the hermeneutical circle. For Derrida there is never a rock-bottom meaning. Rather, texts continually careen off one another in a complex process known as **intertextuality** which has no end point. Final meaning is always deferred.

Deconstruction in Derrida's work is the means by which the instability of meaning within a text is revealed. Through very detailed and scrupulous readings of a variety of written works – from Plato's *Phaedrus* to the US Constitution – Derrida uses deconstruction to show that all texts in spite of their best intentions are inherently contradictory and open to multiple interpretations. Here Derrida's strategy is to isolate pivotal distinctions, distinctions that provide the basis of meaning for the entire logocentric system of the text, and then to show that they are inconsistent and paradoxical because of the rhetorical moves that are necessary to construct them. In this way, 'the text is seen to fall by its own criteria' (Lawson, 1985, page 93). Note that such inconsistencies arise not as a result of *ad hoc* bits of sloppy writing, but subsist in the very logic of language itself. Mirror representations always fail because there is nothing outside the text; our truths are only those that we write to ourselves.

On metaphor

Finally, linked but not necessarily related to post-structuralism is the recent interest by social scientists, including human geographers, in the role of **metaphor**, raising as it does suspicions about the adequacy of naive realism. For, although in literal terms they are nonsensical, metaphors take on a persuasive force in influencing the direction and nature of discourse. That is, metaphors work precisely by decoupling words and things. The persuasiveness of metaphors is not just confined to their role within individual sentences, though, but extends to their influence on broad contours of intellectual inquiry. Richard Rorty (1979, page 12) writes, for example, 'it is pictures rather than propositions, metaphors rather than statements, which determine most of our philosophical convictions'. Much of the recent work carried out on metaphor makes the point that users of 'large' metaphors are often not aware of all the intellectual freight that such metaphors carry, and as a consequence are unconsciously led to positions that they would otherwise not accept. For this reason it is necessary to 'exhume' the dead metaphors that we unwittingly employ, and inspect them critically for their coherence, consistency, and compatibility with the other things that we might also want to say. Doing so often means scrutinising the historical and material origins of the original metaphor which shaped their meaning, and, in this sense, taking metaphor seriously means taking the world seriously too. Metaphors require 'worlding' (see the Introduction), indicating again the complex relationship between words and things.

Texts on texts

Gunnar Olsson's 'Lines of power' is the first of the essays in this section. Although a difficult paper, it is characteristically witty and stylish. Although not his native idiom, Olsson as well as anyone else in the discipline plays with the ambiguities, ironies, and richness of the English language. In many ways, this paper is a retrospective on Olsson himself and is organised according to his long-time fascination with three sets of lines: the equals sign of logic, denoting identity; the slash of the dialectic, representing the merging of opposites; and the Saussurian bar, separating signified and signifier. In each case, Olsson argues, what is on one side of the line 'wants to merge with what stands on the other' (page 146). This is the 'mimetic desire'; the desire for one thing to be the same as another thing. For Olsson this desire is best represented in the English language by the word 'is' which connotes the equivalence of two things: Olsson is a man; war is peace; 'is' is 'is'. Olsson's intention, however, is to deconstruct 'is', and to show that the two entities on either side of the lines are only ever similar, never the same. Moreover, by prising apart the lines, Olsson also thinks that it is possible to see fleetingly what has kept them together: power. His argument is that because perfect translation or mimesis is impossible our conviction that 'is' really means 'is' derives not from logic or reason but from the crust of social convention which is made solid by relationships of social power. We believe $a = b$, or thesis/antithesis produces synthesis, because we have already been socialised into those respective vocabularies and instantiated relations of power such that we could not conceive them otherwise.

Olsson asks tongue-in-cheek at the end of his article, 'Is this geography? Of course it is! For what is geography, if it is not the drawing and interpretation of lines' (page 154). This is only half a joke because one of the most stimulating substantive debates around language in human geography has been precisely about geography's lines, its maps. Brian Harley's essay 'Deconstructing the map' is now already a classic (albeit not without criticism; see Belyea, 1992). He brings together French post-structural theory with his immense knowledge of map-making and the history of cartography. After showing that the supposed 'science of map making' is yet only another culturally constructed and specific discourse, he uses both Derrida's idea of deconstruction to problematise the very representational strategy embedded within maps, and Foucault's work to exemplify the network of power in which they are constructed.

The last essay, by Geraldine Pratt, is on metaphor, and examines the tensions that attend the use of various spatial metaphors by writers within cultural studies. She argues that in trying to overcome the problem of speaking positions – that is, taking into account, and, in part, mitigating the various biases of judgment, politics, and knowledge that subsist in a particular social location – writers in cultural studies have invoked a series of spatial metaphors such as 'mobility', 'marginality', and 'borderland'. In each case use of the metaphor supposedly 'displaces' the normal speaking position allowing the speaker if not to escape her or his social location at least to see the biases inherent within it. Pratt's argument, however, is that such a strategy is unsatisfactory because proponents have not been sensitive enough to the hidden geographical entailments of the metaphors themselves. As she illustrates, while proponents of the metaphor say one thing, the metaphor itself says something quite different. More generally, Pratt carries out a form of deconstruction in that various texts in cultural studies are shown to unravel from within because of inconsistencies in their use of central metaphors.

References

Belyea, B. 1992: Images of power: Derrida/Foucault/Harley. *Cartographica*, **29**; 1–19.

Eagleton, T. 1983: *Literary theory: An introduction.* Minneapolis, MN: University of Minnesota Press.

Gregory, D. 1978: *Ideology, science and human geography.* London: Hutchinson.

Lawson, H. 1985: *On reflexivity: The post-modern predicament.* London: Hutchinson.

Olsson, G. 1975: *Birds in egg.* Michigan Geographical Publications, no. 15 (Ann Arbor, MI: Department of Geography, University of Michigan).

Olsson, G. 1980: *Birds in egg/Eggs in bird.* London: Pion.

Philip, M. 1985: Michel Foucault. In Skinner, Q. (ed.), *The return of grand theory in the human sciences.* Cambridge: Cambridge University Press, 67–81.

Ricoeur, P. 1971: The model of the text: meaningful action considered as text. *Social Research,* **38**; 529–62.

Rorty, R. 1979: *Philosophy and the mirror of nature.* Princeton, NJ: Princeton University Press.

Rorty, R. 1985: Texts and lumps. *New Literary History.* **17**; 1–16.

Suggested reading

Gunnar Olsson's *Birds in egg–Eggs in bird* (London, Pion, 1980) is still the best general treatment of issues of language and geography, although it requires dedicated reading. An exceptionally clear account of Derrida's work is found in Hilary Lawson's book *On reflexivity: The postmodern predicament* (London, Hutchinson, 1985), and Mark Philips provides a good overview of Foucault in his chapter 'Michel Foucault', in Quentin Skinner (ed.) *The return of grand theory in the human sciences* (Cambridge, Cambridge University Press, 1985). A lively and reader-friendly discussion of metaphors of every stripe is found in *Metaphors we live by* by George Lakoff and Mark Johnson (Chicago, IL, Chicago University Press, 1980).

8 Gunnar Olsson
Lines of Power

Reprinted in full from: G. Olsson, *Lines of power/limits of language*, pp. 167–81. Minneapolis, MN: University of Minnesota Press (1992)

For Jacques Lacan the unconscious was structured like a language. For me power is structured like a knowledge.

As I now move from this double beginning, I quickly find myself in the company of René Girard and his theory of mimetic desire. I am driven there by my conviction that knowledge by definition is an exercise in translation. But just as desire desires not to be satisfied, so a translation can never be perfect. And in this light, communication shows itself to be what it really is: a form of collective violence designed to neutralize the deviant by sacrificing it on the altar of social cohesion. The unknown must submit to the known just as the world is divided into reasonable and unreasonable. Subjects are produced through subjection, objects through objection.

Boiled down to its essentials, telling the truth is to claim that something is something else and be believed when you do it. Others trust me, when I say that this is thus or that a = b. To succeed in this tricky business is not so easy, however, for after Nietzsche everybody knows that this is this, not thus, that a equals a, not b. The similar and the same are similar, not same; there is never presence, only proximity. And with this reminder, I have already demonstrated how perfect knowledge is as impossible as perfect translation, how truth has more to do with trust and conviction than with what here-and-now happens to be the case. In addition, I have suggested that even though a and b are striving to become doubles of one another, this desire can never be fulfilled. Both knowledge and language require a defect or a fault, an alter ego who keeps his distance. The dynamics of power are rooted in difference.

*

What remains is a play of make-believe. To this end, I shall attempt a deconstruction of the word IS – the epistemological marker par excellence. Nobody can do without this sign of mimetic desire in concentration, for without it there would be neither understanding nor communication. Without IS nothing would be, not even Nothing.

At the same time, IS would not be IS if it did not have many meanings. It is indeed this ambiguity that turns the IS into a key concept in the vocabulary of power. Its nature is to be supplementary and constantly shifting; if captured in one context, it promptly escapes to another. As it performs its juggling tricks at Polity Fair, power stays in power because of its evasive anonymity; by manifesting itself it conceals itself; in hiding it shows. The word is the manifest nonexistence of what it designates.

To keep track of myself, I shall concretize some various meanings of the IS by constructing a set of straight lines. It is in the turning and twisting of these objective correlates that I shall then catch a glimpse of power with its pants down. A blue eye meets a brown, analyzing and experiencing in the same glance; the law itself as an instance of Epimenides' paradox of the right to say I: all Cretans are liars, for power is in that minuscule yet insurmountable space between speech and silence.

My first drawing of the IS consisted of two parallel lines, an equal sign of the type

=

During the first half of the 1970s, I spent some of my best years in the company of this symbol. Every evening we went to bed, every morning we had breakfast, every day we conversed. Bewitched, bothered and bewildered. And yet. The emptiness between the two lines eventually melted away, the two became one, turned on the side, transcended itself, and became the slash of

/

In the mid-eighties, the potent angle changed once more as the line found a resting place in the Saussurean Bar of

——

While '=' demarcates what is identical to what and '/' stands for the penumbra of a mutual relation, the '——' is the rendezvous of signifier and signified. While '=' has its roots in logics and '/' comes from dialectics, '——' is central to semiotics. Regardless of these differences, however, the three signs are all possessed by the same kind of mimetic desire; what stands on one side of the line wants to merge with what stands on the other. But, here as elsewhere, the desire is defined by its impossibility.

The story of the straight line contains everything I know and everything I have not yet understood. This includes fragments of power. To me it is in the richness of these symbols – signs for the simultaneous splitting apart and joining together – that power reveals itself in its most clear, most elementary, most beautiful form. In comparison, the king's scepter, the general's baton, and the bishop's crosier appear as vulgar as the Danish emperor in his nakedness. It was Georges Bataille (*Visions of Excess*, page 5) who noted that 'the *copula* of terms is no less irritating than the *copulation* of bodies'.

*

If the equation is the wheel of positive science, then the equal sign is the nave of that wheel. The statements on each side of the sign should represent the same quantity, express the same amount, or say the same thing in different words. Clear enough. Yet there is a serious problem: since every reformulation by necessity is supplementary and therefore strictly speaking a distortion, a given supplement can sometimes be accepted as a truth, sometimes rejected as a lie. The issue is further complicated by the requirement that the reformulation should not only be true but also informative; whereas a tautology by definition is true but not informative, a metaphor is often informative but never true. Once again, it is not sufficient to note that something is the case. I must also be believed when I point it out: truth is not a private conviction, but a social convention. Trust is not truth unless shared. Hence, the statement a = b becomes credible when it refers to phenomena that are open to common inspection, a condition that legitimizes the power of logical empiricism and grounds the metaphysics of presence. It is nevertheless an argument of post-modern deconstruction that such abilities to share the world must not be taken for granted. The pronouns I and he are radically different. It is not surprising that truth often is invoked as an excuse: 'Objectively, Comrade!'–'Unconsciously, Sister!'

Indeed it now seems clear that every truth contains an element of distortion, of lying, of them overruling me. Every truth is inevitably formulated around a nucleus of difference, for knowing what something *is* involves knowing what it is *not*. In the public congregation of categorization new truths are being blessed and old prejudices sacrificed. Ecstasy is that brief moment of conversion in which one set of taken-for-granted is replaced by another. But what is it to change one's taken-for-granted?

To change one's taken-for-granted is to change one's central beliefs. It is in the middle of this leap that epistemology turns into ontology, problems of existence into issues of being, reasoning into power. And thus it is that both knowledge and power, resistance and legitimation, rest on a foundation of convention and agreement. The frightening alternative is madness, for Wittgenstein's remark cannot be repeated too often: 'If I were sometime to see quite new surroundings from my window instead of the long familiar ones, if things, humans, and animals were to behave as they never did before, then I would say something like "I have gone mad"; but that would merely be an expression of giving up the attempt to know my way about'.

Does this mean that to be mad is to approach the limits of meaning? Perhaps! For in such situations I am left completely alone, for then I am not like anybody else; if 'a' denotes myself, then there is no 'b' to go with it. Not knowing my way about is another way of saying that I am completely lost, with no fixes to keep me steady, with no contexts to share. It is exactly at this moment of horror that the equal sign shows itself in its imperial nakedness; a blessing in disguise, an instrument of socialization, a standardized tool for making you and me normal, predictable and interchangeable; logic is not merely a matter of form, even though the copula is the real subject of speculative thought.

The basic assumption of this reasoning mode is in the Leibnizian principle of salva veritatae; truth is preserved when two propositions are interchangeable in every context. Put differently, what I say about an object is assumed to be true regardless of what the object is called. In reality, however, two contexts are never exactly the same, for then there would not be two contexts but one. Likewise, two synonyms are never totally overlapping. My credibility is therefore a function of how I distinguish one context from another and of how I name the resulting categories. As expected, this is a privilege held by God himself. Let there be! And there was!

Even the naming is nameable. And this possibility poses again the Epimenidean question of whether a particular Order of Context is to be trusted. Is God – the key figure of power – logical? Is God himself true to the principles of salva veritatae and to the laws of consistency that are symbolized by the equal sign? Of course not! Especially in the Old Testament, God instead exercises his power by creating paradoxes, predicaments, and double binds; first he demands obedience and then he changes his commands. Abraham provides the paradigmatic case, and this is why he was so crucial both to Søren Kierkegaard and to Franz Kafka. But the man in question was already ninety-nine years old, when the Lord made the covenant with him that he was to be the father of a multitude of nations. This altered context was then codified in a decree that proclaimed that his name be changed from Abram to Abraham. New worlds require new labels, for what is true about an object is not independent of what it is called.

But the story continues. Genesis 22. The Almighty designs a test of Abraham's obedience. As a sign, he must sacrifice Isaac, his only son, whom he loves. No regrets, everything ready, the son tied to the altar, the knife on his throat. Then, suddenly, from the sky, the voice of the Lord's angel: 'Stop! Stop! For Heaven's sake. Don't kill. Your master knows that you fear him. He changes his mind'. The murder was no longer necessary, for in everything Abraham had already demonstrated his submission. Yet it was at this very moment of apparent relief that Abraham experienced limitless terror. God – the incarnation of power – is not to be trusted, for if he has changed his mind once, he can do so again. Only logic is predictable and God is not logical. Indeed 'God' is nothing but a proper name for everything we sense is too important to ignore and too evasive to specify. 'God' is a pseudonym of power.

To logic, paradox is an enemy; to power, predicament is an ally. Later dictators have become experts in the same form of institutionalized double bind and uncertainty. The torturer who by mistake kills his victim has eliminated this sense of uncertainty and thereby the whole point of his business; the torture is for the audience, not for the victim. The only defense may indeed be in the deployment of irony, satire, humor, and poetry; even though revolt is often necessary but ineffectual, refusal is always possible and irritating. As a consequence, dictators fear laughter more than tears, lonely poets more than organized protesters, happiness more than sorrow. But whereas humor and poetry engage in a play of logical types, deduction is characterized by consistency and a form of reasoning that moves from an axiomatic beginning to an inevitable end. Under the pressure of salva veritatae, even the equal sign transforms itself into an arrow, the key symbol of mimetic desire.

The best example is of course in the causal model. Here the equal sign serves as a selective filter that lets through more influences in one direction than in the other. At the same time, causal relations are often dressed up as logical relations, even though the two types of implication are drastically different; whereas cause-and-effect involves time, power, and responsibility, logic is without time, without freedom, and without guilt. In both cases, however, the argument strives for acceptance, for like all philosophy also logic is inherently rhetorical; the trick is to be believed, to reason in a manner that increases one's credibility. Belief is the instrument of power par excellence.

In this context, the most interesting rhetorical trope is the metonymy. Its strategy is to create trust by making the reader or listener recognize himself; its tactic is to let the concrete rule over the abstract, the specific over the general. This practice is itself part of the metaphysics of presence that permeates all empiricism. There is a high correlation between credibility and concretization, power and thingification, order and communication; the anatomy of power says that the way to the mind goes via the body, that discipline is anchored in details. Privilege and exploitative power have always masked themselves as duty and responsibility.

Perhaps there is nothing more powerful than the power of the example; speech is not only indicative but also imperative. The current challenge is to subvert that power. Indeed I have come to believe that our very survival depends upon improved abilities to be abstract enough; the most radical point is the point of insolvability. And with this call to abstractness I have already begun to move from the ordering parallels of the equal sign to the devious slant of the slash.

*

As the equal sign belongs to logic, so the slash stems from dialectics. While the former dominates positive science, the latter permeates deconstruction. It is in the former to symbolize knowledge as the restatement of identities, in the latter to denote the inseparability of identity and difference. The former searches for the certain in the ambiguous, the latter for the ambiguous in the certain.

Even though the two tropes are thoroughly intertwined, the equal sign gets its convincing power mainly from metonymy, the slash mainly from metaphor. It follows that the equal sign points the way to standardization and thingification, whereas the slash eludes all attempts to catch it. In both cases, however, the structures of substitution reflect the forms of power and the forms of power the structures of substitution. The ensuing problem has its roots in Plato: what makes us see must itself be invisible, what makes us understand must itself be nonunderstandable; the mirror is a mirror not for the mirrored but for the mirroring, not for the object in front of the glass but for the tain behind it. It is with the writings of Jacques Derrida that the discussion now is returning to a level of abstractness worthy of Hegel, Kant, and Descartes: What is reflection and what makes reflection possible? What can I know about what, except that epistemology never left the mirror stage and that thirsting for knowledge is an instance of unquenchable narcissistic desire?

It is this desire for the invisible that itself becomes visible in the slash, for to me the slanting line serves as a concrete symbol of an abstract relation;

with T. S. Eliot, I conceive of symbols as objective correlates of human feelings. This may sound as if I try to reenlist the forces of metonymy, but then I seek consolation in Jacques Lacan's observation that the symbol manifests itself as the killing of the thing. Lacan's remark is important, because it is the very essence of a relation to be extremely abstract, invisible and untouchable; like silence, every attempt to capture it fails, for every attempt destroys it. It follows that relations cannot be defined, only experienced. Relations should not be confused with what is related, just as the desire for love should not be confused with the loved one.

Through the slash everything hangs together with everything else in a maze of internal and self-referential relations. Even algebra may serve as an example, for three times three would not be nine unless three times seven were twenty-one and seven times seven forty-nine. And so it is easier to understand both why dialectics has been called the algebra of revolution and why all revolutions fail. The concept of implication is a subjunctive that in empirical operationalization is perverted into an indicative. In the need to speak there is nevertheless a desire to silence, an idea of infinity, a sense of neither identity nor difference. The future is future because it is ungraspable, not because it is manipulable.

It follows that dialectics is not a predictive mode of reasoning or a metonymic system of causal models. It is rather a form of epistemology specialized in the discovery of the hidden in the apparent. As a consequence, it can never look into the future, only into the past, an insight that was born with Hegel, lived with Kierkegaard, and died with Stalin: dialectics is not a language of commands but a language of understanding, especially of the relations between repression and submission, master and slave.

The dialectic nature of the slash becomes especially evident in confrontations between the individual self on the one hand and the collective taken-for-granted on the other. Thus there are many who are shocked when they experience how society issues orders that to them seem unfair, or how it makes claims that to them are false. In these situations, the analyst usually asks whether he should put more trust in the individual or in the collective. The dialectician instead answers that neither is to be trusted, for the two words are defined in terms of one another and are hence in constant and unresolvable conflict.

It is nevertheless the void between individual and society that constitutes the realm of political power. The deep insights that fill this social space of silence must not be revealed, just as God the Father could not permit his children to eat from the tree of knowledge or to mention his name. It is indeed through the automatic recital of Commandments that society protects itself from its members and individuals guard against the collective. It is this that the slash symbolizes, that which the angels fear: relations beyond the related.

But how can I know that I as a subject conceive of the world in the same fashion as you do as an object? Assuming that this question lacks an answer (for there is no Other of the Other), by which right do you then engage in political action, an activity that in practice always means that the will of the Other is subordinated to your own? Is sociality to desire your neighbor? If so, how do you handle the problem that the collective 'we' never can be anything but a majority; 'social democracy' is an oxymoron. If the personal is political, then perhaps the political is pathological. For how small must a minority be before it is forbidden? And how large must a majority be before it becomes such an integral part of the taken-for-granted that it turns silent? For the State to wither away, it must be everywhere. Fascism is a tendency to homogeneity. After Gulag and Auschwitz critical thought can be nothing but Plural.

Political power is structured as a slash: internal, self-referential, insatiable as the silence of desire itself. Of all political concepts, intentionality is the most central and the least understood. Thingification is the price we pay for not accepting that there is a beyond beyond the beyond of expression, an absence inscribed without a trace in every discourse. The slash tries to be a symbol of this excluded third, which is neither either-or nor both-and, but something entirely outside the realm of naming; the slash is not what it first might have seemed – a bridge between opposites – but the void of categorical limits itself. And through the silence breaks the voice of Samuel Beckett's Unnamable (*Beckett Trilogy*, page 352): 'perhaps that's what I feel, an outside and an inside and me in the middle, perhaps that's what I am, the thing that divides the world in two, on the one side the outside, on the other the inside, that can be as thin as foil, I'm neither one side nor the other, I'm in the middle, I'm the partition, I've two surfaces and no thickness, perhaps that's what I feel, myself vibrating, I'm the tympanum, on the one hand the mind, on the other the world, I don't belong to either'.

And so it is that the slash perhaps can serve as the signifier of that constellation in which nothing takes place except the place. Maybe it is even at this Mallarméan place of silence that truth and power hold their secret meetings. To stave off all trespassing into this sanctuary, God himself spoke these words to Moses, for him to bring down to the priests and the people (Exodus 20:4–6): 'You shall not make for yourself a graven image, or any likeness of anything that is in heaven or above, or that is in the earth beneath, or that is in the water under the earth; for I the Lord your God am a jealous God, visiting the iniquity of the fathers upon the children to the third and the fourth generation of those who hate me, but showing steadfast love to thousands of those who love me and keep my commandments'.

And so it is that issues of representation and the sublime may have more to do with ethics than with logics and aesthetics. Power is structured as a knowledge: 'And the Lord said to Moses, "Go down and warn the people, lest they break through to the Lord to gaze and many of them perish"' (Exodus 19:21). And in doing as he was told, Moses became the first *politruk*. The eye shows itself to be more powerful than the voice, the gaze more violent than the word.

*

Inherent in the issues of representation is a heightened awareness of the fundamental difference between word and object. This awareness is itself a part of that crisis of the sign which became acute in the second half of the nineteenth century, especially with writers like Baudelaire, Rimbaud, and Mallarmé. Once again, these artists were driven by a desire for the words they did not possess. In realizing that every utterance by necessity is indirect, they experienced how conventional language did not furnish the means to express what they most urgently wanted to express. Even when they said 'stone' and meant 'stone', they were not stone. Even when they said 'you' and meant 'you', they were not you. One can share what one has, not what one is. As a speaking subject, I have no choice but to live in a language that is common and social. The limits of my language mean the limits of my world.

It is this unfathomable problem of representation that to me finds an objective correlate in the Saussurean bar, i.e., in that horizontal line that simultaneously splits and unites the two fractions of the sign of

$$\frac{\mathrm{S}}{\mathrm{s}}$$

where 'S' stands for 'signifier' and 's' for signified'.

It is of course tempting to tie the capital S to the touchable physicality of the sign and the small s to its nontouchable meaning. Such an interpretation would be a serious oversimplification, however, for signs are always threaded together into braids of desire and justification. In a rhythmic dance of creativity, nominator and denominator constantly change positions.

Illustrations of such turnabouts are already in the *Odyssey*, especially in Odysseus's tale of how he was caught in the Cyclopes' cave (Ninth Song). When Polyphemos demanded the name of the intruder, he got the answer Ovtis, *Nohbdy.* The beast in turn replied that 'Nohbdy the last one will be that I eat', upon which he fell asleep, full of red wine and human meat. It was at this crucial moment that the trickster saw his opportunity, rammed the red-hot pike of an olive tree into the drunkard's eye, leaned on it, and turned it as a shipwright turns a drill. And the pierced eyeball hissed broiling and the roots popped. The Cyclops howled in pain, and when his likes gathered outside the cave to learn what was wrong, Polyphemos roared in reply: 'Nohbdy, Nohbdy's tricked me, Nohbdy's ruined me!' To this his friends responded: 'Ah well, if nobody has played you foul there in your lonely bed, we are no use in pain given by great Zeus'. And Odysseus was filled with laughter to see how like a charm the name deceived them.

Thus sounds the original tale of the crisis of the sign, of power structured as a knowledge, of identity statements as lies, of thingification and the sense of meaning. But this ur-tale has interesting connections also with the Greek concept of mimesis and thereby both with postmodernism and with René Girard. Inherent in this notion is the idea that art is an imitation – a re-presentation – of reality. But neither the true artist nor the hegemonic ruler is satisfied with the mere copying of the outer appearance of a phenomenon. They also wish to touch its inner soul. Only with the sublime do they become full-fledged forgers. The bar eliminated, word and object united, mimetic desire completed.

Constitutive of the concept of mimesis is the assumption of a privileged original, a Holy Script, a Book of Nature. With this text as a starting point, the truth-teller's task is to provide a perfect translation, a mirror image, a narcissistic reflection. At the same time, I have already noted that perfect translation is impossible. The supplementary nature of language in fact implies that any reference to an original is highly misleading; the copy is not a copy of an original but a simulacrum, a copy that lacks an original. The Book of Nature is itself a simulacrum. As such it is untranslatable.

And thus it must be repeated: meaning does not emerge from the identity of signifier and signified, but from the difference between them. Indeed it is difference alone that allows a signifier to signify. Without signs no thoughts, without splits no signs. Referring to a once-and-for-all beginning is therefore not to refer at all, for without tenses and cases we would be literally lost. It follows that the ideal of total and perfect representation does not guarantee the stability of truth. Instead it vouches for its indeterminacy.

Filling the ontological gap between signifier and signified is the Saussurean Bar of Power. But whereas power is power because it transforms categories, analysis is analysis because it keeps them apart. As a consequence, the —— can function not only as the wand of the ruler but also as the crowbar of the critic. This explains why also Marxian thought may be considered an early response to the crisis of the sign; the use-value of the commodity corresponds to the signifier, the exchange-value to the signified. To fetishize is in fact to see only the physical and be blind to its meaning; in reality, though, a commodity is a very queer thing abounding in metaphysical subtleties and theological niceties. No society can do without its religion of ontological transformations.

Similar attitudes permeate the theories of psychoanalysis and the practices of psychotherapy. Since the unconscious after Lacan is structured as a language, human crisis can be interpreted as a crisis of the sign, more specifically a crisis locatable to the bar. The therapeutic strategy is first to discover a repressed signified and then to kill it with an explicit signifier. The patient speaks herself well, for, in the power-filled act of naming, anguish becomes graspable. The horrible and noncommunicable loses its frightening grip once it is caught in shared categories and domesticated in common expressions. Desire is tamed when subverted into desired; the phallus becomes a symbol of lack, desire thingified. Penis envy speaks clearly, the castration complex as well. But just as Saussure's bar blocks the road to the fulfillment of desire, so the way to self-understanding always goes via the Other. Fort-da. Fort-da. Fort-dada. But what does a woman want? What does it mean to be a woman in an age of iconography, at once a reflexive consciousness and a social product?

Here, as before, the processes of thingification and alienation seem to be governed by a mimetic desire. This drift to imitation expresses itself less in metaphoric condensations and more in metonymic displacements. It lies within René Girard's theory that we are striving to make signifier and signified into doubles of one another. But when this desire approaches satisfaction, the search for identity turns back on itself and changes into hostility. The desire can be kept alive only through the sacrifice of one of the antagonists. A scapegoat is selected, burdened with the collective guilt, and driven out of the community.

To me, the scapegoat carries many of the same traits as the rejected hypothesis, the power of the example, the prohibition against graven images. The desire is a desire for perfect communication, a desire impossible to satisfy. Put differently: power is a desire, not a need. It is this desire that I have symbolized as lines of power. It is the truth of this desire that no ruler can bear to hear.

*

And now, toward the end of the beginning, it should be clear what fascinates me. It is that strange transition point at which the light or sound waves of my textual performance hit your eyes or ears, move into the gray substance inside your head, stir around and become meaningful words, powerful enough to change both our understanding and our action. This abstract penumbra is to me the intersection of a set of lines, all of which serve as condensations of knowledge, hence of communication. It is not sufficient to say how something is, I must also be believed when I say it. This is not to argue, however, that anything goes. On the contrary, for to be believed is to tread a dance with the taken-for-granted. To be believed is to have power. Power is almighty, the apparatchik its prophet. Not to philosophize is still to philosophize. Power is a desire to control meaning. The prime symbol of meaning is the copula IS, a verb designating an event.

Finally: Is this geography? Of course it is! For what is geography, if it is not the drawing and interpretation of lines. The only quality that makes my geography unusual is that it does not limit itself to the study of visible things. Instead it tries to foreshadow a cartography of thought. To practice this art, however, is incredibly difficult, for any attempt must face the challenge of being abstract enough.

In this context, it is with some dismay that I note how my thoughts of power have been unthinkable without the correlates of the straight lines. Even though this technique of fixing ideas is a minimalist approach, it is an approach nevertheless. But how do I write a theory à la Brancusi, a theory in which form and content are indistinguishable? And how do I practice a writing à la Beckett, in which I do not write *about* something, but in which the writing *is* that something itself? Mallarmé's example was to paint not the thing but the effect it produces.

Perhaps I enter this social space of silence by living in the world as I found it. A world where the unconscious is structured as a language, a world where power is structured as a knowledge, a world where lines are taken to their limits.

Mondrian.

References

Bataille, G. 1985: *Visions of excess: Selected writings 1927–39*. Minneapolis, MN: University of Minnesota Press.

Beckett, S. 1979: *Beckett Trilogy*. London: Picador.
Wittgenstein, L. 1967: *Zettel*. Anscombe, G. E. M. and von Wright, G. H. (eds). Oxford: Blackwell.

9 J. Brian Harley
Deconstructing the Map

Excerpts from: *Cartographica* 26, 1–20 (1989)

This chapter has three points of departure. The first is a belief in the value of maps – both old and new – for the practice of human and historical geography. The second is a frustration with many of the academic cartographers of today, who operate in a tunnel created by their own technologies without reference to the social world. The third is a desire to intervene. Epistemic time has largely stood still in cartography. One effect of accelerated technological change – as manifest in digital cartography and geographical information systems – has been to strengthen its positivist assumptions and it has bred a new arrogance in geography about its supposed value as a mode of access to reality. If it is true that 'New fictions of factual representation' (White 1978) are daily being foisted upon us, then the case for inserting a social dimension into modern cartography is especially strong. Maps are too important to be left to cartographers alone.

One consequence of these developments is that it is increasingly difficult to conduct a dialogue between the practice of cartography as it is now constituted and [various] interpretive strategies . . . . While we can talk freely of our different discourses, cartographers seem unable to situate their maps within the discourse of cartography. Along with other aspects of critical theory, I suspect that the notion of discourse, as defined by Foucault and others, would be alien and bizarre to most cartographers. The orthodox words in their vocabulary are 'impartial', 'objective', 'scientific' and 'true', but we seldom catch the resonances of class, gender, race, ideology, power and knowledge, or myth and ritual. A conceptual vacuum lies between cartography and human geography.

What I am seeking to do in this and related historical essays (Harley 1988a, 1988b, 1989, 1996) is to show how cartography also belongs to the terrain of the social world in which it is produced. Maps are ineluctably a cultural system. Cartography has never been an autonomous and hermetic mode of knowledge, nor is it ever above the politics of knowledge. My key metaphor is that we should begin to deconstruct the map by challenging its assumed autonomy as a mode of representation. From the viewpoint of human geography, maps are perhaps better understood – and used – not so much as discrete or 'unique' images but as accents within a wider theory of representation. Rather than accepting what cartographers tell us maps are supposed to be, the thrust of my deconstruction is to subvert the apparent naturalness and innocence of the world shown in maps both past and present. In the sense of Barthes, I believe we should demystify both cartographic process and the resulting images we call maps.

The notion of deconstruction (Derrida 1976; Norris 1982; Eagleton 1983; Norris 1987) is, of course, a password for the postmodern enterprise. Deconstructionist strategies can now be found not only in philosophy but also in localized disciplines, especially in literature, and in other subjects such as architecture, planning and, more recently, geography (Gregory 1987; Dear 1988; Knox 1988). I shall specifically use a deconstructionist tactic to break the assumed link between reality and representation which has dominated cartographic thinking, has led it in the pathway of 'normal science' since the Enlightenment, and has also provided a ready-made and taken-for-granted epistemology for cultural studies of maps as geographical or historical records. The objective is to suggest that an alternative epistemology, rooted in social theory rather than in scientific positivism, is more appropriate if we are to reassimilate cartography within human geography. It will be shown that even 'scientific' maps are a product not only of 'the rules of the order of geometry and reason' but also of the 'norms and values of the order of social . . . tradition' (Marin 1988: 173). Our task is to search for the social forces that have structured cartography and to locate the presence of power – and its effect – in all map knowledge.

The ideas that follow owe most to writings by Foucault and Derrida. My approach is deliberately eclectic, although the theoretical positions of these two authors are sometimes incompatible. Foucault anchors texts in socio-political realities and constructs systems for organizing knowledge of the kind that Derrida loves to dismantle (Skinner 1985). But even so, by combining different ideas on a new terrain, it may be possible to devise a sketch of social theory with which we can begin to interrogate the hidden agendas of cartography. Such a scheme offers no 'solution' to a humanistic interpretation of the cartographic record, nor a precise method or set of techniques, but as a broad strategy it may help to locate some of the fundamental forces that have driven map making in both European and non-European societies. From Foucault's writings the key revelation has been the omnipresence of power in all knowledge, even though that power is invisible or implied, including the particular knowledge encoded in maps and atlases. While I do not accept Derrida's view that nothing lies outside the text – it clearly defeats the idea of a social history of cartography – his notion of the rhetoricity of all texts offers a provocative challenge. It demands a search for metaphor and rhetoric in maps where previously scholars had found only measurement and topography. Its central question is reminiscent of Korzybski's much older dictim, 'The map is not the territory' (1948), but deconstruction goes further to bring the issue of how the map represents place into much sharper focus.

Deconstruction urges us to read between the lines of the map – 'in the margins of the text' – and through its tropes to discover the silences and contradictions that challenge the apparent honesty of the image. We begin to learn that cartographic facts are facts only within a specific cultural perspective. We start to understand how maps, like art, 'far from being a transparent opening to the world', are but 'a particular human way . . . of looking at the world' (Blocker 1979: 43).

In pursuing this strategy I shall develop three threads of argument. First I will examine the discourse of cartography in the light of some of Foucault's ideas about the play of rules within discursive formations. Second, drawing on one of Derrida's central positions, I will examine the textuality of maps and, in particular, their rhetorical dimension. Third, returning to Foucault, I will consider how maps work in society as a form of power–knowledge.

The rules of cartography

One of Foucault's primary units of analysis is the discourse. A discourse has been defined as 'a system of possibility for knowledge' (Philip 1985: 69). Foucault's method was to ask, it has been said:

> what rules permit certain statements to be made; what rules order these statements; what rules permit us to identify some statements as true and others as false; what rules allow the construction of a map, model or classificatory system . . . what rules are revealed when an object of discourse is modified or transformed . . . Whenever sets of rules of these kinds can be identified, we are dealing with a discursive formation or discourse (Philip 1985: 69).

The key question for us then becomes 'What type of rules govern the development of cartography?'

Cartography I define as a body of theoretical and practical knowledge that map makers employ to construct maps as a distinct mode of visual representation. The question is, of course, both culturally and historically specific: the rules of cartography vary in different societies. Here I refer particularly to two distinctive sets of rules that underlie and have dominated the history of Western cartography, first in Europe and later in its overseas colonial territories, since the seventeenth century. One set may be defined as governing the technical production of maps and are made explicit in cartographic treatises and writings. The history of these technical rules has been extensively written about in the history of cartography (Crone 1978), though not in terms of their social implications nor in Foucault's sense of discourse. The other set of rules relates to the cultural production of maps. These rules must be understood in a broader historical context than either scientific procedure or technique. They are, moreover, rules that are usually ignored by cartographers, so that they form a hidden aspect of their discourse.

The first set of rules for mapping can thus be defined in terms of a positivistic epistemology. From at least the seventeenth century onward there was an epistemic break in activities such as cartography and architecture (Peréz-Gomez 1983), and European map makers increasingly promoted what we would describe today as a standard scientific model of knowledge and cognition. The object of mapping is to produce a 'correct' relational model of the terrain. Its assumptions are that the objects in the world to be mapped are real and objective, and that they enjoy an existence independent of the cartographer; that their reality can be expressed in mathematical terms; that systematic observation and measurement offer the only route to cartographic truth; and that this truth can be independently verified. The procedures of both surveying and map construction came to share strategies similar to those in science in general. Cartography also documents a history of more precise instrumentation and measurement; increasingly complex classifications of its knowledge and a proliferation of signs for its representation; and, especially from the nineteenth century onward, the growth of institutions and a 'professional' literature designed to monitor the application and propagation of the rules (Wolter 1975). Moreover, although cartographers have continued to pay lip service to the 'art and science' of map making (Meynen 1973; Wallis and Robinson 1987), art has been progressively edged off the map. It has often been accorded a cosmetic rather than a central role in cartographic communication (Morris 1982). Even philosophers of visual communication – such as Arnheim, Eco, Gombrich and Goodman (Goodman 1968; Gombrich 1975; Eco 1976; Arnheim 1986) – have tended to categorize maps as a type of congruent diagram – as analogues, models or 'equivalents' creating a similitude of reality – and, in essence, different from art or painting. A 'scientific' cartography (so it was believed) would be untainted by social factors.

In cases where the scientific rules are invisible in the map we can still trace their play in attempting to normalize the discourse. The cartographer's 'black box' has to be defended and its social origins suppressed. The hysteria among leading cartographers at the popularity of the Peters projection (Peters 1983; Loxton 1985a, 1985b; Robinson 1985; Porter and Voxland 1986; Snyder 1988; Vujakovic 1989), or the recent expressions of piety among Western European and North American map makers following the Russians' admission that they had falsified their topographic maps to confuse the enemy give us a glimpse of how the game is played according to these rules. What are we to make of the 3 September 1988 newspaper headlines such as 'Russians caught mapping' (*Ottawa Citizen*), 'Soviets admit map paranoia' (*Wisconsin State Journal*) or (in the *New York Times*) 'In West, map makers hail "Truth" ' and ' "The rascals finally realized the truth and were able to tell it", a geographer at the Defense Department said' ? The implication is that Western maps are value-free. According to the spokesman, our maps are not ideological documents, and the condemnation of Russian falsification is as much an echo of Cold War rhetoric as it is a credible cartographic criticism.

This timely example also serves to introduce my second contention, that the scientific rules of mapping are, in any case, influenced by a quite different set of rules, those governing the cultural production of the map. To discover these rules, we have to read between the lines of technical procedures or of the map's topographical content. They are related to values, such as those of ethnicity, politics, religion or social class, and they are also embedded in the map-producing society at large, and in its other forms of representation. Cartographic discourse operates a double silence towards this aspect of the possibilities of map knowledge. In the map itself social structures are often disguised beneath an abstract, instrumental space, or incarcerated in the coordinates of computer mapping. And in the technical literature of cartography social values are also ignored, notwithstanding the fact that they may be as important as surveying, compilation, or design in producing the statements that cartography makes about the world and its landscapes. Such an interplay of social and technical rules is a universal feature of cartographic knowledge. In maps it produces the 'order' of its features and the 'hierarchies of its practices' (Foucault 1973: xx). In Foucault's sense the rules may enable us to define an *episteme* and to trace an archaeology of that knowledge through time (Foucault 1973: xxii).

[One example] of how such rules are manifest in maps will illustrate their force in structuring cartographic representation. [It] is the well known adherence to the 'rule of ethnocentricity' in the construction of world maps. This has led many historical and modern societies to place their own territory at the centre of their cosmography or world map. While it may be dangerous to assume universality, and there are exceptions, such a 'rule' is as evident in cosmic diagrams of pre-Columbian North American Indians as it is in the maps of ancient Babylonia, Greece or China, or in the medieval maps of the Islamic world or Christian Europe (Harley and Woodward 1987, 1987–1994). Yet what is also significant in applying Foucault's critique of knowledge to cartography is that the history of the ethnocentric rule does not march in step with the 'scientific' history of map making. Thus while the scientific Renaissance in Europe gave modern cartography co-ordinate systems, Euclid, scale maps and accurate measurement, the epistemic break was only partial. Philosophers continued to believe that mathematical thought constituted a privileged channel of communication between human minds and the divine mind, while the new geometrical cartography served to reinforce a new myth of Europe's ideological centrality (Peters 1983). Throughout the history of cartography ideological 'Holy Lands' are frequently centred on maps. Such centricity, a kind of 'subliminal geometry' (Harley 1988a), adds geopolitical force and meaning to representation. Though the link between actual mapping, as the principal source of our world vision, and *mentalité* still has to be thoroughly explored, it is also likely that such maps have in turn helped to codify, to legitimate and to promote the Eurocentric world views that have been prevalent in so much of modern world history (Henrikson 1987; Saarinen 1988).

[More generally,] the point I am making is that [social] rules operate both within and beyond the orderly structures of classification and measurement. They go beyond the stated purposes of cartography. Much of the power of the map, as a representation of social geography, is that it operates behind a mask of a seemingly neutral science. It hides and denies its social dimensions at the same time as it legitimates. Yet whichever way we look at it the rules of society will surface. They have ensured that maps are at least as much an image of the social order as a measurement of the phenomenal world of objects.

Deconstruction and the cartographic text

To move inward from the question of cartographic rules – the social context within which map knowledge is fashioned – we have to turn to the cartographic text itself. The word 'text' is deliberately chosen in an awareness that it offers no simple set of techniques for reading maps. Some cartographers have resisted the metaphor of map as language (Robinson and Petchenik 1976), but it is now generally accepted that the model of text can have a much wider application than to literary texts alone. To non-book texts such as landscapes, musical compositions and architectural structures we can confidently add the graphic texts we call maps (McKenzie 1986; Duncan and Duncan 1988). It is true that literally they have no grammar and lack the temporal sequence of a syntax but 'what constitutes a text is not the presence of linguistic elements but the act of construction' so that maps, as 'constructions employing a conventional sign system' (McKenzie 1986: 35), become texts. With Barthes we could say they 'presuppose a signifying consciousness' that it is our business to uncover (Barthes 1986: 110). 'Text' is certainly a better metaphor for maps than the mirror of nature. Maps are a cultural text: not one code but a collection of codes, few of which are unique to cartography.

By accepting the textuality of maps we are able to embrace a number of different interpretive possibilities. Instead of just the transparency of clarity we can discover the pregnancy of the opaque. To fact we can add myth, and instead of innocence we may expect duplicity. Rather than working with a formal science of communication, or with a cognitive psychology saying nothing about the social world, or perhaps, worse still, with a sequence of loosely related technical processes, our concern is redirected to a history and anthropology of the image. We learn to recognize the narrative qualities of cartographic representation (Wood 1987) as well as its claim to provide a synchronous picture of the world. All this, moreover, is likely to lead to a rejection of the neutrality of maps as we come to define their intentions rather than the literal face of representation, and as we begin to accept the social consequences of cartographic practices.

What, therefore, does deconstruction have to offer? Most important is that it demands a closer and deeper reading of the cartographic text. Deconstruction does not solve the problem: but it aims at as many meanings as possible, even if some aspects of those meanings are undecidable (Hoy 1985). It may be regarded as a search for alternative meanings. 'To deconstruct', it is argued:

> is to reinscribe and resituate meanings, events and objects within broader movements and structures; it is, so to speak, to reverse the imposing tapestry in order to expose in all its unglamorously dishevelled tangle the threads constituting the well heeled image it presents to the world (Eagleton 1986: 80).

The published map also has a 'well heeled image' and our reading has to go beyond the assessment of geometric accuracy, beyond the fixing of location, and beyond the recognition of topographical patterns and geographies. Such interpretation begins from the premise that the map text may contain 'unperceived contradictions or duplicitous tensions' (Hoy 1985: 540) that undermine the surface layer of standard objectivity. Maps are slippery customers. In the words of W. J. T. Mitchell (1986: 8), writing of languages and images in general, we may need to regard them more as 'enigmas, problems to be explained, prison-houses which lock the understanding away from the world'.

We should regard them 'as the sort of sign that presents a deceptive appearance of naturalness and transparence concealing an opaque, distorting, arbitrary mechanism of representation' (Mitchell 1986: 8). Throughout the history of modern cartography in the West, for example, there have been numerous instances of where maps have been falsified, of where they have been censored or kept secret, or of where they have surreptitiously contradicted the rules of their proclaimed scientific status (Harley 1988b).

Taking practices such as these, map deconstruction would focus on aspects of maps that many interpreters have glossed over. Writing of 'Derrida's most typical deconstructive moves', Christopher Norris (1987: 19) noted that:

> deconstruction is the vigilant seeking-out of those 'aporias', blind spots or moments of self-contradiction where a text involuntarily betrays the tension between rhetoric and logic, between what it manifestly *means to say* and what it is nonetheless *constrained to mean.* To 'deconstruct' a piece of writing is therefore to operate a kind of strategic reversal, seizing on precisely those unregarded details (casual metaphors, footnotes, incidental turns of argument) which are always, and necessarily, passed over by interpreters of a more orthodox persuasion. For it is here, in the margins of the text – the 'margins', that is, as defined by a powerful normative consensus – that deconstruction discovers those same unsettling forces at work.

A good example of how we could deconstruct an early map – by beginning with what have hitherto been regarded as its 'casual metaphors' and 'footnotes' – is provided by recent studies reinterpreting the status of decorative art on the European maps of the seventeenth and eighteenth centuries. Rather than being inconsequential marginalia, the emblems in cartouches and decorative title pages can be regarded as basic to the way such maps convey their cultural meaning, and they help to demolish the claim of cartography to produce an impartial graphic science (Harley 1984; Clarke 1988; Harley 1988a, 1996).

Such a strategy need not be limited to historic 'decorative' maps. A recent essay by Wood and Fels on the official state highway map of North Carolina (Wood and Fels 1986) indicates a much wider applicability for a deconstructive strategy by beginning in the 'margins' of the contemporary map. Wood and Fels also treat the map as a text and, drawing on the ideas of Roland Barthes (1986) of myth as a semiological system, develop a forceful social critique of cartography which though structuralist in its approach is deconstructionist in its outcome. They begin, deliberately, with the margins of the map, or rather with the subject matter that is printed on its verso:

> One side is taken up by an inventory of North Carolina points of interest – illustrated with photos of, among other things, a scimitar horned oryx (resident in the state zoo), a Cherokee woman making beaded jewelry, a ski lift, a sand dune (but no cities) – a ferry schedule, a message of welcome from the then governor, and a motorist's prayer ('Our heavenly Father, we ask this day a particular blessing as we take the wheel of our car . . .'). On the other side, North Carolina, hemmed in by the margins of pale yellow South Carolinas and Virginias, Georgias and Tennessees, and washed by a pale blue Atlantic, is represented as a meshwork of red, black, blue, green and yellow lines on a white background, thickened at the intersections by roundels of black or blotches of pink. . . . To the left of . . . [the]

> title is a sketch of the fluttering state flag. To the right is a sketch of a cardinal (state bird) on a branch of flowering dogwood (state flower) surmounting a buzzing honey bee arrested in midflight (state insect) (Wood and Fels 1986: 54).

What is the meaning of these emblems? Like the contents of the Chinese encyclopedia in Borges's short story, referred to by Foucault in the preface to *The Order of Things* (1973), they are bizarre until we've cracked the code. Are they merely a pleasant ornament for the traveller or can they inform us about the social production of such state highway maps? A deconstructionist might claim that such meanings are undecidable, but it is also clear that the state highway map of North Carolina is making other dialogical assertions behind its mask of innocence and transparence. I am suggesting not that these elements hinder the traveller in getting from point A to point B, but that there is a second text within the map. No map is devoid of an intertextual dimension that involves an essentially plural and diffuse play of meanings across the boundaries of individual maps. It has been remarked that 'in the concept of "text" the boundaries which enclosed the "work" are dissolved; the text opens continually into other texts, the space of intertextuality' (Burgin 1988: 51). To read the map we have to dismantle first of all the frame that the cartographer has placed around it.

So it is with our state highway map. The discovery of intertextuality enables us to scan the image as more than a neutral picture of a road network (Bakhtin 1981). It shares the intertextuality of all discourse. The 'users' of the map are not only the ordinary motorists but also the State of North Carolina that has appropriated its publication (distributed in millions of copies) as a promotional device. The map has become an instrument of state policy and an instrument of sovereignty (Wood and Fels 1986). At the same time it is more than an affirmation of North Carolina's dominion over its territory. It also constructs a mythic geography, a landscape full of 'points of interest', with incantations of loyalty to state emblems and to the values of a Christian piety. The hierarchy of towns and the visually dominating highways that connect them have become the legitimate natural order of the world. The map finally insists 'that roads really *are* what North Carolina's all about' (Wood and Fels 1986: 60). The map idolizes our love affair with the automobile. The myth is believable.

[More broadly], the issue in contention is not whether some maps are rhetorical, or whether other maps are partly rhetorical, but the extent to which rhetoric is a universal aspect of all cartographic texts. Thus for some cartographers the notion of 'rhetoric' would remain a pejorative term. It would be an 'empty rhetoric' which was unsubstantiated in the scientific content of a map. 'Rhetoric' would be used to refer to the 'excesses' of propaganda mapping or advertising cartography or an attempt would be made to confine it to an 'artistic' or aesthetic element in maps as opposed to their scientific core. My position is to accept that rhetoric is part of the way all texts work and that all maps are rhetorical texts. Again we ought to dismantle the arbitrary dualism between 'propaganda' and 'true', and between modes of 'artistic' and 'scientific' representation as they are found in maps. All maps strive to frame their message in the context of an audience. All maps state an argument about the world, and they are propositional in nature. All maps employ the common device of rhetoric such as invocations of authority. This is *especially* so in topographical maps, with their reliability diagrams, multiple referencing grids and magnetic error diagrams, or in thematic maps, with their 'trappings of F-scaled symbols and psychometrically divided greys' (Wood and Fels 1986, 99). Maps constantly appeal to their potential readership through the use of colour, decoration, typography, dedications or written justifications of their method (Marin 1988). Rhetoric may be concealed but it is always present, for there is no description without performance.

Maps and the exercise of power

For the final stage in the argument I return to Foucault. In doing so I am mindful of Foucault's criticism of Derrida, that he attempted 'to restrict interpretation to a purely syntactic and textual level' (Hoy 1985: 60), a world where political realities no longer exist. Foucault, on the other hand, sought to uncover 'the social practices that the text itself both reflects and employs' and to 'reconstruct the technical and material framework in which it arose' (Hoy 1985: 60). Though deconstruction is useful in helping to change the epistemological climate, and in encouraging a rhetorical reading of cartography, my final concern is with its social and political dimensions, and with understanding how the map works in society as a form of power–knowledge. This closes the circle to a context-dependent form of cartographic understanding.

We have already seen how it is possible to view cartography as a discourse, a system which provides a set of rules for the representation of knowledge embodied in the images we define as maps and atlases. It is not difficult to find for maps – especially those produced and manipulated by the state – a niche in the 'power/knowledge matrix of the modern order' (Philip 1985: 76). Especially where maps are ordered by government (or are derived from such maps) it can be seen how they extend and reinforce the legal statutes, territorial imperatives and values stemming from the exercise of political power. Yet to understand how power works through cartographic discourse and the effects of that power in society further dissection is needed. A simple model of domination and subversion is inadequate, and I propose to draw a distinction between *external* and *internal* power in cartography. This ultimately derives from Foucault's ideas about power–knowledge, but this particular formulation is owed to Joseph Rouse's book *Knowledge and Power* (1987), where a theory of the internal power of science is in turn based on his reading of Foucault.

The most familiar sense of power in cartography is that of power *external* to maps and mapping. This serves to link maps with the centres of political power. Power is exerted *on* cartography. Behind most cartographers there is a patron; in innumerable instances the makers of cartographic texts were responding to external needs. Power is also exercised *with* cartography. Monarchs, ministers, state institutions, the Church, have all initiated programmes of mapping for their own ends. In modern Western society maps quickly became crucial to the maintenance of state power – to its boundaries, to its commerce, to its internal administration, to control of the population, and to its military strength. Mapping soon became the business of the state: cartography is early nationalized. The state guards its knowledge carefully: maps have been universally censored, kept secret and falsified. In all these cases maps are linked with what Foucault called the exercise of 'juridical power' (Foucault 1980: 88; Rouse 1987). The map becomes a 'juridical territory': it facilitates surveillance and control. A recent textbook on geographical information systems even boasts that it is 'a science of surveillance'. Maps are still used to control our lives in innumerable ways. A mapless society, though we may take the map for granted, would not be politically unimaginable. All this is power *with* the help of maps. It is an external power, often centralized and exercised bureaucratically, imposed from above, and manifest in particular acts or phases of deliberate policy.

I come now to the important distinction. What is also central to the effects of maps in society is what may be defined as the power *internal* to cartography. The focus of enquiry therefore shifts from the place of cartography in a juridical system of power to the political effects of what cartographers do when they make maps. Cartographers manufacture power: they create a spatial panopticon. It is a power embedded in the map text. We can talk about the power of the map just as we already talk about the power of the word or about the book as a force for change. In this sense, just as with other artefacts and technologies, maps do have politics (Winner 1980). It is a power that intersects and is embedded in knowledge. It is universal. Foucault (1978, 73) writes of:

> The omnipresence of power: not because it has the privilege of consolidating everything under its invincible unity, but because it is produced from one moment to the next, at every point, or rather in every relation from one point to another. Power is everywhere; not because it embraces everything, but because it comes from everywhere.

Power comes from the map and it traverses the way maps are made. Maps are a technology of power, and the key to this internal power is cartographic process. By this I mean the way maps are compiled and the categories of information selected; the way they are generalized, a set of rules for the abstraction of the landscape; the way the elements in the landscape are formed into hierarchies; and the way various rhetorical styles that also reproduce power are employed to represent the landscape. To catalogue the world is to appropriate it (Barthes 1980; Wood and Fels 1986), so that all these technical processes represent acts of control over its image which extend beyond the professed uses of cartography. The world is disciplined. The world is normalized. We are prisoners in its spatial matrix. For cartography as much as other forms of knowledge, 'All social action flows through boundaries determined by classification schemes' (Darnton 1984: 192–3). An analogy is what happens to data in the cartographer's workshop and what happens to people in the disciplinary institutions – prisons, schools, armies, factories – described by Foucault (Rouse 1987): in both cases a process of normalization occurs. Or similarly, just as in factories we standardize our manufactured goods, so in our cartographic workshops we standardize our images of the world. Just as in the laboratory we create formulaic understandings of the processes of the physical world, so too, in the map, nature is reduced to a graphic formula. Indeed, cartographers like to promote this metaphor of what they do. One textbook claims:

> Geography thrives on cartographic generalization. The map is to the geographer what the microscope is to the microbiologist, for the ability to shrink the earth and generalize about it . . . The microbiologist must choose a suitable objective lens, and the geographer must select a map scale appropriate to both the phenomenon in question and the 'regional laboratory' in which the geographer is studying it (Monmonier and Schnell 1988: 15).

The power of the map maker was generally exercised not over individuals but over the knowledge of the world made available to people in general. Yet this is not consciously done, and it transcends the simple categories of 'intended' and 'unintended' altogether. I am not suggesting that power is deliberately or centrally exercised. It is a local knowledge which at the same time is universal. It usually passes unnoticed. The map is a silent arbiter of power.

Conclusion

The interpretive act of deconstructing the map can serve three functions in the way we view maps in geographical culture. First, it allows us to challenge the epistemological myth (created by cartographers) of the cumulative progress of an objective science always producing better delineations of reality. Second, deconstructionist argument allows us to redefine the social importance of maps. Rather than invalidating their study, it enhances it by adding different nuances to our understanding of the power of cartographic representation as a way of building order into our world. If we can accept intertextuality then we can start to read our maps for alternative and sometimes competing discourses and they can overflow into a new range of problems. Third, a deconstructive turn of mind may allow geographical cartography to take a fuller place in the interdisciplinary study of text and knowledge. Intellectual strategies such as those of discourse in the Foucauldian sense, the Derridian notion of metaphor and rhetoric as inherent to scientific discourse, and the pervading concept of power–knowledge are shared by many subjects. As ways of looking at maps they are equally enriching. They are neither inimical to hermeneutic enquiry nor anti-historical in their thrust. The possibility of discovering new meanings in maps is enlarged. Postmodernism offers a challenge to read maps in ways that could reciprocally enrich the reading of other texts.

Selected references

Arnheim, R. 1986: The perception of maps. In Arnheim, R. (ed.), *New essays on the psychology of Art.* Berkeley, CA: University of California Press.

Bakhtin, M. M. 1981: *The Dialogical Imagination: Four Essays.* In Holquist, M. (ed./trans.) and Emerson, C. (trans.). Austin, TX: University of Texas Press.

Barthes, R. 1980: The plates of the *Encyclopedia.* In *New critical essays.* New York: Hill and Wang.

Barthes, R. 1986: *Mythologies.* Lavers, A. (trans.). New York: Hill and Wang.

Blocker, H. G. 1979: *Philosophy and art.* New York: Scribner.

Burgin, V. 1988: Something about photographic theory. In Rees, A. L. and Borzell, F. (eds), *The new art history.* Atlantic Highlands, NJ: Humanities Press, 41–62.

Clarke, C. N. G. 1988: Taking possession: the cartouche as cultural text in eighteenth-century American maps. *World and Image*, **4**; 455–74.

Crone, G. R. 1978: *Maps and their makers; An introduction to the history of cartography*, fifth edition. Folkestone: Dawson; Hamden, CT: Archon Books.

Darnton, R. 1984: *The great cat massacre and other episodes in French cultural history.* New York: Basic Books.

Dear, M. 1988: The postmodern challenge: reconstructing human geography. *Transactions of the Institute of British Geographers: New Series*, **13**; 262–74.

Derrida, J. 1976: *Of grammatology*, Spivak, G. C. (trans.). Baltimore, MD: Johns Hopkins University Press.

Duncan, J. S. and Duncan, N. 1988: (Re)reading the landscape. *Environment and Planning D: Society and Space*, **6**; 117–26.

Eagleton, T. 1983: *Literary theory: An introduction.* Minneapolis, MN: University of Minnesota Press.

Eagleton, T. 1986: *Against the grain.* London: Verso.

Eco, U. 1976: *A theory of semiotics.* Bloomington, IN: Indiana University Press.

Foucault, M. 1973: *The order of things: An archaeology of the human sciences.* New York: Vintage.

Foucault, M. 1978: *The history of sexuality 1: An introduction.* New York: Random House.

Foucault, M. 1980: *Power/knowledge: Selected interviews and other writings, 1972–77.* Gordon, C. (ed.). Gordon, C., Marshall, L., Mepham, J. and Sopher, K. (trans.). New York: Pantheon.

Gombrich, E. 1975: Mirror and map: theories of pictoral representation. *Philosophical Transactions of the Royal Society of London*, Series B, Biological Sciences, **270**; 119–49.

Goodman, N. 1968: *Languages of art: An approach to a theory of symbols.* Indianapolis and New York: Bobbs-Merrill.

Gregory, D. 1987: Postmodernism and the politics of social theory. *Environment and Planning D: Society and Space*, **5**; 245–8.

Harley, J.B. 1984: Meaning and ambiguity in Tudor cartography. In S. Tyacke (ed.) *English Mapmaking, 1500–1650: Historical Essays.* London: British Library Reference Division.

Harley, J. B. 1988a: Maps knowledge and power. In Cosgrove, D. and Daniels, S. (eds), *The iconography of landscape.* Cambridge: Cambridge University Press.

Harley, J. B. 1988b: Secrecy and silences: the hidden agenda of cartography in early modern Europe. *Imago Mundi*, **40**; 111–30.

Harley, J. B. 1989: Historical geography and the cartographic illusion. *Journal of Historical Geography*, **15**; 80–91.

Harley, J. B. 1996: Power and legitimation in the English geographical atlases of the eighteenth century. In Wolter, J. A. (ed.), *Images of the world: The atlas through history.* New York: McGraw Hill.

Harley, J. B. and Woodward, D. 1987: *The history of cartography 1: Cartography in prehistoric ancient and medieval Europe and the Mediterranean.* Chicago: University of Chicago Press.

Harley, J. B. and Woodward, D. 1987–1994: *The History of Cartography 2*, Book 1, *Cartography in the traditional Islamic and south Asian societies*, and Book 2, *Cartography in the traditional East Asian societies.* Chicago: University of Chicago Press.

Henrikson, A. K. 1987: Frameworks for the world. Preface to Ehrenberg, R. E., *Scholars' guide to Washington D.C. for Cartography and Remote Sensing Imagery.* Washington DC: Smithsonian Institution Press.

Hoy, D. 1985: Jacques Derrida. In Skinner, Q. (ed.), *The return of grand theory in the human sciences.* Cambridge: Cambridge University Press.

Knox, P. L. (ed.) 1988: *The design professionals and the built environment.* London: Croom Helm.

Korzybski, A. 1948: *Science and sanity: An introduction to non-Aristotelian systems and general semantics*, third edition. Lakeville, CT: International non-Aristotelian Library.

Loxton, J. 1985a: The Peters phenomenon. *Cartographic Journal*, **22**; 106–8.

Loxton, J. 1985b: The so-called Peters projection. *Cartographic Journal*, **22**; 108–10.

McKenzie, D. F. 1986: *Bibliography and the sociology of texts.* London: British Library.

Marin, L. 1988: *Portrait of the king.* Houle, M. H. (trans.). Minneapolis, MN: University of Minnesota Press.

Meynen, E. (ed.) 1973: *Multilingual dictionary of technical terms in cartography.* Wiesbaden: Steiner.

Mitchell, W. J. T. 1986: *Iconology: Image, text, ideology.* Chicago, IL: University of Chicago Press.

Monmonier, M. S. and Schnell, G. A. 1988: *Map appreciation.* Englewood Cliffs, NJ: Prentice-Hall.

Morris, J. 1982: The magic of maps: the art of cartography. Unpublished MA dissertation, University of Hawaii, Hawaii.

Norris, C. 1982: *Deconstruction: Theory and practice.* London: Methuen.

Norris, C. 1987: *Derrida.* Cambridge, MA: Harvard University Press.

Peréz-Gomez, A. 1983: *Architecture and the crisis of modern science.* Cambridge: Cambridge University Press.

Peters, A. 1983: *The new cartography.* New York: Friendship Press.

Philip, M. 1985: Michel Foucault. In Skinner, Q. (ed.), *The return of grand theory in the human sciences.* Cambridge: Cambridge University Press.

Porter, P. and Voxland, P. 1986: Distortions in maps: the Peters projection and other devilments. *Focus*, **36**; 22–30.

Robinson, A. H. 1985: Arno Peters and his new cartography. *American Cartographer*, **12**; 103–107.

Robinson, A. H. and Petchenik, B. B. 1976: The map as a communication system. *Cartographic Journal*, **12**(11); 7–15.

Rouse, J. 1987: *Knowledge and power: Toward a political philosophy of science.* Ithaca, NY: Cornell University Press.

Saarinen, T. F. 1988: Centering of maps of the world. *National Geographic Research*, **4**(1); 112–27.

Skinner, Q. (ed.) 1985: *The return of grand theory in the human sciences.* Cambridge: Cambridge University Press.

Snyder, J. P. 1988: Social consciousness and world maps. *Christian Century*, **24** (February); 190–2.

Vujakovic, P. 1989: Arno Peters' cult of the 'new cartography': from concept to world atlas. *Bulletin of the Society of University Cartographers*, **22**(2); 1–6.

Wallis, H. M. and Robinson, A. H. (eds) 1987: *Cartographical innovations: An international handbook of mapping terms to 1900.* Tring: Map Collector Publications and International Cartographic Association.

White, H. 1978: *Tropics of discourse: Essays in cultural criticism.* Baltimore, MD: Johns Hopkins University Press.

Winner, L. 1980: Do artefacts have politics? *Daedalus*, **109**(1); 121–36.

Wolter, J. A. 1975: The emerging discipline of cartography. Unpublished PhD. dissertation, University of Minnesota, Minneapolis, MN.

Wood, D. 1987: Pleasure in the idea/The atlas as narrative form. In Carswell, R. J. B., de Leeuw, G. J. A., and Waters, N. M., (eds), *Atlases for schools: design principles and curriculum perspectives. Cartographica*, **24**(1); 24–25 (Monograph 36).

Wood, D. and Fels, J. 1986: Designs on signs/Myth and meaning in maps. *Cartographica*, **23**(3); 54–103.

10 Geraldine Pratt

Spatial Metaphors and Speaking Positions

Reprinted in full from: *Environment and Planning D: Society and Space* 10, 241–4 (1992)

The task of writing an editorial has been, for me, vaguely embarrassing and difficult; 'editorialising' conjures an image of intellectual centrality seemingly at odds with the times. Although it is now accepted as an ethical and political responsibility to position oneself so as to betray the partiality of one's perspective, establishing the grounds for taking a position and the right to speak – for oneself and certainly *about* others – is by no means unproblematic. The stakes surrounding the politics of presence are high. They offer the potential for more democratic, inclusive public discussions and institutions. Donna Haraway (1991) argues that 'situated knowledge' also comprises a middle course between relativism and objectivism (a type of limited objectivity). At the same time, a commitment to positioning each speaker can be used to put people in their place, to shut them down. In practice, the diversification of perspectives opens up the possibility of a pluralisation of antagonisms, an essentialising of some differences (for example, within feminism, the subtlety of the theorising of the category of gender is frequently attended by a remarkably static treatment of other social categories, such as race or sexual orientation), and a trivialisation of intellectual debate to ad femina/hominem insults.

As a means of thinking about positionality and knowledge, spatial metaphors are very much in vogue. Although geographers hold no patent on these metaphors, our disciplinary location could lead us to think more critically about them. My view is that this is a matter that readers of *Society and Space* can and should attend to; although some of the spatial metaphors circulating through contemporary academic discussions are useful aids for problematising positionality, others dress up and potentially reproduce some very conventional intellectual subject positions (for example, that of the distanced observer), underwrite new sets of dividing practices, and promote a remarkable arrogance or naivety towards the construction and destruction of and caring for places.

There are three sets of spatial metaphors in popular use: ones that draw upon the rhetoric of mobility (for example, 'nomadism', 'travelling', 'migration', the 'flaneur'); others that emphasise the position of marginality and exile; and a third that represents the borderland as a place. Each opens up and closes down thought in different ways. A discussion of the potentials and limitations of each is complicated by the slipperiness of the ways that these metaphors are used: they are often employed to represent the author's own and others' subject positions (sometimes metaphorically, sometimes realistically), in some cases they are used to draw the contours of disciplinary practices (for example, Clifford, 1992); and in others they are applied as literal descriptions of contemporary cultures. Given my concerns about the politics of presence – the grounds for finding a speaking position and the possibilities for speaking across differences – I concentrate my comments on the first use.

The rhetoric of mobility and detachment from place runs through much of the contemporary postmodern/poststructuralist/postcolonial cultural studies literature. Trinh Minh-ha (1990, page 334) describes the postcolonial feminist as 'a permanent sojourner walking bare-footed on multiply de/re-territorialized land'. This movement is tied to a desire continually to displace the boundaries between centre and margins, as a neverending strategy to displace controlling reference points. In a similar gesture, Edward Said ties a 'scrupulous subjectivity', one that entails a consistent questioning of fixed identities and categories, to a detachment from place. He quotes Hugo St Victor, a twelfth-century Saxon monk: 'The tender soul has fixed his love on one spot in the world; the strong man has extended his love to all places; the perfect man has extinguished his' (Said, 1990, page 365). Reflecting the same concern, Gayatri Spivak notes that, 'As far as I can tell, one is always on the run, and it seems I haven't really had a home base – and this may have been good for me. I think it's important for people not to feel rooted in one place. So wherever I am, I feel I'm on the run in some way . . .' (1990, page 37). Bell hooks describes how travelling can disrupt reifications of 'the Other' and offer critical insights into one's own and others' social locations: by leaving one's place, one comes to know one's place. James Clifford's ability to portray travel playfully reflects, she notes, his own social position and travel experiences as a white American male academic (see also Haraway, 1988, page 147). She contrasts her own travel experiences as a black American woman, during which she has been detained and harassed by customs officials, characterising these experiences as 'encounter[s] with terrorism' (hooks, 1992, page 343). Clifford (1992) extends the use of travel metaphorically, beyond this self-discovery of social location. He refigures culture *as* travel, in part to mitigate the process of 'representational essentialising', the fixing of certain people as 'natives' or around a few unchanging attributes.

This talk about movement, the representation of self-knowledge and the vigilant deconstruction of conceptual stasis and cultural hierarchy in terms of mobility, is immensely suggestive. At the same time the limitations of the rhetoric of mobility need to be spelt out and considered carefully. Clifford (1992, page 110) outlines some of the limitations of the travel metaphor, acknowledging that it is tainted by associations with 'gendered, racial bodies, class privilege, specific means of conveyance, beaten paths, agents, frontiers, documents, and the like'. He chooses to use the term precisely because of the obviousness of these associations; they offer a good reminder that all translation terms (terms used for comparison in strategic ways) 'get us some distance *and* fall apart' (page 110). I think that this translation term falls apart in ways beyond those that Clifford acknowledges and I wonder about the point at which the metaphor clouds as much as liberates our thinking about subjectivity, categories, and positionality.

A focus on mobility and the fluidity of identity and social categories runs the risk of reproducing the privilege of the unsuited observer. The view on the move can potentially shade into the pretensions of a view from nowhere, an ungrounded detached gaze. Nancy Fraser (1991) raises this issue in relation to Judith Butler's (brand of) poststructuralist reading of subjectivity; her privileging of deconstructive over reconstructive, normative critique; and her location of women's liberation in the dislocation of dereification of identity. Fraser argues that a privileging of fluidity and the destabilisation of identities allows social theorists to escape some hard political questions, in the case of feminists 'whether there are real conflicts of interest among women of different classes, ethnicities, nationalities, and sexual orientations, conflicts so intractable as not to be harmonizable, or even finessable, within feminist movements' (1991, page 175). One has to acknowledge the stability and social reification of some identities to tackle these questions.

When fluid readings of subjectivity are interwoven with spatial metaphors of mobility this has some worrying implications for the preservation, caring for, and creation of places. As Stuart Hall notes in the discussion that followed Clifford's (1992) presentation of his ideas about 'Traveling cultures', we also have to think about the meaning of dwelling and to acknowledge, not only the dangers of reactionary forms of dwelling (for example, some types of 'community' or nationalism), but the legitimacy and value of peoples' struggles to create their own places and memories. This may seem a self-evident and somewhat 'banal' observation to readers of *Society and Space*, but it is one that must be placed insistently alongside the rhetoric of movement that privileges detachment from place; we must do this in order to break down a new hierarchy of difference created through the seemingly fashionable mobility–dwelling duality. [See also Morris (1990) for a critique of any 'thematics of place' that is articulated around the binary opposition of propriety and mobility.]

The metaphor of exile or living at the margins is used to represent some of the same ideals underlying the rhetoric of mobility: the desire to disrupt categories and the authority of dominating hegemonic viewpoints. Teresa de Lauretis (1990) lists 'marginality as location' as one of the key characteristics of a postcolonial feminism that pushes the concerns of white, straight, middle-class Western feminists from centre stage. For bell hooks (1990), marginality is a site of resistance, a position from which to resist colonisation by the dominant (white) culture, one that allows a space to imagine alternative ways of existing and the opportunity to create counter hegemonic cultures.

The vision of being located outside or at the margins shares some of the attractions and limitations of the image of continual movement. Although the theorist is situated *somewhere* and the metaphor of marginality *can* be used effectively to problematise positionality, it can *also* encourage the rigidification of difference and the pretension of being outside dominant society. Spivak (1989) captures the arrogance of this pretension in a typically revealing anecdote. She tells the story of walking up to an upper-class African gay male friend at a party, asking him to tell a group of listeners his name. 'He gave us a string of his names, the sixth name, as he said, was the name of his ancestor. Undoubtedly somewhat troubled by the burden of hyperbolic admiration because of the color of his skin, the other side of racism, he added quietly, ''A slave trader''. Collaboration with the enemy does not depend on the color of your skin or on your gender.' Spivak's point is that we *all* collaborate with some enemies and that the points of difference are situated within nets of interwoven similarities and shared complicities. The metaphor of exile captures difference; it encourages us to lose sight of commonalities.

The pretension of being 'outside' can also lead to a naivety concerning one's effect in place and on places. The metaphor does at least problematise individuals' relations to place insofar as those living at the margins stay in place long enough for the sceptic to detect their effect on places. For example, in a widely cited autobiographical narrative, Minnie Bruce Pratt (1984) describes her coming to consciousness of her own complicity with racism, heterosexism, and antisemitism. At the end of the narrative, Pratt is living as a solitary white woman in a black neighbourhood in Washington DC – self-consciously very much the outsider – choosing this location, in part, as a vehicle for continually destabilising her sense of identity. Focused on her sense of exile, Pratt seems oblivious to her *locatedness* and her potential role as first-wave gentrifier, an impact immediately perceptible to an observer attuned to the production of urban spaces.

The third metaphor, that of borderland, seems to me the most provocative for thinking through the politics of difference and presence, perhaps because the focus is shifted from the individual to a socially constructed place in which difference and conflict is constructed and lived. Borders are saturated with inequality, domination, forced exclusion (Giroux, 1992); they are social and political constructions that are used to construct differences. But they are also relational places where individuals live and construct themselves in relation to each other. We can redraw borders; we recognise that different types of boundaries operate at different scales.

We should, however, recognise the limits of any metaphor and resist being seduced by geographical and spatial metaphors that are ultimately aspatial and insensitive to place. We need ways to envision places and situations in which people are forced into dialogues that admit conflict and real differences *and* the social construction and permeability of these differences. We need to hold change (mobility, deconstruction) and limited stasis (dwelling, placement, normative ideals) in tension. Finally, as distinct from Said and Hugh St. Victor, I find perfection of self a dubious ideal, both in relation to subjectivity and place; tenderness and strength may offer more workable (and less individualistic) guides to action.

References

Clifford J, 1992, 'Traveling cultures', in *Cultural Studies* Eds L Grossberg, C Nelson, P Triechler (Routledge, Chapman and Hall, New York) pp 96–112

de Lauretis T, 1990, 'Eccentric subjects: feminist theory and historical consciousness' *Feminist Studies* **16** 115–150

Fraser N, 1991, 'False antitheses: a response to Seyla Benhabib and Judith Butler' *Praxis International* **11**(2) 166–177

Giroux M, 1992, 'Resisting difference: cultural studies and the discourse of critical pedagogy' in *Cultural Studies* Eds L Grossberg, C Nelson, P Triechler (Routledge, Chapman and Hall, New York) pp 199–212

Haraway D, 1988, 'Panel discussion 3' *Inscriptions* **3/4** 145–152

Haraway D, 1991, 'Situated knowledges: the science question in feminism and the privilege of partial perspectives', in *Simians, Cyborgs, and Women: The Reinvention of Nature* (Routledge, Chapman and Hall, New York) pp 183–201

Hooks B, 1990, 'Marginality as site of resistance', in *Out There: Marginalization and Contemporary Culture* Eds R Ferguson, M Gever, M T Trinh, C West (The New Museum of Contemporary Art, New York, and MIT Press, Cambridge, MA) pp 341–343

Hooks B, 1992, 'Representing whiteness in the black imagination', in *Cultural Studies* Eds L Grossberg, C Nelson, P Triechler (Routledge, Chapman and Hall, New York) pp 338–346.

Morris M, 1990, 'Banality in cultural studies', in *Logics of Television: Essays in Cultural Criticism* Ed. P Mellencamp (Indiana University Press, Bloomington, IN) pp 14–43

Pratt M B, 1984, 'Identity: skin blood heart', in *Yours in Struggle: Three Feminist Perspectives on Anti-semitism and Racism* Eds E Burkin, M B Pratt, B Smith (Firebrand Books, Ithaca, NY) pp 10–63

Said E, 1990, 'Reflections on exile', in *Out There: Marginalization and Contemporary Culture* Eds R Ferguson, M Gever, M T Trinh, C West (The New Museum of Contemporary Art, New York, and MIT Press, Cambridge, MA) pp 357–366

Spivak G C, 1989, 'A response to "the difference within: feminism and critical theory"', in *The Difference Within* Eds E Meese, A Parker (John Benjamin, Amsterdam) pp 207–220

Spivak G C, 1990, 'Strategy, identity, writing', in *The Postcolonial Critic: Interviews, Strategies, and Dialogues* Ed S Harasym (Routledge, Chapman and Hall, New York) pp 35–49

Trinh M T, 1990, 'Cotton and iron', in *Out There: Marginalization and Contemporary Culture* Eds R Ferguson, M Gever, M T Trinh, C West (The New Museum of Contemporary Art, New York, and MIT Press, Cambridge, MA) pp 327–335

Geography and the Politics of Nature

GEOGRAPHY AND THE POLITICS OF NATURE

The nature of nature

The British literary critic Raymond Williams (1976, page 184) says that 'nature is perhaps the most complex word in the language'. Certainly the term has had a complex history in geography. In his tome *Traces on the Rhodian shore* Clarence Glacken (1967) identifies three pre-19th-century Western intellectual responses to nature: something to dominate, something to be dominated by, and something to live in harmony with. Although examples of all three approaches can be found during geography's formative period of institutionalisation of the late-19th and early-20th centuries, it was primarily the second that held sway, taking the form of a crude **environmental determinism**: we are what nature makes us. Or, as the early US geographer Ellen Churchill Semple (1911, page 1) put it, 'Man [*sic*] is a product of the earth's surface. This means not merely that he is a child of the earth, dust of her dust; but that the earth has mothered him, fed him, set him tasks, [and] directed his thoughts . . . .'

As a number of recent writers have noted (Hudson, 1977; Livingstone, 1992), the persuasiveness of environmental determinism was clearly linked to the frenetic pace of European imperialism occurring at the same time (for example, the 'scramble for Africa' of the 1880s). Environmental determinism provided an intellectual legitimation for colonisation: if because of their natural environment a people could not 'develop' then it was 'the white man's burden', as Kipling put it, to intervene and show them how.

By the 1920s, the shallowness, shoddiness, and ethnocentrism of environmental determinism was increasingly challenged on a number of related fronts: by **possibilism**, an idea most associated with the French regional school of geography and which suggested that nature provided a range of opportunities for human action among which individuals could choose to varying degrees; by **human ecology**, the idea of a mutual adjustment between humans and their natural environment; and **cultural landscape**, an idea associated with Carl Sauer (1925) (see Section Six) that emphasised the interactive relationship between culture and nature, the product of which was a distinctive landscape morphology. Clearly, improvements over environmental determinism, each of these three responses nevertheless continued to conceive nature as neutral and external, whose fixed characteristics and regulatory laws could be understood only through the canons of natural science.

After the end of the Second World War, nature as a topic of discussion in human geography was increasingly marginalised as **spatial science** took centre stage (including, ironically, in physical geography). By emphasising isotropic plains and uniform resource distributions, spatial science effectively assumed nature away; space and location were the key terms, and nature and environment were at best minor complicating factors. By the end of the 1960s, though, partly as a result of wider concerns about the environment prompted by disasters such as the Torrey Canyon oil spill off the Cornish coast, and partly as a result of the desire for geographers to be socially relevant ('Earthday', a day given over to celebrating the global environment, was in 1970), nature came back on to the human geographer's agenda. Moreover, as a result of the then emerging critique of spatial science's objectivism, and a wider suspicion about the power of natural science, the older objectivist view of nature was also called into question (in human geography seen in Harvey, 1974). Subsequent responses have varied, but many claimed that the nature of nature had been seriously misunderstood; indeed for many there was now nothing natural about nature at all.

Politicising nature

With this definitional questioning that occurred both in the academy and outside, nature as a term became highly contested and politicised. A variety of groups wanted to lay claim to it for their own ends. The result was a set of unusual alliances and enmities among different classes, social movements, schools of intellectual thought, and environmental and political stakeholders as each sought to make their version of nature the official version: workers and capitalists joined together against middle-class environmentalists; deep ecologists contested various aboriginal practices such as baby seal hunting while claiming a common environmental philosophy; and the political left was splintered in every which way – into reds, greens, ecosocialists, radical environmentalists, dialectical biologists, and more beside. Nature is indeed a complex word.

Technocentrism

Because of this involved pattern of political fissures, we provide here only a highly abbreviated account of the main positions. The orthodox view is what O'Riordan (1989) calls technocentrism, and derives from the first of Glacken's responses by humans to nature: something to dominate. In this approach nature is taken as malleable and manipulatable through both human ingenuity and technological prowess. Moreover, through that intervention both nature and human life are improved. Implicit within this position are a number of assumptions. First, there is the supposition of **anthropocentrism**: nature exists to satisfy *human* purposes. All life on planet earth is not equally equal; humans are the most important species with the right to alter nature in order to satisfy their various ends. Second, there is the presumption that nature is external. Nature lies outside of human life both in the sense that there is something special about us that sets us apart from it, but also in the sense that nature possesses a brute reality, an 'out thereness', that must be confronted and subdued. Finally, there is the assumption that the best method to do that subduing is science. Scientific methods allow for the control of nature, to squeeze and mold it in ways useful to humans. Science forms the basis of the 'techno' part of technocentrism.

After at least two decades of critique, technocentrism is not propounded very much in human geography, although it is making a come back through the back door of geographical information systems (GIS) and the associated movement of applied geography (examples are found in the journal of the same name). Criticisms of technocentrism are legion and take the form of everything from moral and **epistemological** arguments (see below) to specific examples of the sometimes catastrophic consequences of humans attempting to control nature.

Ecocentrism and deep ecology

At the opposite pole to technocentrism is ecocentrism with its origins in Glacken's third response by humans to nature: the quest for harmony. Such a desire has clearly lain behind the most significant environmental movements of the past hundred years. According to Merchant (1992), there have been three main waves of environmentalism in North America: the first came in the late 19th century with the founding of the Sierra Club and Audobon society; the second arose after the Second World War and was associated with various government and quasi-government bodies struck to manage environmental conservation; and the third arose during the 1960s as part of a more general political unrest in both Europe and North America and which culminated in the formation of official political parties, such as the Green Party in Germany (established in 1979), that attempted to implement through the parliamentary process a wide-ranging environmental political platform.

One much discussed element of the 'third wave' is **deep ecology**. Originating with the Norwegian philosopher Arne Naess's **biocentric** approach, the gist is that nature should be accorded in and of itself a deep moral respect by humans. The richness and diversity of *all* life, both human and non-human, is inherently valuable; it is simply immoral for one kind of life to expunge another kind. Some have further argued that the moral imperative of deep ecology is already found in, or can be easily integrated with, existing religious systems, for example, Buddhism, Native American shamanism, and Druidism [but not Christianity; see White's (1967) famous essay]. Combinations of deep ecology with one or more of these existing religions, in turn, form new transcendental philosophies that are the bases of various spiritual ecologies. Although this spiritualism is at the extreme tip of Merchant's third wave, there is no denying that the values of deep ecology have already seeped deeply into contemporary culture through the influence of all kinds of 'New Age' movements – from health to music to food. The political consequences have also been impressive – from abandoning the sinking of the Brent Spar in the North Sea to abandoning logging sites in US Pacific Northwest forests to save the spotted owl.

Marxist conceptions of nature

One of the criticisms of deep ecology, as least as manifest in its more transcendental form, is its political naivity. Coming particularly from Marxists, they argue that much of the environmentalist creed is ideological, obscuring what should be the real focus of enquiry – capitalist social relations. Social relations are so critical because for post-war **Western Marxists** they form the very basis of the **production of nature** (the classic statement is by Schmidt, 1971; and in geography by Smith, 1984). We need to be careful here with the term 'produce'. It doesn't mean that capitalism's social relations are able to defy the principles of physics, chemistry, and biology and create something where there was nothing before. But it does mean that when capitalism emerged on to the scene it utterly transformed pre-existing nature (sometimes called 'first nature'). Nature still exists, but how it is exploited, used, thought about, and represented is now completely different. Castree (1995, page 20), following Cronon (1991), provides a useful example: 'the "natural" regions of . . . the midwestern United States, cannot be understood simply as pre-existent natural grasslands, as the traditional notion of "first nature" would imply. Instead . . . they must be seen as *constructed natural environments* evolving out of decades of intensive, profit-driven conversion into what they presently are.'

One of the difficulties with this orthodox Western Marxist position, though, is that it seems to give too much away to social relations; nature comes across as infinitely malleable, capable of taking any form dictated by capitalism's Promethean impulses. Against this social–productionist view, other Marxists have recently attempted to retheorise nature allowing it some power of resistance. Benton's work (1989) is probably the best known, and argues for a dialectical relationship between society and nature where capitalism produces nature, and nature as a produced entity both enhances and limits capitalism. For example, the capitalist production of chloroflurocarbons (CFCs) in the manufacture of aerosol sprays has produced a nature in which there is now a large hole in the ozone layer, but that hole reacts back and influences capitalism itself – most immediately through some country's governments banning CFCs, and, more catastrophically, if left unchecked, through precipitating the demise of capitalism itself as the produced human world becomes uninhabitable.

Ecofeminism

Ecofeminism as a movement is a reaction against both the anthropocentrism of the technocentrist approach, and the **androcentrism** of much of deep ecology and Marxism, one that privileges men over women. At the heart of ecofeminism is the idea that there are strong parallels between the domination and exploitation of nature and the domination and exploitation of women. Ecofeminists argue that both are a consequence of a particular form of reasoning within Western societies which is termed **masculinist** (see Section Eight). Characterised by the use of hierarchical, dualistic, and supposedly universal categories, masculinist thought aims to subject the world to its authority. But under scrutiny that authority, say ecofeminists, is held by, and is in the interests of, only one of earth's creatures, the male member of *Homo sapiens*. This is why deep ecology never goes far enough. Although recognising the domination of nature, it fails to recognise similar forms of the same fundamental relationship, and as a result also fails to identify root causes which are bound up with power and knowledge. It follows that a better world both for women and for nature is possible only by changing relations of power so that they are non-**patriarchical** and non-exploitive, and rethinking knowledge so that it is contextual, blurred, non-hierarchical, and decentred.

The social construction of nature

Strands of both the Marxist emphasis on the production of nature and the ecofeminist stress on knowledge and power are found in the final approach, social constructionism. There are a number of different variants, but the one on which we concentrate originates in the sociology of scientific knowledge (SSK), and in its most recent form combines two distinct meanings of social constructionism: that our knowledge about the world reflects the social origins of its production, and that nature and society bleed into one another in complicated ways, rendering the very notion of a separate nature and society problematic.

As it emerged in the 1970s, SSK was a reaction against the traditional view of science that assumed scientists possessed a 'God's eye view' (Haraway, 1992); a view of nature that is truthful and without blemish. In contrast, proponents of SSK argued that what counts as nature is actively constructed, shaped according to the context of its manufacture. For example, if 20th century medical researchers believed that sperm were intrepid, forceful, and resolute in attempting fertilisation it was because of the power of masculinist thought (Harding, 1976). In this case, nature is constructed not as it 'really is', but according to the dictates of the social interests of those putting forward such representations. Note that this view is not **solipsistic**, that the world is entirely in our heads. It means only that we can never know nature on its own terms; that the language nature speaks is one we provide, with all the biases, contradictions, and hidden claims that this implies.

In more recent work associated particularly with Bruno Latour (1993) and Donna Haraway (1992), the first sense of social constructionism is joined with the second. In part this is necessary because of the dramatic changes over the past 25 years in such fields as robotics, bioengineering, artificial intelligence, and cybernetics. Both Latour and Haraway argue that the traditionally drawn line between nature and culture, organism and machine, the natural and the artificial is being systematically erased. We are increasingly living in a world where these old kind of dualisms are losing their purchase. It is not only the old idea of nature as pristine and natural that is 'dead' (Merchant, 1983), but also the traditional conception of 'culture' as something human made. Instead, we are living in a 'cyborg' nature/culture defined as one where the distinctions among humans, animals, machines, and the non-physical are progressively blurred.

Although cyborg life has its problems, it also offers new possibilities, both political and epistemological. The political is the potential to disrupt the old dualistic categories on which oppressive and discriminatory practices are based: man–woman, white–coloured, straight–gay, and, of course, nature–culture. The epistemological is the possibility of understanding scientific knowledge of nature as a product or an effect of a network of heterogeneous materials, both animate and inanimate. This requires a little elaboration. Known as actor-network theory, the argument is that knowledge, which itself is often embodied in material forms (such as this book), is the end product of a lot of hard work in organising, controlling, and speaking for bits and pieces of the world, and then patterning them into an ordered whole. Those bits and pieces can be technical (the computer on which this text is being written), physical (the coffee we are sipping while talking about what to write next), textual (the books and articles we've read to write this introductory essay), institutional (the privileged setting of the university in which we work and which allows us uncluttered August mornings in which to write), and conceptual (discussions we've had with Bruce, Maureen, and Noel about this essay). A great variety of things, then, go into making knowledge, and distinguishing the different constituents by, say, either their human or machine character is neither always possible nor useful. It is cyborg knowledge.

Geographies of nature

The first of the readings is by Margaret FitzSimmons, and it is a plea for radical geographers to take nature more seriously in their work (it was one of a series of papers written in the late 1980s in *Antipode* on 'What's left to do'). She argues nature's neglect as a topic is a result of, first, the kind of theorists on whom radical geographers have relied, such as Weber, Marx, and Foucault, who emphasise time and space over the natural environment; and, second, the urban bias of much of radical research, which not only sets itself against nature but also makes use of a positivist methodology that although acceptable within the sociology of the academy is unacceptable as a means to confront nature's 'social **ontology**'. The neglect of nature by radical geographers has been politically costly, though. For the standard view of nature as 'externalized, abstracted [and] . . . primordial provide[s] a source of authority to a whole language of domination' (page 186). By conceiving nature in different terms, FitzSimmons resists such language. Arguing that the present Western formulation of nature arose with capitalism, she avers that nature is 'a material, practical and conceptual reconstitution and reification of what are essentially social relationships' (page 185). The point is that once nature is conceived as a social product, then any justification of the language of domination as natural is undermined; it must be seen for what it is, a consequence of social relations.

Nature also gets short shrift, according to Cathy Nesmith and Sarah Radcliffe, in feminist geography. Moreover, the cause of the neglect is the same as in radical geography: an overemphasis on space. Furthermore, this disregard of nature is not some minor omission but hinders the very political project of feminism itself. As from the ecofeminist position, the domination of nature and the domination of women are seen to be linked, and so the aims of feminism can never be met unless the movement critically theorises the relationship between gender and nature. It is here, argue Nesmith and Radcliffe, that feminist geographers offer a potentially important contribution, situated, as they are, in a discipline that historically focuses on nature. In particular, Nesmith and Radcliffe stress the need to analyse the cultural coding of nature which in the West, at least, is posited as feminine, and is made to contrast with and be subservient to a masculine culture. The political imperative of such an analysis, they argue, is to break the links between nature/woman–culture/man, thus unshackling both women and nature. For there is nothing essential about such pairings; they exist only because of historically, culturally and gender specific constellations of power and interest.

Although not all ecofeminists would agree with their analysis – spiritual ecofeminists, for example, would likely want to maintain the essential naturalness of women, seeing it as a source of strength and power – Nesmith and Radcliffe maintain that the alternative is only a naive **essentialism**; one that is the cause of the problem, and not its solution.

The final paper in this section, by David Demeritt, directly addresses an issue that cuts across much of the more recent literature on nature, and is implicitly found in the previous two papers: the complicated relationship between social constructionist and realist views of nature. The problem is that many of the recent approaches to nature, such as orthodox Marxism and ecofeminism, want to have it both ways. On the one hand, they argue that the other side, whoever they maybe – for example, technocentrists – are wrong because their view of nature is only a social construction, a product of, say, class location or **phallocentric** thought. On the other hand, those same critics abandon that social constructionist argument by later suggesting that once the ideology of class or patriarchy is peeled away *real* nature is revealed.

Demeritt shows that this same inconsistency is found in the recent debates in ecology and environmental history. He suggests that by mobilising socially constructionist arguments first proposed by Thomas Kuhn, ecocentrist environmentalists very effectively criticised the technocentrist's realist claims about nature. But having done so, ecocentrists then couched their arguments against the further ravaging of nature by humans in precisely the same realist terms that they earlier disparaged. Clearly this is contradictory, but is there a way out? Is there a balance between realism and social constructionism? Drawing upon Latour and Haraway, Demeritt argues that there is. It involves opening up the conversation to all interested parties, while at the same time denying that any one of them has sole purchase on the truth. As he writes, 'Exclusive knowledge claims serve only to divide, to empower a few anointed experts, and to exclude the many different voices that can and should speak' (page 223). This doesn't mean that anything goes. Rather, we should see claims by, say, scientists as potentially useful pieces of information, but also ones that need to be set alongside other pieces of information that are not scientific in origin. Only then can we make our moral choice. But to suppose that scientists always possess the right answers is to abandon that choice; it is to allow them to speak for us, and to create nature in their image and not ours.

References

Benton, T. 1989: Marxism and natural limits: an ecological critique and reconstruction. *New Left Review*, No. 178, 51–81.

Castree, N. 1995: The nature of produced nature: materiality and knowledge construction in Marxism. *Antipode*, **27**; 12–48.

Cronon, W. 1991: *Nature's metropolis: Chicago and the Great West*. New York: W. W. Norton.

Glacken, C. 1967: *Traces on the Rhodian shore. Nature and culture in Western thought from ancient times to the end of the eighteenth century*. Berkeley, CA: University of California Press.

Haraway, D. 1992: *Simians, cyborgs and women: The reinvention of nature*. New York: Routledge.

Harding, S. 1976: *The science question in feminism*. Ithaca, NY: Cornell University Press.

Harvey, D. 1974: Population, resources and the ideology of science. *Economic Geography*, **50**; 256–77.

Hudson, B. 1977: The new geography and the new imperialism: 1870–1918. *Antipode*, **9**; 12–19.

Latour, B. 1993: *We have never been modern*. Porter, C. (trans.). Cambridge, MA: Harvard University Press.

Livingstone, D. N. 1992: *The geographical tradition: Episodes in the history of a contested enterprise*. Oxford: Blackwell.

Merchant, C. 1983: *The death of nature: Women, ecology and the scientific revolution*. New York: Harper & Row.

Merchant, C. 1992: *Radical ecology: The search for a liveable world*. New York: Routledge.

O'Riordan, T. 1989: The challenges for environmentalism, in Peet, R. Thrift, N. J. (eds), *New models in geography*, volume 1, London: Unwin Hyman, pp. 77–102.

Sauer, C. 1925: The morphology of landscape. *University of California Publications in Geography*, **2**(2); pp. 19–54.

Schmidt, A. 1971: *The concept of nature in Marx*. London: New Left Review Books.

Smith, N. 1984: *Uneven development: Nature, capital and the production of space*. Oxford: Blackwell.

Semple, E. C. 1911: *Influences of geographic environment on the basis of Ratzel's anthropo-geography*. New York: Henry Holt.

White, L. 1967: The historic roots of our ecological crisis. *Science*, 10 March.

Williams, R. 1976: Nature. *Kewords: A vocabulary of society and nature*. London: Fontana.

Suggested reading

A very readable discussion around general issues of nature is found in Neil Everden's *The social creation of nature* (Baltimore, MD, John Hopkins University Press, 1992). The classic Marxist account of nature in human geography remains Neil Smith's *Uneven development* (Oxford, Blackwell, 1982), but also very useful is David Pepper's *Eco-socialism: From deep ecology to social justice* (London, Routledge, 1993). A clearly written and grounded examination of the relationship between feminism and nature, and especially with environmentalism, is *Earth follies* by Joni Seager (New York, Routledge, 1993). Finally, there have been a slew of essays and monographs exploring the social construction of nature. Latour's (1993) and Haraway's (1992) works are probably the best known, but Steve Woolgar's *Science: The very idea* (London, Sage, 1988) is one of most accessible and comprehensive.

11 Margaret FitzSimmons
The Matter of Nature

Reprinted in full from: *Antipode* 21, 106–20 (1989).

In its origins, the term 'geography' refers to the earth and to formal, socially-constructed knowledge of the material reality of the world – not to space and spatial relations. If we ask 'what's left to do' in the reconstruction of human geography, we ought not overlook this essential aspect of the geographical project. Yet, in the twenty years of development in radical geography . . . the question of the material reality of what I will call 'social Nature' has received relatively little theoretical attention.

Most work by geographers in the radical tradition has continued a peculiar silence on the question of social Nature: the geographical and historical dialectic between societies and their material environments. I argue that this question of the social production of nature has deep roots and a necessary place in any reconstructed human geography. Nature, I suggest, is a 'concrete abstraction' (Harvey, 1985, following Marx) that is equivalent and parallel to Space in its mystifying role and power in our social and intellectual life.[1] Radical geography seems an appropriate arena within which to engage this abstraction but, as Peet suggested more than ten years ago:

> [w]ithin radical geography, the theoretical base is increasingly strong in one of the traditional areas of geographic interests (spatial relations) and conspicuously weak in another area (environmental relations). The Marxist theory of spatial relations is becoming more and more sophisticated, especially in the area of underdevelopment processes. In that area a coherent body of theory, developed outside geography, already existed; in addition, there is a condition of crisis in spatial relationships between the center and the periphery of capitalism, marked by a series of successful wars of Third World liberation, which has spurred on theoretical inquiry. But there is also an *environmental* crisis of monumental proportions, and the materialist approach of Marxist geography can easily be applied to environment–man [*sic*] relations; yet this area of geography remains largely untouched by radical geographers . . . This would seem to indicate that radical geography is still trapped in the areas of emphasis of the 'new geography' which stressed spatial rather than environmental questions. If so, it is time to break the last links with liberalism, for a truly mature radical geography must cover all capitalist contradictions of a geographic type, be holistic within itself, and be integrated with both social and physical academic disciplines to contribute fully to a whole radical science (1977: 26).

Why have radical geographers been unable to come to grips with the theoretical problem of Nature, even though we have made substantial progress in addressing the theoretical problem of Space? What is it about our social practice, our intellectual culture, and our institutional environment that has contributed to the relative underdevelopment of this area of the geographical project?

Our sense of Nature, I suggest, arises from real history and geography. The pervasive view that Nature is external and primordial is unconsciously confirmed by our placement as intellectuals in a spatially organized society in which 'intellectual work' and 'intellectual life' are urban. It is further supported by our particular insertion into the institutions of that intellectual life, institutions dominated by academic disciplines and by the growing power of science (for which the model is 'natural' science) as the privileged form of social knowledge. But where we accept this view unchallenged we remain subject to the mystifying power of this abstraction of Nature, blind to substantial aspects of the real human geography of capitalism.

Urbanization, the urban question, and the social construction of nature and space

First, let me make a gesture of solidarity – we are all, those of us concerned with the human geography of contemporary life, *urban* geographers to some degree. Urbanization is a process of reconstitution of the relationship between humans and the material world of everyday life, just as it is a process of reconstitution of social space. I refer here to that great historical and geographical differentiation of the city and the countryside, a differentiation which is, particularly in capitalism, a crucial aspect of the movement of history and the construction of geography as we know it.

But the urban is just one aspect of urbanization, an epiphenomenon in some sense, a viewpoint into the process but (if dominant) a myopic viewpoint. We see Nature through the geographical and historical experience of the urban; in that the urbanists are correct. Nonetheless, *the urban question is neither ontologically prior nor ontologically complete.* The city is merely the characteristic capitalist form of human settlement in which our lives are embedded, but we are agreed that capitalism is not a transhistorical form.

I suggest an ontological rupture, a rupture which would simultaneously engage two aspects of our current ontology: the retrospective extension of history which we receive from the Enlightenment, and the separation of nature from culture which we perceive as urban beings. Consider the proposition that *Nature as we know it was invented* in the differentiation of city and countryside, in the differentiation of mental and manual labor, and in the abstraction of contemporary culture and consciousness from the necessary productive social work of material life. Set aside the conception of nature as external, as primordial, as historically prior to the development of humans and human society. Try to see Nature, like History, Geography, and Space, as a material, practical and conceptual reconstruction and reification of what are essentially social relationships. If you do this you will overcome a conception which is deeply written into the unconscious presuppositive ontology of our culture – even more deeply hidden than the unconscious historicism (absent geography) with which Ed Soja is concerned (Soja, 1989).

This deconstruction of received ontology is very difficult for us to achieve. Nature provides one pole of all the great Enlightenment antinomies – nature and society, nature and culture, nature and nurture (Williams, 1980). This ontological and historical separation of nature coincided with the transition from feudal geographical relationships – the movement of the primary locus of human economic and intellectual activity from dispersed settlements to early modern urban (and rural) forms – and with the division of the labor of those who work *with* nature from those (scientists) who work *on* Nature. In the course of this great geographical transition to capitalism, the Enlightenment invented a new fundamental cultural code. As Foucault suggests in *The Order of Things* (1970), the intellectuals and ideologists of this transition reworked nature, which the Middle Ages believed had been made by God for the perfection of man, a Nature suffused with a system of God's signs. They constructed a new 'epistemological space', invented external Nature through 'rules of formation' as a mechanism subject to analysis, and pursued, rather than perused, a Nature which mattered not for its meaning but for its laws.[2] This invention of primordial Nature reformed and restructured our sense of the world. It also redefined our place in it, our human nature (Ollmann, 1971).

But primordial nature is just as consequential (and as inconsequential) to human geographers as is primordial space. *We must recognize that externalized, abstracted, Nature-made-primordial provides a source of authority to a whole language of domination.* This is the domination *of* nature, but also the domination of human reality *by* Nature. We are shown one side of this – the domination *of* nature – by Marx: 'Capitalist production, therefore, develops technology, and the combining together of various processes into a social whole, only by sapping the original sources of all wealth – the soil and the labourer' (1967, 506–7). But Marx was not free of the ontology of his time; though he hints at the social production of nature, he often slips into a language which implies that nature is external (Smith, 1984). In fact, Marx's materialism at times seems contaminated by an almost Darwinian struggle for survival, where the historical role of capitalism is finally to overcome the grubby realm of material necessity which until then has ruled human life. A better critique of the assumption of domination *by* Nature appears in the feminist critique of science (Harding, 1986) and in the radical critique of biology (Levins & Lewontin, 1985; Rose and Rose, 1980). Nature in this sense is used to tell us who we are, meeting Therborn's classic short definition of ideology as the accepted answer to the questions: What exists? What is good and bad? and What is possible? (Therborn, 1980). Raymond Williams draws our attention to the ideological role of this metaphor:

> [W]hat is usually apparent [when reference is made to nature] is that it is selective, according to the speaker's general purpose. 'Nature is . . .' – What? Red in tooth and claw; a ruthlessly competitive struggle for existence; an extraordinary interlocking system of mutual advantage; a paradigm of interdependence and cooperation (1980: 70).

Urbanization as a process has constituted the city and the countryside, society and nature, a 'unity of opposites' constructed from the integrated, lived world of human social experience. At the same time, the 'urbanization of consciousness' constitutes Nature as well as Space. Urbanization is relational – a process which contains, constructs and conceptualizes both the city and the countryside. It ought not be reduced, in our theory or our practice, to the categorical 'urban'.

Disciplining nature

In his book, *Postmodern Geographies*, Ed Soja (1989) reviews the subordination of the geographical imagination, associating this with the rise of a hegemonic historicism. This followed, for the left, on Marx's discovery of transformative history and, for the right, on the gradualism and progressivism of nineteenth-century British natural philosophy. He points to the involution of Modern Geography, its retreat from the exploration of the world to the 'explanation of geography in terms of geography', to a real differentiation as catalogues of empirical moments assembled by propinquity within regions, and finally to a terminal mathematical empiricism of what he calls 'space invaders' – 'more theoretically-minded [urban] geographers [who] wandered into every disciplinary location they could find' (Soja, 1989: 16), carrying with them a mathematicized space fully abstracted from material content. It is this last phase of Modern Geography which Peet referred to as the 'new geography', and only urban geography managed this level of abstraction.

But this Modern Geography involved three deconstructions of the geographical project under the pressure of an insistent, hegemonic, increasingly scientistic modernism. The first schism deconstructed geography into human and physical dominions, as the rising power of natural science disciplines broke the original geographical bond. The second deconstruction separated social nature and social space within human geography, in line with the great differentiation of social theory between core and periphery: cultural geography was drawn by anthropology into the holistic, historical, ecological primitive, taking as its question the intimate production of the now natural landscapes of everyday (often rural) life; while urban geography took up with sociology, becoming systematic, urban, and increasingly formalized toward the interactions of roles and places within the city. Despite Hartshorne's attempt to weld these two factions with a reconstructed regional metaphor, we had instead two regional geographies, one looking to history, one to science, for its epistemological model.

At the critical moment when human geographers from both camps revolted against the limiting a- or anti-theorism of their predecessors, in concord with the renaissance of radical thought across the social sciences, both cultural and urban geographers were equally represented (Peet, 1977). And both persist today, though work in these two traditions remains separated and unequally represented in the limited academic space the discipline of geography has managed to defend.

The reasons for this unbalanced representation of Nature and Space in academic geography are complex. Some arise in the relation between geography and the other disciplines in general, some in the persistent institutions of modern intellectual life as these have been differentially achieved within human geography. I use the device of the three deconstructive schisms to suggest an outline of this problem.

The first is the separation of human and physical geographies. This was the capitulation within geography to the abstraction of Nature as external to society, nature existing without social meaning. When brought to full flower, in the early years of this century, it left geography as a divided discipline with two discrete phenomenal foci: external nature and the human world. Early attempts to heal this breach led to the excesses of environmental determinism (given the ontological priority of nature already widely accepted). Reaction against these excesses increased the intellectual distance between these two domains, leaving physical geography and human geography uncomfortably housed together, finding common ground only in the 'space' they shared.

The philosophy of naturalism, of a single overarching science, complicated this separation. Nineteenth-century philosophies of science carried forward Lyell's uniformitarianism and the methods of Cartesian reductionism into a general ontological position on the unity of science: in the end, when Nature was known and all its laws revealed, complex phenomena would be explained in terms of the essential laws of organization of their simpler components. With the increasing economic, social and institutional power of natural science, Descartes' hierarchy of knowledge seemed demonstrated; so the questions of geography would eventually be dissolved into the successful discoveries of physics. Human geographers were forced on the defensive. Few could argue, as Gregory later did, that

> [t]he integration of human and physical systems, I suggest, is not so much an epistemological problem as an ontological one. In these terms it is resolved every day that men [sic] appropriate their material universe in order to survive. The two worlds are necessarily connected by social practice, and there is nothing in this which requires them to be connected through a formal system of common properties and universal constructs. In reifying this sort of system, human geography must inevitably represent social structures and space economies as parts of the fabric of nature, which man [sic] can regulate only within limits which he transcends at his peril (1978: 75).

When, in the early decades of the Twentieth Century, even biology capitulated to reductionism, human geography was left without allies to support a position that in fact the world was complex and not simply additive: that higher-order systems (be they organism, ecosystem or human society) were organized by their own distinguishing laws and tendencies.

This set the ground for the second schism within human geography: the ontological separation of Nature and Space. I have dealt with this at length elsewhere (FitzSimmons, 1985), so will state here only that cultural geography and urban-economic geography express two different responses to the confrontation between the traditional questions of geography and the increasing formalization of science. Urban-economic geography took Space as its unique object of analysis; but it was Space devoid of nature, a 'featureless plain' about which it could theorize in increasingly mathematical terms, discovering empirical regularities in proper scientific fashion (the gravity model, the rank-size rule, etc.). Urban geographers turned to economics and sociology for the social theories that structured the spaces they described. Cultural geography addressed Nature but eschewed formal theory, working with metaphor and narrative at the uncomfortable margins of proper science, flirting with anthropology and, like anthropology, retreating from the increasing complexities of the capitalist core. This was cultural geography's own form of *historical* reductionism: the assumption that cultural geography (with anthropology) must fully understand the 'anthropological (and geographical) primitive' before it could address the more complex processes of 'our Western Commercial Culture' (Wagner & Mikesell, 1962, 23).

What was an ontological differentiation thus also became an epistemological difference. Urban economic geography took up the methods of positivist social science, becoming finally the 'new geography'; cultural geography resisted these. The third schism was thus an epistemological and sociological development, as well as an ontological one.

It seems to me that we have inadvertently carried over to this putative reconstructive period in human geography the old divisions of the second and third (and, of course, the first) deconstructions. Though the continuing distance arises out of the intellectual habits of both sides, it does not do so symmetrically. There has been a relative closure of the urban-economic school against the 'nature' question – a subtle walling off which arises out of a number of circumstances and practices.

The first barrier is the theoretical authorities to which many radical geographers turn. That intellectual life is necessarily urban is a continental European view; it is less hegemonic in England (Williams, 1973) and, until recently, was not characteristic of the United States (Worster, 1977). It is there in Marx, though rural idiocy was not his primary bête-noire. Weber's bureaucracies made cities, managed cities and lived in them. An urban bias is clear, more recently, in Althusser, Foucault, Derrida and Lacan. The association of Western Marxism with the various Latinate Communist Parties, which Perry Anderson (1976) points out, extended an urban bias (which he does not point out), since these Parties extended the anti-ruralism of the Second International to a complete disregard of farmers and the working class of the primary section.

Furthermore, too much theoretical Francophilia accentuates this division. With all due deference to Lefebvre's broader vision, most French left thought has left the question of nature to the technocrats of the Grandes Écoles (Pincetl, 1988). French left intellectual culture, with its ties to the Communist Party and the industrial unions, acquiesces with enthusiasm to the technocratic, instrumental and externalized conception of nature which complements their obsession with *la vie urbaine, la vie parisienne*, as the only civilized manifestation of *la vie quotidienne.*

The second barrier appears in the sociology of institutional and intellectual power.[3] Urban economic geography constituted itself as a positivist science in a way that those concerned with human environment relations did (perhaps could) not. Positivist urban geography constructed for itself all of the modern disciplinary structures and forms, and its critical inheritors have made use of these, which are well fitted to the structures of discipline of the academic world. This provides, for those concerned with critical urban geography, a certain disciplinary protection in a period of deconstruction of whole university departments and of a significant number of non-tenured faculty.

This has, inevitably, had some effect on the loyalties and commitments of a number of geographers who began their careers in the human-environment tradition. If the gravity model obtains anywhere, this is where we can see it. Furthermore, the proclivity of this community to condense around a limited problematic – urbanism, industrialism, flexible accumulation – and to tolerate positivism and empiricism with more enthusiasm than historical work has usually shut out of the discourse those of us who took up the question of nature.

It is not the disciplinary practices alone, but the social circumstances in which they are embedded, that is at issue. Set within the modern university – a dominating and hierarchical intellectual system – the institutions of urban (and radical urban) geography become dominating inasmuch as they are closed to a broader tradition. But at the same time they also defer to the hegemony, and the Comptean hierarchy, of nineteenth-century conceptions of 'science' (Bauman, 1978). It is safer, after all, for human geographers not to take up the question of the social ontology of nature. Otherwise, we directly confront the natural sciences, and lose access to their scientistic authority and the complicity of scientific power.

So what has happened to that initial community of critical human geographers concerned with the question of the social environment? Most of them are working in the third world, watching the production of the city and the countryside as capital reworks the geography of underdeveloped countries with not-so-silent violence. Some have been driven out of the universities, and perhaps out of formal intellectual work, by a disregard, even disdain, for the significance of this question. And some of us still struggle to contribute to the theorization of social nature and to convince others of what a thorough understanding of that issue might mean to us all.

Sources for a theory of the social production of nature

We have the basis, in theory, for an understanding of the social production of nature. We have it first in Marx, through his analysis of the labor process, in which 'Man opposes himself to Nature *as one of her own forces* . . .' (Marx, 1967: 177). In fact, Marx's geography is much better developed as an analysis of social nature than social space: 'The exploitation of the worker is simultaneously the exploitation of the soil . . .' (Marx, 1967: 506). It is in social labor that we encounter the material world, and this material moment in its peculiar geographical and historical expression is the entrypoint of a materialist philosophy. The question of the labor theory is hardly metaphysical, as Joan Robinson (1960) believed – it is only unconscious accommodation to the abstraction of Nature that makes it seem so. What Marx offers is the metaphor of *metabolism*, of the material production of the worker by the work, as well as the work by the worker.

We have the contributions of Kropotkin and Reclus on the social differentiation of the city and the countryside and the implications of a division of labor which fits the worker to a single task – an urban task. We might be further with this theoretical work if the Communist Parties had not followed Lenin's sense of the revolutionary proletariat as the manufacturing working class. Imagine the world if we had had the IWW instead.

More recently, within geography, we have a number of good starts. First, consider this passage:

> . . . a 'thing' cannot be understood or even talked about independently of the relations it has with other things. For example, [NATURE] can be defined only in relationship to the *mode of production* which seeks to make use of [IT] and which simultaneously 'produces' [IT] through both the physical and mental activity of the users. There is, therefore, no such thing as [NATURE] in abstract or [NATURE] which exists as a 'thing in itself' (Harvey, 1980: 212 – I have substituted the word *nature* for the word *resources*).

We have other maps to this new terrain. One, which reviews the initial ontological rupture, is Kenneth Olwig's book *Nature's Ideological Landscape* (1984). One may not think to read this, because its historical focus is the preservation of the Danish heathlands. But in this text is a compelling argument for the abstraction of nature as a concomitant of urbanization, in the political and ideological reworking of everyday life toward the urbanization of consciousness which has disguised and excused the development of capitalism.

Another map, overlapping Olwig's in some sections but delving further into the general construction of geography as nature *and* space, is Neil Smith's book, *Uneven Development* (1984). Smith's map has some sketchy moments, but he travels farther than anyone else toward an integrated general theory of how capitalism makes geography.

We have a large collection of critical work on the production of nature, and of the antinomy of city and countryside, in the Third World. Some examples are the works of Michael Watts (1983),[4] Michael Redclift (1984, 1987), Susanna Hecht (1985), Cindi Katz (1988), Kirsten Johnson (1977) and Ben Wisner (1978). We also have a new, critical attention to technological hazards within geography, drawing together people such as Ken Hewitt, Andrew Kirby and Sallie Marston, who are concerned with the social construction of risk and who are extending the radical reanalysis of 'natural' hazards (Hewitt, 1983) to the social hazards of technology.

We have a growing body of work within and without geography on the production of nature in the First World. In my own work on agriculture, I have been aided substantially by the thinking of the new generation of rural sociologists and agricultural economists, who have taken Marxian political economy into the primary sector with significant intellectual results (see, for example, Buttel and Newby, 1980; Friedland, Barton & Thomas, 1981; Mann & Dickenson, 1978, 1980, 1987; Mooney, 1983). After all, agriculture is an industry, too (FitzSimmons, 1986; Pudup, 1987). We have the whole critical literature on environmental politics and planning, including Buttel's contribution to this area (Buttel, 1975) and the early work of Walker and Storper (1978). We have also the new social ecology, which includes James O'Connor (1987) and Rudolf Bahro (1982), among many others.

And finally, we are not without allies in our confrontation with science, allies within science itself (Young, 1975). There is a growing radical critique within biology, whose participants include Hilary and Stephen Rose (1980), Richard Levins and Richard Lewontin (1985) and Stephen Jay Gould (1987), and in physics, with the rise of the new realism that has intrigued Bhaskar (1986) and Harré (1986). This is not to argue for a simplistic naturalism, but to suggest that the radical critique of science in natural science has gone much further than many in the social sciences know, and that if we accommodate the question of the social relation with nature – the natural basis of social life – within geography, we will not do so in intellectual isolation.

Why this is to be done

This development of an effective theoretical understanding of Nature as social nature is urgent. It is a necessary contribution that we, as left geographers, can make to a whole complex of current struggles throughout the world. Some of these struggles are ideological, struggles over the meaning of Nature and the restrictions and possibilities open to human beings and human societies (Harvey, 1974, offered an example of this in his reanalysis of the population-resources question). Many are immediately political, as geographically-defined communities in the home, workplace, community and region struggle to control the way in which capitalism is poisoning the material environment in which they are embedded (FitzSimmons & Gottlieb, 1988), or as Third-World groups such as the Brazilian rubber-tappers fight to contain and redress the ecological destruction which threatens their worlds (Beck, 1989). If we do not engage in these struggles, we abandon them to those who use Nature to justify not only the domination of nature by humans, but also the domination of humankind itself. Further, if we do not critically engage the Nature that capitalism constructs, which is as much a social product of capitalist production, organization and technology as is class exploitation, we may accept forms of socialism which simply repeat this Nature under alternative forms of social domination.

This agenda is also intellectually necessary. Without theoretical consideration of the social construction of Nature, in its geographical and intellectual manifestations, we restrict ourselves to a partial view of the real geography of capitalism. Localities do not end at city boundaries, and the south of England was not a featureless plain before the reconstruction of British industrial capitalism in this decade. The new economic geographies of capitalism may be urban in form, but they are written over and by means of a complex world which includes city and countryside, Space and Nature. Silicon Valley arose as much out of the surplus labor of the Mexican-American agricultural workers of San Jose as it did out of the technological ingenuities of Stanford professors (Saxenian, 1985).

What Soja has done for space, drawing our attention to the trenchant but constructed and integrally embedded spatiality of everyday life, can be done for nature as well. If we reconstruct human geography from both sides, we will all gain a more symmetrical understanding of the geography of capitalism. If you look closely you will find that those of us concerned with the social production of nature are reading and criticising, but also using and extending, the work of radical geographers in the urban tradition. We are working toward a new integrated geography of Nature and Space. Let us join together.

> We have mixed our labour with the earth, our forces with its forces too deeply to be able to draw back and separate either out. Except that if we mentally draw back, if we go on with the singular abstractions, *we are spared the efforts of looking*, in any active way, at the whole complex of social and natural relationships which is at once our product and our activity. (Williams, 1980: 83)

If what we seek is not just to understand the world, but to change it, we must address capitalism in its hidden moments, its reproduction of disguising abstractions such as 'nature' and 'space'. Urban (*and rural*) society (*and nature*) as we know them are capitalist forms; the domination of one by the other echoes and continues the structural domination of capitalism. We cannot avoid this, even though we may continue to choose to deny it. If we work together, however, we may participate in its transformation.

Notes

1 I use the capitalised form of Nature to differentiate explicit reference to Nature as a concrete abstraction from a more general use of the term. Space is capitalised in parallel fashion. With Harvey, I would argue, 'the abstractions are not of my making. They are embodied in a social process that creates abstract forces that have concrete and personal effects in daily life' (Harvey, 1985: 1). The genesis and effect of this abstraction on the discipline of geography are the focus of this argument.

2 Smith (1984) verifies Foucault's examination of these changes in the French Enlightenment in his discovery of the parallel arguments of the major early figures of modern British natural science – Roger Bacon, Isaac Newton, even Charles Darwin. Smith also draws our attention to the differentiation of two modern concepts of Nature, Nature as external and as universal. However, in his critical review of Schmidt's (1971) analysis of Marx's concept of Nature, Smith seems to fall into the trap by which Schmidt (and even perhaps Marx) have also been captured – that is, he treats 'first Nature', primordial Nature, as historically and analytically prior to 'second Nature', social Nature. I argue that 'first Nature' is itself a social construction, an abstraction constructed under the particular social conditions of early urban capitalism.

3 Bordieu (1988) and Latour (1987) offer fascinating examples of the insights which can be developed from examining intellectual life, or science and technological change, from an anthropological method. Dickson (1984) reviews the changing political economy of science. Unfortunately, none of these studies pays particular attention to geography, which only Bordieu mentions, as the discipline at the bottom of the French intellectual hierarchy.

4 References are to major or characteristic works.

References

Anderson, P. (1976) *Considerations on Western Marxism.* London: New Left Books.

Bahro, R. (1982) *Socialism and Survival.* London: Heretic Books.

Bauman, Z. (1978) *Hermeneutics and Social Science.* London: Hutchinson.

Beck, M. (1989) Chronicle of a death foretold: Murder and ecological tragedy in the Amazon. *Newsweek* (January 9, 1989): 62.

Bhaskar, R. (1986) *Scientific Realism and Human Emancipation.* London: Verso.

Bodieu, P. (1988) *Homo Academicus.* Stanford, CA: Stanford University Press.

Buttel, F. (1975) The environmental movement: Consensus, conflict and change. *Journal of Environmental Education* 7: 53–56.

Buttel, F. and Newby, H. (Eds.) (1980) *The Rural Sociology of the Advanced Societies: Critical Perspectives.* Montclair, N.J.: Allanheld Osmun.

Dickson, D. (1984) *The New Politics of Science.* New York: Pantheon.

FitzSimmons, M. (1985) Hidden philosophies: How geographical thought has been limited by its theorctical models. *GeoForum* 16(2): 139–149.

FitzSimmons, M. (1986) The new industrial agriculture: *Economic Geography*, 62: 334–353.

FitzSimmons, M. and R. Gottlieb (1988) A new environmental politics. In M. Davis and M. Sprinker (Eds.) *Reshaping the U.S. Left: Popular struggles in the 1980s.* Volume 3 of 'The Year Left'. London: Verso, pp. 114–130.

Foucault, M. (1970) *The Order of Things: An Archaeology of the Human Sciences.* New York: Vintage Books.

Freidland, W., A. Barton and R. Thomas (1981) *Manufacturing Green Gold.* Cambridge: Cambridge University Press.

Gould, S. J. (1987) *An Urchin in the Storm.* New York: W. W. Norton.

Gregory, D. (1978) *Ideology, Science and Human Geography.* New York: St. Martins.

Harding, S. (1986) *The Science Question in Feminism.* Ithaca, N.Y.: Cornell University Press.

Harré, R. (1986) *Varieties of Realism.* Oxford: Basil Blackwell.

Harvey, D. (1974) Population, resources, and the ideology of science. *Economic Geography* 50: 256–277.

Harvey, D. (1980) *Limits to Capital.* Chicago: University of Chicago Press.

Harvey, D. (1985) *Consciousness and the Urban Experience.* Oxford: Basil Blackwell.

Hecht, S. (1985) Environment, development and politics: Capital accumulation and the livestock sector in eastern Amazonia. *World Development* 13(6): 663–684.

Hewitt, K. (Ed.) (1983) *Interpretations of Calamity.* Boston: Allen & Unwin.

Katz, C. (1988) Children's environmental learning, knowledge and interactions under conditions of socio-economic transformation: The possibilities of change. Paper presented at the Waren Nystrom Award Special Session, 85th annual meeting of the Association of American Geographers, Phoenix, Arizona.

Johnson, K. (1977) *Do As the Land Bids.* Unpublished Ph.D. dissertation, Clark University.

Latour, B. (1987) *Science in Action.* Cambridge: Harvard University Press.

Levins, R. and R. Lewontin (1985) *The Dialectical Biologist.* Cambridge, Mass: Harvard University Press.

Lewontin, R., S. Rose and L. Kamin (1984) *Not in our genes.* New York: Pantheon.

Mann, S. and J. Dickenson (1978) Obstacles to the development of a capitalist agriculture. *Journal of Peasant Studies* 5: 466–481.

Mann, S. and J. Dickenson (1980) State and agriculture in two eras of American capitalism. In Buttel, F. & Newby, H. (Eds.) *The Rural Sociology of the Advanced Societies: Critical Perspectives.* Montclair, N.J.: Allanheld Osmun, pp. 283–325.

Mann, S. and J. Dickenson (1987) One furrow forward, two furrows back: A Marx–Weber synthesis for rural sociology? *Rural Sociology*, 52(2): 264–285.

Marx, K. (1967) *Capital.* Volume I. New York: International Publishers.

Mooney, P. (1983) Toward a class analysis of midwestern agriculture. *Rural Sociology*, 48(4): 563–584.

Ollman, B. (1971) *Alienation: Marx's Conception of Man in Capitalist Society.* Cambridge: Cambridge University Press.

Olwig, K. (1984) *Nature's Ideological Landscape.* London: George Allen & Unwin.

O'Connor, J. (1987) *The Meaning of Crisis.* Oxford: Blackwell.

Peet, R. (1977) 'The development of radical geography in the United States.' In R. Peet, (Ed.) *Radical Geography: Alternative Viewpoints on Contemporary Social Issues.* Chicago: Maaroufa Press, pp. 6–30.

Pincetl, S. (1988) 'Cooptation or disinterest: Environmental movements and politics in France'. Paper presented at the annual meetings of the Association of American Geographers, Phoenix, Arizona, April 1988.

Pudup, M. B. (1987) From farm to factory: Structuring and location of the U.S. farm machinery industry. *Economic Geography* 63: 203–222.

Redclift, M. (1984) *Development and the Environmental Crisis: Red or Green Alternatives?* London: Methuen.

Redclift, M. (1987) *Sustainable Development.* London: Methuen.

Robinson, J. (1960) *An Essay on Marxian Economics.* New York: St. Martin's Press.

Rose, H. and S. Rose (Eds.) (1980) *Ideology of/in the natural sciences.* Boston: G. K. Hall.

Saxenian, A. L. (1985) Let them eat chips. *Environment and Planning D: Society and Space* 3: 121–127.

Schmidt, A. (1971) *The Concept of Nature in Marx.* London: New Left Books.

Smith, N. (1984) *Uneven Development: Nature, Capital and the Production of Space.* Oxford: Basil Blackwell.

Soja, E. (1989) *Postmodern Geographies: The Reassertion of Space in Critical Social Theory.* London: New Left Books.

Therborn, G. (1980) *The Ideology of Power and the Power of Ideology.* London: Verso.

Wagner, P. and M. Mikesell (1962) The themes of cultural geography. In P. Wagner and M. Mikesell (Eds.) *Readings in Cultural Geography*. Chicago: University of Chicago, pp. 1–24.

Walker, R. (1979) Editor's Introduction. Special issue on Natural Resources and Environment. *Antipode* 11(2): 1–16.

Walker, R. and M. Storper (1978) Erosion of the clean air act of 1970: a study in the failure of government regulation and planning. *Boston College Environmental Law Review* 7(2): 189–258.

Watts, M. (1983) *Silent Violence: Food, Famine and Peasantry in Northern Nigeria.* Berkeley: University of California Press.

Williams, R. (1973) *The Country and the City.* New York: Oxford University Press.

Williams, R. (1980) Ideas of nature. In *Problems of Materialism and Culture.* London: Verso, pp. 67–85.

Wisner, B. (1978) Does radical geography lack an approach to environmental relations? *Antipode* 10(1): 84–95.

Worster, D. (1977) *Nature's Economy.* San Francisco: Sierra Club Books.

Young, B. (1975) Science is social relations. *Radical Science Journal* 5: 65–129.

12 Cathy Nesmith and Sarah A. Radcliffe

(Re)mapping Mother Earth: A Geographical Perspective on Environmental Feminisms

Excerpts from: *Environment and Planning D: Society and Space* 11, 379–94 (1993)

Introduction

Given the centrality of the concepts of nature and environment to geography, the theoretical aspects of the connection between women, gender, and nature, particularly in relation to landscape, have been of some interest to geographers (Ford 1991; Kolodny, 1975; 1984; Monk, 1984; Norwood and Monk, 1987; Porteous, 1986; Rose, 1991; Schaffer, 1988). Overall, however, the woman–nature connection has merited little more than passing notice in the discipline (for example, see Bowlby et al, 1989; Smith, 1984, pages 13–14). Indeed, the theorisation by radical geographers of nature itself has lagged behind theorisation of space (Fitzsimmons, 1989; Wisner, 1978). This situation is most likely linked to the privileged position held by space as the theoretical focal point of geography in recent years. As environmental conditions now approach crisis proportions in many regions, the time seems right for a new emphasis on nature by geographers. As we will argue, such an orientation necessitates a sustained dialogue with existing environmental feminist arguments. Here, the discussions of environmental feminisms are used to articulate a geographical approach to issues around gender and environment, while at the same time utilising recent advances in geographical feminisms and sociocultural geography to overcome absences within previous feminist writings on the environment.

The emerging theory or discourse around women, nature, and the environment can be identified as environmental or ecological feminisms (also termed 'ecofeminism' and 'feminist ecology'). Often comprising an innovative mixture of academic and nonacademic forms (for example, poetry, fiction, journalism), environmental feminisms are varied, extending to Third World development issues (Shiva, 1988), women's spirituality (termed spiritual environmental feminisms in this paper) (Spretnak, 1989; Starhawk, 1989), contemporary environmental concerns (Biehl, 1991; Peterson and Merchant, 1986), and philosophical debates (Plumwood, 1988; Salleh, 1984; Warren, 1987; 1988; 1990; Zimmerman, 1987). In attempting a geographical approach to environmental issues, we suggest a critical engagement with these current feminist theories of environment, as these approaches provide a broad feminist vision of ecological work. In the environmental feminist approaches, the relations between men and women, nature and culture, industrialised and nonindustrialised societies are rearticulated in theory and practice. A critique of these environmental feminisms creates a space for the generation of new theories and practices, informed by recent feminist and cultural geographical writing.

Ecological perspectives: the need for a feminist vision

The fundamentals of environmental feminisms need to be outlined in order to provide background to new, gender-aware approaches to environments. In developing a perspective on the putative relationships between men and 'culture' on the one hand, and women and 'nature' on the other, environmental feminisms have argued for a fundamental reconceptualisation of these categories. Environmental feminist positions have brought four specific points to the foreground of environmental analysis: first, there are important connections between the oppression and domination of women and the exploitation of nature; second, an understanding of the nature of these connections is necessary for any adequate theorisation of the oppression of women and the exploitation of nature; third, because of this connected and concurrent domination, women have a particular interest in ending the domination of nature; and, last, this connected domination entails a linking of environmentalism and feminism.

Writing with a gender-aware approach, Zimmerman notes that, although deep ecology or radical environmentalism may on the surface appear to be 'in almost complete agreement with the feminist view that abstract, dualistic atomistic, and hierarchical categories are responsible for the domination of nature, on another level the deep ecologist critics of anthropocentrism do not address their own androcentrism' (1987, page 37; also, see Salleh, 1984). According to environmental feminists, many advocates of deep ecology continue to work within male-dominant social relations. For example, feminists point to the deep ecologists' use of the generic term 'man' as a cause of environmental crises, a lack of recognition of connections between oppression of women and nature, and a prescription of birth-control regimes without regard for the desires and needs of women who would be affected by such policies (Salleh, 1984). Identification of the discrepancies between deep ecology and feminist perspectives highlights how feminism can define a new perspective. As Warren puts it, 'without the addition of the word *feminist*, one presents environmental ethics as if it has no bias, including male-gender bias, which is just what [ecological] feminists deny: failure to notice the connections between the twin oppressions of women and nature *is* male-gender bias' (1990, page 144, emphasis in original). The development of feminisms of the environment has been concerned largely with addressing the conceptual issues of dualism, hierarchy, domination, which underlie Western notions of nature, environment, and gender difference (see Warren, 1991). It is these themes which unify what in practice is often a diversity of philosophies and theoretical grounding in feminist writings on the environment.

Environmental feminisms are thus not only critical of ecological perspectives, but also of previous Western feminist theories, arguing that they in turn are ecologically blind. After assessing the utility of the four dominant branches of Western feminist thought (liberal, traditional Marxist, radical, and socialist)[1] from an environmental feminist viewpoint, Warren (1987) argues that none provides a sufficient basis for the theoretical development of ecological feminism. The alternative of a 'transformative feminism', Warren asserts, offers a synthesis of feminist and ecological approaches, bound by a politics of 'naturism', a politics against the domination of nature, and also against the domination of other subalterns (those oppressed by gender, race, class, etc), as 'an integral part of any feminist solidarity' (1990, page 132). Warren (1987; 1990) concludes with a call for rethinking what it means to be 'human', a point which will be taken up later.

Arguing from the perspective of a postcolonial Third World position, Shiva (1988) enlarges the fields of metaphors with which to challenge Western notions of male–female, nature–culture. Drawing from Indian (specifically Hindu)[2] cosmology as an alternative, symbolic system of understanding gender and nature, she argues against the dualism dominant in Western philosophy. According to Shiva, the Hindu notions of person–nature are encapsulated through the concepts of masculine *purusha* and feminine *prakriti* principles. These are not opposing forces as in Western thought – they are a 'duality in unity' (page 40). She advocates a nondichotomous understanding of masculine and feminine, to further break down hierarchical categorisations:

> In this non-gender based philosophy the feminine principle is not exclusively embodied in women, but is the principle of activity and creativity in nature, women and men. One cannot really distinguish the masculine from the feminine, person from nature, *Purusha* from *Prakriti*. Though distinct, they remain inseparable in dialectical unity, as two aspects of one being (page 52).

However, the notion of 'duality in unity' advocated by Shiva as an alternative appears to replicate many characteristics of Western gender and nature categories, in which maleness cannot be understood except as that which is not female, as the nonfemale: the two are thus intimately and inextricably bound up in one another, resembling a unit with two interconnected parts.

In relation to the interests of geographers, feminist writing on the environment brings into view three sets of ideas, related respectively to environment–human relations, to development and environment, and to gendered landscapes. First, the notion that human–environment relations are gendered is, according to environmental feminisms, explicitly constructed as such by various writers in the Western tradition . . . Second, environmental feminisms argue that industrialising, First World 'maldevelopment' is both an ecological and a gendered process . . . , and third, the ecological feminisms offer new perspectives on issues of identities and landscapes . . . .

As environmental feminisms argue overall, an understanding of a gendered hierarchical and environmentally exploitative society necessitates a blend of feminisms and environmentalisms (Agarwal, 1992; Warren, 1987). It is proposed, then, that environmental feminisms articulate specific philosophies, politics, and identities that may contribute to the development of a new critical geography of the environment. In the sections following, the strengths and weaknesses of environmental feminist positions are considered in light of recent geographical work, which can offer solutions to the problems raised by essentialist and Western-centric environmental feminisms.

Gender and nature: the gendering of human–nature relations

Feminist rewriting of historical and contemporary discourses around notions of 'natural' femininity and masculine mastery over nature has been developing in recent years (MacCormack and Strathern, 1980). As a response to persistent patriarchal notions of the 'female', environmental feminist writings have uncovered the gender notions around the dichotomy of nature versus culture. Feminist writings on the environment are thus of great interest to contemporary debates in geography, precisely because the underlying notions of geographical work on the environment are related to what are historically gendered concepts. Environmental feminisms challenge such notions as the patriarchal (a)symmetry set up between women and nature, which can be traced back, at least to Bacon's rape metaphor for the scientific method (Harding, 1986; Merchant, 1980). Environmental feminisms highlight the close ties between nature and gender in Western thought, and have developed a critique of the same through the deployment of Ortner's work. Ortner's (1974) assertion that women in all societies, because of their biology, are *thought* to be closer to nature, made a first step in the deconstruction of these categories. Nature, Ortner suggested, is universally devalued compared with culture, the creation of men. As environmental feminists point out, the equation of femaleness or womanhood with nature has historically been used to oppress women. Women's childbearing ability and menstrual cycle were used to keep them out of schools in the 19th century (Sayers, 1982), and, more recently, to belittle their intelligence and deny them opportunities in the work place (on China, see Wolf, 1987; on the West, see Sayers, 1982). It has been a short step, for some, to suggest that, clearly, women's vocation is in the home, bearing and rearing children.

Not surprisingly, feminisms strove historically to shatter the patriarchal ideological link between women and nature, particularly nature in the form of female biology (Firestone, 1971; Wollstonecraft, 1792). Anthropological work also demonstrated that the link was by no means universal over all cultures (MacCormack and Strathern, 1980). It is from this recognition and representation of a common oppression of women and nature that present environmental feminist perspectives have developed.

Environmental feminisms have also pointed to the historicity of the supposedly fixed relations between gender, nature, and culture. From within a Western tradition, environmental feminisms have attempted to trace, and in turn question, why oppositions around nature and gender are so fundamental to our understanding of the environment. The history of Western notions of nature, gender, and civilisation provides evidence both of the mutability of the political activism around humanity's relation to 'nature' and of its grounding in particular spaces. For example, throughout history, social and political activists have employed particular notions of nature, culture, and humanity in moral debates concerning social change and political priorities. As various feminist versions have suggested, Western hegemonic ideologies of nature and culture, often rooted in Enlightenment thought (Bloch and Bloch, 1980), are predicated upon notions of tension and struggle, whereby culture imposes itself and dominates the force of nature: man separates himself from nature by subordinating and conquering it, including 'woman'. More recently these ideas have been rearticulated by environmental feminists, as discussed above, who, in opposition to Enlightenment social models, are attempting to formulate a feminist version of nature–culture–gender relations, which does not necessarily equate woman and nature.

However, some spiritual environmental feminists have not rejected the woman–nature connection. Rather, the common heritage of oppression by patriarchy experienced by women and nature is taken as a common bond (Merchant, 1980), and others, in an inversion of historic gendered notions of nature, choose to celebrate the perceived unique connection between women, as life-giving beings, and the 'natural' environment (Daly, 1979; Griffin, 1978). Through claims to a special sustainable relationship between women, the earth, and its life forces, or a natural alliance between feminism and ecology (for example, King, 1989), certain ecological feminists articulate an alternative conception of nature–culture relations, and male–female interaction. As with some Enlightenment writers, these spiritual ecological feminists tend to speak from what is constructed as a privileged position of those (women) who are assumed to be closer to 'nature'. Whereas Enlightenment thinkers minimally addressed the value-laden issue of gender difference in relation to 'nature', spiritual environmental feminisms challenge the supremacy of the cultural and the male by making a virtue of women's putative closeness to the (disdained) 'natural'. Such feminist affirmations of a 'female' is at times realised through the reversal of values associated with nature, such that 'closeness to nature, previously used to put women down, [is] . . . recast as a virtue' (Plumwood, 1989, page 20). In a similar vein, Reynolds (1989) suggests that spiritual ecological feminists distinguish between 'feminine' and 'masculine' approaches to the environment, whereby the feminine views the world as a home, and the masculine as a resource and raw material to be made into something else. This gives rise to ecological political practice in which treatment of the world as a 'home' is more sustainable and ecologically sensitive.

However, from a geographical perspective, such essentialisms are problematic. Recent geographical work has highlighted the problematic around issues of attributing or claiming essentialised relationships (for instance, Bondi and Domosh, 1992). From this cultural geographical perspective, feminist thought on the environment is closely connected, or should be, to the question of essentialism (Fuss, 1989). Essentialism in the context of environmental feminism translates into an irreducible essence, an innate quality in women that connects them to nature, which is not present in or available to men. A hazard of ecological feminist approaches, then, is that they can be reduced to essentialising women's characteristics to explain or substantiate women's connection to the environment. By making the (strategic) 'risk of essence' (Fuss, 1989), feminisms have risked reinscribing the symmetry between women and nature which has so harmed women in the past. An essentialising tendency is widespread across the spiritual ecological feminist literature, although expressed with varying degrees of subtlety and in some cases rejected entirely. Fundamental to the essentialising themes in these feminisms is the notion of women as 'life givers', as capable of giving birth. For example, the following quote is an example of a privileged and recurring image in environmental feminist writings:

> Women's monthly fertility cycle, the tiring symbiosis of pregnancy, the wrench of childbirth and the pleasure of suckling an infant, these things already ground women's consciousness in the knowledge of being coterminous with Nature (Salleh, 1984, page 340).

Female images articulated by environmental feminisms include discourses around women's essential ability to empathise or connect with 'nature'. Such effects are found in environmental feminisms as superficially varied as Shiva's and the British campaign group, Women's Environmental Network (WEN). The underlying philosophy of the WEN, an active lobbying and information group in Britain, is that women are linked to nature in a way men cannot possibly be (Vallely, WEN director, personal communication, 1990). Shiva (1988) too adopts similar positions. The overall tone of Shiva's book is essentialising – the feminine principle is nature sensitive, and women have it. Although she says that men have it too, or have the potential to develop it, the book is not written from that ontological viewpoint, so she does not follow through on her nongender perspective. Similarly, although attempting to form an 'antidualistic' ecological politics and feminism, King (1989) in the end falls back into the trap of seeing an essentialised connection between women and nature which will provide an alternative to the masculine war-machine currently in operation.

These images, and similar essentialisms, do much 'ideological work' (Poovey, 1990) in environmental feminism. First, they skim over very real differences between women, thereby calling into being a 'given' political constituency and 'common' agenda. Second, and in related vein, they assume an essential meaning around bodies and their biological capacities, what they do, and how they differ from other bodies. Those associations or expectations and meanings around such bodily groundings have built up over years from within Western patriarchal thought. They have thereby excluded women of colour and Third World women, whose interpellation of Western discourses is highly variable. Questions which have not been addressed by ecological feminists include questions raised increasingly in cultural and feminist geography regarding issues of 'race' and difference between women. By discounting the diverse lived realities of women in the South and North, do ecofeminisms not reinscribe racisms and lack of knowledge about non-Western women found in other forms of feminism?[3]

An argument against this essentialising tendency and the need to classify the world in masculine and feminine terms is presented by Plumwood (1988; also, Warren, 1987; 1990). Rethinking what it means to be 'human', Plumwood (1988, page 19) points out that the Western notion of human is not, 'as it pretends to be, gender neutral, but instead coincides or converges with that of masculine character, while the ideals of womanhood diverge'. In order to overcome such a masculinised subject [one recognisable from other feminist work in related fields (for example, Fraser, 1989; Pateman, 1988)], Plumwood calls for a reexamination and redefinition of humanity in which the masculinising trend (that of science, technology, and superiority over nature) and the feminising trend (the 'genuinely feminine' of difference theorists; simple inversions of virtue by some ecological feminists) are both rejected. She argues that to free the concept of masculinity which underlies our notions or knowledges of humanity does not necessarily entail the wholesale adoption of a 'rival feminine ideal' (1988, page 22). Rather she criticises both masculinity and femininity for their links with the 'traditionally associated dualisms of mind/body, rationality/emotionality, public/private, and so on, which [must] also [be] rejected as false choices' (pages 22–23). As a formal alternative, she suggests that what is required is a 'transcendence' of the traditional dichotomous gender categories and the associated set of dualisms which underpin our notions of humanity (page 23; see also, Plumwood, 1989). She calls for a 'degendering' of humanity's characterisation, in the sense that human characteristics would not be connected with one gender over the other, but on the basis of independent considerations (1988, page 23). Although suggesting that this does not necessarily entail an androgynous humanity or a lack of sexual difference, Plumwood argues that this notion could lead to a reassessment of 'the characteristics that we as humans share with the non-human world' (1988, page 24). For Plumwood, then, it is not necessary to use essentialised gendered constructions to develop a sustainable relationship between humans and nature.

In summary, an opposition between the spheres of 'nature' and of culture is not essential to critical geographical perspectives on the environment. Rather than retrenching and deepening the gulf between masculine and feminine, 'nature' and 'culture', in simplistic dichotomous systems, other environmental feminisms argue that an alternative approach to the issues of gender, environment, politics and morals can be developed. Such an approach is fruitful for developing geographical insights on the environment.

The global politics of environment and development

In common with recent geographical feminist writings (for example, Momsen and Townsend, 1987; Nesmith, 1991; Radcliffe and Westwood, 1993), environmental feminisms address global and local issues related to development and environment. Engaging unique political practices and political identities, Western environmental feminisms are concerned with global issues such as nuclear arms and loss of tropical rainforests, and local concerns such as community health, and air and water quality (see Biehl, 1991; Diamond and Orenstein, 1990; Peterson and Merchant, 1986; Plant, 1989; WEN, various dates). Starting from a Western, domestic work focus (see Hurtado, 1989), they have taken action on issues as varied as restricting the use of tampons because of the threat to women's health, consciousness-raising among women about the waste of food packaging, and campaigning for alternatives to disposable diapers (nappies). These political actions and concerns have not yet been analysed in geographical writings as a politics or as a set of political identities which are located historically and spatially.

Of particular interest to geographers is the fact that environmental feminisms have made innovative links between the previously separate issues of the body and environmental politics in the processes of First World development. Activist environmental feminist politics is termed by Peterson and Merchant (1986) the 'politics of reproduction', as it examines how women's involvement in environmental issues has been guided by their gender role. They suggest that women ground their environmental political agendas in those environmental issues which relate to reproduction and women's activities in daily life. However, according to Peterson and Merchant, the ideological bases for women's involvement in a politics which challenges their socially constructed relation to their bodies and reproduction remain wide; they range across the spectrum from 'acceptance of women's role as nurturant social mother . . . to strategies designed to change the distribution of patriarchal and capitalist power' (1986, page 472). Concerns uniting issues of bodies, global development, and the environment are identified: 'biological (intergenerational) reproduction', which contributes to concern for ecological and nuclear issues; 'family (intragenerational) reproduction' which leads to a concern for health and 'local' environmental quality; 'the welfare state', where female wage workers in health care, social work, and education lobby for improved working conditions and standards; and 'the liberation of women as reproduction workers', which refers to women's equality with men and to changes in environmental professions and sciences.

In these ways, environmental feminisms attempt a fundamental rethinking and reexpression of the relation between women's bodies, broad narratives of 'development', and the gender-hierarchical categories of the Enlightenment. This brings into focus the importance of an analysis of bodies and their relation to the environment. Bodies, despite the broad distinctions of gender, race, and age, are unique in terms of physical ability, colours, form, and cultural imprints (scars, bound feet, shaved legs, etc), while also holding different relations with one another and with normative notions of the natural and cultural (MacCormack and Strathern, 1980). Grounded in physical bodies, people's social relations with environments are mediated through these same bodies, a fact which ecological feminisms consistently draw attention to. In their formulation, all organic bodies are perceived as 'opposing to the technological' (Haraway, 1991, page 174). Developing their oppositional imaginations (Cocks, 1989), environmental feminisms articulate a heightened awareness of beings delicately reproduced through the genetic material of DNA. As Haraway suggests (1991), it is perhaps only in the late 20th century, with its fixations on industrialising development models and the ubiquity of machinery, that such 'organicist' discourses could arise. However, by attributing agency to (Mother) Earth, environmental feminism 'makes room for some unsettling possibilities' (page 199). No longer constrained by human agency over products, environmental feminisms suggest that alternative politics and personal relations exist in an oppositional, organicised, and gendered sphere.

Writing from a distinctly postcolonial perspective, the environmental feminist Shiva articulates a distinct set of issues, also of interest to geography. Shiva's book *Staying Alive: Women, Ecology and Survival in India* (1988), foregrounds domination and imperialism as central issues in women's environmental politics. Her book has been influential in the development of environmental feminist ideas because of its sustained critique of the global and gendered forces shaping environments and the politics around them. Central to this is the concept of 'maldevelopment', defined as the ecological breakdown and social inequality inherent in the current dominant development paradigm, as advocated and practiced by the industrialising North. With the devaluation of nature and its destruction through maldevelopment, argues Shiva, women who are closely linked to nature through their sustaining, life-giving work (providing food, fuel, and water for their families) are also being destroyed. From this, she argues, 'the women's and ecology movements are therefore one, and are primarily counter-trends to a patriarchal maldevelopment' (page 47). Environmental feminist writing, such as Shiva's, illustrates a desire to give voice to silent groups. Feminists' interests in the environment can lead to an appreciation of the perspective of aboriginal people, whose cultural attitudes to the land tend to have been nonexploitative and partnership-oriented (Means, 1980). As feminism has been accused of being exclusionary and generally derived from a white middle-class perspective, this incorporation of aboriginal concerns can enhance the broadening of feminism which is already taking place (hooks, 1984; Spelman, 1988). At a conference on Women and the Environment, held in Vancouver, Canada, in 1991, for example, an overwhelming number of the speakers were women from aboriginal groups from the Americas. (Unfortunately, the balance of speakers was such that other views on the relationship between women and the environment were seriously underrepresented or unheard.)

Shiva has argued that Third World women retain a sustainable progressive politics towards environments, that 'third world women, whose minds have not yet been dispossessed or colonized, are in a privileged position to make visible the invisible oppositional categories that they are the custodians of' (page 46). Tribals and peasants 'who have been left out of the processes of maldevelopment' are accorded a similar privileged status (page 46). These tropes raise a critique of environmental feminisms' tendency to romanticise women's and aboriginal people's relationship to the environment. Geographers have become increasingly sensitive to the mechanisms of creating 'imaginative geographies' (Said, 1978), attributing specific characteristics to people in a particular location (see Slater, 1992). Geography thus provides the tools for a critique of the romanticisation of women's relation to the environment. This is especially relevant in light of the use made of the Indian Chipko movement (in which rural women hugged trees to save them from commercial felling) in ecological feminist writings. For example, Warren (1988) cites the Chipko movement as a 'reality' of a practically based environmental feminist ethic. Women in Chipko are represented as ideologically committed to a feminist ecology, rather than committed to sustaining their households and communities which rely directly on the surrounding environment (see Agarwal, 1992; Ahmed, 1985; Guha, 1989; Jain, 1984). In this way, environmental feminists risk making the move, recognised from other strands of feminism (Mohanty, 1991), of appropriating other women's struggles as part of Western Feminism, as a monolithic and unproblematic category. In other words, environmental feminisms have not addressed issues of difference between women. For example, Warren's analysis of the gendering of 'race' and class relations is formulaic, not questioning the particularities of differences between women:

> Eco[logical] feminists insist that the sort of logic of domination used to justify the domination of humans by gender, racial or ethnic, or class status is also used to justify the domination of nature (Warren, 1990, page 132).

The Chipko example, so often cited in ecofeminist texts, introduces questions about the political role played by this example in environmental feminisms. In these texts, are not Indian women in Chipko playing the role for ecological feminists of suffering Third World women in struggle? Such constructions are recognisable as imperialist–colonialist constructions of Third World women evident in Western feminist texts (Mohanty, 1991).

Tendencies towards romanticisation in environmental feminisms are diverse, not uniform. Shiva's work demonstrates this romanticising characteristic in a different light, when she presents rural Indian women as being spiritually attuned to nature and consciously seeing themselves as linked to nature. For example, she suggests that '*prakriti*, far from being an esoteric abstraction, is an everyday concept which organises daily life' (1988, page 40). In the terms that Shiva writes, this is an authorising claim, unsubstantiated with oral confirmation from the women concerned. It is an advance in feminist work to start from Third World women's viewpoints. However, such an analysis must also take into account the fact that women have sometimes been marginalised by non-Western societies' distribution of resources, and that women have often not been awarded the status and/or attributes of the fully 'human', however that has been defined in each local culture (MacCormack and Strathern, 1980). As it stands, the romanticisation move occurs when Shiva suggests that rural women, tribals, and peasants are *necessarily* living in a sustainable or ecologically conscious way, rather than that their management of the environment is guided by their needs for survival (for a general critique of Shiva's version of ecological feminism, see Agarwal, 1992). Raghunandan (1987) has criticised this attitude toward tribal life-styles, and Dietrich (1988) has questioned the extent to which we can expect poor, overburdened, rural women to come up with ways to save the world from ecological disaster. Such romanticising moves are perhaps an authorising technique. In a perceived need to present alternative politics, a utopian vision of future society is presented by certain environmental texts; a narrative of progression and development towards an ideal society reappears as a recurring metaphor in post-Enlightenment thought (Boyne and Rattansi, 1990).

Related to these points is the politically charged issue of the privileged perspective. Certain ecological feminist writings suggest that one's view is not credible unless one is a woman, or, better still, a Third World or aboriginal woman. From the perspective being developed here, the privileging of certain groups risks false separation of people into categories that may not be most relevant to the formulation of their views and practice. It also returns to the dilemma of essentialism, in that such a perspective on individuals and groups of people assumes a single unfractured identity; a simplistic assumption in this world of multiple sites of oppression and identity formation (see Harding, 1986; Smith, 1987). Moreover, it is generally true that only white, middle-class women can afford the time and the financial and emotional commitments to promote environmental feminisms, including spiritual feminisms, which in their perspectives and practices assumes a socioeconomic position of privilege. In other words, there has been little attempt in environmental feminisms to acknowledge the situatedness of knowledges and perspectives, such that coalitions between partial perspectives can be developed (Haraway, 1988).

Notions of a gendered environmental relationship forming part of a globalising development pattern thus represent a tool for a new perspective in geographical approaches to the environment, by bringing into focus the interactions between global forces of maldevelopment and the politics of reproduction, through a gender-aware analysis. Such a focus foregrounds the geographical specificities of gendered relationships and practices with the environment, and the politics of identity related to these practices which act to reaffirm humanity, notions of 'culture', and the priorities of gendered use of resources, in a nonessentialising way. In a practical sense, environmental feminism points geography more clearly towards an analysis of women's access to environmental resources, and raises the question of how policies, meanings, and practices of the environment and gender affect this access. Questions to be asked include the following. What does a heightened awareness of ecological issues and striving for recycling and the use of unprocessed products mean for women's time and energy? Who is/will be spending extra hours in food preparation and recycling? How do these relate to 'development' agendas in, and for, the Third World? In what ways, and with what effects, are the politics of environmentalism gendered? How do 'masculine' and 'feminine' activisms vary across space? This approach avoids the simplistic and damaging viewpoint that only one gender (or other identifiable group) has the prioritised identity and authority in relation to the environment, a philosophy which historically has excluded the (natural) female gender from (cultured, male) humanity in Western history. Non-Western views on environments suggests that the dialogue around gender, nature, and culture is spatially and temporally creative of distinct coalitions of identities, action, and agendas.

Conclusions: spatial inscriptions on a gendered landscape

In what new directions can this combination of feminism and environmentalism lead? What positions can critical social and feminist geographies take in relation to it? In the above discussion we have suggested various ways in which current feminist approaches to the environment are problematic and we have hinted at some future directions for research. Although in geography there is a long-standing concern for the relationship(s) between people and environments, this field has not been well developed within the new social and cultural geography: it is in this field of geography that we place the need for critical engagement with environmental feminisms. Environmental feminist thought contains ideas which contribute to geographical debates in a number of fields, concerned as it is not only with environmentalism *tout court* but also with issues of landscape, identities, environmental practices, and politics. Geographers concerned with understanding the gendered global relations of capitalism and environmental dynamics in the First and Third Worlds can gain insights from these writings. The linking of ecology and feminisms can thus lead to a greater geographical awareness of the links between gender relations and the environment, at the levels of practices, ideologies, and identities. As discussed, geography can learn that there is nothing inherently feminist or egalitarian (with reference to humans) in an ecological perspective. The viewpoint of women, as well as other minority groups, needs to be heard, and the special needs of these groups in relation to environmental resources must be addressed. Similarly, feminisms, engrossed with analysis of gender relations, are not necessarily attuned to environmental issues, regardless of the putative woman–nature connection. Ecological feminists usefully link these two arguments and suggest various strands of a new geographical approach to the environment.

By contrast, geography has elements to contribute to ecological feminist approaches. A geographical perspective on environmental relations highlights aspects of global and local issues which have not gained environmental feminists' attention. For example, a geographical perspective on gendered environments would of necessity incorporate analyses of technologies and of urban areas, both often overlooked in standard environmental feminist accounts. A critical geography would thus not divide the environment into nature and *man*ufactured things, as the environment we live in is a mixture of *man*made things and things which have a capacity to recreate themselves. In this respect, certain environmental feminist perspectives, encompassing as they do campaigns against nonbiodegradable materials and the dominant development model of industrialisation, would be similar to geographical approaches. Moreover, the new geographical debates over postcolonialism and difference represent important contributions to the environmental feminism literature. The argument developed here suggests that the relations and ideologies of *gendered* environments are worked out over space and in specific landscapes. Social differentiation, ideologies of human–environment interactions, and moral geographies (Philo, 1991) are inscribed in landscapes: environmental feminisms remind us that such landscapes and their inscriptions are gendered, in imagery and through daily practices. Such features could occupy a central position in new research agendas in social and cultural geographies, grounded as they are in practical and theoretical concerns. At a theoretical level, a critical geographical approach to the environment would analyse gender differences in practice, ideology, agendas, socialisation, and (self-)representation in relation to the environment. To take into account such factors means recognition of racialisations, class positions, and other processes of differentiation between women, without reinscribing the hierarchy between nature and culture and between men and women. The reassessment of the relationships between human and nonhuman environmental features, and, by implication the definition of humanity, articulated by environmental feminisms, therefore contributes to a constructive, feminist, non-Eurocentric, nonhierarchical, and sustainable vision and practice for environments and landscapes.

Notes

1 In addressing the debate over the relation between capitalism and patriarchy, so important in the history of geographical feminisms (see WGSG, 1984), environmental feminism usefully highlights the historical connection between the two in the form of their (capitalist patriarchal) domination of women and nature. Ecological feminisms presents an *environmental* critique of both capitalism and patriarchy. Pioneered by Merchant (1980), environmental feminisms link the early development of capitalism and 'modern science' to the oppression of nature and women (see Driver and Rose, 1992). Shiva (1988) picks up this form of analysis with a critique of the ecological impact of patriarchal capitalist maldevelopment on agriculture, water, and forest resources in India, in light of women's relationships to these resources.

2 Shiva discusses 'Indian' philosophy yet refers only to Hinduism, not other philosophical and religious traditions in India such as Islam, Buddhism, Jainism, Sikhism, etc.

3 From this critique of essentialism, it is necessary to identify some caveats. Fuss (1989) argues that pure essentialism or nonessentialism is virtually unattainable and demonstrates how each position can be found embedded in the other. Further, despite the current antiessentialism furore, we do live in bodies, female and male bodies, which although very similar are also different. As authors, we take a kind of materialist and constructivist viewpoint in relation to bodies, in which bodies are understood within specific geographical and historical contexts (see Ortner and Whitehead, 1981).

Selected references

Agarwal B, 1992, 'The gender and environment debate: lessons from India' *Feminist Studies*, **18** 119–158

Ahmed S, 1985, 'The socio-political economy of deforestation in India', OP–29, School of Development Studies, University of East Anglia, Norwich

Biehl J, 1991, *Finding our way: Rethinking Ecofeminist Politics* (Black Rose Books, Montreal)

Bloch M, Bloch J E, 1980, 'Women and the dialectics of nature in eighteenth century French thought', in *Nature, Culture and Gender* Eds C MacCormack, M Strathern (Cambridge University Press, Cambridge) pp 25–41

Bondi L, Domosh M, 1992, Other figures in other places: on feminism, postmodernism and geography, *Environment and Planning D: Society and Space*, **10** 199–213

Bowlby S, Lewis J, McDowell L, Foord J, 1989, 'The geography of gender', In *New*

Models in Geography. Volume 2 Eds R Peet, N Thrift (Unwin Hyman, London) pp 157–175

Boyne R, Rattansi A (Eds), 1990 *Postmodernism and Society* (Macmillan, London)

Cocks J, 1989 *The Oppositional Imagination: Feminism, Critique and Political Theory* (Routledge, Chapman and Hall, Andover, Hants)

Daly M, 1979 *Gyn/Ecology: The Metaethics of Radical Feminism* (Women's Press, London)

Diamond I, Orenstein G (Eds), 1990 *Reweaving the World: The Emergence of Ecofeminism* (Sierra Club Books, San Francisco, CA)

Dietrich G, 1988, 'Development, ecology and women's struggles' *Social Action*, **38**(1) 1–14

Driver F, Rose G (Eds), 1992 *Nature and Science: Essays in the History of Geographical Knowledge* Historical Geography Research Series 28, Historical Geography Research Group; available from C Withers, Department of Geography, Cheltenham and Gloucester College of Higher Education, Cheltenham. Glos

Firestone S, 1971 *The Dialectic of Sex* (Bantam Books, New York)

Fitzsimmons M, 1989, 'The matter of nature' *Antipode* **21**(2) 106–120

Ford S, 1991, 'Landscape re-visited: a feminist reappraisal', in *New Words, New Worlds: Reconceptualising Social and Cultural Geography* compiled by C Philo, Social and Cultural Geography Study Group, Institute of British Geographers; copy available from C Philo, Department of Geography, St David's University College, Lampeter, SA48 7ED, pp 151–155

Fraser N, 1989 *Unruly Practices: Power, Discourse and Gender in Contemporary Social Theory* (Polity Press, Cambridge)

Fuss D, 1989 *Essentially Speaking: Feminism, Nature and Difference* (Routledge, Chapman and Hall, New York)

Griffin S, 1978 *Woman and Nature: The Roaring Inside Her* (Harper and Row, New York)

Guha R, 1989 *The Unquiet Woods: Ecological Change and Peasant Resistance in the Himalaya* (Oxford University Press, New Delhi)

Haraway D, 1988, 'Situated knowledges: the science question in feminism and the privilege of partial perspective' *Feminist Studies* **14** 575–599

Haraway D, 1991 *Simians, Cyborgs and Women: The Reinvention of Nature* (Free Association Books, London)

Harding S, 1986 *The Science Question in Feminism* (Open University Press, Milton Keynes)

hooks b, 1984 *Feminist Theory: From Margin to Centre* (South End Press, Boston, MA)

Hurtado A, 1989, 'Relating to privilege: seduction and rejection in the subordination of white women and women of color' *Signs* **14** 833–855

Jain S, 1984, 'Women and people's ecological movement: a case study of women's role in the Chipko movement in Uttar Pradesh' *Economic and Political Weekly* **23** 1233–1237

King Y, 1989, 'The ecology of feminism and the feminism of ecology', in *Healing the Wounds: The Promise of Ecofeminism* Ed. J Plant (Between the Lines Press, Toronto) pp 18–28

Kolodny A, 1975 *The Lay of the Land* (University of North Carolina Press, Chapel Hill, NC)

Kolodny A, 1984 *The Land Before Her* (University of North Carolina Press, Chapel Hill, NC)

MacCormack C, Strathern M, 1980 *Nature, Culture and Gender* (Cambridge University Press, Cambridge)

Means R, 1980, 'Fighting words; on the future of the earth' *Mother Jones* (December) 25–38

Merchant C, 1980 *The Death of Nature: Women, Ecology and the Scientific Revolution* (Harper and Row, New York)

Mohanty C T, 1991, 'Under Western eyes: feminist scholarship and colonial discourses' *Feminist Review*, **30**; reprinted in *Third World Women and the Politics of Feminism* Eds C T Mohanty, A Russo, L Torres (Indiana University Press, Bloomington, IN) pp 51–80.

Momsen J, Townsend J (Eds), 1987 *Geography of Gender in the Third World.* (Hutchinson Education, London)

Monk J, 1984, 'Aproaches to the study of women and landscape' *Environmental Review* **8**(1) 23–33

Nesmith C, 1991, 'Gender, trees and fuel: social forestry in West Bengal, India' *Human Organization* **50** 337–348

Norwood V, Monk J, 1987 *The Desert is No Lady: Southwestern Landscapes in Women's Writing and Art* (Yale University Press, New Haven, CT)

Ortner S, 1974, 'Is female to male as nature is to culture?' in *Woman, Culture and Society* Eds M Rosaldo, L Lamphere (Stanford University Press, Standford, CA) pp 67–87

Ortner S, Whitehead M (Eds), 1981 *Sexual Meanings: The Cultural Construction of Gender and Sexuality* (Cambridge University Press, Cambridge)

Pateman C, 1988 *The Sexual Contract* (Polity Press, Cambridge)

Peterson A, Merchant C, 1986, '"Peace with the earth": women and the environmental movement in Sweden' *Women's Studies International Forum*, **9** 465–479

Philo C (compiler), 1991 *New Words, New Worlds: Reconceptualising Social and Cultural Geography* Social and Cultural Geography Study Group, Institute of British Geographers: copy available from C Philo, Department of Geography, St David's University College, Lampeter SA48 7ED

Plant J (Ed.), 1989 *Healing the Wounds: The Promise of Ecofeminism* (Between the Lines Press, Toronto)

Plumwood V, 1988, 'Women, humanity and nature' *Radical Philosophy* **48** 16–24

Plumwood V, 1989, 'Do we need a sex/gender distinction?' *Radical Philosophy* **51** 2–11

Poovey M, 1990 *Uneven Developments* (Routledge, Chapman and Hall, Andover, Hants)

Porteous J D, 1986, 'Bodyscape: the body-landscape metaphor' *Canadian Geographer* **30**(1) 2–12

Radcliffe S, Westwood S (Eds), 1993 *Viva! Women and Popular Protest in Latin America* (Routledge, Chapman and Hall, Andover, Hants)

Raghunandan D, 1987, 'Ecology and consciousness' *Economic and Political Weekly* **22** 545–554

Reynolds F, 1989, 'Ecofeminism' *Ecos* **10**(2) 2–8

Rose G, 1991, 'Geography as a science of observation: the landscape, the gaze and masculinity', unpublished seminar paper presented in London, January; copy available from the author, Department of Geography, Queen Mary and Westfield College, University of London, London

Rose G, 1993 *Feminism and Geography* (Polity Press, Cambridge)

Said E, 1978 *Orientalism* (Penguin Books, Harmondsworth, Middx)

Salleh A K, 1984, 'Deeper than deep ecology: the ecofeminist connection *Environmental Ethics* **6** 339–345

Sayers J, 1982 *Biological Politics: Feminist and Anti-feminist perspectives* (Tavistock Publications, Andover, Hants)

Schaffer K, 1988 *Women and the Bush: Forces of Desire in the Australian Cultural Tradition* (Cambridge University Press, Cambridge)

Shiva V, 1988 *Staying Alive: Women, Ecology and Survival in India* (Kali for Women, New Delhi)

Slater D, 1992, 'On the borders of social theory: learning from other regions' *Environment and Planning D: Society and Space* **10** 307–328

Smith D, 1987 *The Everyday World as Problematic: A Feminist Sociology* (University of Toronto Press, Toronto)

Smith N, 1984 *Uneven Development* (Basil Blackwell, Oxford)

Spelman E, 1988 *Inessential Woman: Problems of Exclusion in Feminist Thought* (Beacon Press, Boston, MA)

Spretnak C, 1989, 'Toward an ecofeminist spirituality', in *Healing the Wounds: The Promise of Ecofeminism* Ed. J Plant (Between the Lines Press, Toronto) pp 127–132

Starhawk, 1989, 'Feminist earth-based spirituality and ecofeminism' in *Healing the Wounds: The Promise of Ecofeminism* Ed. J Plant (Between the Lines Press, Toronto) pp 174–185

Warren K, 1987, 'Feminism and ecology: making connections' *Environmental Ethics* **9** 3–20

Warren K, 1988, 'Toward an ecofeminist ethic' *Studies in the Humanities* **15** 140–156

Warren K, 1990, 'The power and the promise of ecological feminism' *Environmental Ethics*, **12** 125–146

Warren K (Ed.), 1991, 'Ecological feminism', special issue *Hypatia: A Journal of Feminist Philosophy* **6**(1) 1–214

WEN, 1989–92 *Newsletter* Women's Environmental Network, 287 City Road, London EC1V 1LA

WGSG, 1984 *Geography and Gender: An Introduction to Feminist Geography.* Women and Geography Study Group. (Hutchinson in association with the Explorations in Feminism Collective, London)

Wisner B, 1978, 'Does radical geography lack an approach to environmental relations?' *Antipode* **10**(1) 84–95

Wolf M, 1987 *Revolution Postponed: Women in Contemporary China* (Methuen, Andover, Hants)

Wollstonecraft M, 1792 *A Vindication of the Rights of Woman* (Joseph Johnson, London)

Zimmerman M, 1987, 'Feminism, deep ecology, and environmental ethics' *Environmental Ethics* **9** 21–44

13 David Demeritt

Ecology, Objectivity and Critique in Writings on Nature and Human Societies

Reprinted in full from: *Journal of Historical Geography* 20, 22–37 (1994)

While the North American environmental movement has won only moderate gains in the legislative arena, it has produced a body of writing remarkable both for its impassioned criticism of the environmental devastation wrought by modern Western society and its sensitive analysis of the historic and geographic dimensions of global change. I wish that I could report historical geographers out in the forefront of this discussion. I cannot. Like other human geographers, we seem to have all but abandoned the once venerated study of human relations with the environment.[1] What little geographic research there is has generally conceived the important questions to be technical ones about natural hazards and their human impacts. Geographers, as a result, have tended to produce socially thin environmental impact statements without enough of the caring and reflexive moral engagement that we must make with the world in this era of global warming.[2] We need a more fully critical effort that both diagnoses the deeper social and economic causes of present environmental problems and points the way forward to some preferable future that we might make for ourselves. To date, geographers writing about North America, in marked contrast to their colleagues interested in less developed countries, have not been up to this critical task.[3]

Instead, historians have provided the most insightful critiques of the environmental crisis in North America. Environmental history emerged only recently as a distinct subfield within the historic profession. 'Its goal', according to Donald Worster, a leading scholar and outspoken publicist for the field, 'is to deepen our understanding of how humans have been affected by their natural environment through time, and conversely and perhaps more importantly in view of the present global predicament, how they have affected that environment and with what results'. From the outset, it embodied 'a strong moral concern . . . [with] some political reform commitments behind it'.[4]

Lately, however, the scientific and epistemological foundations of this critical project have been severely eroded. Revisions within ecological science have discredited the holistic concepts of equilibrium, ecosystem and climax used by environmental historians to measure anthropogenic damage to nature.[5] I would like to start by considering the new empirical claims of ecology and their implications for the critique offered by environmentalists.[6] From there, I will discuss two further challenges to this critique that move us farther and farther away from authoritative claims to know the world as it really is. Historians and philosophers of science have impugned the ideas of objectivity and observer neutrality that legitimate the exclusive power of science to represent nature. But if ecologists cannot represent nature as it really is, then surely environmental historians cannot do so either. Barbara Leibhardt and Elizabeth Ann R. Bird have responded to this challenge from [the] history of science by rethinking the relationship between scientific inquiry and its objects of study. The extension of a diverse linguistic turn has also destabilized history and the other human sciences. William Cronon has offered an extensive response to the epistemological difficulties posed by various postmodern theories of language and knowledge to historians constructing and evaluating historical narratives and environmental critiques.

In the final part of the essay, I will discuss the possibilities and practice of environmental critique without foundational authority, be it in science or elsewhere. By dispensing with questions of truth and falsity, we can evaluate different representations of nature in terms of their likelihood to produce the kind of world we want to live in.

The challenge of revision in ecological science

At the moment, ecologists are busy rethinking their discipline. There is now a hollow ring to the bold claims of the 1960s and 1970s that information theory, systems analysis or mathematical modelling would provide the unifying metatheory to transform ecology into a mature, nomothetic science like physics. Whether the present ferment marks the 'death of the old ecology', a tolerant and creative 'pluralism in ecology', or perhaps the rise of some new 'ecology of chaos' depends on the observer.[7] For the time being I would like to bracket discussion of whether these scientific claims can provide a true account of nature. I will consider the actual substance of the revisions in ecology before moving on later in the essay to question their foundations in a theory of knowledge.

Contemporary debate fractures the discipline along several ancient and intersecting faultlines of institutional, methodological, and philosophical difference. These divisions were held steady through the 1960s and 1970s by a set of hegemonic ideas about the science of ecology and the nature that it studied. In this period, plant and animal ecologists of many different philosophical persuasions accepted – and idiographically minded field ecologists could not effectively challenge – the idea that ecology was a nomothetic science whose highest goal was to produce broad, context-independent generalizations about nature.[8] To this end, ecologists directed their research to produce and modify statements about niches, trophic levels, and ecosystems. These statements quickly took on an existence of their own. Holistically inclined systems theorists and reductionist mathematical modelers may have disagreed about the causes of the natural regularities they called ecosystems, but they certainly agreed that the word ecosystem referred to a real natural entity that could be described mathematically in terms of an atemporal dynamic equilibrium. Their shared vocabulary of ecosystem and equilibrium not only bridged their differences but also lent itself to holistic readings by those outside the discipline.[9]

Of course, there were also very real and often bitter disputes in ecology during this period. Donald Worster has described the very important debate over the fundamental level of biological organization.[10] E. P. Odum and other system theorists insisted that ecosystems have a real ontological status, while population biologists and mathematical modelers like Robert MacArthur held that the properties of ecosystems emerge from the interactions of individual species. However, it is my contention that this debate occurred within the context of a much broader consensus about ecology as a nomothetic science. Ecological discourse could accommodate these apparently incommensurable statements because both groups used mathematics to articulate the broad context-independent generalizations about nature that defined ecology as a true science.

The consensus about ecology as a law-finding science, which held the discipline together through the 1960s and 1970s, has eroded with a renewed belief that 'pesky ''biological details'' matter a lot'.[11] Ecologists are not as confident that their statements about ecosystems actually represent context-independent generalizations about nature or even name real objects.[12] Population biologists and fisheries managers no longer hold that the Lotka-Volterra equations, which focus exclusively on predator–prey interactions, provide an adequate basis to predict population fluctuations.[13] Plant ecologists and park managers, who once saw a dynamic equilibrium in the species composition of forests like the Boundary Waters Canoe Area on the Minnesota–Ontario border, now believe that many species and natural systems are disturbance dependent and require disturbance at some temporal and spatial scale to reproduce themselves.[14] In short, recent revisions in ecological science have not proceeded independent of problems and concerns outside of the discipline.

Although ecologists remain riven by institutional, epistemological, and ontological differences too complicated to consider here in any depth, many agree that the integration of different spatial and temporal scales is the key problem for their discipline. Ecologists hope that greater attention to scale will allow them to produce context-independent generalizations about nature.[15] What appears as a homogeneous forest environment at one spatial scale, is, at another, a mosaic of many different environments. Similarly, paleoecologists interested in very long term vegetation dynamics describe a constant flux of different species and communities where those working at shorter time scales see only stability and equilibrium. Succession is now widely conceived in terms of small scale gap processes caused by senescence and larger scale patch dynamics caused by natural disturbances like windfall rather than in the once dominant terms of the autogenic development of relatively larger and more homogeneous landscape units like ecosystems or vegetation formations.[16] Analysis of fossil pollen from plant species returning to a deglaciated North America makes visible an independence in the migration of individual species that also undermines the idea of forest communities as organic landscape units whose species composition is fixed by unchanging and unilineal successional pathways to some natural climax.[17]

Recognition of temporal and spatial scale also requires a redefinition of many other ecological concepts. Whereas once ecologists spoke rather loosely in terms of a generalized sort of equilibrium or balance of nature, they now unpack this idea more carefully and break out several different ways of thinking about ecological stability.[18] Although the terminology varies, ecologists often distinguish resilience (the rate at which a population returns to its former population density following a decline), persistence (how long the composition of a community persists without change), variability (the temporal variability of a community's population densities), and resistance (the degree to which species' densities change following a change in the population density of a single species in the community).[19]

More generally, recognition of the importance of scale in ecology signals a greater respect for description and data collection, a more humble approach to ecological complexity, and a renewed skepticism towards mathematical models and abstract theories.[20] Ecologists reject any off-handed dismissal of their discipline's present descriptive bent as somehow being 'unscientific'. Many still hold out the hope that ecology will eventually mature enough to transcend its descriptive status and become a nomothetic science like physics. For the moment, though, the complexities of natural systems seem to overwhelm ecological theory. Comparing their young discipline to nineteenth century cellular biology, ecologists Lawrence Pomeroy, Eugene Hargrove, and James Alberts urge their colleagues to continue descriptive studies that, like Gregor Mendel's work, may one day provide the basic data for experimentally validated predictions about natural systems.[21]

These revisionary developments in ecology would be of only passing interest to environmental historians if they were not so reliant upon metatheories borrowed from ecological science. Warren Dean attributed the problems of Brazilian rubber producers to the coevolution of specially adapted pathogens in the rainforest.[22] Others, on still thinner evidence, have explained the structure of the contact-era forests of eastern North America in terms of fire ecology theory.[23] Explanations culled from ecological science may be appropriate in particular situations, but ecology cannot provide general covering laws. In the case of Indian-set fires, historians have paid scant attention to the pollen record or to the contention of botanist Emily Russell that anthropogenic fires were much less common than environmental historians have suggested.[24] For the most part, it would seem, environmental historians have approached ecological science as a mine of potential theories and data about nature, the tools and raw materials with which to construct their own narratives.

The reasons for this reliance are clear enough. Ecologists' authoritative claims to understand nature as well as their apparent independence from the economic forces of environmental destruction have impressed historians and other environmental critics. In this new science, they find a familiar romanticism. As Donald Worster and others have noted, the organicism of the Romantic movement leavened the early development of ecological thought.[25] Strains of this nineteenth century romanticism persist in scientific concepts like ecosystem, climax, and equilibrium. When environmental historians press their critique of contemporary society and its relations with nature, they can appropriate the scientific authority of ideas infused with the romantic reaction to capitalism. This, of course, is what makes them so attractive to environmentalists. By their very nature these ecological concepts are ready-made to highlight human disturbance of the natural order from which modern humanity is, by definition, alienated.

Donald Worster has been the most enthusiastic advocate of this use of ecological theory. In an otherwise favorable review of William Cronon's *Changes in the land*, Worster chided Cronon for failing to use 'the idea of nature as a yardstick for comparing different adaptations . . . It would have been far more clear and compelling to have said with the ecologist that nature, left alone, demonstrates a marvellous system of organization, that the Indians survived for so long by adapting their lives to that order, and that the English newcomers [to New England] generally did not'.[26] The problem is that ecologists no longer say this in such unequivocal terms, and so environmental historians cannot rely on the authority of ecology to legitimate their condemnation of modern society.

Not surprisingly, therefore, Worster is also the most prominent and trenchant critic of the revisions in ecology. The new ideas about natural disturbance, succession and the relatively ephemeral composition of contemporary ecosystems deprive environmentalists of the scientific grounds to construct hard and fast distinctions between human and natural disturbance on the landscape and thus between natural and disturbed landscapes and geographies. Without these distinctions it is impossible to argue that old growth forests are somehow more natural than cutover forests and thus best preserved from logging and other human disturbances which, by definition, are unnatural. Worster complains that the new uncertainty about nature, equilibrium, and stability can 'serve to justify the destruction wrought by contemporary industrial societies'.[27] It blunts the scientific authority of his calls to respect nature, and he fears, leads to an 'environmental relativism' whereby it is impossible to distinguish 'between the balance achieved by nature and that contrived by man'.[28] 'What,' he asks, 'does the phrase "environmental damage" mean in a world of so much natural chaos?'[29]

This is a very important question, but his response is to savage the messenger rather than confront the issues of ontology and epistemology posed by revision in ecological science. Worster seems content to dismiss the science he dislikes, on the grounds that it is a social construction, while still appealing to the authority of Clements' and Odum's ecology as 'a scientific check' on contemporary society.[30] The fundamental questions posed by ecologists about nature and our ability to know it demand a more direct response than head in the sand conservatism. With his intransigence, Worster risks being ignored at a time when we need to hear his critique of modern society more than ever. Worster tells his readers to respect nature, but as I read the contemporary literature in ecology, it says that recognizing a nature actually 'out there' to respect is much more complicated than it may seem. In a world where the stamp of humanity is more extensive than ever and distinguishing the natural from the artificial is, at best, highly problematic, we must act to shape the best world we can, rather than appealing to some extra-social 'nature' to authorize someone to arbitrate between the moral from the immoral. As Donna Haraway has observed, the kind of nature that Worster speaks for and tells us to respect is not so much a thing in itself as a rhetorical place in our language. By displacing the represented object (nature) from the surrounding relationships that constitute it, and relocating it in the authorial domain of the representative, this strategy authorizes the representative, be it scientists or environmentalists, to speak exclusively for a nature rendered inarticulate by this very act of distancing.[31] Insofar as Worster and environmental critics like him present us with only two options, either heed their exhausted romantic creed – 'respect nature' – or fall off the precipice of relativism, they have failed, it seems to me, to provide a useful response to the contemporary challenges posed by ecological science.

The challenge from history of science

History of science poses a second challenge to the critical project of environmental history. Ironically enough, environmental historians are themselves responsible for much of their own discomforting skepticism about ecology's claims to represent nature accurately. They were quick to deploy the work of Thomas Kuhn and others who argued that scientific knowledge is socially constructed. They cheered as their critique discredited the objectivity claims of foresters, nuclear engineers, and other scientists opposed by the environmental movement. Now, however, environmental historians and their allies rest uneasily as they realize that ecology, their scientific darling child, is no different from any other science; its knowledge is also socially constructed and cannot claim to be a mirror of nature. More disorienting still is the realization that even their own critiques of science and our society can no longer claim a privileged vantage point from which to represent the world as it really is out there.

Donald Worster's *Nature's economy*, a history of ecological ideas, was in the forefront of the historicist critique of science. His systematic assault on scientific objectivity claims proceeded in two ways. First, Worster distinguished between science and ideology by documenting the influence of outside, 'non-scientific' ideas on particular scientists. A. G. Tansley's attack on Clementian climax ecology was not motivated by an objective re-evaluation of the data, but by a fervent wish 'to put down the threat to the legitimacy of human empire posed by the natural climax theory'. Second, Worster measured the inaccuracy of scientific knowledge by charting the shifting fortunes of the reductionist and holistic metaphors in ecology against the implicitly unchanging ontological status of nature. Since nature does not change while the predominant scientific representations of it do, scientific knowledges must be 'valid relatively, suited to or at least rooted in their times'.[32] This tack smuggles an epistemological realism about its own claims to know an unchanging nature into its rendering of socially constructed science.

By following this general two pronged strategy, other environmental historians have critiqued the sciences of nature. John Perkins outlined the effects on scientific thinking of the public uproar over DDT use. He also applied Kuhn's theory of scientific revolutions to explain the internal paradigm shifts in agronomy and economic entomology.[33] In *The death of nature*, Carolyn Merchant described the eclipse of an older tradition of animism and analogical, holistic thought by the mechanistic, Cartesian paradigm of the Enlightenment.[34] More recently, she has attempted to map such scientific paradigm changes onto larger ecological revolutions that encompass environmental changes as well as shifts in social and economic practice.[35] Similarly, Arthur McEvoy documented the interpenetration of changing ecological conditions, legal practice and economic power with science in the California fishery.[36]

These attacks on scientific objectivity erode the epistemological foundation that many environmental historians seek to build beneath their wider critique of modern society and its relations with the biogeochemical environment. If scientists cannot claim to represent nature truly, then environmental historians cannot rely upon ecology to provide the truth about nature. Ecology may be a preferable science to nuclear engineering, but it cannot claim to occupy a privileged vantage point from which to represent nature more accurately than other sciences. This realization is particularly troubling because environmental historians have relied so heavily upon the authority of favored scientists like Rachel Carson to unveil the truth about human devastation of the environment.

The social constructivist approach to knowledge goes still further in undermining the foundations of environmental history. As practiced by environmental historians and others, the social constructivist critique of science relies upon an instrumental irony in which historic actors are unaware of the underlying social influences on their thinking to which only the historian and reader are privy. In unveiling the contextual, partial nature of scientific knowledge, this approach displays a 'fairly inflexible commitment to epistemological realism'.[37] If scientists cannot claim to represent nature as it really is, then neither can historians claim to represent scientists, or nature, as they really are. Donald Worster is not the only one to worry about the practical effects of a historicist view of scientific knowledge. A number of other environmental historians have also called for the development of new theories of 'historical change and causation'.[38] But, for the most part, environmental historians continue to rely on a realist epistemology and to lean heavily on the authority of ecology in writing environmental histories. They remain blissfully ignorant of the unsettling implications of their own social constructivist approach to knowledge.

The social constructivist approach to knowledge does not have to lead to a relativism wherein there are no reasonable grounds to choose between different scientific knowledges or environmental histories. Barbara Leibhardt and Elizabeth Ann R. Bird have each turned to anthropology as a guide to evaluating different accounts of nature without relying on a realist epistemology. Leibhardt appealed to Clifford Geertz who 'steered clear of relativism by separating good interpretations from bad according to how well they capture and made sense of the complexities of human action'. She would apply similar criteria to environmental histories, evaluating them 'on the ability to describe the "thickness" of the relationships at issue'.[39] Unfortunately, Geertz's ethnography tends to ignore power relations and subsume social conflicts to a shared symbolic vocabulary, so is not a terribly good model of rich social analysis.[40] Certainly his recognition of complexity in social and ecological relationships is valuable, and in tune with much of the present thinking in ecological science. However, his notion of thickness cannot overcome the problems of judgement and evaluation facing environmental historians. In practice, it turns out to be as difficult to distinguish the thick from the thin as it is to separate the natural from the artificial. Rather than resolving this pressing dilemma, Geertz's terminology simply restates it in a new way.

James Clifford's ideas about authoritative modes of ethnographic representation guide Elizabeth Ann R. Bird's understanding of the relationship between science and nature. An ethnography, Clifford argues, does not faithfully reproduce its object of study; instead, it textualizes a previous multisubjective encounter in which the ethnographer(s) and native subjects mutually constitute one another. Bird has applied this idea to scientific inquiry.[41] Like proponents of the strong sociology of knowledge, she reverses the traditional relationship between reality and representations of it: scientific practice actually constitutes the facts of nature it purports to represent. Furthermore, as Bruno Latour and Steve Woolgar have argued, this 'process of construction involves the use of certain devices whereby all traces of production are made extremely difficult to detect'.[42] However, Bird goes beyond these sociologists of science in asserting the active role of nature 'in negotiating reality . . . Nature's role in that negotiation takes the form of actively creating something materially new and of resisting or accommodating the range of metaphorical and theoretical imaginings with which it is approached'.[43] For her, scientific texts, like Clifford's ethnographies, are textualizations of earlier, multisubjective, processes whereby nature and scientists mutually construct one another.

Many environmental historians find the social constructivist approach to knowledge unsettling. It seems to undermine any secure foundation for legitimate critique of our society and its environmental policy. Quite the contrary in fact. It is possible to 'argue against environmentally destructive technologies, but not on the grounds that they are anti-natural'. Such distinctions are impossible to make. 'A better argument', Bird replies, 'should rest on the grounds that those technologies do not foster the nature we want to exist'.[44] Here environmental historians can make a valuable contribution: not by directing us to true laws of nature that we must respect – a chimerical wish at best – but by providing a historical context for our discussions of the kind of nature we hope to make.

The challenge of the linguistic turn

The epistemological realism of environmental historians faces a third challenge in the recent 'linguistic turn' in history. This rather unfortunate label refers to the ascendence of many different ideas about language and knowledge that directly assault the ideal of objectivity which, as Peter Novick has shown, serves as the founding myth of the profession.[45] Historians and sociologists of science, like Thomas Kuhn, emphasize the theory-ladenness of observation. They disavow the scientific method as an Olympian gateway to singular truth about the world, and by implication discredit the natural sciences as a model for an objective history.[46] Philosophers like Hilary Putnam and Richard Rorty reject the correspondence theory of truth. They insist that there can be no 'God's Eye point of view' secure above language and cultural values from which to hold a mirror to nature.[47] Post-structuralist literary critics like Jacques Derrida deny that language is a transparent medium for the transmission of content; on the contrary, they argue, language actually produces meaning. Their emphasis on narrative form and the rhetorical practice blurs the sacred distinctions between history and fiction.[48] The result of these many different linguistic turns has set Ranke's historic ideal, telling the past as it really was, adrift from its moorings in a foundational theory of knowledge.

Although these moves shake the foundations of their discipline, historians have been rather slow to respond. A few have enthusiastically embraced the post-foundational situation and the concomitant focus on language and the construction of meaning.[49] Others have redoubled their commitment to the traditional justifications of objective history in some form of realism, be it empiricism, Hempel, or simple 'common sense'.[50] Many more historians have been bewildered by the whole controversy. The fulminations of arcane philosophers, literary critics, and their various disciplines seem to say precious little to the familiar vexations of archival research or the actual business of writing history. Discussions of postmodernism and the foundations of historical knowledge are only just beginning to move beyond the programmatic to influence historians' actual work.[51]

William Cronon is the first environmental historian to respond at any length to the linguistic turn and its implications for environmental critique. Because of their avowedly critical aims, environmental historians stake a great deal on epistemological realism. They aspire to do more than just tell good stories about people and nature; they want their diagnosis of our society and its environmental problems to lead to social action. 'Without a clear demonstration of causality' and the authority of truth, Richard White and his colleagues worry that their critique will be unconvincing.[52] In this context, Cronon's response to the linguistic turn is particularly important because it will help set the terms of future discussion about the problems and possibilities of historical knowledge and environmental critique.

In Cronon's analysis of the many different and incommensurable histories of the Great Plains, he acknowledges the centrality of narration to the environmental history project. Like other humans, environmental historians tell stories about nature, but 'their narrative form has less to do with nature than with human discourse'. Narratives organize human reality in terms of beginnings, middles, and ends. This structure orders an otherwise messy chronicle of past events by linking connected facts, themselves determined by the narrative structure, and excluding extraneous details from the story.[53]

Historians have imposed one of two basic plot lines on the history of the Great Plains. Historians like Frederick Jackson Turner and Walter Prescott Webb depicted sagas of frontier progress, in which white settlers transformed an empty or savage wilderness into a prosperous region of productive farms, civilized towns and American democracy. On the other hand, New Deal reformers saw in the Great Plains a tragic story of human ignorance resulting in environmental disaster and miserable poverty. As Cronon shows, the meaning of these very different stories is prefigured by their representation of the beginning and its juxtaposition with the ending. The difference between the two not only sets progressive histories apart from tragic ones, it also 'gives us a chance to extract a moral from the rhetorical landscape'.[54] This moral framework dictates, in large measure, how the reader will judge the people, places, and events of the story.

Through their writing then, historians exercise considerable power over the objects they represent. In addition to configuring meaning, the placement of the beginning and ending also determines who and what will be included and excluded in the story. This process is never innocent. Turner and Webb's histories of frontier progress, like the New Dealer's stories of tragic ignorance and human misery, were centrally concerned with the development of white society. They systematically ignored both pre-contact aboriginal history (called, tellingly enough, *prehistory*) and the many stories of death and destruction, resistance and revival of native peoples relegated to reservations. There are countless other erasures at work in the history of the American West; women, ethnic and racial minorities, and class struggles have been systematically silenced as well. Environmental historians are familiar with the prominent erasure of nature and make it their business to recover 'the earth itself as an agent and presence in history'.[55] As they do so, however, they must also acknowledge the tyranny of the narrative and appreciate that storytelling is inevitably an exercise of power. With the flick of a pen or the stroke of a key, historians can erase the struggles and aspirations of entire cultures from the landscape or enframe them in unfavorable or unflattering ways.

In these different theories of narrative and language, environmental historians confront the troubling artifice in their representations of nature. Traditionally, historians have maintained that their craft, despite the important elements of creativity and imagination, is at some fundamental level a reflection of an external reality against which individual histories can be measured and evaluated. Cronon's analysis of the many different and incommensurable histories of the Great Plains points to the very artificial, perspectival, and constructed nature of history. If so many different histories of the same thing are possible, how then, Cronon asks, 'are we to choose among the infinite stories that our different values seem capable of generating?' He is afraid that acknowledgement of the indeterminacy of meaning and the plurality of possible histories leads to relativism whereby historians have no rational means to choose between stories. Unfortunately, Cronon's attempt to 'accommodate the lessons of critical theory without giving in to relativism' does not escape the terms of an objectivist/relativist dichotomy whereby relativism is the only alternative to a realist epistemology. Rejecting relativism, Cronon clings to the idea of 'the past (and nature) as *real* things to which our storytelling must somehow *conform* lest it cease being history altogether (emphasis added)'. He points to a tripartite structure of rules and institutions that ensures the correspondence of history to this external reality. First, historians agree that 'our stories cannot contravene known facts about the past'. To this stricture, environmental historians would add the further requirement that 'our stories must make ecological sense'. Finally, Cronon says, historians 'write as members of communities, and we cannot help but take those communities into account as we do our work'. Simply put, 'community, past reality and nature itself' cannot provide the 'ultimate justification of history'. They cannot secure a foundation for the establishment of historic truth through correspondence with external reality. They cannot make the past and nature 'real things to which our storytelling must somehow conform'. The correspondence theory of truth is dead. To hope otherwise seems completely at variance with the first twenty-five pages of Cronon's essay in which he described the many different, indeed incommensurable, stories that different historians have written about the Great Plains and what they each took to be its nature and its past. As Cronon so astutely observed, what passes for nature and history in any particular story is an effect of its narrative structure.[56]

But Cronon is on to something important when he points to the power that the rules and structures of historical writing exercise over the stories historians tell. More than simply an effect of narrative structure, what pass for nature, the past, and the facts about them are also an effect of the rules and structures of the historical discipline. History is a discourse with its own particular politics of truth.[57] Cronon seems close to saying as much when he describes the process of criticism that shaped his own essay. The profession evaluates individual texts, enforcing these rules, sanctioning truth claims and disciplining wayward accounts of the past. The discourse works to shape the range of possible stories, but not by making them converge on an improved or somehow more accurate representation of the past, as Cronon is left arguing with his insistence on the past and nature as 'real things to which our storytelling must somehow conform'.[58] Historical texts conform to past reality only insofar as historical discourse actually *produces* the past reality to which they conform.

Historic facts, after all, are constructed by this particular discursive community. Under the right circumstances, facts can unravel and lose their truth within this discourse. I think, for instance, of the now discredited, but at the turn of the century near unanimous denunciation by American historians of the abolitionists as irresponsible agitators and of the Reconstruction governments imposed on the South by the federal government as criminal outrages.[59] Some might respond that these racist statements were interpretations and therefore of a different order than the historic facts from which they were constructed. Such facts, like the date of American entry into the Second World War, may seem somehow more concrete, more real, but the difference lies not in the nature of December 7, 1941, but rather in the difficulty of challenging the consensus that has developed around its status as a fact.[60]

American warships had been skirmishing with German submarines since September 1941, but these anonymous encounters in the North Atlantic do not live in infamy.[61] The attack on Pearl Harbor, unlike the anti-submarine operations, is so widely acknowledged, that I do not need to assert its status as a fact with a footnote. Some facts are so well established and taken for granted that they become like second nature to us, inhering in our very language. Garry Wills has recently argued that Lincoln's Gettysburg Address transformed the United States from a plural concept, requiring a plural verb in the antebellum period, to a singular idea that necessitated a singular verb.[62]

Just as historic facts are constructed by historians, facts about nature are constructed by natural scientists. With Cronon's second requirement that environmental history must make ecological sense, he does not enter a privileged realm of truth about nature. Rather he simply requires himself to conform the facts as set down by a second discursive community, ecological science. Ecological discourse has its own particular discursive rules and disciplinary structures that help produce (and police) facts about nature. And at the moment, environmental historians find the politics of truth about nature changing rapidly in ecological discourse.

Cronon and many other environmental historians react with great fear and anxiety to the suggestion that the past and even nature itself are things we construct. They are caught in the terms of the objectivist/relativist dichotomy whereby relativism is the only alternative to a realist epistemology. Relativism is logically self-refuting and morally repugnant, but it is not the only game in town. We do not need to retreat to the ruined redoubt of realism to keep relativism at bay.

Though facts, history, and nature may have no more concrete claim on us than that they work for us, we can use them all the same. Their constructed nature does not make them any less of a guide to our world as it is and as we want it to be. As long as we agree to live by them, constructed or not, facts, history and nature can help explain the world we make for ourselves to ourselves. This process of construction always involves the negotiation of relationships between many different actors, not all of them human. But, a recognition of the discursive construction of knowledge does not preclude making moral choices about the world as we happen to understand it, despite what some environmental historians may fear. Such anti-foundationalism simply undermines exclusive claims to legitimate knowledge about the world, claims that both environmentalists and their political opponents have used entirely too much.

Moving beyond foundationalism

Moving beyond foundationalism and the objectivist/relativist dichotomy will require a new approach to environmental critique and a new politics of participation.[63] The North American environmental history project has lost the firm, authoritative foundation that underwrote its critique of our society and our relations with the other organisms and material objects in the world. No longer can environmental historians call upon ecology to propound holistic concepts that measure anthropogenic damage. No longer can they rely upon Science and its claims of objectivity to authorize their representations of nature and prescriptions of social action. No longer can environmental historians claim objectivity for themselves and arbitrate between different accounts by appealing to the facts alone.

Of course the challenges posed by this sweeping anti-foundationalism affect much more than this select group of academics. In its wake, the entire Green movement finds the diagnostic element of its critique decoupled from its utopian vision. Greens, like many other social critics, have advanced their normative program by positioning it as the only logical response to an objective diagnosis of the planet's environmental woes. Like environmental historians, Greens have relied upon ecological theory to describe human devastation to nature and upon the scientific authority of ecology to provide what Donald Worster enthusiastically calls 'a scientific check' on social action.[64] Without an epistemological foundation on which to make these claims, Green critics will have to advance their utopian vision of a sustainable world by some other means. They are not alone. The idea of a singular, transcendental truth about the world underwrites the entire Enlightenment project. The consequences of this anti-foundationalism are enormous.

Let me restrict myself then to some brief comments about the possibilities and practice of an anti-foundational environmental critique. The arguments against foundationalism are not just anti-epistemological; they are also political. The authoritarian strategies of the environmental movement, though perhaps once helpful as counterarguments to the claims of a totalitarian science of technocrats and a narrow instrumentalism serving chiefly the interests of capital, have now outlived their utility. Exclusive knowledge claims serve only to divide, to empower a few anointed experts, and to exclude the many different voices that can and should speak – and that we need to hear – in the ongoing conversation about nature.[65] Green critics, therefore, should end their efforts to build an epistemological Mount Sinai atop science and the facts from which to commune with nature and bring its commandments down to the children of Israel. Instead, they should be satisfied with a more humble, but ultimately less malevolent, and more attainable goal: simple participation in the ongoing conversation about the nature we have made and the one we hope to make.

Ultimately, of course, environmental narratives are not legitimated in the lofty heights of foundational epistemology but rather in the more approachable and more contested realm of public discussion and debate. But in understanding the dynamics of the public sphere, pragmatic philosophers like Richard Rorty who celebrate the possibilities of open conversation in liberal society are unnecessarily optimistic about the possibilities of free and open discussion. In practice, the 'we' in conversations about nature is not an especially inclusive one. It tends to include far more white, middle class, male academics (like me) than other categories of people, to say nothing of reaching out beyond the boundary between humans and other earthlings to allow them to participate as well.

In this conversation, we should look to science not as a mirror to nature but as a useful tool for engaging our world critically. Ecology, like every other science, is a discourse with its own particular rules and disciplinary structures that produce representations of nature. These representations involve the exercise of power, and we should treat them as such. This situation does not rule out appropriations from ecological science or other fields of knowledge where they prove useful and convincing. Science can still provide an important way to make our relationships with the world visible to us. These knowledges are necessarily perspectival, situated ones, but this fact makes the atmospheric carbon dioxide measurements at Mauna Loa no less important in helping us reevaluate anthropogenic carbon releases.[66]

The political response to global warming has been slow thus far, as the recent summit in Rio de Janiero sadly demonstrated. Even so, the knowledge produced by scientists at an observatory 11 000 feet above the tropical Pacific has helped bring into focus a whole complex of different relationships connecting environmental groups, atmospheric scientists, rubber tappers, petroleum companies, international bankers, and politicians to each other and to tropical rain forests, computer models, carbon dioxide, water vapor in clouds, and a host of other trace compounds in the atmosphere that absorb infrared radiation.[67] The wide range of these relationships suggests the great power and utility of science to make visible our relationships with the world and the other creatures, things, and miscellaneous earthlings with whom we share it.

Of course, the very act of representing the world scientifically establishes a particular kind of relationship between representatives and the objects they represent. As we consider this process we should always bear in mind that different actors bring different resources to these encounters and that the relations of power that result are rarely, if ever, equal ones. Here, historians and geographers can play a crucial role by helping to situate competing knowledges about the world.[68] This effort can make visible the material and discursive effects of different environmental narratives. It provides us with the means to move beyond foundationalism, to evaluate competing accounts, not in terms of their truth or falsehood, but in terms of their likelihood to produce the kind of world we hope to live in and leave behind us.

Notes

1 For a similar assessment of our limited contribution to this conversation, see Robert W. Kates, The human environment: the road still beckoning *Annals of the Association of American Geographers* **77** (1987) 525–34; Terry G. Jordan, Preadaptation and European colonization in rural North America *Annals of the Association of American Geographers* **79** (1989) 489–500. The discipline has not really progressed beyond programmatic statements like Margret FitzSimmons, Reconstructing nature *Environment and Planning D: Society and Space* **7** (1989) 1–3; Chris Philo, Delimiting human geography: new social and cultural perspectives, in C. Philo (ed.), *New worlds, new words* (Aberystwyth 1991) 25.

2 Further justification for this remark would itself require an extended review of the geographic literature which neither space nor time will allow. I submit the following, however, as symptomatic of the disengaged and excessively technical bent of

geographic scholarship on global change: R. J. Palm, *Natural hazards* (Baltimore 1990); B. L. Turner II *et al.* (Eds.), *The Earth as transformed by human action* (New York 1990); A. M. Mannion, *Global environmental change* (New York 1991). Some of the reasons for this are suggested by Margaret FitzSimmons, The matter of nature *Antipode* **21** (1989) 106–20.

3 See, for example, M. Watts, *Silent violence* (Berkeley 1983); S. Hecht and A. Cockburn, *The fate of the forest* (London 1990); M. W. Martin, *Wagering the land* (Berkeley 1992).

4 Donald Worster, Transformations of the earth: toward an agroecological perspective in history *Journal of American History* **76** (1990) 1089. For reviews of the environmental history project, see Richard White, American environmental history: the development of a new historical field *Pacific Historical Review* **54** (1985) 297–335; Donald Worster (Ed.), *The ends of the Earth* (New York 1988); and the roundtable discussion of Worster, Transformations of the earth, by Alfred W. Crosby, Richard White, Carolyn Merchant, William Cronon, and Stephen J. Pyne in *Journal of American History* **76** (1990) 1087–1147. As the nationality of these authors and their publication outlets would indicate, environmental history is much better established in the United States than in Canada.

5 I use the word 'nature' here and throughout this essay with some ambivalence. In *Keywords* (rev. ed., London 1983) 219, Raymond Williams called it 'perhaps the most complex word in the language'. It is also one of the most potent. Nature suggests a very troubling distinction between humans and the other organisms and material objects in the world. In this discursive position as 'other', nature has also helped constitute many different sorts of racism, colonialism, sexism, and class domination. In the sense of the inherent force directing the world, nature provides a silent, but transcendent, authorization for scientific and other discourses that are legitimated by appeal to the way the world works. Through such appeals to nature, science has replaced religion as the pre-eminent form of social legitimation. See, W. Wright, *Wild knowledge* (Minneapolis 1992). Unfortunately, the English language provides few substitutes for the word nature, and so its use becomes difficult to avoid. I have tried to use it with care, but the reader should take note of these different effects sneaking into my text.

6 While I am sympathetic to the many local varieties of environmental activism and the many different shades of Green in the environmental movement, I would still maintain that, for all their many differences, environmental critics have all depended on ecological discourse to claim an exclusive and privileged knowledge about nature. As I shall explain, revision in ecological science challenges the environmentalists' reading of nature. For a discussion of different national histories of the Green movement, see S. P. Hays, *Beauty, health, and permanence* (New York 1987); A. Bramwell, *Ecology in the 20th century* (New Haven 1989); J. McCormick, *Reclaiming paradise* (Bloomington 1989); P. Alphandery, Pierre Bitoun, and Yves Dupont, *Léquivoque écologique* (Paris 1991).

7 Robert K. Cowell, What's new? community ecology discovers biology, in P. W. Price *et al.* (Eds.), *A new ecology* (New York 1984) 392; Robert P. McIntosh, Pluralism in ecology *Annual Review of Ecology and Systematics* **18** (1987) 321–41; Donald Worster, The ecology of order and chaos *Environmental History Review* **14** (1990) 1–18.

8 The development of the ideographic/nomothetic distinction in neo-Kantian philosophy is described in J. N. Entriken, *The betweenness of place* (Baltimore 1991), 93–102.

9 The best histories of recent ecological science are M. Nicolson, The development

of plant ecology, 1790–1960 (unpubl. Phd. diss., University of Edinburgh 1984); R. P. McIntosh, *The background of ecology* (New York 1985); S. E. Kingsland, *Modelling nature* (Chicago, 1985); T. Söderqvist, *The ecologists* (Stockholm 1986); Worster, Ecology of order and chaos, 1–18; G. Mitman, *The state of nature: ecology, community and American social thought, 1900–1950* (Chicago 1992); J. B. Hagen, *An entangled bank: the origins of ecosystem ecology* (New Brunswick 1992); R. A. Overfield, *Science with practice: Charles E. Bessey and the nurturing of American botany* (Ames 1993). Daniel Botkin winds his own idiosyncratic and highly stimulating way across much of this territory as well. D. B. Botkin, *Discordant harmonies* (New York, 1990).

10 Worster, Ecology of order and chaos, 1–18. Questions about methodological individualism and the fundamental level of biological organization were not unrelated to wider social currents. See Evelyn Fox Keller, Demarcating public from private values in evolutionary discourse *Journal of the History of Biology* **21** (1988) 195–211.

11 Colwell, *op cit.* 389.

12 Botkin, *Discordant harmonies* reviews much of the recent uncertainty in ecology about the ontological status of their analytical units.

13 Kenneth H. Mann, Towards predictive models for coastal marine ecosystems, in L. R. Pomeroy and J. L. Alberts (Eds.). *Concepts of ecosystem ecology* (New York 1988) 291–316; A. F. McEvoy, *The fishersman's problem* (New York 1986).

14 M. L. Heinselman, Fire in the virgin forests of the Boundary Waters Canoe Area, Minnesota *Quarternary Research* **3** (1973) 329–82.

15 John A. Wiens, On understanding a non-equilibrium world: myth and reality in community patterns and processes, in D. R. Strong *et al.* (Eds.), *Ecological communities* (Princeton 1984) 439–57; Robert M. May, Levels of organization in ecology, in J. M. Cherrett (Ed.), *Ecological concepts* (Princeton 1988) 339–63; H. L. Shugart and D. L. Urban, Scale, synthesis, and ecosystem dynamics, in *Concepts of ecosystem ecology*, 279–89.

16 See, for example, David E. Hibbs, Gap dynamics in a hemlock-hardwood forest *Canadian Journal of Forest Research* **12** (1982) 522–27; S. T. A. Pickett and P. S. White (Eds.), *The ecology of natural disturbance and patch dynamics* (Orlando 1985).

17 The landmark paper here is George L. Jacobson, Thom Webb III, and Eric C. Grimm, Patterns and rates of vegetation change during the deglaciation of eastern North America, in W. F. Ruddiman and H. E. Wright, Jr (Eds.), *North America and adjacent oceans during the last deglaciation* (Boulder, CO: Geological Society of America, The Geology of North America, vol k–3, 1987). Of course, neither the use of phytolith data, nor the argument that species behave individually was new or unique to these authors. In his dispute with Clements, Herbert Gleason used phytolith data. Gleason, Vegetational history of the Middle West *Annals of the Association of American Geographers* **12** (1922) 78–85. For a recent review of these techniques, see Glen M. MacDonald and Kevin J. Edwards, Holocene palynology: I principles, population, and community ecology *Progress in Physical Geography* **15** (1991) 261–89 and their, Holocene palynology: II human influence and vegetation change *Progress in Physical Geography* **15** (1991) 364–91.

18 For a brief outline of ecologists' uses of the idea of a balance of nature, see Frank N. Egerton, Changing concepts in the balance of nature *Quarterly Review of Biology* **48** (1973) 322–50.

19 These particular definitions of resilience, persistence, variability, and resistance come from S. L. Pimm, *The balance of nature?* (Chicago 1991). There are others.

See, for example, C. S. Holling, Resilience and stability of natural ecosystems *Annual Review of Ecology and Systematics* **4** (1973) 1–23.
20 See, for example, the essays in J. M. Cherret (Ed.), *Ecological concepts*, (Princeton 1988).
21 Laurence R. Pomeroy, Eugene C. Hargrove, and James J. Alberts, The ecosystem perspective, in L. R. Pomeroy and J. J. Alberts (Eds.), *Concepts of ecosystem ecology* (New York 1988) 17.
22 W. Dean, *Brazil and the struggle for rubber* (New York 1987) 53–66.
23 S. J. Pyne, *Fire in America* (Princeton 1982); W. Cronon, *Changes in the land* (New York 1983); M. Williams, *Americans and their forests* (New York 1989); T. Silver, *A new face on the countryside* (New York 1990).
24 Emily W. B. Russell, Indian-set fires in the forests of the northeastern United States *Ecology* **64** (1983) 78–88. Also see the exchange between R. T. T. Forman and E. W. B. Russell, Evaluation of historical data in ecology *Bulletin of the Ecological Society of America* **64** (1983) 4–5 and R. L. Myers and P. A. Peroni, Approaches to determining aboriginal fire use and its impact on vegetation *Bulletin of the Ecological Society of America* **64** (1983) 217–218. For an analysis of the fossil pollen data, see, William A. Patterson III and Kenneth E. Sassaman, Indian fires in the prehistory of New England, in G. P. Nicholas (Ed.), *Holocene human ecology in northeastern North America* (New York 1988); William A. Patterson III and Andrew E. Backman, Fire and disease history of forests, in B. Huntley and T. Webb III (Eds.), *Vegetation history* (Dordecht 1988) 603–32; John H. McAndrews, Human disturbance of North American forests and grasslands: the fossil pollen record, in *ibid.*, 673–98.
25 Worster, *Nature's economy* (San Francisco 1977); R. Tobey, *Saving the prairies* (Berkeley 1981).
26 Donald Worster, Review of *Changes in the land* by William Cronon *Agricultural History* **58** (1984) 508–509.
27 Worster, History as natural history: an essay on theory and method *Pacific Historical Review* **53** (1984) 13.
28 Worster, *Nature's Economy*, 242, 241.
29 Worster, Ecology of order and chaos, 16.
30 Worster, *Nature's Economy*, 241.
31 Very similar politics of representation are at work in the debate about women's reproductive rights. By claiming the exclusive right to speak for the unborn, anti-abortionists silence pregnant women who are discursively reconstituted as beings with opposing interests. Donna Haraway, The promises of monsters: a regenerative politics for inappropriate/d Others, in L. Grossberg *et al.* (Eds.), *Cultural studies* (New York 1992) 311–15.
32 Worster, *Nature's economy*, 241, 345.
33 J. Perkins, *Insects, experts, and the insecticide crisis* (New York 1982).
34 C. Merchant, *The death of nature* (San Francisco 1980). For a rather different view of these changes, see Denis Cosgrove, Environmental thought and action: pre-modern and post-modern *Transactions of the Institute of British Geographers*, new series **15** (1990) 344–58.
35 Merchant, *Ecological revolutions.*
36 McEvoy, *op. cit.*
37 Steve Woolgar, Irony in the social study of science, in K. D. Knorr-Cetina and Michael Mulkay (Eds.), *Science observed* (London 1983) 262. For a fuller critique of this instrumental irony, see S. Woolgar, *Science, the very idea* (Chichester 1982) 89–111. Richard White has also expressed some ambivalence about this use

of irony in western American history. White, Trashing the trails, in P. N. Limerick *et al.* (Eds.) *Trails* (Lawrence, KA 1991) 34–35.

38 Richard White, Environmental history, ecology, and meaning *Journal of American History* **76** (1990) 1114. For similar concerns, see, Richard White, American environmental history: the development of a new historical field *Pacific Historical Review* **54** (1985) 334; John Opie, Environmental history: pitfalls and opportunities *Environmental Review* **7** (1983) 15.

39 Leibhardt, Interpretation and causal analysis: theories in environmental history *Environmental Review* **12** (1988) 26.

40 For critiques of Geertz in human geography and anthropology, see C[arolyn] Mills, 'Life on the Upslope': the postmodern landscape of gentrification *Environmental and Planning D: Society and Space* **6** (1988) 171–72; Bob Scholte, The charmed circle of Geertz's hermeneutics: a neo-marxist critique *Critique of Anthropology* **6** (1986) 261–79.

41 J. Clifford, *The predicament of culture* (Cambridge, Mass. 1988); Elizabeth Ann R. Bird, The social construction of nature: theoretical approaches to the history of environmental problems *Environmental Review* **11** (1987) 255–64.

42 B. Latour and S. Woolgar, *Laboratory life* (Beverley Hills 1979), 176.

43 Bird, *op. cit.*, 25.

44 *ibid*, 261.

45 P. Novick, *That noble dream* (New York 1988).

46 T. Kuhn, *The structure of scientific revolutions* (2nd edn. Chicago 1970).

47 H. Putnam, *Reason, truth and history* (Cambridge 1981) 50; R. Rorty, *Philosophy and the mirror of nature* (Princeton 1979). C. G. Prado, *The limits to pragmatism* (Atlantic Highlands, N.J., 1987) provides a sympathetic introduction to Rorty.

48 See, for example, J. Derrida, *Of grammatology*, trans. G. C. Spivak (Baltimore 1976).

49 See, for example, J. W. Scott, *Gender and the politics of history* (New York 1988).

50 J. H. Hexter, *The history primer* (New York 1971) 270, quoted in Novick, *op. cit.*, 594. Novick's book provides a useful roadmap of the debate over realism in the American profession, although he is not especially sensitive to the political implications of these challenges. For all their other differences, both the Left and the Right defend realism by indiscriminate assaults on postmodern theories and theorists. In tone, the fulminations of B. Palmer, *Descent into discourse* (Philadelphia 1990) read much like D. Harvey, *The condition of postmodernity* (Oxford 1989). On the Right, see G. Himmelfarb, *The new history and the old* (Cambridge, MA. 1987).

51 Certainly it would be possible to qualify this broad-brush assessment somewhat. Many quarters of the discipline, particularly feminist, post-colonial, and intellectual historians, have taken up these questions about language and the social construction of knowledge. But in the main, historians have continued their research undisturbed by these esoteric and seemingly irrelevant debates.

52 White, Environmental history, ecology, and meaning, 1114.

53 William Cronon, A place for stories: nature, history, and narrative *Journal of American History* **78** (1992) 1347–76 The quotation is from page 1352.

54 *Ibid.* 1370.

55 Donald Worster, Doing environmental history, in *Ends of the earth*, 289.

56 Cronan, *loc. cit.* 1370–74.

57 The phrase, of course, belongs to Michel Foucault, quoted in L. Kritzman, (Ed.), *Michel Foucault* (New York 1988) 118. His extensive writings provide a diverse

and insightful guide to the workings of discursive formations and the relations between power and knowledge.

58 Cronon, *loc. cit.* 1372.

59 On the changing interpretations of the Civil War, see Thomas J. Pressly, *Americans interpret their Civil War* (2nd edn. New York 1965); Novick, *op. cit.*, 72–80.

60 B. Latour, *Science in action* (Cambridge, Mass. 1987) describes the tremendous resources aligned behind facts and the great difficulty of challenging them.

61 S. E. Morrison, *History of US naval operations in World War II, Vol. 1* (Boston 1947); D. Van Der Vat, *The Atlantic Campaign* (New York 1988).

62 G. Wills, *Lincoln at Gettysburg* (New York 1992).

63 R. J. Bernstein, *Beyond objectivism and relativism* (Philadelphia 1983) provides a very useful guide in the effort to escape the trap of this dichotomy

64 Worster, *Nature's economy*, 241.

65 For a discussion of this very danger, see Peter J. Taylor, Technocratic optimism, H. T. Odum, and the partial transformation of the ecological metaphor after World War II *Journal of the Hisjtory of Biology* **21** (1988) 213–44.

66 Some of the first discussions of these measurements took place at a 1972 conference at the Brookhaven National Laboratory. G. M. Woodwell and E. V. Pecan (Eds.), *Carbon and the biosphere* (Oak Ridge, TN. 1973).

67 One of the best guides to new and different relationships created by the deforestation and development of the tropical rainforests is Hecht and Cockburn *Fate of the forest.*

68 Both the phrase and the vision of situated knowledges belong to Donna Haraway, Situated knowledges: the science question in feminism and the privilege of partial perspective, reprinted in her, *Simians, cyborgs, and women* (New York 1991) 183–201.

Space, Spatiality and Spatial Structure

SPACE, SPATIALITY AND SPATIAL STRUCTURE

Spaces in question

Human geography has been interested in questions of space (in one form or another) since its inception. Before the formalisation of **spatial science** in the middle decades of the 20th century, two of the discipline's most significant debates over its conceptions of space were staged within regional geography and political geography. Traditional regional geography had to find ways of integrating accounts of physical space with accounts of social space; traditional political geography was racked by arguments over the ways in which state power was inscribed in space, most fiercely in discussions over geopolitics and territoriality. These two axes – the relations between physical space and social space, and what we might call more generally the politics of space – were carried over into controversies about spatial science, and they run through much of the discussion that follows.

Geography and spatial science

'Spatial science' is a school of geographical thought directed towards spatial analysis, the modelling of spatial systems, and the geometric conceptualisation of spatial structure. According to Allen Scott (1980), its formation as an **empirical–analytic science** was closely connected to the planning ideologies of post-war capitalism: as such, it emerged in the 1950s, accelerated in the 1960s, and reached its first summit in the early 1970s. Since then it has been the object of vigorous critical discussion, but the tradition is by no means a relic of the recent past. Spatial science continues to make important contributions to spatial statistics and **epidemiological** forecasting, and it is pursued most ambitiously by those scholars who see the junction between a revivified spatial science and the vastly enhanced data handling and displaying capacities of geographical information systems (GIS) as the lodestar of future geographical inquiry.

In its original form, spatial science rested on four main pillars: geometry, models of physical science, a revisionist historiography, and philosophy of science.

Geometry

Many of the principal architects of spatial science were critical of what they called geography's *exceptionalist* tradition. That term was derived from a distinction made towards the end of the 18th century by the German philosopher Immanuel Kant, between knowledge that classifies the world on the basis of formal or functional similarities (a *logical* classification) and knowledge that classifies the world on the basis of coexistence in time or space (a *physical* classification). Kant had argued that the sciences depended on logical classifications, which in principle allow for generalisation and universalisation, whereas history and geography were alone – hence 'exceptional' – in their dependence on physical classifications which captured unique configurations in time and in space respectively:

'Science'	*'History and Geography'*
logical classification	physical classification
similarities	coexistence in time and space
generalisation	uniqueness

If traditional regional geography were to remain the centre of the discipline, the very core of geographical inquiry, studying those elements of the biophysical and cultural–historical landscapes that are found together within the unique space of a particular region, then – in Kantian terms, at any rate – geography could not aspire to properly scientific status.

For geography to relinquish its exceptionalist position and to undertake a search for generalisations and perhaps even scientific laws, the architects of spatial science believed it necessary to establish a 'logical' rather than a 'physical' basis for the analysis of such spaces. They found a solution in the universals of geometry: as William Bunge put it in *Theoretical geography* (1962), 'The science of space [geography] finds the logic of space [geometry] a sharp tool'. It was also suggested that conceiving spatial structure in abstract geometric terms might well provide a foundation for an integrated discipline, because human geography and physical geography would then be able to draw on common analytical techniques (such as network analysis) to identify common spatial structures (in topologies of drainage systems and road networks, for example: see Haggett and Chorley, 1969).

Models of physical science

The advocates of spatial science modelled many of their claims on the physical sciences, perhaps most rhetorically when Chorley and Haggett (1967) tacitly invoked Einstein to foreshadow the construction of 'a general theory of locational relativity', but more prosaically through a series of models derived from the physical sciences. Some of these were explicit, such as the use of a gravity model to predict spatial interaction, or the use of a mechanical Varignon frame to calculate the moments of Weber's model of industrial location. Others were less obvious (but no less powerfully present). Many of the most significant contributions to modern location theory, including Christaller's central place theory and Lösch's general equilibrium model of the space-economy, were based on marginalist economics and its 20th-century heir, neoclassical economics, discourses that crucially depended on concepts drawn directly and deliberately from statistical mechanics (see Gregory, 1994).

A revisionist historiography

Haggett (1965) urged that spatial science simply reinstated the geometric tradition that 'was basic to the original Greek conception of the subject', a view that conveniently promoted Ptolemy and ignored Strabo's much earlier claims for geography as regional description. But most proponents were much less interested in reawakening classical traditions of geography than in constructing a *discontinuist* history that represented spatial science as a distinctively *new* intellectual formation: a truly 'New Geography'. In order to do so, they often invoked T. S. Kuhn's controversial account of scientific revolutions. Kuhn had argued that what distinguished the intellectual history of the physical sciences from that of many other sciences, and most obviously from the social sciences and humanities, was the existence of long-term consensus ('normal science') governed by a shared set of assumptions and procedures (a scientific 'paradigm') punctuated by intense periods of radical change (a 'scientific revolution').

These ideas were used by many advocates of spatial science to represent traditional geography as a pre-paradigmatic discipline, one that was informal, unsystematic, and without coherent conceptual structure. They insisted that it was only through the application of formal mathematical and statistical methods during the discipline's so-called 'Quantitative Revolution' of the 1960s that the geography had achieved paradigmatic – and hence truly scientific – status. As Kuhn's model applied specifically to the physical sciences, the plausibility of its polemical extension to geography plainly depended on situating spatial science within the tradition of the physical sciences (for a critical discussion, see Mair, 1986).

Philosophy of science

When spatial science reached its height there was a belated attempt to underwrite and so legitimise its models and methods through a formal philosophy of science: usually, the philosophy (or, rather, Philosophy) of **logical positivism** that we described at the beginning of this book. This was codified most elegantly in David Harvey's *Explanation in geography*, published in 1969, although it has to be said that many practitioners of spatial science were not greatly interested in philosophical arguments (and were thus largely unmoved by the subsequent philosophical critique), and that Harvey himself soon moved away from many of the positions he set out in that book.

Critiques and continuities

The critique of spatial science has been many-stranded, but in general it has followed two closely connected paths: a critique of both the *conception of science* and the *conception of space* involved in the formation of spatial science. These find common ground in their dissent from the ways in which spatial science typically abstracts its *procedures* and its *explanation*s from the social (Sheppard, 1995).

As we argued in our introduction to Section Two, post-positivist geographies typically acknowledge that they are grounded in particular social settings: that their knowledges are always situated and that there is a politics behind their claims to truth. This has involved human geographers thinking very carefully about their own representations of space and the representations of space that inhere in the conduct of everyday life. In doing so, they have reflected on the various spatial **metaphors** that have become a commonplace within the late-20th-century academy, and indeed late-20th-century culture more generally, and they have drawn attention to the problematic nature of many of them. They argue that most spatial metaphors – such as 'region', 'landscape', or 'place' – are used too lightly, as though they are such familiar and everyday terms we know almost instinctively what they mean. But the critics urge that in fact 'space' is never innocent but always shot through with relations of power, so that metaphorical spaces must be articulated with the dense frictions of material spaces (Smith and Katz, 1993).

Inquiries of this kind inevitably transform one's sense of scientific practice, especially once the significance of Livingstone's 'historical geographies of science' is properly appreciated (Chapter 1), but they also have a direct effect on the ontological basis presupposed by science. This is a difficult idea to summarise simply, but **positivist** philosophy assumes an event-**ontology**: in other words, the leading edge of any scientific inquiry conducted under its sign is the observation of *events*, which are thus the ultimate objects of its knowledge. In elementary diffusion models, for example, events of one kind (telling someone about an innovation) are coupled to events of another kind (adopting or rejecting the innovation). Explanation then resides in linking a map of one set of events (the circulation of information) via some sort of distance function (usually a matrix of distance-bound interactions) to a map of another set of events (the diffusion of innovations). Post-positivist geographies, in contrast, typically depend upon an ontology of *positions and places, sites and structures* which are the very conditions for human existence and, by extension, for the formation of the human sciences themselves (Pickles, 1985; Schatzki, 1991). This substantially alters the analysis of our spatial diffusion example, focussing attention on the ways in which positions and sites within a sociospatial structure open up and close off access to resources in a complex and variable geography of power and domination (see Gregory, 1985).

As this simple example implies, post-positivist geographies have been instrumental in what Neil Smith (1990) describes as the 20th-century discovery of 'deep space'. 'Deep space,' he declares, 'is quintessentially social space; it is physical extent fused through with social intent'. In the course of these explorations, spatial science has been criticised for what was, in effect, its ambitious limitation: its attempt to explain spatial structure in terms of purely *spatial processes*, a project that Robert Sack (1974) criticised, in a mischievous mirroring of Kant's exceptionalism, as 'spatial separatism', and that Edward Soja (1989) lambasted as an introverted, defensive 'geographical involution'.

When, in the wake of these objections, human geographers began to turn outwards, they first turned to political economy, to social and political theory, and to cultural anthropology in order to theorise the substantive processes by which spatial structures are produced. But these were all disciplines whose spatial implications are at best oblique, and it was not long before the projection of their aspatial formulations onto the plane of human geography was seen to be problematic. The need to reconstruct critical social theory to capture what Soja (1989) termed the 'sociospatial dialectic' was widely recognised within the discipline (see Chapter 14). So, for example, Harvey (1982) claimed that the production of space is a largely untheorised but nonetheless necessary condition for Marx's critique of the capitalist **mode of production**, and worked to establish what he came to call an '**historico-geographical materialism**' that clarifies the spatial foundations of Marx's political economy. Or again, a '**structurationist** school' argued that the spatiality of social life is intrinsic to any theorisation of the intersections between 'human **agency**' and 'social structure' that has for so long bedevilled orthodox social theory (see Chapter 22). These examples could and no doubt will be multiplied, and there are often sharp disagreements between their proponents, but all of them accentuate the production of social space – or **spatiality** – by *social processes* and accept that the spatial structures of social life cannot adequately be explicated by the abstract logics of geometry or the abstract operations of mathematical processes. Indeed, one of the most invigorating contributions to post-positivist geography has been an exploration of the multiple ways in which so many areas of modern life have been colonised through the social production of exactly these abstract spaces (see Lefebvre, 1991; Gregory, 1994).

These objections strike at the very heart of spatial science, and they have profoundly (and properly) politicised spatial analysis. But it is important not to lose sight of the *continuities* that remain within the critical project. These extend beyond the use of mathematical and statistical techniques, particularly in those economic geographies that draw upon analytical political economy, though this is more widespread than many commentators are prepared to acknowledge. Many of those who have proposed explicitly socialised conceptions of spatial structure have retained an unambiguous commitment to scientific forms of inquiry. They draw on non-positivist philosophies of science – including **phenomenology** and **realism** – but they continue to work with those 'depth models' that are central to modern science: this is particularly clear, we think, in attempts to disclose the generative structures that underlie the uneven particularities of the capitalist space-economy. In addition, some of those contributions continue to be fascinated by – and to learn from – the insights of modern physics. Sometimes this occurs at several removes, as in Harvey's (1989) metaphor of '**time–space compression**' (see Chapter 15) to dramatise what would once have been called a reduction in the 'friction of distance'. Sometimes it is much more direct, as in Doreen Massey's (1994) 'view from physics' to reveal the inseparability of space and time, and to urge upon human geography a conception of space–time as being produced through the changing connections and dislocations between objects. And both of these authors continue to think, at least in part, in terms of geometries: the 'structured coherence' that Harvey (1982) detects in the regional production systems of a capitalist economy, and the 'power geometry' that Massey (1993) sees embedded in time–space compression.

Even more radical views can retain some of these inflections. Those feminist geographies that draw upon Donna Haraway's work, for example, are involved in the production of a recognisably scientific tradition since the point of her project, as we explained in our introduction to Section One, is not to dismiss the relevance of science but to redescribe and learn from its practices. In seeking to establish a different way of thinking about '**objectivity**', Haraway provides a series of opportunities for a critical analysis of spatiality: perhaps most of all by emphasising the particularity and embodiment of vision, since visuality plays such an important part in conceptualising spatiality and conducting spatial analysis (see Gregory, 1994). Within the discipline itself, Gillian Rose (1992) has developed one of the most sustained critiques of the optics of spatial science and, even more important, of many of its successor projects. In her view, many of these analyses are predicated on a **masculinist** will-to-power that desires to render space as 'infinitely knowable' and to make space transparent to its (critical) scientific gaze; yet her own explorations of 'a politics of paradoxical space' – of the ambiguities, pluralities and mobilities of a non-masculinist spatial imaginary – rest on a principled recognition of what she calls, after Haraway, 'a geometrics of difference'.

None of this implies that nothing has changed or that behind the critical façade it is business as usual. But we do think it is extremely important to reflect on the ways in which the predicament of spatial science continues to mark the contemporary discipline. That said, there have been a number of significant *departures*. Foremost among them, we suggest, is a determination to recognise that, just as our knowledges are embodied and implicated in relations of **power**, so a critical human geography must address the multiple ways in which relations between bodies and spaces are refracted through grids of power and knowledge. Investigations of this sort are in their infancy, and many of them have been inspired by versions of **post-structuralism**. Post-structuralism is a diverse group of ideas and arguments, but common to most of them is an insistence on the importance of *difference* (see also our introduction to Section Eight). This may not seem much of a departure: both regional geography and spatial science revolved around systems of difference. But these geographical traditions both organised their constitutive differences around what post-structuralist critics would call an *essential unity*. Within traditional regional geography, for example, the world was assumed to have a fundamentally regional structure that could be disassembled and reassembled like some vast and intricate jigsaw puzzle. What is more, each piece of the puzzle – each region – had its own internal unity: thus the significant elements found within a particular region contributed towards a sense of regional identity, a distinctive 'sense of place', that gave the region its coherence and distinguished it from other regions. Within spatial science the world was assumed to have a fundamentally geometric structure that could be dissected or replicated by the modelling of superimposed patterns of varying wavelengths. The discovery of common spatial patterns, repeated from one place to another, was taken to be evidence of an underlying and invariant order – a deep 'spatial structure' – beneath the surface mosaic of spatial variation. It is exactly that conception of a single order, an 'essential unity', often supposedly produced by some central generating mechanism, that post-structuralism disavows.

Post-structuralism was not developed with geography in mind; in most cases, its criticisms were directed against the essential unity of the singular, knowing, and rational subject that is at the heart of **humanism**. This does have important implications for human geography, however, because similar conceptions occupy the core of the mainstream discipline: the 'culture groups' of traditional cultural geography, for example, the 'rational economic man' of spatial science and location theory, and the 'human subject' of **humanistic geography**. There is thus no doubt, as we suggest in our introduction to Section Seven, that post-structuralism's emphasis on the complex, contradictory ways in which subjectivities are constructed has invited the construction of an altogether different and explicitly post-humanist geography. But it turns out that post-structuralism bears much more directly than this on the analysis of space, spatiality, and spatial structure. For many of the most prominent post-structuralist philosophers have made extensive use of a spatialised vocabulary – most vividly, Michel Foucault and Jacques Lacan – and in some cases this is much more than a spatial metaphoric. Foucault once remarked that 'a whole history remains to be written of spaces', for example, 'which would at the same time be the history of powers', and several geographers have started to work with Foucault's writings on power, space, and subjectivity (Driver, 1985; Philo, 1992). Yet this affirmation of the historicity of spatial structure, which was a conspicuous absence from spatial science, is not altogether unproblematic. Soja (1989) has cautioned that Foucault's 'spatialised historiography' is emphatically not an **historicism**, for example, and those geographers who have been drawn to the writings of Jacques Lacan in order to explore a psychoanalytics of space are likely to have an even more reserved attitude to historicism (see Copjec, 1994). Be that as it may, psychoanalytic theory introduces an emphasis on the corporeality and spatiality of *desire* that is altogether foreign to the analytics of spatial science (Pile, 1996; see also Chapter 23).

Space, spatiality, and spatial structure

All three essays that follow represent attempts to go beyond the formulations of modern spatial science. In the first essay, Edward Soja notes that many of the early critiques of spatial science were informed by radical political economy. It was during these debates that Marx's objections to 'commodity fetishism' were extended into the domain of what came to be called 'spatial fetishism'. Just as Marx cautioned against reducing relations between people in capitalist society (class) to relations between things (commodities) – thereby making a 'fetish' of the commodity – so radical critics of spatial science urged the need to avoid reducing relations between people in society to relations between points in space: making a 'fetish' of space. Soja's central argument is that this analogy holds *only if the essentially dialectical nature of Marx's thought is appreciated*: after all, Marx's critique of political economy did not displace the commodity, but instead *placed it at the very centre of analysis in order to disclose the social relations that were inscribed within and constituted through its various forms and productions.* In much the same way, Soja argues, it is necessary to place the concept of *spatiality,* 'the created space of social organization and production', at the very centre of critical human geography, *in order to disclose the social relations that are inscribed within and constituted through its various forms and productions.*

Although this space has a dense materiality and physicality, Soja insists that it is not the space of the physicists ('space in its generalized and abstracted physical form'); it is, rather, both the medium and the outcome of social processes that are bound in to the operation of the capitalist mode of production. Hence Soja invokes the French Marxist philosopher and sociologist Henri Lefebvre, to insist that 'Space is not a scientific object removed from ideology and politics; it has always been political and strategic'. Soja knows very well that Lefebvre's interest in these questions placed him on the periphery of mainstream Marxism, and he claims that **historical materialism** has typically prioritised time and history over space and geography. The reasons for this are, in part, matters of intellectual history – an over-reliance on what Soja sees as 'the largely aspatial, closed-system theorizing' of the published volumes of *Capital* and a neglect of Hegel's writings on the state and spatiality – but they are also matters of historical geography. For Soja believes that the construction and clarification of the concept of spatiality is a profoundly political intervention, one of decisive contemporary significance, because the social organisation of space has assumed a much greater significance in advanced capitalism than it had in Marx's 19th-century world.

As it is set out here, Soja's argument places him very close to David Harvey's projected historico-geographical materialism, but Soja subsequently developed his version of the sociospatial dialectic to construct an avowedly 'postmodern geography' that tried to retain many of Marx's central analytical insights while also moving some considerable way beyond ('post') them (Soja, 1989; 1996). David Harvey himself has a much more reserved attitude to **postmodernism**, however, and in the second essay of this section he argues that the representations of time and space which are contained within late-20th-century postmodern culture have been brought about by – and are in large measure reflections of – transformations in the modern capitalist economy and, most particularly, in that transition from Fordism to flexible accumulation that he describes in detail elsewhere (Harvey, 1989; see also Chapter 4). For Harvey, it follows that celebrations of postmodernism can all too easily become uncritical endorsements of capitalism.

Harvey treats time and space as social constructions which, in their dominant forms, acquire an objectivity – a 'matter-of-factness' – that provides tacit guarantees of continuity, stability, and social order. These constructions are by no means stable; they differ between societies and are often contested within them. They have a special volatility within capitalist societies, Harvey explains, because capitalism is relentlessly revolutionary: its path of 'creative destruction' has always involved 'radical reorganizations of time and space'. These reorganizations have taken multiple forms – turnpikes and canals, railways and steamships, telegraph and radio, aircraft and satellites, television and telecommunications – but they are all bound into a dramatic acceleration in the turnover time of capital: what Harvey calls 'the elimination of spatial barriers and the struggle to "annihilate space by time"' (see also Kirsch, 1995). The fulcrum of Harvey's argument is that these economic processes have a series of complex impacts on cultural productions, and that these time–space displacements (in economic and cultural registers) have intensified in the course of the 19th and 20th centuries. In making these claims Harvey seems to be trembling on the edges of what, in aesthetic theory, would be called the 'sublime' – a sense of being overwhelmed by the scale and sheer power of the world – that was a persistent motif in 19th-century Western thought and which reappears in late-20th-century postmodernism (Jameson, 1991). When Harvey talks about 'time–space compression', he means to accentuate exactly this 'sense of overwhelming change in space–time dimensionality', a sense of disorientation, bewilderment, even fear that has been questioned by subsequent writers (Massey, 1993; Thrift, 1995).

In the final essay of this section, Felix Driver presents a preliminary engagement with some of the ideas of Michel Foucault. He begins by emphasising Foucault's 'powerful sense of history', and here his remarks dovetail with his own insistence on the *historicity* of human geography (Driver, 1988). Foucault's was a very particular sense of history, however, one that rejected 'a continuous history, centred around the human subject' and installed in its place a 'decentred history in a space of dispersion'. This is exactly what we described in our discussion of post-structuralism (above, page 238). Like us, what interests Driver is Foucault's 'new kind of cartography', his 'language of space', his 'spatial fixation'. These obsessions are revealed with a special clarity in one of Foucault's most influential texts, *Discipline and punish*, ostensibly an inquiry into the history of the prison in France but in fact a sustained consideration of changing modalities of power. Until Foucault's intervention, most studies of power were directed towards what he called juridico-political or *sovereign power*, which is to say power that is exercised from a centre outwards and from a summit downwards; these are the terms which are usually used to analyse state power. But Foucault believed that power takes many forms, that it 'gains hold of human beings' in different ways. What most interested him was another modality of power, which he called disciplinary power, which was located not at the centre but at the margins; it was capillary, localised, and 'ascending'. Foucault's analysis of modern regimes of surveillance and punishment, which were directed towards the production of what he called 'docile and obedient subjects', showed that the production and regulation of space was constitutively implicated in the production and regulation of bodies. This has turned out to be a provocative claim, but Foucault's project was not a finished and fixed set of propositions, and Driver does not treat it as such. As we have set them out here, sovereign power and disciplinary power appear to be antinomies, for example, but Driver draws attention to the complex ways in which the state traded on both these regimes and thus he invites an analysis of the connections between different modalities of power and their spaces. Other writers have responded to these suggestions by exploring the ways in which sovereign and disciplinary power were implicated in the production of colonial spatialities (Said, 1978; Mitchell, 1988).

References

Bunge, W. 1962: *Theoretical geography*. Lund: Gleerup.

Chorley, R.J. and Haggett, P. (eds) 1967: *Models in geography*. London: Methuen.

Copjec, J. 1994: *Read my desire: Lacan against the Historicists*. Cambridge, MA: MIT Press.

Driver, F. 1985: Power, space, and the body: a critical assessment of Foucault's *Discipline and Punish*. *Environment and Planning D: Society and Space* **3**; 425–446.

Driver, F. 1988: The historicity of human geography. *Progress in Human Geography*, **12**; 497–506.

Gregory, D. 1985: Suspended animation? The stasis of diffusion theory, in Gregory, D. and Urry, J. (eds), *Social relations and spatial structures*. London: Macmillan, 296–336.

Gregory, D. 1994: *Geographical imaginations*. Oxford, UK, and Cambridge, MA: Blackwell.

Haggett, P. 1965: *Locational analysis in human geography*. London: Arnold.

Haggett, P. and Chorley, R.J. 1969: *Network analysis in geography*. London: Arnold.

Harvey, D. 1969: *Explanation in geography*. London: Arnold.

Harvey, D. 1982: *The limits to capital*. Oxford, UK, and Cambridge, MA: Blackwell.

Harvey, D. 1989: *The condition of postmodernity: an enquiry into the origins of cultural change*. Oxford, UK, and Cambridge, MA: Blackwell.

Jameson, F. 1991: *Postmodernism, or the cultural logic of late capitalism* Durham, NC: Duke University Press.

Kirsch, S. 1995: The incredible shrinking world? Technology and the production of space. *Environment and Planning D: Society and Space*. **13**; 529–56.

Lefebvre, H. 1991: *The production of space*. Oxford, UK, and Cambridge, MA: Blackwell.

Mair, A. 1986: Thomas Kuhn and understanding geography, *Progress in Human Geography*, **10**; 345–70.

Massey, D. 1993: Power-geometry and a progressive sense of place. In Bird, J. Curtis, B. Putnam, T. Robertson, G. and Tickner, L. (eds), *Mapping the futures: local cultures, global change*. London and New York: Routledge, 59–69.

Massey, D. 1994: Politics and space/time. In *Space, place and gender*. Cambridge: Polity Press; Minneapolis, MN: University of Minnesota Press, 249–72.

Mitchell, T. 1988: *Colonising Egypt*. Cambridge: Cambridge University Press.

Philo, C. 1992: Foucault's geography. *Environment and Planning D: Society and Space*, **10**; 137–61.

Pickles, J. 1985: *Phenomenology, science and geography: spatiality and the human sciences*. Cambridge: Cambridge University Press.

Pile, S. 1996: *The body and the city: Psychoanalysis, space and subjectivity*. London and New York: Routledge.

Rose, G. 1992: *Feminism and geography: The limits of geographical knowledge*. Cambridge, UK: Polity Press; Minneapolis, MN: University of Minnesota Press.

Sack, R.D. 1974: The spatial separatist theme in geography, *Economic Geography*, **50**; 1–19.
Said, E. 1978: *Orientalism*. Harmondsworth, Middx: Penguin (reprinted 1995).
Schatzki, T. 1991: Spatial ontology and explanation. *Annals of the Association American Geographers*, **81**; 650–70.
Scott, A. 1980: The meaning and social origins of discourse on the spatial foundations of society. In Gould, P. and Olsson, G. (eds), *A search for common ground*. London: Pion, 141–56.
Sheppard, E. 1995: Dissenting from spatial analysis. *Urban geography* **16**; 283-303.
Smith, N. 1990: *Uneven development: Nature, capital and the production of space*. Oxford, UK, and Cambridge, MA: Blackwell.
Smith, N. and Katz, C. 1993: Grounding metaphor: towards a spatialized politics. In Keith, M. and Pile, S. (eds), *Place and the politics of identity*. London and New York: Routledge, 67–83.
Soja, E. 1989: *Postmodern geographies: The reassertion of space in critical social theory*. London: Verso.
Soja, E. 1996: *Thirdspace: a journey through Los Angeles and other real-and-imagined places*. Oxford, UK and Cambridge, MA: Blackwell.
Thrift, N. 1995: A hyperactive world? In Johnston, R. J. Taylor, P. and Watts, M. (eds), *Geographies of global change: Remapping the world in the late twentieth century*. Oxford, UK, and Cambridge, MA: Blackwell, 18–35.
Wolin, R. 1990: *The politics of being: The political thought of Martin Heidegger*. New York: Columbia University Press.

Suggested reading

For a discussion of spatial science and abstract space, see Gregory (1994, pp. 53–69); for a post-structuralist critique of some of the central assumptions of spatial science, see Trevor Barnes, 'Probable writing: Derrida, deconstruction and the Quantitative Revolution in economic geography', in *Logics of dislocation: models, metaphors and meanings of economic space* (New York: Guilford, 1996), pp. 161–84.

Attempts to socialise the analysis of spatial structure are provided by the essays in Derek Gregory and John Urry (eds), *Social relations and spatial structures* (London, Macmillan, 1985). An excellent summary discussion is Sheppard (1995). For general theorisations of the production of space under capitalism, see Neil Smith (1990), and Harvey (1982). For perspectives on the spatiality of late-20th-century capitalism in particular, see Harvey (1989), Soja (1989), and Scott Lash and John Urry, *Economies of signs and space* (London, Sage, 1994).

For some of the possibilities of a post-structuralist geography, see Pile (1996) and Derek Gregory, *Power, knowledge and geography* (Oxford, UK, and Cambridge, MA: Blackwell, forthcoming).

14 Edward Soja
The Socio-spatial Dialectic

Excerpts from: E. Soja, *Postmodern geographies,* Chapter 3. London: Verso (1989)

> Space and the political organization of space express social relationships but also react back upon them . . . Industrialization, once the producer of urbanism, is now being produced by it . . . When we use the words 'urban revolution' we designate the total ensemble of transformations which run throughout contemporary society and which bring about a change from a period in which questions of economic growth and industrialization predominate to the period in which the urban problematic becomes decisive.

These observations are drawn from a postscript to *Social Justice and the City* (1973, 306) in which David Harvey presented a brief appreciation and critique of the ideas of Henri Lefebvre on urbanism, the organization of space, and contemporary Marxist analysis. But Harvey's interpretation accomplished something more than a sympathetic introduction of Lefebvre to anglophonic Marxist geography. It recapitulated the pattern of response to Lefebvre's theory of space that had already appeared in French through Manuel Castells's important work, *La Question urbaine* (1972). Harvey praised Lefebvre but dissented from his insistence on the 'decisive' and 'pre-eminent' role of spatial structural forces in modern capitalist society. Both Harvey and Castells recognized Lefebvre's contribution in dealing brilliantly with the organization of space as a material product, with the relationship between social and spatial structures of urbanism, and with the ideological content of socially created space. But surely Lefebvre had gone too far? They both insinuated that he had elevated the urban spatial 'problematic' to an intolerably central and apparently autonomous position. The structure of spatial relations was being given an excessive emphasis while the more fundamental roles of production (versus circulation and consumption), social (versus spatial) relations of production, and industrial (versus finance) capital were being submerged within an overinterpreted alternative – what Lefebvre called the 'urban revolution', *La Révolution urbaine* (1970). In his conceptualization of urbanism, Lefebvre appeared to them to be substituting spatial/territorial conflict for class conflict as the motivating force behind radical social transformation.

The key question to Harvey in 1973 was whether the organization of space (in the context of urbanism) was '*a separate structure* with its own laws of inner transformation and construction', or 'the *expression* of a set of relations embedded in some broader structure (such as the social relations of production)'. To Harvey – as to Castells previously – Lefebvre seemed to be a 'spatial separatist' and was thus succumbing to what might be called a fetishism of space. Struggling to be serious and rigorous in their application of Marxism, pioneers of Marxist geography like Harvey and Castells thus began to establish certain boundaries beyond which radical spatial analysis must not reach.

This pattern of response pervaded the new Marxist analysis of space that developed in the 1970s, significantly blunting its impact and weakening its accomplishments. The reaction to Lefebvre and the misunderstanding on his ideas was one manifestation of this rigidifying tendency. Indeed one can go a step further and argue that the first generation to develop a spatially explicit form of Marxist analysis – exemplified best in the pioneering works of Harvey and Castells but also in the rapidly expanding literature on radical urban and regional political economy . . . – was built upon an unnecessarily limited conceptualization of spatial relations. Thus what should have been the far-reaching implications of Marxist spatial analysis were unnecessarily blunted through the well-intended but short-sighted efforts of radical scholars to avoid the presumed dangers of spatial fetishism.

Rather ironically, the primary source of misunderstanding seemed to lie in the failure of Marxist analysts to appreciate the essentially dialectical character of social and spatial relationships as well as that of other structurally linked spheres like production and consumption. As a result, instead of sensitively probing the mix of opposition, unity, and contradiction which defines a socio-spatial dialectic, attention was too often drawn to empty categorical questions of causal primacy. Within this rigidly categorical logic, it was difficult to see that the socio-spatial dialectic fitted neither of the two alternatives pressed upon Lefebvre by David Harvey. The structure of organized space is not a separate structure with its own autonomous laws of construction and transformation, nor is it simply an expression of the class structure emerging from social (and thus aspatial?) relations of production. It represents, instead, a dialectically defined component of the general relations of production, relations which are simultaneously social and spatial.

To establish this simultaneity, it must be clearly demonstrated that there exists a corresponding spatial homology to traditionally defined class relations and hence to the contingencies of class conflict and structural transformation. As I will attempt to demonstrate, such a space-to-class homology can be found in the regionalized division of organized space into dominant centres and subordinate peripheries, socially created and polarized spatial relations of production which are captured with greater precision in the concept of geographically uneven development. This conceptualization of the links between social and spatial differentation does not imply that the spatial relations of production or the centre–periphery structure are separate and independent from the social relations of production, from class relations. On the contrary, the two sets of structured relations (the social and the spatial) are not only homologous, in that they arise from the same origins in the mode of production, but are also dialectically inseparable.

That there exists such a dialectical association between what might be called the vertical and horizontal dimensions of the mode of production, is suggested in the writings of Marx and Engels: in discussions of the antithesis between town and countryside, the territorial division of labour, the segmentation of urban residential space under industrial capitalism, the geographical unevenness of capitalist accumulation, the role of rent and private ownership of land, the sectoral transfer of surplus value, and the dialectics of nature. But one hundred years of Marxism have not been enough to develop the logic and scope of these insights.

The atrophy of the geographical imagination in the intervening epoch helps explain why the rebirth of Marxist spatial analysis was so arduous and burdened with unfounded fears of spatial fetishism. The long hiatus also explains why there was so much controversy over terminology, emphasis, and credentials; as well as why divisions persisted between urban, regional, and international political economies rather than leading to the creation of a more unified spatial political economy. Finally it helps us understand why, except for Lefebvre, there was such a lack of audacity – that is, why amid claims that the resurgence of spatially explicit, radical political economy represented a 'new' urban sociology, a 'new' economic geography, a 'new' urban politics, or a 'new' planning theory, no one else was ready to grasp the really radical implication that what was emerging was a dialectical materialism that is simultaneously historical and spatial. What follows is an attempt to recapture the initial assertion of the socio-spatial dialectic and the need for a historico-geographical materialism as originally developed in Soja (1980) and Soja and Hadjimichalis (1979).

Spatiality: the organization of space as a social product

It is necessary to begin by making as clear as possible the distinction between space *per se*, space as a contextual given, and socially-based spatiality, the created space of social organization and production. From a materialist perspective, whether mechanistic or dialectical, time and space in the general or abstract sense represent the objective form of matter. Time, space, and matter are inextricably connected, with the nature of this relationship being a central theme in the history and philosophy of science. This essentially physical view of space has deeply influenced all forms of spatial analysis, whether philosophical, theoretical or empirical, whether applied to the movement of heavenly bodies or to the history and landscape of human society. It has also tended to imbue all things spatial with a lingering sense of primordiality and physical composition, an aura of objectivity, inevitability, and reification.

Space in this generalized and abstracted physical form has been conceptually incorporated into the materialist analysis of history and society in such a way as to interfere with the interpretation of human spatial organization as a social product, the key first step in recognizing a socio-spatial dialectic. Space as a physical context has generated broad philosophical interest and lengthy discussions of its absolute and relative properties (a long debate which goes back to Leibniz and beyond), its characteristics as environmental 'container' of human life, its objectifiable geometry, and its phenomenological essences. But this physical space has been a misleading epistemological foundation upon which to analyse the concrete and subjective meaning of human spatiality. Space in itself may be primordially given, but the organization, and meaning of space is a product of social translation, transformation, and experience.[1]

Socially-produced space is a created structure comparable to other social constructions resulting from the transformation of given conditions inherent to being alive, in much the same way that human history represents a social transformation of time. Along similar lines, Lefebvre distinguishes between Nature as naively given context and what can be termed 'second nature', the transformed and socially concretized spatiality arising from the application of purposeful human labour. It is this second nature that becomes the geographical subject and object of historical materialist analysis, of a materialist interpretation of spatiality.

> Space is not a scientific object removed from ideology and politics; it has always been political and strategic. If space has an air of neutrality and indifference with regard to its contents and thus seems to be 'purely' formal, the epitome of rational abstraction, it is precisely because it has been occupied and used, and has already been the focus of past processes whose traces are not always evident on the landscape. Space has been shaped and molded from historical and natural elements, but this has been a political process. Space is political and ideological. It is a product literally filled with ideologies (1976b, 31).

Organized space and the mode of production: three points of view

Once it becomes accepted that the organization of space is a social product – that it arises from purposeful social practice – then there is no longer a question of its being a separate structure with rules of construction and transformation that are independent from the wider social framework. From a materialist perspective, what becomes important is the relationship between created, organized space and other structures within a given mode of production. It is this basic issue that divided Marxist spatial analysis in the 1970s into at least three distinctive orientations.

First, there were those whose interpretations of the role of organized space led them to challenge prevailing Marxist approaches, especially with regard to definitions of the economic base and superstructure. Again, Lefebvre offered a key argument:

> Can the realities of urbanism be defined as something superstructural, on the surface of the economic base, whether capitalist or socialist? No. The reality of urbanism modifies the relations of production without being sufficient to transform them. Urbanism becomes a force in production, rather like science. *Space and the political organization of space express social relationships but also react back upon them.*[2]

Here we have opened the possibility of a complex socio-spatial dialectic operating within the structure of the economic base, in contrast with the prevailing materialist formulation which regards the organization of spatial relations only as a cultural expression confined to the superstructural realm. The key notion introduced by Lefebvre in the last sentence becomes the fundamental premise of the socio-spatial dialectic: that social and spatial relations are dialectically inter-reactive, interdependent; that social relations of production are both space-forming and space-contingent (at least insofar as we maintain, to begin with, a view of organized space as socially constructed).

Within a regional as opposed to urban frame, similar ideas were developed by Ernest Mandel. In his examination of regional inequalities under capitalism, Mandel (1976, 43) declared that 'The unequal development between regions and nations is the very essence of capitalism, on the same level as the exploitation of labour by capital'. By not subordinating the spatial structure of uneven development to social class but viewing it as 'on the same level', Mandel identified for the regional and international scale a spatial problematic that closely resembled Lefebvre's interpretation of urban spatiality, even to the point of suggesting a powerful revolutionary force arising from the spatial inequalities which he clearly claimed were necessary for capitalist accumulation. In his major work, *Late Capitalism* (1975), Mandel focused upon the crucial historical importance of geographically uneven development in the accumulation process and thereby in the survival and reproduction of capitalism itself. In so doing, he presented one of the most rigorous and systematic Marxist analyses of the political economy of regional and international development ever written.

Neither Lefebvre nor Mandel, however, fully succeeded in defining a cross-scalar synthesis of the socio-spatial dialectic and their formulations thus remained incomplete. Nevertheless, in their attribution to the structure of spatial relations of a significant transformational potential in capitalist society comparable to that which has conventionally been associated with the 'vertical' class struggle, the direct social conflict between labour and capital, both Lefebvre and Mandel presented a point of view which provoked strong resistance from other Marxists who saw the spectre of spatial determinism rising again.

This resistance to the suggestion that organized space represents anything more than a reflection of the social relations of production, that it can engender major contradictions and transformational potential with regard to the mode of production, that it is some way homologous to class structure and relations, defined another, much larger group of radical scholars. Included here was the growing cadre of critics seeking to maintain some form of Marxist orthodoxy by persistent screening of the 'new' urban and regional political economy. Characteristic to this group was the belief that neo-Marxist analysis adds little that is inherently new to more conventional Marxist approaches, that the centrality of traditional class analysis is inviolable, and thus that neo-Marxist urban and regional analyses, while interesting, are too often unacceptably revisionist and analytically muddled. Needless to say, the conceptualization (or non-conceptualization) of space adhered to by this group deviated little from the traditional historicism of Marxism after Marx.

A third approach can be identified, however, falling somewhere in between these two extremes. Its practitioners appeared, implicitly at least, to be adopting much the same formulation as described by Lefebvre and Mandel. Yet when pushed to an explicit stance, they maintained the pre-eminence of aspatial social class definitions, sometimes to the point of tortuously trying to resist the implications of their own analyses. In this group were Manuel Castells, David Harvey, Immanuel Wallerstein, André Gunder Frank, and Samir Amin, all of whom contributed insightful depictions of the socio-spatial dialectic as I have defined it. Each, however, backed off from an open recognition of the formative significance of spatiality into analytically weak and vulnerable positions on the role of spatial structure in the development and survival of capitalism. Whereas the first group mentioned occasionally overstated the socio-spatial dialectic, this group retreated from it without effectively capturing its meaning and implications, creating a confusing ambivalence – which was reacted to in turn by the more orthodox Marxist critics.

To take a prominent example, consider Castells's conceptualization of space in *The Urban Question* (1977), a book purposefully titled to contrast with *The Urban Revolution*, written by his former teacher, Lefebvre.

> To consider the city as the projection of society on space is both an indispensable starting point and too elementary an approach. For, although one must go beyond the empiricism of geographical description, one runs the very great risk of imagining space as a white page on which the actions of groups and institutions are inscribed, without encountering any other obstacle than the trace of past generations. This is tantamount to conceiving of nature as entirely fashioned by culture, whereas the whole social problematic is born by the *indissoluble union* of these two terms, through the *dialectical process* by which a particular biological species (particularly because divided into classes), 'man', transforms himself and transforms his environment in his struggle for life and for the differential appropriation of the product of his labour.
>
> *Space is a material product*, in relation with other elements – among others, men, who themselves enter into particular social relations, which give to space (and to the other elements of the combination) a form, a function, a social signification. It is not, therefore, a mere occasion for the deployment of social structure, but a concrete expression of each historical ensemble in which a society is specified. It is a question, then of establishing, in the same way as for any other real object, the *structural and conjunctural laws that govern its existence and transformation*, and the specificity of its articulation with the other elements of a historical reality. This means that there is no theory of space that is not an integral part of a general social theory, even an implicit one (page 115, emphasis added).

This complex passage encompasses a socio-spatial dialectic but is presented as an alternative to the rejected Lefebvrean view. No wonder the readers of the English translation were confused.

Castells clearly presented space as a material product emerging dialectically from the interaction of culture and nature. Space was thus not simply a reflection, a 'mere occasion for the deployment', of the social structure, but the concrete expression of a combination of instances, an 'historical ensemble' of interacting material elements and influences. How then can one understand and interpret this created space? The route was through what Castells described as the 'structural and conjunctural laws that govern its existence and transformation', a hard indication of the Althusserian structuralism that was then governing Castells's approach to the urban question.

What appeared to separate Castells and Lefebvre was the former's unqualified argument that 'particular social relations' give form, function, and significance to the spatial structure and all other 'elements of the combination'. One 'structure' – the supposedly aspatial social relations of production (which somehow include property rights while ignoring their spatial/territorial dimension) – was thus given a determinant role. But it is precisely this determinative relationship which Lefebvre began to qualify and amend by associating class formation with both social and spatial relations of production and embedding the 'social problematic' in a simultaneously social and spatial division of labour, a vertical and horizontal dimension. There did not yet exist in the 1970s a rigorous formulation of these spatial relations of production and spatial divisions of labour, certainly not one which matched the depth and persuasiveness of Marxist analyses of the social relations of production and the social divisions of labour. But there was no reason to reject the formulation of a socio-spatial dialectic on the grounds that a century of Marxism failed to incorporate a materialistic interpretation of spatiality to match its materialist interpretation of history.

The origins of the neglect of spatiality in Western Marxism

It became common practice amongst Marxist geographers and urban sociologists in the 1970s to argue that within the classic works of Marx, Engels, and Lenin, there are powerful geographical and spatial intuitions but that these emphases and orientations remained weakly developed by successive generations. Many thus approached the task of Marxist spatial analysis largely in terms of drawing out and elaborating upon these classical observations in the context of contemporary capitalism. David Harvey's analysis of the geography of capitalist accumulation (1975) and Jim Blaut's work on imperialism and nationalism (1975) are excellent examples, while larger projects aimed at extracting the geographical implications of Marx's writings were established under the direction of the contributors to *Antipode* and members of its linked organization, the Union of Socialist Geographers.

Relatively little attention, however, was given to explaining why spatial analysis remained so weakly developed for so long a time. Indeed, until recently, Western Marxism paralleled the development of bourgeois social science in viewing the organization of space as either a 'container' or an external reflection, a mirror of the social dynamic and social consciousness. In an almost Durkheimian manner, the spatiality of social life was externalized and neutered in terms of its impact on social and historical processes and seen as little more than backdrop or stage. Explaining this submergence of spatial analysis in Marxism remains a major task. It is possible, however, to establish some initial theses.

1. *The late appearance of Grundrisse*. Marx's *Grundrisse*, which was not disseminated widely in translation until well after the Second World War, probably contains more explicit geographical analysis than any of his writings. Its two volumes were first published in Russian in 1939 and 1941. The

first German edition appeared in 1953, the first English edition in 1973. In addition, as is now well-known, Marx never completed his plans for subsequent volumes of *Capital* dealing with world trade and the geographical expansion of capitalism, only hinting at their possible content in the late-appearing *Grundrisse*. In the absence of these sources, heavy emphasis was placed upon the largely aspatial, closed-system theorizing of the published volumes of *Capital*. Although Marx never fails to illustrate his arguments with specific historical and geographical examples, volumes I and II of *Capital* in particular remain encased in the simplifying assumptions of a closed national economy, an essentially spaceless capitalism systematically structured almost as if it existed on the head of a pin. Volume III and the proposed additional volumes were to represent concretizations of Marx's theory, projections outward into the historical and geographical analysis of world markets, colonialism, international trade, the role of the state, etc. – in essence, toward an analysis of the uneven development of productive sectors, regions, and nations.

Through the contributions of Bukharin, Lenin, Luxembourg, Trotsky and others, the theory of imperialism and associated conceptualizations of the processes of uneven development became the major context for geographical analysis within Western Marxism. There was an implied spatial problematic in these theorizations of imperialism, but it rested primarily on the simple recognition of an ultimate physical limitation to the geographical expansion of capitalism. For most of the major theoreticians, these geographical limits to capital were unlikely ever to be reached because social revolution would intervene long before the entire world became uniformly capitalist. Nevertheless, the processes of geographically uneven development were recognized and placed on the theoretical and political agenda, to be revived by a new generation of Marxist scholarship, led by such figures as Wallerstein, Amin, Emmanuel, Palloix, Hymer, and, especially, Ernest Mandel. How much this new generation was influenced by the post-war translations of *Grundrisse* remains an interesting but still open question.

2. *Anti-spatial traditions in Western Marxism.* Failure to develop the spatial emphases inherent in Marx's and later works on the geographical expansion of capitalism and in the equally spatial interpretations of the town–countryside antithesis that appear so vividly in *The German Ideology* and elsewhere in Marx's writings can also be linked to a deep tradition of anti-spatialism. Paradoxically perhaps, this tradition of rejecting geographical explanations of history originates with Marx himself, in his response to the Hegelian dialectic.

In many ways, Hegel and Hegelianism promulgated a powerful spatialist ontology and phenomenology, one which reified and fetishized space in the form of the territorial state, the locus and medium of perfected reason. As Lefebvre argues in *La Production de l'espace* (1974, 29–33), for Hegel historical time became frozen and fixed within the imminent rationality of space as state-idea. Time thus became subordinated to space, with history itself being directed by a territorial 'spirit', the state. Marx's anti-Hegelianism was not confined to a materialist critique of idealism. It was also an attempt to restore historicity – revolutionary temporality – to primacy over the spirit of spatiality. From this project arose a powerful sensitivity and resistance to the assertion of space into a position of historical and social determination, an anti-Hegelian anti-spatialism which is woven into virtually all of Marx's writings.

The possibility of a 'negation of the negation', a non-prioritized recombination of history and geography, time and space, was buried under subsequent codifications of Marx's theory of fetishism. A historical materialistic dialectic was embraced as human beings were contextualized in the making of history; but a spatial dialectic, even a materialist one, with human beings making their geographies and being constrained by what they have made, was unacceptable. This form of anti-spatialism may have been most rigidly codified by Lukacs in *History and Class Consciousness*, wherein spatial consciousness is presented as the epitome of reification, as false consciousness manipulated by the state and by capital to divert attention away from class struggle.

This anti-spatial armour served well to resist the many attacks on Marxism and the working class based on uncontestable spatial reification – Le Corbusier's choice between 'Architecture or Revolution' being among the most innocuous of these attacks, fascism by far the most vicious – but it also tended to associate all forms of spatial analysis and geographical explanation with fetishism and false consciousness. Not only does this tradition continue to interfere with the development of Marxist spatial analysis, it has also been partially responsible for the characteristic confusion surrounding the formulation of a sufficiently concretized Marxist theory of the state, nationalism, and local politics.

Mention must also be made of the anti-spatial character of Marxist dogmatism as it emerged from the Second International and was consolidated under Stalinism. Spatial questions, among many other aspects of Marxist theory and practice, were treated by the Second International and its leaders through the scope of a sterile economic reductionism. Marxism was turned into positivistic scientism under Stalin, emphasizing a belief in technocratic thought and strictly economic causality in the links between base and superstructure. Culture, politics, consciousness, ideology, and, along with them, the production of space were reduced to simple reflections of the economic base. Spatiality became absorbed in economism as its dialectical relationship with other elements of material existence was broken.

3. *Changing conditions of capitalist exploitation.* The early neglect and recent revival of interest concerning the spatial problematic in Marxism may, in the end, be primarily a reflection of changing material conditions. Lefebvre has argued, in *La Pensée marxiste et la ville* (1972), that during the nineteenth century and into the early twentieth the spatial problematic was simply less important than it is today with respect to both the exploitation of labour and the reproduction of the essential means of production. Under the conditions of competitive industrial capitalism, machines, commodities, and the labour force were reproduced under specific social legislation (labour contracts, civil laws, technological agreements) and an oppressive state apparatus (police, the

military, colonial administration). The production of space was accommodative, conformal, and directly shaped by the market and state power. The spatial structure of the industrial capitalist city, for example, repeated itself over and over again in its functional concentricity and segregated sectors of social class.

Exploitation and social reproduction were primarily embedded in a manipulative matrix of time. The rate of exploitation, Marx's ratio between surplus value and variable capital, is, after all, an expression derived from the labour theory of value and its fundamental measurement of socially necessary labour time. Like the formulas for the organic composition of capital and the rate of profit, its derivation assumes a closed system view of capitalist production relations, devoid of significant spatial differentiation and unevenness. In addition, given the massive urbanization associated with expanding industrialization, the reproduction of the labour force was much less crucial an issue than the process of direct exploitation through a system of subsistence wages and the domination of capital over labour at the point of production. In the extraction of absolute surplus value, the social organization of time appeared to be more important than the social organization of space.

In contemporary capitalism (setting aside for the moment the question of transition and restructuring, its causes, timing, and so on) the conditions which underlie the continuing survival of capitalism have changed. Exploitation of labour time continues to be the primary source of absolute surplus value, but within increasing limits arising from reduction in the length of the working day, minimum wage levels and wage agreements, and other achievements of working-class organization and urban social movements. Capitalism has been forced to shift greater and greater emphasis to the extraction of relative surplus value through technological change, modifications in the organic composition of capital, the increasingly pervasive role of the state, and the net transfers of surplus associated with the penetration of capital into not fully capitalist spheres of production (internally, through intensification, as well as externally, through uneven development and geographical 'extensification' into less industrialized regions around the world). This has required the construction of total systems to secure and regulate the smooth reproduction of the social relations of production. In this process, the production of space plays a crucial role. It is this switch in significance between the temporality and spatiality of capitalism that provoked Lefebvre to argue that 'industrialization, once the producer of urbanism, is now being produced by it'.

The spatial problematic and the survival of capitalism

Lefebvre's writings are marked by a persistent search for a political understanding of how and why capitalism has survived from the competitive industrial form of Marx's time to the advanced, state-managed and oligopolistic industrial capitalism of today. . . . [H]e presented a series of increasingly elaborated 'approximations' beginning with his conceptualization of everyday life in the modern world and passing through revolutionary urbanization and urbanism to his major thesis on the social production of space. This thesis is pointedly condensed in *The Survival of Capitalism* (1976a: 21).

> Capitalism has found itself able to attenuate (if not resolve) its internal contradictions for a century, and consequently, in the hundred years since the writing of *Capital*, it has succeeded in achieving 'growth.' We cannot calculate at what price, but we do know the means: *by occupying space, by producing a space.*

Lefebvre links this advanced capitalist space directly to the reproduction of the social relations of production, that is, the processes whereby the capitalist system as a whole is able to extend its existence by maintaining its defining structures. He defines three levels of reproduction and argues that the ability of capital to intervene directly and affect all three levels has developed over time, with the development of the productive forces. First, there is biophysiological reproduction, essentially within the context of family and kinship relations; second is the reproduction of labour power (the working class) and the means of production; and third is the still broader reproduction of the social relations of production. Under advanced capitalism the organization of space becomes predominantly related to the reproduction of the dominant system of social relations. Simultaneously, the reproduction of these dominant social relations becomes the primary basis for the survival of capitalism itself.

Lefebvre grounds his argument in the assertion that socially produced space (essentially urbanized space in advanced capitalism, even in the countryside) is where the dominant relations of production are reproduced. They are reproduced in a concretized and created spatiality that has been progressively 'occupied' by an advancing capitalism, fragmented into parcels, homogenized into discrete commodities, organized into locations of control, and extended to the global scale. The survival of capitalism has depended upon this distinctive production and occupation of a fragmented, homogenized, and hierarchically structured space – achieved largely through bureaucratically (that is to say, state) controlled collective consumption, the differentiation of centres and peripheries at multiple scales, and the penetration of state power into everyday life. The final crisis of capitalism can only come when the relations of production can no longer be reproduced, not simply when production itself is stopped (the abiding strategy of *ouvrièrisme*).

Thus, class struggle (yes, it still remains class struggle) must encompass and focus upon the vulnerable point: the production of space, the territorial structure of exploitation and domination, the spatially controlled reproduction of the system as a whole. And it must include all those who are exploited, dominated, and 'peripheralized' by the imposed spatial organization of advanced capitalism: landless peasants, proletarianized petty bourgeoisies, women, students, racial minorities, as well as the working class itself. In advanced capitalist countries, Lefebvre argues, the struggle will take the form of an 'urban revolution' fighting for *le droit à la ville* and control over *la vie quotidienne* within the territorial framework of the capitalist state. In less industrialized countries, it will also focus upon territorial liberation and reconstruction, on taking control over the production of space and its polarized system of dominant cores and dependent peripheries within the global structure of capitalism.

With this chain of arguments, Lefebvre defines an encompassing spatial problematic in capitalism and raises it to a central position within class struggle by embedding class relations in the configurative contradictions of socially organized space. He does not argue that the spatial problematic has always been so central. Nor does he present the struggle over space as a substitute for, or alternative to, class struggle. Instead, he argues that no social revolution can succeed without being at the same time a consciously spatial revolution. In much the same way as other 'concrete abstractions' (such as the commodity form) have been analysed in the Marxist tradition to show how they contain within them, mystified and fetishized, the real social relations of capitalism, so too must we now approach the analysis of space. The demystification of spatiality will reveal the potentialities of a revolutionary spatial consciousness, the material and theoretical foundations of a radical spatial praxis aimed at expropriating control over the production of space. Berger's claim comes back again: 'prophesy now involves a geographical rather than historical projection; it is space, not time, that hides consequences from us'.

Notes

1 The dominance of a physicalist view of space has so permeated the analysis of human spatiality that it tends to distort our vocabulary. Thus, while such adjectives as 'social', 'political', 'economic', and even 'historical' generally suggest, unless otherwise specified, a link to human action and motivation, the term 'spatial' typically evokes a physical or geometrical image, something external to the social context and to social action, a part of the 'environment', a part of the setting for society – its naively given container – rather than a formative structure created by society. We really do not have a widely used and accepted expression in English to convey the inherently social quality of organized space, especially since the terms 'social space' and 'human geography' have become so murky with multiple and often incompatible meanings. For these and other reasons, I have chosen to use the term 'spatiality' to specify this socially-produced space.

2 This observation, with added emphasis, comes from Harvey's (1973) translation of a segment of *La Révolution urbaine* (1970, 25). At this point in the development of his ideas on the production of space, Lefebvre had fastened on to urbanism as a summative conceptualization of capitalist spatiality. Unfortunately, this explicitly urban metaphor prevented readers from seeing the much more general spatial emphasis that lay behind his developing argument and provoked responses to a perceived reification of the urban. Castells would crystallize this view by describing Lefebvre's conceptualization of the urban revolution as the left-wing version of the 'urban ideology' promulgated by the bourgeois theoreticians of the Chicago School of urban ecology, which he considered an equally mystifying overspecification of the urban as theoretical object.

Selected references

Blaut, James 1975: Imperialism: the marxist theory and its evolution. *Antipode* 7, 1–19.

Castells, Manuel 1972: *La question urbaine*. Paris: Maspero. Trans. 1975: *The urban question: a marxist approach*. London: Edward Arnold.
Harvey, David 1973: *Social justice and the city*. London: Edward Arnold. Reprinted 1988: Oxford, UK: Blackwell.
Harvey, David 1975: The geography of capitalist accumulation: a reconstruction of marxian theory. *Antipode* 7, 9–21.
Lefebvre, Henri 1970: *La revolution urbaine*. Paris: Gallimard.
Lefebvre, Henri 1972: *La pensée marxiste et la ville*. Paris: Casterman.
Lefebvre, Henri 1974: *La production de l'espace*. Paris: Anthropos.
Lefebvre, Henri 1976a: *The survival of capitalism*. London: Allison and Busby.
Lefebvre, Henri 1976b: Reflections on the politics of space. *Antipode* 8, 30–37.
Mandel, E. 1975: *Late capitalism*. London: Verso.
Mandel, E. 1976: Capitalism and regional disparities. *Southwest Economy and Society* 1, 41–47.
Soja, Edward 1980: The socio-spatial dialectic. *Annals of the Association of American Geographers* 70, 207–25.
Soja, Edward and Hadjimichalis, C. 1979: Between geographical materialism and spatial fetishism: some observations on the development of marxist spatial analysis. *Antipode* 11, 3–11.

15 David Harvey

Between Space and Time: Reflections on the Geographical Imagination

Reprinted in full from: *Annals of the Association of American Geographers* 80, 418–434 (1990).

The question I wish to consider is the construction of a historical geography of space and time. Since that sounds and indeed is a double play on the concepts of space and time, the idea requires some initial elaboration. I shall then explore the implications of the idea in relation to the historical geography of everyday life and the social practices of those who call themselves geographers.

The spaces and times of social life

Durkheim pointed out in *The Elementary Forms of the Religious Life* (1915) that space and time are social constructs. The writings of anthropologists such as Hallowell (1955), Lévi-Strauss (1963), Hall (1966) and, more recently Bourdieu (1977) and Moore (1986) confirm this view: different societies produce qualitatively different conceptions of space and time (see also Tuan 1977). In interpreting this anthropological evidence, I want to highlight two features.

First, the social definitions of space and time operate with the full force of objective facts to which all individuals and institutions necessarily respond. For example, in modern societies, we accept clock time, even though such time is a social construct, as an objective fact of daily life; it provides a commonly held standard, outside of any one person's influence, to which we turn again and again to organize our lives and in terms of which we assess and judge all manner of social behaviors and subjective feelings. Even when we do not conform to it, we know very well what it is that we are not conforming to.

Secondly, the definitions of objective space and time are deeply implicated in processes of social reproduction. Bourdieu (1977) shows, for example, how in the case of the North African Kabyle, temporal and spatial organization (the calendar, the partitions within the house, etc.) serve to constitute the social order through the assignment of people and activities to distinctive places and times. The group orders its hierarchies, its gender roles and divisions of labor, in accordance with a specific mode of spatial and temporal organization. The role of woman in Kabyle society is, for example, defined in terms of the spaces occupied at specific times. A particular way of representing space and time guides spatial and temporal practices which in turn secure the social order.

Practices of this sort are not foreign to advanced capitalist societies. To begin with, space and time are always a primary means of both individuation and social differentiation. The definition of spatial units as administrative, legal or accounting entities defines fields of social action which have wide-ranging impacts on the organization of social life. Indeed, the very act of naming geographical entities implies a power over them, most particularly over the way in which places, their inhabitants and their social functions get represented. As Edward Said (1978) so brilliantly demonstrates in his study of Orientalism, the identity of variegated peoples can be collapsed, shaped, and manipulated through the connotations and associations imposed upon a name by outsiders. Ideological struggles over the meaning and manner of such representations of place and identity abound. But over and beyond the mere act of identification, the assignment of place within a sociospatial structure indicates distinctive roles, capacities for action, and access to power within the social order. The when and where of different kinds of social activity and of different manners of relating convey clear social messages. We still instruct children, for example, in the idea that there is 'a time and a place for everything' and all of us, at some level of meaning, know what our place is (though whether or not we feel comfortable with it is another question). We all know, furthermore, what it means to be 'put in one's place' and that to challenge what that place might be, physically as well as socially, is to challenge something fundamental in the social order. Sit-ins, street demonstrations, the storming of the Bastille or the gates of the US embassy in Teheran, the striking down of the Berlin Wall, and the occupation of a factory or a college administration building are all signs of attack against an established social order.

Sufficient accounts of these phenomena exist to render further proof of their generality superfluous, though the exact manner in which concepts of space and time operate in social reproduction is so subtle and nuanced as to require, if we are to read it right, the most sophisticated apparatus of enquiry we can muster. But the evidence is solid enough to support the following proposition: *each social formation constructs objective conceptions of space and time sufficient unto its own needs and purposes of material and social reproduction and organizes its material practices in accordance with those conceptions.*

But societies change and grow, they are transformed from within and adapt to pressures and influences from without. Objective conceptions of space and time must change to accommodate new material practices of social reproduction. How are such shifts in the public and objective conceptions of time and space accomplished? In certain instances, the answer is simply given. New concepts of space and time have been imposed by main force through conquest, imperial expansion or neocolonial domination. The European settlement of North America imposed quite alien conceptions of time and space upon the Plains Indians for example, and in so doing altered forever the social framework within which the reproduction of these peoples could, if at all, take place. The imposition of a mathematically rational spatial order in the house, the classroom, the village, the barracks and even across the city of Cairo itself, Mitchell (1988) shows, were centerpieces of a late nineteenth-century project to bring Egypt into line with the disciplinary frameworks of European capitalism. Such impositions are not necessarily well received. The spread of capitalist social relations has often entailed a fierce battle to socialize different peoples into the common net of time discipline implicit in industrial organization and into a respect for partitions of territorial and land rights specified in mathematically rigorous terms (see Sack 1986). While rearguard actions against such impositions abound, it is nevertheless true that public definitions of time and space throughout much of the contemporary world have been imposed in the course of capitalist development.

Even more interesting problems arise when the public sense of time and space is contested from within. Such contestation in contemporary society in part arises out of individual and subjective resistance to the authority of the clock and the tyranny of the cadastral map. Modernist and postmodernist literature and painting are full of signs of revolt against simple mathematical and material measures of space and time, while psychologists and sociologists have revealed, through their explorations, a highly complicated and often confused world of personal and social representations which departs significantly from dominant public practices. Personal space and time do not automatically accord with the dominant public sense of either and, as Tamara Hareven (1982) shows, there are intricate ways in which 'family time' can be integrated with and used to offset the pressing power of the 'industrial time' of deskilling and reskilling of labor forces and the cyclical patterns of employment. More significantly, the class, gender, cultural, religious and political differentiation in conceptions of time and space frequently become arenas of social conflict. New definitions of what is the correct time and place for everything as well as of the proper objective qualities of space and time can arise out of such struggles.

A few examples of such conflict are perhaps in order. The first comes from the chapter in *Capital* on 'The Working Day' in which Marx (1967, 233–35) sets up a fictitious conversation between capitalist and worker. The former insists that a fair day's work is measured in relation to how much time a worker needs to recuperate sufficient strength to return to work the next day and that a fair day's wage is given by the money required to cover daily reproduction costs. The worker replies that such a calculation ignores the shortening of his life which results from unremitting toil and that the measure of a fair day's work and wage looks entirely different when calculated over a working life. Both sides, Marx argues, are correct from the standpoint of the laws of market exchange, but different class perspectives dictate different time horizons for social calculation. Between such equal rights, Marx argues, force decides.

The gendering of 'Father Time' yields a second example. It is not only that time gets construed quite differently according to gender roles through the curious habit of defining working time as only that taken up in selling labor power directly to others. But, as Forman (1989) points out, the reduction of a woman's world to the cyclical times of nature has had the effect of excluding women from the linear time of patriarchal history, rendering women 'strangers in the world of male-defined time'. The struggle, in this case, is to challenge the traditional world of myth, iconography and ritual in which male dominion over time parallels dominion over nature and over women as 'natural beings'. When Blake, for example, insisted that 'Time and Space are Real Beings. Time is a Man, Space is a Woman, and her masculine Portion is Death' (quoted in Forman, page 4), he was articulating a widespread allegorical presumption that has echoes even unto the present day. The inability to relate the time of birthing (and all that this implies) to the masculine preoccupation with death and history is, in Forman's view, one of the deeper psychological battlegrounds between men and women.

The third example derives from a conversation between an economist and a geologist over the time horizon for optimal exploitation of a mineral resource. The former holds that the appropriate time horizon is set by the interest rate and market price, but the geologist, holding to a very different conception of time, argues that it is the obligation of every generation to leave behind an aliquot share of any resource to the next. There is no logical way to resolve that argument. It, too, is resolved by main force. The dominant market institutions prevailing under capitalism fix time horizons by way of the interest rate and, in almost all arenas of economic calculation (including the purchase of a house with a mortgage), that is the end of the story.

We here identify the potentiality for social conflict deriving entirely from the time horizon over which the effect of a decision is held to operate. While economists often accept the Keynesian maxim that 'in the long run we are all dead' and that the short-run is the only reasonable time horizon over which to operationalize economic and political decisions, environmentalists insist that responsibilities must be judged over an infinite time horizon within which all forms of life (including that of humans) must be preserved. The opposition in the sense of time is obvious. Even when, as in Pigouvian economics, longer time horizons are introduced into economic calculation, the effective means is through a discount rate which is set by economic rather than ecological, religious or social calculation (see, for example, the report by Pearce, Markandya and Barbier (1989) on a *Blueprint for a Green Economy*, which insists that all environmental impacts can be monetized and that the discount rate is a perfectly adequate means by which to take account of long-term environmental impacts). The whole political-economic trajectory of development and change depends upon which objective definition we adopt in social practice. If the practices are capitalistic, then the time horizon cannot be that to which environmentalists cleave.

Spatial usages and definitions are likewise a contested terrain in both practical and conceptual realms. Here, too, environmentalists tend to operate with a much broader conception of the spatial domain of social action, pointing to the spillover effects of local activities into patterns of use that affect global warming, acid rain formation and global despoliation of the resource base. Such a spatial conception conflicts with decisions taken with the objective of maximizing land rent at a particular site over a time horizon set by land price and the interest rate. What separates the environmental movement (and what in many respects makes it so special and so interesting) is precisely the conception of time and space which it brings to bear on questions of social reproduction and organization.

Such deep struggles over the meaning and social definition of space and time are rarely arrived at directly. They usually emerge out of much simpler conflicts over the appropriation and domination of particular spaces and times. It took me many years, for example, to understand why it was that the Parisian communards so readily put aside their pressing tasks of organizing for the defense of revolutionary Paris in 1871, in order to tear down the Vendôme column. The column was a hated symbol of an alien power that had long ruled over them; it was a symbol of that spatial organization of the city that had put so many segments of the population 'in their place', by the building of Haussmann's boulevards and the expulsion of the working class from the central city. Haussmann inserted an entirely new conception of space into the fabric of the city, a conception appropriate to a new social order based on capitalistic (particularly financial) values. The transformation of social relations and daily life envisaged in the 1871 revolution entailed, or so the communards felt, the reconstruction of the interior spaces of Paris in a different non-hierarchical image. So powerful was that urge that the public spectacle of toppling the Vendôme column became a catalytic moment in the assertion of communard power over the city's spaces (Ross 1988). The communards tried to build an alternative social order not only by reoccupying the space from which they had been so unceremoniously expelled but by trying to reshape the objective social qualities of urban space itself in a nonhierarchical and communitarian image. The subsequent rebuilding of the column was as much a signal of reaction as was the building of the Basilica of Sacré Coeur on the heights of Montmartre in expiation for the Commune's supposed sins (see Harvey 1985).

The 1989 annual convention of the Association of American Geographers in Baltimore likewise took place in what is for me, a resident of that city for some eighteen years, alien territory. The present carnival mask of the inner harbor redevelopment conceals a long history of struggle over this space. The urban renewal that began in the early 1960s was led by the property developers and financial institutions as they sought to recolonize what they saw as a strategic but declining central city core. But the effort was stymied by the unrest of the 1960s that had the downtown dominated by anti-war demonstrations, counter-cultural events and, most devastating of all for investor confidence, street uprisings mainly on the part of impoverished African-Americans. The inner city was a space of disaffection and social disruption. But in the wake of the violence that rocked the city after Martin Luther King's assassination in 1968, a coalition sprang to life to try and restore a sense of unity and belonging to the city. The coalition was broad; it included the churches (the Black Ministerial Alliance in particular), community leaders of all kinds, academics and downtown lawyers, politicians, trade unionists, bureaucrats, and, bringing up the rear in this instance, the business community, which was plainly at a loss as to what to do or where to turn. The struggle was on to try and put the city back together again as a cohesive social entity, as a working and living community alert to racial and social injustice.

One idea that emerged from that effort was to create a city fair in the inner city, a fair that would celebrate 'otherness' and difference by being based on the city's distinctive religious, ethnic and racial composition but which would also celebrate the theme of civic unity within that diversity. In 1970 the first fair took place, bringing a quarter of a million people over a weekend, from all neighborhoods of the city, into the inner city space of disaffection. By 1973, nearly two million came and the inner harbor was reoccupied by the common populace in ways which it had been impossible to envisage in the 1960s. It became a site of communal affirmation of unity within difference.

During the 1970s, in spite of considerable popular opposition, the forces of commercialism and property development recaptured the space. It became the site of a public–private partnership in which vast amounts of public moneys were absorbed for purposes of private rather than civic gain. The Hyatt-Regency Hotel, headquarters for the AAG meetings, was built with $5 million of private money, a $10 million Urban Development Action Grant, and a complicated deal of city investment in infrastructures and shell which took some $20 million of a city bond issue. The inner city space became a space of conspicuous consumption, celebrating commodities rather than civic values. It became the site of 'spectacle' in which people are reduced from active participants in the appropriation of space to passive spectators (Debord 1983). This spectacle diverts attention from the awful poverty of the rest of the city and projects an image of successful dynamism when the reality is that of serious impoverishment and disempowerment (Levine 1987). While all that money was pouring into the inner city redevelopment, the rest of the city gained little and in some instances lost much, creating an island of downtown affluence in a sea of decay (Szanton 1986). The glitter of the inner harbor diverts the gaze from the gathering trajedy of injustice in that other Baltimore, now safely (or so it seems) tucked away in the invisible neighborhoods of despair.

The point of these examples is to illustrate how social space, when it is contested within the orbit of a given social formation, can begin to take on new definitions and meanings. In both Paris and Baltimore, we see the struggle for command over strategic central city spaces as part of a broader struggle to replace a landscape of hierarchy and of pure money power with a social space constructed in the image of equality and justice. While both struggles were unsuccessful, they do illustrate how dominant and hegemonic definitions of social space (and time) are perpetually under challenge and always open to modification.

Materialist perspectives on the historical geography of space and time

If space and time are both social and objective, then it follows that social processes (including social conflicts of the sort already outlined) have a role to play in their objectification. How then, would we set out to study the ways in which social space and time get shaped in different historical and geographical contexts? There is no answer to that independent of the explicit character of our ontological and epistemological commitments. My own are, as is well known, explicitly Marxist, which means the organization of enquiry according to the basic principles of historical geographical materialism. The objective definitions must in the first instance be understood, not by appeal to the world of thoughts and ideas (though that study is always rewarding), but from the study of material processes of social reproduction. As Smith (1984, 77) puts it, 'the relativity of space (is) not a philosophical issue but a product of social and historical practice'.

Let me illustrate such a principle at work. I often ask beginning geography students to consider where their last meal came from. Tracing back all the items used in the production of that meal reveals a relation of dependence upon a whole world of social labor conducted in many different places under very different social relations and conditions of production. That dependency expands even further when we consider the materials and goods used in the production of the goods we directly consume. Yet we can in practice consume our meal without the slightest knowledge of the intricate geography of production and the myriad social relationships embedded in the system that puts it upon our table.

This was the condition that Marx (1967, 71–83) picked upon in developing one of his most telling concepts – *the fetishism of commodities*. He sought to capture by that term the way in which markets conceal social (and, we should add, geographical) information and relations. We cannot tell from looking at the commodity whether it has been produced by happy laborers working in a cooperative in Italy, grossly exploited laborers working under conditions of apartheid in South Africa, or wage laborers protected by adequate labor legislation and wage agreements in Sweden. The grapes that sit upon the supermarket shelves are mute; we cannot see the fingerprints of exploitation upon them or tell immediately what part of the world they are from. We can, by further enquiry, lift the veil on this geographical and social ignorance and make ourselves aware of these issues (as we do when we engage in a consumer boycott of nonunion or South African grapes). But in so doing we find we have to go behind and beyond what the market itself reveals in order to understand how society is working. This was precisely Marx's own agenda. We have to get behind the veil, the fetishism of the market and the commodity, in order to tell the full story of social reproduction.

The geographical ignorance that arises out of the fetishism of commodities is in itself cause for concern. The spatial range of our own individual experience of procuring commodities in the market place bears no relationship to the spatial range over which the commodities themselves are produced. The two space horizons are quite distinct, and decisions that seem reasonable from the former standpoint are not necessarily appropriate from the latter. To which set of experiences should we appeal in understanding the historical geography of space and time? Strictly speaking, my answer will be both because both are equally material. But it is here that I insist we should deploy the Marxian concept of fetishism with its full force. We will arrive at a fetishistic interpretation of the world (including the objective social definitions of space and time) if we take the realm of individual experience (shopping in the supermarket, traveling to work and picking up money at the bank) as all there is. These latter activities are real and material, but their organization is such as to conceal the other definitions of space and time set up in accordance with the requirements of commodity production and capital circulation through price-fixing markets.

A pure concern for the material base of our own daily reproduction ought to dictate a working knowledge of the geography of commodity production and of the definitions of space and time embedded in the practices of commodity production and capital circulation. But in practice most people do without. This also raises important moral issues. If, for example, we consider it right and proper to show moral concern for those who help put dinner on the table, then this implies an extension of moral responsibility throughout the whole intricate geography and sociality of intersecting markets. We cannot reasonably go to church on Sunday, donate copiously to a fund to help the poor in the parish, and then walk obliviously into the market to buy grapes grown under conditions of apartheid. We cannot reasonably argue for high environmental quality in the neighborhood while still insisting on living at a level which necessarily implies polluting the air somewhere else (this is, after all, the heart of the ecologists' argument). Our problem is indeed precisely that in which Marx sought to instruct us. We have to penetrate the veil of fetishisms with which we are necessarily surrounded by virtue of the system of commodity production and exchange and discover what lies behind it. In particular, we need to know how space and time get defined by these material processes which give us our daily bread. It is to this world that I now turn.

The historical geography of space and time in the capitalist epoch

Consideration of the historical geography of space and time in the era of Western capitalism illustrates how conceptions and practices with respect to both have changed in accordance with political–economic practices. The transition from feudalism to capitalism, Le Goff (1980, 1988) argues, entailed a fundamental redefinition of concepts of space and time which served to reorder the world according to quite new social principles. The hour was an invention of the thirteenth century, the minute and the second became common measures only as late as the seventeenth. While the first of these measures had a religious origin (illustrating a deep continuity between the Judeo-Christian view of the world and the rise of capitalism), the spread of adequate measures of time-keeping had much more to do with the growing concern for efficiency in production, exchange, commerce and administration. It was an urban-based revolution 'in mental structures and their material expressions' and it was 'deeply implicated', according to Le Goff (1980, 36), 'in the mechanisms of class struggle.' 'Equal hours' in the city, Landes (1983, 78) confirms, 'announced the victory of a new cultural and economic order'. But the victory was partial and patchy, leaving much of the western world outside of its reach until at least the mid-nineteenth century.

The history of cartography in the transition from feudalism to capitalism has, like the history of time-keeping, been very much about refinement of spatial measurement and representation according to clearly defined mathematical principles. Here, too, the interests of trade and commerce, of property and territorial rights (of the sort unrecognizable in the feudal world) were of paramount importance in reshaping mental structures and material practices. When it became clear that geographical knowledge was a vital source of military and economic power, then the connection between maps and money, as Landes (1983, 110) shows, followed not far behind. The introduction of the Ptolemaic map into Florence in 1400 and its immediate adoption there as a means to depict geographical space and store locational information, was arguably the fundamental breakthrough in the construction of geographical knowledge as we now know it. Thereafter it became possible in principle to comprehend the world as a global unity.

The political significance of this cartographic revolution deserves consideration. Rational mathematical conceptions of space and time were, for example, a necessary condition for Enlightenment doctrines of political equality and social progress. One of the first actions of the French revolutionary assembly was to ordain the systematic mapping of France as a means to ensure equality of political representation. This is such a familiar constitutional issue in the democracies of the world (given the whole history of gerrymandering) that the intimate connection between democracy and rational mapping is now taken for granted. But imagine attempting to draw up an egalitarian system of representation armed only with the Mappa Mundi! The Jeffersonian land system, with its repetitive mathematical grid that still dominates the landscape of the United States, likewise sought the rational partitioning of space so as to promote the formation of an agrarian democracy. In practice this proved admirable for capitalist appropriation of and speculation in space, subverting Jefferson's aims, but it also demonstrates how a particular definition of objective social space (in this case strictly interpreted in rationalistic Enlightenment terms) facilitated the rise of a new kind of social order.

Accounts of the sort which Le Goff and Landes provide illustrate beyond doubt that concepts of space and time and the practices associated with them are far from socially neutral in human affairs. Precisely because of such political and economic implications, the sense of space and time remains contested and more problematic than we are wont to admit. Helgerson (1986) points out, for example, the intimate connection between the Renaissance maps of England (by Speed, Nordon, Caxton, and the others), the fight with dynastic privilege and the latter's ultimate replacement by a politics in which the relation between individual and nation became hegemonic. Helgerson's point is that the new means of cartographic representation allowed individuals to see themselves in terms that were more in accord with these new definitions of social and political relations. In the colonial period, to take a much later example, the maps of colonial administrations had very distinctive qualities that reflected their social purposes (Stone 1988).

Since I have taken up the above themes elsewhere (Harvey 1985, 1989a), I shall here merely assert that the construction of new mental conceptions and material practices with respect to space and time were fundamental to the rise of capitalism as a particular socioeconomic system. These conceptions and practices were always partial (though they became more hegemonic as capitalism evolved), and they were, in any case, always subject to social contestation in specific places and times. But social reproduction of the capitalist sort required their deep implantation in the world of ideas as well as in the realm of social practices.

Capitalism is, however, a revolutionary mode of production, always restlessly searching out new organizational forms, new technologies, new lifestyles, and new modalities of production and exploitation. Capitalism has also been revolutionary with respect to its objective social definitions of time and space. Indeed, when compared with almost all other forms of innovation, the radical reorganizations of space relations and of spatial representations have had an extraordinarily powerful effect. The turnpikes and canals, the railways, steamships and telegraph, the radio and the automobile, containerization, jet cargo transport, television and telecommunications, have altered space and time relations and forced us to new material practices as well as to new modes of representation of space. The capacity to measure and divide time has been revolutionalized, first through the production and diffusion of increasingly accurate time pieces and subsequently through close attention to the speed and coordinating mechanisms of production (automation, robotization) and the speed of movement of goods, people, information, messages, and the like. The material bases of objective space and time have become rapidly moving rather than fixed datum points in human affairs.

Why this movement? Since I have explored its roots in greater detail elsewhere (Harvey 1982, 1989a) I simply summarize the principal argument. Time is a vital magnitude under capitalism because social labor time is the measure of value and surplus social labor time lies at the origin of profit. Furthermore, the turnover time of capital is significant because speed-up (in production, in marketing, in capital turnover) is a powerful competitive means for individual capitalists to augment profits. In times of economic crisis and of particularly intense competition, capitalists with a faster turnover time survive better than their rivals, with the result that social time horizons typically shorten, intensity of working and living tends to pick up and the pace of change accelerates. The same sorts of proposition apply to the experience of space. The elimination of spatial barriers and the struggle to 'annihilate space by time' is essential to the whole dynamic of capital accumulation and becomes particularly acute in crises of capital overaccumulation. The absorption of surpluses of capital (and sometimes labor) through geographical expansion into new territories and through the construction of a completely new set of space relations has been nothing short of remarkable. The construction and reconstruction of space relations and of the global space economy, as Henri Lefebvre (1974) acutely observes, has been one of the main means to permit the survival of capitalism into the twentieth century.

The general characteristics (as opposed to the detailed where, when and how) of the historical geography of space and time which results are not accidental or arbitrary, but implicit in the very laws of motion of capitalist development. The general trend is towards an acceleration in turnover time (the worlds of production, exchange, consumption all tend to change faster) and a shrinking of space horizons. In popular terms, we might say that Toffler's (1970) world of 'future shock' encounters, as it were, Marshall McLuhan's (1966) 'global village'. Such periodic revolutions in the objective social qualities of time and space are not without their contradictions. It takes for example, long term and often high cost fixed capital investments of slow turnover time (like computer hardware) to speed up the turnover time of the rest, and it takes the production of a specific set of space relations (like a rail network) in order to annihilate space by time. A revolution in temporal and spatial relations often entails, therefore, not only the destruction of ways of life and social practices built around preceding time–space systems, but the 'creative destruction' of a wide range of physical assets embedded in the landscape. The recent history of deindustrialization is amply illustrative of the sort of process I have in mind.

The Marxian theory of capital accumulation permits theoretical insights into the contradictory changes that have occurred in the dimensionality of space and time in Western capitalism. If, as is the case, the temporal and spatial world of contemporary Wall Street is so very different from that of the nineteenth century stock exchange and if both depart from that of rural France (then and now) or of Scottish crofters (then and now), then this must be understood as a particular set of responses to a pervasive aggregate condition shaped by the rules of commodity production and capital accumulation. It is the contradictions and tensions implied therein that I want to examine.

Cultural and political responses to the changing dimensionality of space and time

Rapid changes in the objective qualities of social space and time are both confusing and disturbing, precisely because their revolutionary implications for the social order are so hard to anticipate. The nervous wonderment at it all is excellently captured in the *Quarterly Review* for 1839:

> Supposing that our railroads, even at our present simmering rate of travelling, were to be suddenly established all over England, the whole population of the country would, speaking metaphorically, at once advance *en masse*, and place their chairs nearer to the fireside of their metropolis . . . As distances were thus annihilated, the surface of our country would, as it were, shrivel in size until it became not much bigger than one immense city (cited in Schivelbusch 1978, 32).

The poet Heine likewise recorded his 'tremendous foreboding' on the opening of the rail link from Paris to Rouen:

> What changes must now occur, in our way of looking at things, in our notions! Even the elementary concepts of time and space have begun to vacillate. Space is killed by the railways. I feel as if the mountains and forests of all countries were advancing on Paris. Even now, I can smell the German linden trees; the North Sea's breakers are rolling against my door (cited in Schivelbusch 1978, 34).

The German theatre director Johannes Birringer (1989, 120–38) records a similar sense of shock in a contemporary setting. On arrival in Dallas and Houston, he felt an 'unforeseen collapse of space', where 'the dispersion and decompositions of the urban body (the physical and cultural representation of community) have reached a hallucinatory stage'. He remarks on:

> the unavoidable fusion and confusion of geographical realities, or the interchangeability of all places, or the disappearance of visible (static) points of reference into a constant commutation of surface images.

The riddle of Houston, he concludes:

> is one of community: fragmented and exploded in all directions . . . The city impersonates a speculative disorder, a kind of positive unspecificity on the verge of a paradoxical hyperbole (global power/local chaos).

I shall call this sense of overwhelming change in space–time dimensionality 'time–space compression' in order to capture something of Heine's sense of foreboding and Birringer's sense of collapse. The experience of it forces all of us to adjust our notions of space and time and to rethink the prospects for social action. This rethinking is, as I have already argued, embedded in political-economic struggles. But it is also the focus of intense cultural, aesthetic and political debate. Reflection on this idea helps us understand some of the turmoil that has occurred within the fields of cultural and political production in the capitalist era.

The recent complex of movements known as 'post-modernism' is, for example, connected in the writings of authors as diverse as Jameson (1984), Berman (1982) and Daniel Bell (1976) to some new experience of space and time. Interestingly, having advanced the idea, none of them tells us exactly what they might mean by it. And the material basis upon which these new experiences of space and time might be built, and its relation to the political economy of capitalist development, remains a topic lost in the shadows. I am particularly interested to see how far postmodernism can be understood simply by relating it to the new experiences of space and time generated out of the political-economic crisis of 1973 (Harvey 1989a).

Much of the advanced capitalist world was at that time forced into a major revolution in production techniques, consumption habits and political-economic practices. Strong currents of innovation have focused on speed-up and acceleration of turnover times. Time-horizons for decision making (now a matter of minutes in international financial markets) have shortened and lifestyle fashions have changed rapidly. And all of this has been coupled with a radical reorganization of space relations, the further reduction of spatial barriers, and the emergence of a new geography of capitalist development. These events have generated a powerful sense of time–space compression which has affected all aspects of cultural and political life. Whole landscapes have had to be destroyed in order to make way for the creation of the new. Themes of creative destruction, of increased fragmentation, of ephemerality (in community life, of skills, of lifestyles) have become much more noticeable in literary and philosophic discourse in an era when restructuring of everything from industrial production techniques to inner cities has become a major topic of concern. The transformation in 'the structure of feeling' which the move towards postmodernism betokens seems to have much to do with the shifts in political-economic practices that have occurred over the last two decades.

Consider, glancing backwards, that complex cultural movement known as modernism (against which postmodernism is supposedly reacting). There is indeed something special that happens to writing and artistic representation in Paris after 1848 and it is useful to look at that against the background of political-economic transformations occurring in that space and at that time. Heine's vague foreboding became a dramatic and traumatic experience in 1848, when, for the first time in the capitalist world, political-economy assumed an unlooked for simultaneity. The economic collapse and political revolutions that swept across the capitals of Europe in that year indicated that the capitalist world was interlinked in ways that had hitherto seemed unimaginable. The speed and simultaneity of it all was deeply troubling and called for some new mode of representation through which this interlinked world could be better understood. Realist modes of representation, which took a simple narrative structure as their model, simply could not do the job (no matter how brilliantly Dickens ranged across space and time in a novel like Bleak House).

Baudelaire (1981) took up the challenge by defining the modernist problematic as the search for universal truths in a world characterized by (spatial) fragmentation, (temporal) ephemerality and creative destruction. The complex sentence structure in Flaubert's novels and the brushstrokes of Manet defined totally new modes of representation of space and time that allowed new ways of thinking and new possibilities for social and political action. Kern's (1983) account of the revolution in the representation of space and time that occurred shortly before 1914 (a period of extraordinary experimentation in fields as diverse as physics, literature, painting and philosophy) is one of the clearest studies to date of how time–space compression generates experiences out of which new conceptions are squeezed. The avant-garde movements in the cultural field in part reflected but in part also sought to impose new definitions of space and time upon a Western capitalism in the full flood of violent transformation.

A closer look at the contradictions built into these cultural and political movements illustrates how they can mirror the fundamental contradictions in capitalist political economy. Consider the cultural response to the recent speed-up and acceleration of capital turnover time. The latter presupposes, to begin with, a more rapid turnover in consumption habits and lifestyles which consequently become the focus of capitalist social relations of production and consumption. Capitalist penetration of the realm of cultural production becomes particularly attractive because the lifetime of consumption of images, as opposed to more tangible objects like autos and refrigerators, is almost instantaneous. In recent years, a good deal of capital and labor has been applied to this purpose. This has been accompanied by a renewed emphasis upon the production of controlled spectacles (of which the Los Angeles Olympic Games was a prime example) which can conveniently double as a means of capital accumulation and of social control (reviving political interest in the old Roman formula of 'bread and circuses' at a time of greater insecurity).

The reactions to the collapse of spatial barriers are no less contradictory. The more global interrelations become, the more internationalized our dinner ingredients and our money flows, and the more spatial barriers disintegrate, so more rather than less of the world's population clings to place and neighborhood or to nation, region, ethnic grouping, or religious belief as specific marks of identity. Such a quest for visible and tangible marks of identity is readily understandable in the midst of fierce time–space compression. No matter that the capitalist response has been to invent tradition as yet another item of commodity production and consumption (the reenactment of ancient rites and spectacles, the excesses of a rampant heritage culture), there is still an insistent urge to look for roots in a world where image streams accelerate and become more and more placeless (unless the television and video screen can properly be regarded as a place). The foreboding generated out of the sense of social space imploding in upon us (forcibly marked by everything from the daily news to random acts of international terror or global environmental problems) translates into a crisis of identity. Who are we and to what space/place do we belong? Am I a citizen of the world, the nation, the locality? Not for the first time in capitalist history, if Kern's (1983) account of the period before World War I is correct, the diminution of spatial barriers has provoked an increasing sense of nationalism and localism, and excessive geopolitical rivalries and tensions, precisely because of the reduction in the power of spatial barriers to separate and defend against others.

The evident tension between *place* and *space* echoes that fundamental contradiction of capitalist political economy to which I have already alluded; that it takes a specific organization of space to try and annihilate space and that it takes capital of long turnover time to facilitate the more rapid turnover of the rest. This tension can be examined from yet another standpoint. Multinational capital should have scant respect for geography these days precisely because weakening spatial barriers open the whole world as its profitable oyster. But the reduction of spatial barriers has an equally powerful opposite effect; small-scale and finely graded differences between the qualities of places (their labor supply, their infrastructures and political receptivity, their resource mixes, their market niches, etc.) become even more important because multinational capital is in a better position to exploit them. Places, by the same token, become much more concerned about their 'good business climate' and inter-place competition for development becomes much more fine-tuned. The image-building of community (of the sort which characterizes Baltimore's inner harbor) becomes embedded in powerful processes of interurban competition (Harvey 1989b). Concern for both the real and fictional qualities of place increases in a phase of capitalist development in which the power to command space, particularly with respect to financial and money flows, has become more marked than ever before. The geopolitics of place tend to become more rather than less emphatic. Globalization thus generates its exact opposite motion into geopolitical oppositions and warring camps in a hostile world. The threat of geopolitical fragmentation in global capitalism – between geopolitical power blocks such as the European Common Market, the North American Common Market, and the Japanese trading empire – is far from idle.

It is for these reasons that coming to terms with the historical geography of space and time under capitalism makes so much sense. The dialectical oppositions between place and space, between long and short-term time horizons, exist within a deeper framework of shifts in time-space dimensionality that are the product of underlying capitalist imperatives to accelerate turnover times and to annihilate space by time. The study of how we cope with time–space compression illustrates how shifts in the experience of space and time generate new struggles in such fields as aesthetics and cultural representation, how very basic processes of social reproduction, as well as of production, are deeply implicated in shifting space and time horizons. In this regard, I find it intriguing, if I may make the aside, that the exploration of the relations between literature and geography that have so far emanated from the geographer's camp have almost without exception concentrated on the literary evocation of place (see, for example, Mallory and Simpson-Housley 1987) when the far more fundamental question of spatiality in, say, the novels of Flaubert and Joyce (a topic of great import for literary historians) has passed by unremarked. I also find it odd that geographers have concentrated so much more upon the importance of locality in the present conjuncture, leaning, as it were, to one side of the contradictory dynamic of space and place, as if they are separate rather than dialectically related concepts.

Geography in relation to social and aesthetic theory

Armed with such epistemological and ontological commitments as historical-geographical materialism provides, we can begin to unravel the theoretical and philosophical conceptions of space and time which sustain (explicitly or implicitly) particular social visions and interpretations of the world. In so doing, it is useful to begin with consideration of a major divide in Western thought between aesthetic and social theory.

Social theory of the sort constructed in the diverse traditions of Adam Smith, Marx, or Weber tends to privilege time over space in its formulations, reflecting and legitimizing those who view the world through the lenses of spaceless doctrines of progress and revolution. In recent years, many geographers have sought to correct that defective vision and to reintroduce the concept of space as not only meaningful but vital to the proper understanding of social processes (see Gregory and Urry 1985; Soja 1989). To some degree that effort has been rewarded by the recognition on the part of some social theorists that space indeed does matter (for example, Giddens 1984). But that task is only partly complete. Getting behind the fetishism of commodities challenges us to integrate the historical geography of space and time within the frame of all our understandings of how human societies are constructed and change. Our interventions in social theory stand to be strengthened even further by the exploration of that theme, though this presupposes, as always, the training of geographers with a powerful command over social theory and seized intellectually by the challenge to explore the difficult terrain of interface between society and the social construction of space and time.

But there is, curiously, another terrain of theoretical intervention which remains largely unexplored, except in that unsatisfactory and partial manner that always comes with nibbling at hidden rather than struggling over overt questions. I refer here to the intersection between geographical work and aesthetic theory. The latter, in direct contrast to social theory, is deeply concerned with 'the spatialization of time', albeit in terms of how that experience is communicated to and received by knowing, sensuous individuals. The architect, to take the most obvious case, tries to communicate certain values through the construction of a spatial form. Architecture, suggests Karsten Harries (1982), is not only about domesticating space, wresting and shaping a livable place out of empty space. It is also a deep defense against 'the terror of time'. The 'language of beauty' is 'the language of a timeless reality'. To create a beautiful object is 'to link time and eternity' in such a way as to redeem us from time's tyranny. The aim of spatial constructs is 'not to illuminate temporal reality so that (we) might feel more at home in it, but . . . to abolish time within time, if only for a time'. Even writing, comments Bourdieu (1977, 156), 'tears practice and discourse out of the flow of time'.

There are, of course, as many varieties of aesthetic theory as there are of social theory (see, for example, Eagleton's (1990) brilliant treatise on the subject). But I quote these comments from Harries to illustrate one of the central themes with which aesthetic theory grapples: how spatial constructs are created and used as fixed markers of human memory and of social values in a world of rapid flux and change. There is much to be learned from aesthetic theory about how different forms of produced space inhibit or facilitate processes of social change. Interestingly, geographers now find even more support for their endeavors from literary theorists (Jameson

1984 and Ross 1988) than from the social theorists. Conversely, there is much to be learned from social theory concerning the flux and change with which aesthetic theory has to cope. Historical geography, insofar as it lies at the intersection of those two dimensions, has an immense potentiality to contribute to understanding them both. By playing these two currents of thought off against each other, we may even aspire to create a more general theoretical framework for interpreting the historical geography of space and time while simultaneously figuring how cultural and aesthetic practices – spatializations – intervene in the political-economic dynamic of social and political change.

Let me illustrate where the political significance of such an argument might lie. Aesthetic judgments (as well as the 'redemptive' artistic practices that attach thereto) have frequently entered in as powerful criteria of political and social action. Kant argued that independent aesthetic judgment could act as a mediator between the worlds of objective science and of subjective moral judgment. If aesthetic judgment gives space priority over time, then it follows that spatial practices and concepts can, under certain circumstances, become central to social action.

In this regard, the German philosopher Heidegger is an interesting figure. Rejecting the Kantian dichotomies of subject and object, and fearing the descent into nihilism that Nietzschen thought seemed to promote, he proclaimed the permanence of *Being* over the transitoriness of *Becoming* and attached himself to a traditionalist vision of the truly aesthetic political state (Chytry 1989). His investigations led him away from the universals of modernism and Judeo-Christian thought and back to the intense and creative nationalism of pre-Socratic Greek thought. All metaphysics and philosophy, he declared (Heidegger 1959), are given their meaning only in relation to the destiny of the people. The geopolitical position of Germany in the interwar years, squeezed in a 'great pincer' between Russia and America, threatened the search for that meaning. 'If the great decision regarding Europe is not to bring annihilation', he wrote, the German nation 'must move itself and thereby the history of the West beyond the center of their future "happening" and into the primordial realm of the powers of being' and 'that decision must be made in terms of new spiritual energies unfolding historically from out of the center'. Herein for Heidegger lay the 'inner truth and greatness of the National Socialist movement' (Blitz 1981, 217).

That a great twentieth-century philosopher, who has incidentally inspired the philosophizing of Karsten Harries as well as much of the geographical writing on the meaning of place (see Relph 1976; Seamon and Mugerauer 1989), should so compromise himself politically and throw in his lot with the Nazis is deeply troubling. But a number of useful points can be made from the standpoint of my present argument. Heidegger's work is deeply imbued with an aesthetic sense which prioritizes *Being* and the specific qualities of *place* over *Becoming* and the universal propositions of modernist progress in universal space. His rejection of Judeo-Christian, values, of the myth of machine rationality, and of internationalism was total. The position to which he subscribed was active and revolutionary precisely because he saw the necessity for redemptive practices which in effect depended upon the restoration of the power of myth (of blood and soil, of race and fatherland, of destiny and place) while mobilizing all of the accoutrements of social progress towards a project of sublime national achievement. The application of this particular aesthetic sense to politics helped alter the historical geography of capitalism with a vengeance.

I scarcely need to remind geographers of the tortured history of geopolitical thinking and practices in the twentieth century and the difficulty geographers have had in confronting the thorny issues involved. I note that Hartshorne's (1939) *The Nature of Geography*, written in Vienna shortly after the *Anschluss*, totally rejects aesthetics in geography and reserves its most vitriolic condemnations for the mythologies of landscape geography. Hartshorne, following Hettner, seems to want to expel any opening for the politicizing of academic geography in an era when geography was suffused with politics and when sentiments of place and of aesthetics were being actively mobilized in the Nazi cause. The difficulty, of course, is that avoiding the problem does not eliminate it, even in academic geography.

This is not to say that everyone who, since Hartshorne, has sought to restore an aesthetic dimension to geography is a crypto-Nazi, for, as Eagleton (1990, 28) points out, the aesthetic has ever been 'a contradictory, double-edged concept'. On the one hand 'it figures as a genuinely emancipatory force – as a community of subjects now linked by sensuous impulse and fellow feeling' while on the other it can also serve to internalize repression, 'inserting social power more deeply into the very bodies of those it subjugates and so operating as a supremely effective mode of political hegemony'. The aestheticization of politics has, for this reason, a long history, posing both problems and potentialities in relation to social progress. There are left and right versions (the Sandinistas, after all, aestheticize politics around the figure of Sandino, and Marx's writings are full of references to an underlying project of liberation of the creative senses). The clearest form the problem takes is the shift in emphasis from historical progress and its ideologies towards practices which promote national (or even local) destinies and culture, often sparking geopolitical conflicts within the world economy. Appeals to mythologies of place, person and tradition, to the aesthetic sense, have played a vital role in geopolitical history.

Herein, I think, lies the significance of conjoining aesthetic with social theoretic perspectives, bringing together understandings that give space priority over time with those that give time priority over space. Historical geography in general, and the study of the historical geography of space and time, lies exactly at that point of intersection and therefore has a major intellectual, theoretical, political and practical role to play in understanding how human societies work. By positioning the study of geography between space and time, we evidently have much to learn and much to contribute.

The geographical imagination

I conclude with a brief commentary on the implications of such a perspective for the study of geography and for that relatively small group of scholars occupying a niche labeled 'geographer' within the academic division of labor.

The latter is a product of late nineteenth-century conditions and concerns. It is by no means self-evident that the disciplinary boundaries then drawn up (and subsequently fossilized by professionalization and institutionalization) correspond to contemporary conditions and needs. Partly in response to this problem, the academy has moved towards an increasing fragmentation in the division of labor within disciplines, spawned new disciplines in the interstices and looked for crosslinks on thematic topics. This history resembles the development of the division of labor in society at large. Increasing specialization of task and product differentiation, increasing roundaboutness of production and the search for horizontal linkages are as characteristic of large multinational corporations as they are of large universities. Within geography this process of fragmentation has accelerated since the mid-1960s. The effect has been to make it harder to identify the binding logic that is suggested by the word 'discipline'.

The turnover time of ideas in academia has also accelerated. Not so long ago, to publish more than two books in a lifetime was thought to be over-ambitious. Nowadays, it seems, leading academics have to publish a book every two years if they are to prove they are still alive. Definitions of productivity and output in academia have become much more strictly applied and career advancement is more and more measured simply in such terms. There is, of course, a certain intersection here between research and corporate/nation state requirements, between academia and the publishing trade, and the emergence of education as one of the big growth sectors in advanced capitalist societies. Speed-up in the production of ideas parallels a general push to accelerate turnover time within capitalism as a whole. But greater output of books and journals must rest on the production of new knowledge, and that implies the much fiercer competitive search for new ideas, a much greater proprietary interest in them. Such frenetic activity can converge upon some consensual and well-established 'truth' only if Adam Smith's hidden hand has all those effects in academia that it plainly does not have in other markets. In practice, the competitive marketing of ideas, theories, models, topic thrusts, generates color-of-the-month fashions which exacerbate rather than ameliorate conditions of rapid turnover, speed-up and ephemerality. Last year it was positivism and Marxism, this year structurationism, next year realism and the year after that constructivism, postmodernism, or whatever. It is easier to keep pace with the changes in Benetton's colors than to follow the gyrations of ephemeral ideas now being turned over within the academic world.

It is hard to see what we can do to resist such trends, even when we bewail their effects. Our job descriptions do not encompass those of 'intellectual geographer' but much more typically specify ever narrower proficiencies in everything from mere command of techniques (remote sensing and GIS) to specialists in transport modeling, industrial location, groundwater modeling, Soviet geography, or flavor of the month topic (sustainable development, chaos theory, fractal geometry or whatever). The best we can do is appoint specialists and hope they have an interest in the discipline as a whole. Our seeming inability or unwillingness to resist fragmentation and ephemerality suggests a condition in which something is being done to us by forces beyond our control. I wish, for example, that those who now so loudly proclaim the power of individual agency in human affairs could demonstrate how their or our specific agencies have produced this macroshift in our conditions of working and living. Are we mere victims of social processes rather than their real progenitors? If here, too, I prefer the Marxian conception of individuals struggling to make history but not under conditions of their own making, it is because most of us have a lifetime of exactly that kind of experience behind us.

This same question comes to mind when we consider the resurgent interests in aesthetics, landscape geography and place as central to the concerns of many human geographers. The claim that the place of geography in academia is to be secured by attaching the discipline to a core concept of place (even understood as a unique configuration of elements) has strengthened in a phase of capitalist development when the particular qualities of place have become of much greater concern to multinational capital and when there has simultaneously been a renewed interest in the politics and image of place as an arena of supposed (even fictional) stability under conditions of powerful time–space compression. The social search for identity and roots in place has reentered geography as a leitmotif and is in turn increasingly used to provide the discipline with a more powerful (and equally fictitious) sense of identity in a rapidly changing world.

A deeper understanding of the historical geography of space and time sheds considerable light on why the discipline might cultivate such arenas of research in this time and place. It provides a critical perspective from which to evaluate our reactions to the social pressures that surround us and suffuse our lives. Do we, in unthinkingly accepting the significance of place to our discipline, run the danger of drifting into subconscious support for a reemergence of an aestheticized geopolitics? The question does not imply avoidance of that issue but a proper confrontation of it through a conception of geography that lies at the intersection between social and aesthetic theory.

The historical geography of space and time facilitates critical reflection on who we are and what it is we might be struggling for. What concepts of space and time are we trying to establish? How do these relate to the changing historical geography of space and time under capitalism? What would the space and time of a socialist or ecologically responsible society look like? Geographers, after all, are contributors (and potentially powerful and important ones at that) to the whole question of spatiality and its meanings. Historical geographers with their potential interests in both space and time have unbounded potentiality to reflect back not only on the history of this or that place and space relations but the whole conundrum of the changing experience of space and time in social life and social reproduction.

Critical reflection on the historical geography of space and time locates the history of ideas about space and time in their material, social and political setting. Hartshorne did not write *The Nature of Geography* in a political vacuum but in post-*Anschluss* Vienna, and that fact (though never mentioned in consideration of that work) is surely present in its manner of construction and intervention in the world of ideas. This text of mine is likewise constructed in the light of a certain experience of time–space compression, of shifting mores of social reproduction and political argument. Even the great Kant did not develop his ideas on space and time, his distinctions between aesthetic, moral and scientific judgments, in a social vacuum. His was the grand attempt to codify and synthesize the evident contradictions inherent in the bourgeois logic of Enlightenment reason as it was then unfolding in the midst of the revolutionary impulses sweeping Europe at the end of the eighteenth century. It was a very distinctive product of that society with its particular and practical interests in commanding space and time with rational and mathematical precision, while experiencing all the frustrations and contradictions of initiating such a rational order given the nascent social relations of capitalism. If Hegel attacked Kant (on everything from aesthetics to his theory of history) and if Marx attacked both Hegel and Kant (again, on everything from aesthetics to basic conceptions of materiality and history), then these debates had everything to do with trying to redefine the paths of social change. If I, as a Marxist, still cling to that quest for an orderly social revolution that will take us beyond the contradictions, manifest injustices and senseless 'accumulation for accumulation's sake' logic of capitalism, then this commits me to a struggle to redefine the meaning of space and time as part and parcel of that quest. And if I am still so much in a minority in an academy in which neo-Kantianism dominates (without, it must be said, most people even knowing it), then this quite simply testifies to the persistence of capitalist social relations and of the bourgeois ideas that derive therefrom, including those defining and objectifying space and time.

Attachment to a certain conception of space and time is a political decision, and the historical geography of space and time reveals it so to be. What kind of space and time do we, as professional geographers, seek to promote? To what processes of social reproduction do those concepts subtly but persistently allude? The current campaign for geographical literacy is laudable, but what language is it that we teach? Do we simply insist that our students learn how many countries border on Chad? Do we teach the static rationality of the Ptolemaic system and insist that geography is nothing more than GIS, the contemporary version of the Hartshornian rule that if it can be mapped, then it is geography? Or do we teach the rich language of the commodity, with all its intricate history of social and spatial relations stretching back from our dinner table into almost every niche of labor activity in the modern world? And can we go on from that to teach the rich and complex language of uneven geographical development, of environmental transformations (deforestation, soil degradation, hydrological modifications, climatic shifts) whose historical geography has scarcely begun to be reconstructed? Can we go even further and create a deep awareness of how social processes can be given aesthetic forms in political debates (and learn to appreciate all the dangers that lurk therein)? Can we build a language – even a whole discipline – around a project that fuses the environmental, the spatial and the social within a sense of the historical geography of space and time?

All such possibilities exist to be explored. But whatever course we take entails a political commitment as to what kind of space and time we wish to promote. We are political agents and have to be aware of it. And the politics is an everyday question. The marketing head of a US communications firm in Europe commented (*International Herald Tribune*, 9 March 1989), on conversations with senior bankers in which he sought to go beyond the banter about it being the warmest January on record and talk seriously about the long-term effects of global warming. His clients all reacted in such a way as to suggest they thought about the environment 'in the same way we practice a hobby, in the comfort of our homes' and at weekends, when we should really think about it all the time 'especially at work'. But how can international bankers think about such things when their time-horizon is minutes? If twenty-four hours is a very long time in financial markets, and if finance capital is today the most powerful force in international development, then what kinds of long-term decisions can we expect from that quarter that make any sense from the standpoint of even long-term planning of investments, let alone of environmental regulation? When the commander of the Vincennes had to make the life-and-death decision on whether an image on a screen was a diving fighter or an Iranian airbus, he was caught in the terror of time–space compression which ultimately dissolves everything into ephemera and fragments such that the devil takes not the hindmost but the global totality, the whole social fabric of an internationalizing society that is more closely linked than ever before and in which the pace of change has suddenly accelerated.

Geographers cannot escape the terrors of these times. Nor can we avoid in the broad sense becoming victims of history rather than its victors. But we can certainly struggle for a different social vision and different futures with a conscious awareness of stakes and goals, albeit under conditions that are never of our own making. It is by positioning our geography between space and time, and by seeing ourselves as active participants in the historical geography of space and time, that we can, I believe, recover some clearer sense of purpose for ourselves, define an arena of serious intellectual debate and inquiry and thereby make major contributions, intellectually and politically, in a deeply troubled world.

References

Baudelaire, C. 1981. *Selected writings on art and artists*. Cambridge: Cambridge University Press.

Bell, D. 1976. *The cultural contradictions of capitalism*. New York: Basic Books.

Berman, M. 1982. *All that is solid melts into air*. New York: Simon and Schuster.

Birringer, J. 1989. Invisible cities/transcultural images. *Performing Arts Journal* 12:33–34, 120–38.

Blitz, M. 1981. *Heidegger's being and time: The posibility of political philsophy*. Ithaca, NY: Cornell University Press.

Bourdieu, P. 1977. *Outline of a theory of practice*. Cambridge: Cambridge University Press.

Chytry, J. 1989. *The aesthetic state: A quest in modern German thought*. Berkeley: University of California Press.

Debord, G. 1983. *Society of the spectacle*. Detroit: Black and Red Books.

Durkheim, E. 1915. *The elementary forms of the religious life*. London: Allen and Unwin.

Eagleton, T. 1990. *The ideology of the aesthetic*. Oxford: Basil Blackwell.

Forman, F. J., with Sowton, C., eds. 1989. *Taking our time: Feminist perspectives on temporality*. Oxford: Pergamon Press.

Giddens, A. 1984. *The constitution of society*. Oxford: Polity Press.

Gregory, D., and Urry, J., eds. 1985. *Social relations and spatial structures*. London: Macmillan.

Hall, E. T. 1966. *The hidden dimension*. Garden City, NY: Doubleday.

Hallowell, A. 1955. *Culture and experience*. Philadelphia: University of Pennsylvania Press.

Hareven, T. 1982. *Family time and industrial time*. Cambridge: Cambridge University Press.

Harries, K. 1982. Building and the terror of time. *Perspecta: The Yale Architectural Journal* 19:59–69.

Hartshorne, R. 1939. *The nature of geography*. Lancaster, PA: Association of American Geographers.

Harvey, D. 1982. *The limits to capital*. Chicago: University of Chicago Press.

Harvey, D. 1985. *Consciousness and the urban experience*. Baltimore: Johns Hopkins University Press.

Harvey, D. 1989a. *The condition of postmodernity*. New York: Basil Blackwell.

Harvey, D. 1989b. From managerialism to entrepreneurialism: The transformation in urban governance in late capitalism. *Geografiska Annaler* 71(Series B):3–17.

Heidegger, M. 1959. *An introduction to metaphysics*. New Haven, CT: Yale University Press.

Helgerson, R. 1986. The land speaks: Cartography, chorography, and subversion in Renaissance England. *Representations* 16:51–85.

Jameson, F. 1984. Postmodernism, or the cultural logic of late capitalism. *New Left Review* 146:53–92.

Kern, S. 1983. *The culture of time and space, 1880–1918*. London: Weidenfeld and Nicholson.

Landes, D. 1983. *Revolution in time: Clocks and the making of the modern world*. Cambridge, MA: Harvard University Press.

Lefebvre, H. 1974. *La production de l'espace*. Paris: Anthropos.

Le Goff, J. 1980. *Time, work and culture in the middle ages*. Chicago: University of Chicago Press.

Le Goff, J. 1988. *Medieval civilisation*. Oxford: Basil Blackwell.

Levine, M. 1987. Downtown redevelopment as an urban growth strategy; a critical appraisal of the Baltimore renaissance. *Journal of Urban Affairs* 9(2):103–23.

Lévi-Strauss, C. 1963. *Structural anthropology*. New York: Random House.

Mallory, W. F., and Simpson-Housley, P., eds. 1987. *Geography and literature: A meeting of the disciplines*. Syracuse, New York: Syracuse University Press.

Marx, K. 1967. *Capital, Volume 1*. New York: International Publishers.

McLuhan, M. 1966. *Understanding media: The extensions of man*. New York: McGraw-Hill.

Mitchell, T. 1988. *Colonising Egypt*. Cambridge: Cambridge University Press.

Moore, B. 1986. *Space, text and gender*. Cambridge: Cambridge University Press.

Pearce, D., Markandya, A. and Barbier, E. 1989. *Blueprint for a green economy*. London: Earthscan Publications.

Relph, E. 1976. *Place and placelessness*. London: Pion.

Ross, K. 1988. *The emergence of social space: Rimbaud and the Commune.* Minneapolis: University of Minnesota Press.
Sack, R. 1986. *Human territoriality: Its theory and history.* Cambridge: Cambridge University Press.
Said, E. 1978. *Orientalism.* New York: Random House.
Schivelbusch, W. 1978. Railroad space and railroad time. *New German Critique* 14:31–40.
Seamon, D., and Mugerauer, R., eds. 1989. *Dwelling, place and environment.* New York: Columbia University Press.
Smith, N. 1984. *Uneven development: Nature, capital and the production of space.* Oxford: Basil Blackwell.
Soja, E. 1989. *Postmodern geographies.* London: Pluto.
Stone, J. 1988. Imperialism, colonialism and cartography. *Transactions, Institute of British Geographers.* N.S. 13, No. 1, 57–64.
Szanton, P. 1986. *Baltimore 2000.* Baltimore: Goldseker Foundation.
Toffler, A. 1970. *Future shock.* New York: Bodley Head.
Tuan, Yi-Fu. 1977. *Space and place.* Minneapolis: Minnesota University Press.

16 Felix Driver

Bodies in Space: Foucault's Account of Disciplinary Power

Excerpts from: C. Jones, and R. Porter (eds), *Re-assessing Foucault,* Chapter 7. London: Routledge (1994)

Foucault and the spaces of history

Amongst academic historians, the work of Michel Foucault has been received with a mixture of indifference, scepticism and downright hostility.[1] A few, to be sure, have been more generous; but in general, Foucault's writings on madness, health, punishment and sexuality have been regarded as beyond the disciplinary pale. Banished from the kingdom of History proper, they have received a warmer welcome amongst philosophers, although professional opinion remains sharply divided. In recent years, debates over modernity and postmodernism have inspired new readings of Foucault's work amongst cultural theorists and geographers, particularly in the United States. Ironically, what they frequently find attractive in Foucault's work is precisely its powerful sense of history; its obsession with discontinuities, practices, concrete details and complex genealogies. If Foucault is now widely seen as a philosopher of modernity, it is important to recall that much of his work was framed by what Peter Dews describes as an 'individual historical vision'.[2] This sense of history sets Foucault apart from other philosophers of his generation; the comparison with Derrida, for example, is instructive. Foucault addresses general questions by analysing specific moments, setting out to create 'philosophical fragments in historical building sites'.[3]

Foucault's sheer enthusiasm for historical documents was plain throughout his writing, in his archaeologies of the human sciences, in his genealogies of power and his studies of the techniques of the self.[4] Foucault's history was a history without nostalgia; there was no longing for a return to a mythical past, no dream of origins, no comforting vision of transcendence. In place of a continuous history, centred around the human subject, he proposed a decentred history, history in a space of dispersion.

> Nothing in man [he insisted], not even his body, is sufficiently stable to serve as the basis for self-recognition or for understanding other men. The traditional devices for constructing a comprehensive view of history and for retracing the past as a patient and continuous development must be systematically dismantled. Necessarily, we must dismiss those tendencies that encourage the consoling play of recognitions.[5]

The Nietzschean rhetoric is perhaps overdone, but the anti-humanist message is clear: how are we to write a history without the transcendental subject? A history, that is, which accounts for the constitution of subjects, as much as of knowledges, discourses and practices, 'without having to make reference to a subject which is either transcendental in relation to the field of events or runs in its empty sameness throughout the course of history'.[6] This suspicion of universals was a theme to which Foucault constantly returned throughout his work. Rather than reducing discourses and practices to simple essences or fundamentals, he wanted to explore their heterogeneity. As he mischievously observed in 1982, 'nothing is fundamental. That is what is interesting in the analysis of society.'[7] Foucault looked to history as a way of challenging the assumption that our conceptions of such things as madness, sexuality, crime and health spring directly and self-evidently from our experiences. Foucault's history disconcerts; it offers a means of criticizing the present, without the possibility of a return to the past.

Foucault's histories do not present us with linear narratives. They are concerned less with the flow of individual intentions, actions and consequences than with discourses, practices and effects. We might characterize many of his historical inquiries as maps rather than stories; a new kind of cartography, to borrow a phrase from Gilles Deleuze.[8] Evidence for such a view may be found throughout Foucault's work, especially in his constant recourse to the language of space; in the emphasis on exclusions and boundaries in *Madness and Civilization*, medical and social spaces in *The Birth of the Clinic* ('This book is about space, about language, and about death; it is about the act of seeing, the gaze'),[9] discursive spaces in *The Archaeology of Knowledge*, and diagrams or figures of power in *Discipline and Punish*. Critics differ in their interpretations of Foucault's spatial fixation. Some have argued that Foucault's sensitivity to space represents a decisive break from the historicism of much social theory, providing a new model for a postmodern geography. Others contend that Foucault's spatial metaphors were all too often used uncritically.[10] My own argument would be that space was indeed central to Foucault's concerns, although geography was not. This can be demonstrated in several ways. First, I would suggest that Foucault's spatial perspective represents a radical version of historical modes of inquiry, rather than an alternative to them. His genealogies attempt to bring grand narratives of truth and reason back to earth, as it were, grounding them in the realm of bodies, powers and spaces: 'Once knowledge can be analysed in terms of region, domain, implantation, displacement, transposition, one is able to capture the process by which knowledge functions as a form of power'.[11] Second, Foucault's constant use of terms such as *dispositif* (usually translated as 'apparatus') was supposed to highlight relations of connection rather than that of causation, simultaneity rather than succession. Power, in Foucault's analysis, does not exist prior to discourses and practices, on some other plane or level; rather, it operates through them. Third, spatial concepts lend themselves to an analysis of power relations in terms of strategic encounters, battles, terrains, colonizations and campaigns. Foucault certainly found these terms useful as metaphors; but their significance was more than metaphorical, because they highlighted the central role of the body, or rather bodies, in power relations. Foucault believed that 'space is fundamental in any exercise of power' mainly because he regarded the control of bodies as fundamental.[12] Fourth, when Foucault wrote about spaces, he often focused on particular institutions – the clinic, the prison, the hospital, the asylum – which help to constitute, or make visible, particular aspects of modern subjectivity. These were the sites within which various forms of knowledge and power were inscribed on bodies and souls. What was perhaps unclear was how these spaces were to be situated within a broader social whole; how the various kinds of spaces were to be related. It has been suggested, for example, that in Foucault's work, 'power and knowledge operate in the space of the body, not of geography'.[13] This is an important point to which I shall return.

In the rest of this chapter, I shall consider aspects of Foucault's account of disciplinary power. It would be misleading to treat Foucault's disparate statements on power as if they made up a coherent and consistent general theory, without recognizing the very different contexts in which they were made. In his books, especially in *Discipline and Punish*, Foucault explored particular aspects of discourses of power in specific settings; in his interviews, on the other hand, he was frequently tempted to express his views on power more freely. In both, he was less concerned with establishing a new theory of power than with challenging those which already existed. As Foucault's writings on power continue to stimulate lively debate, I should emphasize that in what follows I am less concerned with the overall logic, or otherwise, of Foucault's work than with its utility for specific kinds of historical research. (I shall also say little about the normative and ethical dimensions of Foucault's account of power.)[14] With this aim in mind, we might apply to our subject what Foucault himself once rather teasingly said of Nietzsche:

> the only valid tribute to thought such as Nietzsche's is precisely to use it, to deform it, to make it groan and protest. And if commentators then say that I am being faithful or unfaithful to Nietzsche, that is of absolutely no interest.[15]

Power and the disciplinary society

It is clear from many of his comments in essays and interviews that Foucault was uneasy about his reputation as a theorist of power. In much of his writing on power, as on other subjects, he was at pains to emphasize the limitations of general theories. The only thing that could be said about power in general, he once remarked, is that it is an open-ended, more or less coordinated 'cluster of relations'.[16] Rather than seeking the essence of power in some simple theoretical formula, Foucault posed apparently more modest questions about *how* power is exercised in particular sites and settings. At the same time, however, Foucault did make some far-reaching criticisms of existing theories of power, and if these did not amount to a coherent alternative, they did at least offer new perspectives on the problem.[17] Instead of portraying power as the property of any particular group or institution, Foucault preferred to describe it as a heterogeneous ensemble of strategies and techniques. He was thus sceptical of any approach which mapped power onto an abstract model of class relations. ('I believe that anything can be deduced from the general phenomenon of the domination of the bourgeois class. What needs to be done is something quite different. One needs to investigate historically, and beginning at the lowest level, how mechanisms of power have been able to function.')[18] Rather than confining his analysis to key institutions such as the state, he emphasized that power took many forms, often at its most effective where it was least visible. ('We must escape from the limited field of juridical sovereignty and state institutions, and instead base our analysis of power on the study of the techniques and tactics of domination.')[19] Finally, rather than seeing power always in negative terms, as prohibitive, Foucault asked how power could function as a positive force, creating, shaping, moulding subjects and subjectivity. ('What makes power hold good . . . is simply the fact that it doesn't only weigh on us as a force that says no, but that it traverses and produces things, it induces pleasure, forms knowledge, produces discourse.')[20] This was the general perspective which inspired Foucault's account of the emergence of modern systems of punishment in *Discipline and Punish*.

In *Discipline and Punish*, Foucault set out to explore the changing rationale and techniques of punishment in modern Europe. His analysis was framed in terms of three distinct 'regimes' of power: the monarchical, the contractual and the carceral. (This approach shared some of the features of Foucault's earlier work on discursive formations in *The Order of Things* and *The Archaeology of Knowledge*, although now he was as interested in the actual techniques of punishment as in their theoretical rationale.) The monarchical regime is exemplified in the opening pages of *Discipline and Punish* by the horrifying spectacle of public torture and execution. The contractual regime is read through the writings of reforming jurists, such as Beccaria, who called for a more efficient and transparent system of criminal justice. The carceral regime is associated with the disciplinary institution, pre-eminently the prison, where the inmate was trained in new ways of behaviour. For Foucault, each of these regimes depended on different mechanisms of power. The public execution, for example, ritually reaffirmed the power of the sovereign. The reforming jurists, in contrast, put their faith in the law itself as a 'technology of representation', a means of transmitting appropriate messages about the calculable consequences of criminal acts. The theorists of the carceral regime, in their turn, looked to disciplinary training as a means of producing docile, obedient individuals. Instead of representing punishment as a kind of exchange value modulated according to the criminal act, the carceral regime was designed to produce reformed individuals. The modern prison, in Foucault's account, was the place where the 'semio-techniques' of the law were superseded by the disciplinary techniques of surveillance.

Foucault interprets the movement between these regimes of punishment as a shift in the way in which power gains hold of human beings. Within the monarchical regime, power was exercised at a distance, exceptionally and capriciously; within its successors, power was supposed to operate routinely and universally, throughout the social body. The contractual regime, according to Foucault, offered a new economy of the power to punish in a universe of juridical subjects. He argues, however, that its semiotic techniques were ultimately superseded by the disciplinary technology of the prison, designed to produce obedient objects. In the carceral regime, individuals were instead to be trained into new habits, new patterns of conduct; their bodies subject to a *dressage* of disciplinary routines, their conduct monitored as closely as possible. 'In the first instance', argues Foucault, 'discipline proceeds from the distribution of individuals in space . . . each individual has his own place; and each place its own individual.'[21] Classifications, timetables and routines organize activity in space and time; prisons are divided by cells, landings and wings, just as schools are managed by classes and hospitals by wards. These arrangements are carefully designed to ensure the increasingly efficient surveillance of individual conduct and the exercise of what Foucault calls normalizing judgment. The new technique[s] of moral regulation depend on a more calculative attitude towards human behaviour; new modes of power go hand in hand with new forms of knowledge.

Discipline and Punish has attracted considerable criticism from historians. Was the movement between punitive regimes as clear-cut as Foucault implies? Did he take into account the gap between the reformers' intentions and actual outcomes? What about the differences between different institutions, in different times and places? What of conflicts between policy-makers, administrators, guards and the prisoners themselves? All of these questions are of course vitally important for historians. To be fair, Foucault has anticipated or responded to many of them, acknowledging, for example, that historical changes are always complex and uneven, that unintended consequences are a chronic feature of all projects of social reform, that there were important differences between different institutions and that conflicts and struggles played a critical role in the development of institutional practices.[22] However, to accept the importance of all these things for the history of prisons and prison reform is not necessarily to undermine Foucault's own project. For Foucault was not attempting to reconstruct the history of the prison system in all its complexity. Instead, his aim was to trace the emergence and consolidation of a set of disciplinary strategies which, he argues, provided the basis for a new regime of punishment. It is important to recall that Foucault was concerned more with practices than with closed institutions *per se*; with techniques of training which are found at their most concentrated within institutions, but are also present in all sorts of other social settings. Rather than offering *Discipline and Punish* as a history of prisons or penal policy, Foucault intended it to be read as a chapter in the history of 'punitive reason'.[23]

The differences between these readings of *Discipline and Punish* may be illustrated by the controversy over Foucault's account of the Panopticon, Bentham's bizarre scheme for a model prison. Some critics have taken Foucault to task for suggesting that the Panopticon, an institution which was never actually built, should be seen as a symbol of a new disciplinary regime. Others have complained that the Panopticon model cannot faithfully express institutional realities, especially given the myriad ways in which inmates may escape the gaze of officials.[24] To some extent, this criticism simply misses Foucault's point. The Panopticon is not presented by Foucault as an 'expression' or a 'reflection' of the reality of institutional life; rather, it is a paradigm, a model, in which many disciplinary strategies are concentrated. In *Discipline and Punish*, for example, Foucault describes the Panopticon as a 'diagram of a mechanism of power reduced to its ideal form . . . a pure architectural and optical system [abstracted from any] obstacle, resistance or friction'.[25] Foucault's account of 'panopticism' is thus to be read as a model of a disciplinary programme, not as a description of actual disciplinary institutions.[26] Why does Foucault regard it as exemplary? Because it embodied principles and techniques which were increasingly influential in the discourses of social policy: a faith in the moral powers of design; an emphasis on the surveillance of individual conduct; and an attempt to cultivate a sense of self-discipline amongst those to be trained – the habitual criminal, the demoralized pauper, the unreformed delinquent.

Although Bentham's Panopticon scheme attracted little support amongst his contemporaries, the principles it embodied were to exert an important influence on projects of moral regulation during the eighteenth and nineteenth centuries. It was in the sphere of institutional policy that strategies of surveillance and reformatory training were developed with most effect. Bentham himself had no doubt about the moral powers of architecture. (As he had proclaimed in drawing up his Panopticon plan: 'Morals reformed – health preserved – industry invigorated – instruction diffused – public burden lightened – economy seated as it were upon a rock – the Gordian knot of the Poor Laws not cut but untied – all by a simple idea in Architecture!')[27] In recent years, historians of prisons, asylums, hospitals and reformatories have confirmed that this faith in institutional design as a technique of moral discipline was remarkably widespread. This is suggested by the very terms they have used – the 'moral architecture' of the asylum, the 'moral geometry' of the prison, the 'moral space' of the reformatory, the 'moral universe' of the hospital, the 'school as machine'.[28] Contemporary reformers made a variety of assumptions about the mechanisms which linked the design of space with the pattern of individual behaviour. Within Bentham's plan, for example, the psychology of associationism provided a materialist framework for understanding the process by which individual sensation, perception and conduct were connected; a means by which the inmate could be trained to be the agent of his own reformation. The Panopticon in a sense combined what Foucault terms the 'semio-techniques' of the contractual regime (which presumed an economy of signs centred on the rational, calculating individual) with the disciplinary techniques of the carceral regime (which presumed a political technology based on disciplinary training).[29] Yet Foucault's general argument in *Discipline and Punish* is that the former were superseded by the latter, as the dreamworld of the Enlightenment gave way to the corporeal mechanisms of the prison. In so far as it consigns the 'semio-techniques' of the reforming jurists to a kind of historical no man's land, this view presents a rather one-dimensional account of the genealogy of Victorian discourses of social policy.[30] It also neglects the continuing significance of religious frames of reference in discourses of moral reformation. As Ignatieff has argued, penal reformers continued throughout the nineteenth century to attach importance to 'symbolic persuasion' (surely a kind of 'semio-technique') as much as to 'disciplinary routinization'.[31]

If Foucault's account of the relationship between these different forms of power is vulnerable to criticism, the task of situating them within some wider social whole is more problematic still. Foucault suggests in *Discipline and Punish* that the simple efficiency and transparency of the disciplinary techniques enabled their diffusion throughout society. At one point, he comes close to suggesting a simple functionalist explanation for the success of the disciplines, pointing to the imperatives of rapid capital accumulation and population growth, the need for reformed legal systems, and the technological revolutions of the eighteenth and nineteenth centuries.[32] More generally, there is little evidence in *Discipline and Punish* of Foucault's concern with resistance and struggle. In response to criticism on this point, Foucault drew an important distinction between the idea of a 'disciplinary society' (in which programmes such as the Panopticon proliferate) and that of a completely 'disciplined society' (in which such programmes are in a sense unnecessary!).[33] Self-evidently, he argued, his concern in *Discipline and Punish* was with the former rather than the latter; there was always something to be disciplined, something beyond the Panopticon. This said, Foucault's emphasis was overwhelmingly on the disciplinary techniques themselves, rather than on the various ways in which they were actually diffused, resisted and deflected in eighteenth and nineteenth-century societies. If the disciplines are to be located in a wider field of social relationships and resistances, then we need to know more about the actual workings of different institutional regimes and their position in the wider world. How were institutional programmes actually put into effect? By what means were disciplinary practices implemented or not implemented? To what extent was there an integrated system of institutions?

In answering these questions, it is necessary to consider the role of the state much more directly than Foucault does in *Discipline and Punish*. The claim that Foucault's work betrays a 'state-centred conception of social order'[34] seems rather odd in this context, because the institutions of the state are conspicuous by their absence from his account. Foucault has consistently rejected the idea that the new regime of power was imposed 'from the top', arguing instead that the disciplines were 'invented and organized from the starting points of local conditions and particular needs'.[35] This position clearly reflects Foucault's antipathy to state-centred theories of power. Yet a more focused account of the role of the state would not necessarily undermine Foucault's account. Bentham, for example, applied his principle of inspection to the regulation of institutions by the state as well as the surveillance of inmates by institutional officials. More generally, the establishment of new methods of government regulation during the eighteenth and nineteenth centuries could also be seen in the context of Foucault's ideas about surveillance.[36] Although Foucault was reluctant to be drawn into an institutional analysis of this sort, he was interested in forms of knowledge and power often associated with the state. His work on notions of governmental power and the idea of police inevitably raised questions about the role of the state in the surveillance of populations. Yet it was not altogether clear how his analysis of these forms of power might be connected with his discussions of discipline.[37]

Discipline and Punish does not offer us a self-contained history of prisons, or anything like it. Nevertheless, it poses a number of problems for historians as well as others to investigate; problems about the changing nature of punishment, the diffusion of disciplinary practices, the technologies of behaviour, the government of subjectivity. Foucault does not, of course, address these questions in the manner of a conventional historian: at the same time, however, he finds it worthwhile to approach them through historical work. Indeed, *Discipline and Punish* is a book absolutely obsessed with historical details, little events and obscure figures. If anything is lacking, it is a sense of society rather than a sense of history. What is left open by Foucault's analysis is how the exemplars of punitive regimes – the Panopticon, for example – might be located in a wider social space. If we wish to situate Foucault's account of disciplinary power in a more general context, we need to consider how various kinds of spaces and powers might be related to each other.

Foucault himself refused to outline these relationships in anything but the most schematic ways, suspicious as he was of the very idea of a theory of society.[38] He left others to ponder what might properly be described as a problem of location – how to situate the disciplines within both the broader flows of history and the complex geographies of the modern world.[39]

Conclusion

> What I say ought to be taken as 'propositions', 'game openings', where those who may be interested are invited to join in; they are not meant as dogmatic assertions that have to be taken or left *en bloc*.[40]

If the test of a good book lies in the stimulus it has given to further research and debate, there is no doubt that *Discipline and Punish* was one of Foucault's greatest successes. In this, as in his other writings, Foucault did not set out to write a definitive history; rather, he found historical work a useful way of posing problems about the nature of particular kinds of power. His hybrid form of analysis in fact took history very seriously; not as a set of grand continuities and totalities, but as a means of thinking about the way different forms of knowledge and power are rooted in particular historical situations. Foucault offered his readers a kind of geopolitics of discourse, a study of the means by which subjects and objects were placed and displaced. Disciplinary power was, perhaps above all else, a colonizing form of power; it cultivated new ways of seeing, calculating and ordering.[41]

Foucault's self-confessed obsession with space and spatiality is evident throughout his work. In his genealogies, in particular, he found spatial terms and concepts useful as a way of bringing questions of power to the fore; power in terms of concrete practices rather than ideologies. Yet I have suggested at several points in this essay that Foucault does not offer much guidance for those interested in developing an account of the way spaces might be articulated with each other; in other words, he presents only one side of what I have called the problem of location. There is little sense in *Discipline and Punish*, for example, of how to place panopticism in relation to other forms of power and resistance. This is not to say that Foucault does not discuss the problem of diffusion; in fact, he places considerable emphasis on the fact that the disciplines do escape from closed institutions, taking very different spatial forms outside the prison. This is an important point, for it demonstrates that disciplinary techniques do not necessarily depend on confinement or even spatial segregation. In his account of the Mettray colony, just as in his discussion of the York Retreat in *Madness and Civilization*, Foucault stresses the way in which normalizing judgement can operate in different contexts to those of the closed institution; through the family and the community, for example.[42] What one needs to know is whether and in what ways the disciplinary techniques themselves were modified, reshaped and resisted in the very process of their 'diffusion'. This remains an open question.

Notes

1 One critic has gone so far [as] to describe the influence of 'theory-mongers' (including Foucault) on historians as a 'cancerous radiation': G. R. Elton, *Return to Essentials: Some Reflections on the Present State of Historical Study*, Cambridge, Cambridge University Press, 1991, p. 41.
2 P. Dews, 'Power and subjectivity in Foucault', *New Left Review*, 144, 1984, p. 73.
3 M. Perrot (ed.), *L'impossible prison: Recherches sur le Système Pénitentiare au XIXe siècle*, Paris, Seuil, 1980, p. 41.
4 M. Foucault, 'The discourse of history' in S. Lotringer (ed.), *Foucault Live: Interviews 1966–1984*, New York, Semiotext(e), 1989, pp. 11–33; M. Foucault, 'The subject and power' in H. Dreyfus and P. Rabinow, *Michel Foucault: Beyond Structuralism and Hermeneutics*, Brighton, Harvester, 1982, p. 208.
5 M. Foucault, 'Nietzsche, genealogy, history' in D. Bouchard (ed.), *Language, Counter-Memory, Practice*, Ithaca, Cornell University Press, 1977, p. 153. Similar

arguments will be found in both Foucault's 'archaeological' and 'genealogical' phases, suggesting that the 'break' between the two can be overdrawn.

6 M. Foucault, 'Truth and power' in *Power/Knowledge: Selected Interviews and Other Writings, 1972–1977*, Brighton, Harvester, 1980, p. 117.

7 M. Foucault, 'Space, knowledge and power', *Skyline*, March 1982, p. 18.

8 G. Deleuze, 'Ecrivain non: un nouveau cartographe', *Critique*, 31, 1975, pp. 1207–27.

9 M. Foucault, *The Birth of the Clinic: An Archaeology of Medical Perception*, London, Tavistock, 1973, p. ix.

10 For comment on various aspects of Foucault's spatial obsession, see E. Soja, *Postmodern Geographies: The Reassertion of Space in Critical Social Theory*, London, Verso, 1989; M. Jay, 'In the empire of the gaze: Foucault and the denigration of vision in twentieth-century French thought' in D. C. Hoy (ed.), *Foucault: A Critical Reader*, Oxford, Blackwell, 1986, pp. 175–204; C. Philo, 'Foucault's geography', *Society and Space*, 10, 1992; R. Diprose and R. Ferrell, *Cartographies: Poststructuralism and the Mapping of Bodies and Spaces*, Sydney, Allen & Unwin, 1991.

11 M. Foucault, 'Questions on geography' in *Power/Knowledge*, p. 69.

12 Foucault, 'Space, knowledge and power', p. 20.

13 C. Lemert and G. Gillan, *Michel Foucault: Social Theory as Transgression*, New York, Columbia University Press, 1982, p. 98.

14 The key question here concerns Foucault's attitude towards humanism. The best critical account is to be found in N. Fraser, *Unruly Practices: Power, Discourse and Gender in Contemporary Social Theory*, Cambridge, Polity, 1989, chs. 1–3. See also the essays in Hoy, *Foucault: A Critical Reader.*

15 M. Foucault, 'Prison talk' in *Power/Knowledge*, pp. 53–4.

16 M. Foucault, 'The confession of the flesh' in *Power/Knowledge*, p. 199. See also M. Foucault, 'Two lectures' in *Power/Knowledge*, pp. 96–108, and *The History of Sexuality 1: An Introduction*, London, Allen Lane, 1979.

17 Cf. F. Driver, 'Geography and power: the work of Michel Foucault', reprinted in P. Burke (ed.), *Michel Foucault: Critical Essays*, Aldershot, Scolar Press, 1992, pp. 147–156.

18 Foucault, 'Two lectures,' p. 100.

19 Ibid., p. 102.

20 Foucault, 'Truth and power', p. 119.

21 M. Foucault, *Discipline and Punish: The Birth of the Prison*, Harmondsworth, Penguin, 1977, pp. 141–9.

22 Foucault, *Discipline and Punish*, pp. 14–16, 139; 'The eye of power' in *Power/Knowledge*, pp. 146–65; 'La poussière et le nuage' in Perrot, *L'impossible prison*, pp. 29–39.

23 Foucault, 'La poussière et le nuage', p. 33.

24 A. Giddens, *The Constitution of Society*, Cambridge, Polity Press, 1984, pp. 153–4.

25 Foucault, *Discipline and Punish*, p. 205.

26 C. Gordon, 'Afterword', in *Power/Knowledge*, p. 246.

27 J. Bowring (ed.), *The Works of Jeremy Bentham*, London, 1843, vol. iv, p. 39.

28 A. Scull, 'Moral architecture: the Victorian lunatic asylum' in A. Scull (ed.), *Social Order/Mental Disorder: Anglo-American Psychiatry in Historical Perspective*, Berkeley, University of California Press, 1989, pp. 213–38; R. Evans, *The Fabrication of Virtue: English Prison Architecture, 1759–1840*, Cambridge, Cambridge University Press, 1982; F. Driver, 'Discipline without frontiers?',

Journal of Historical Sociology, 3, 1990, pp. 272–93; C. Rosenberg, 'Florence Nightingale on contagion: the hospital as moral universe' in C. Rosenberg (ed.), *Healing and History*, New York, Folkestone, Dawson, 1979, pp. 116–36; T. Markus, 'The school as machine' in T. Markus (ed.), *Order in Space and Society: Architectural Form and its Context in the Scottish Enlightenment*, Edinburgh, Edinburgh University Press, 1982, pp. 201–56. See also D. Rothman, *The Discovery of the Asylum*, Toronto, Little, Brown & Co., 1971; A. D. King, *Buildings and Society*, London, Routledge & Kegan Paul, 1980; M. Donnelly, *Managing the Mind: A Study of Medical Psychology in Early Nineteenth-Century Britain*, London, Tavistock, 1983; J. Thompson and G. Goldwin, *The Hospital: A Social and Architectural History*, New Haven, Yale University Press, 1975; M. Foucault, B. Barret-Kriegel, A. Thalamy, F. Béguin and B. Fortier, *Les machines à guérir: Aux origines de l'hôpital moderne*, Brussels, Pierre Mardaga, 1976.

29 These regimes are also associated with two different kinds of subjectivity: Fraser, *Unruly Practices*, pp. 45–6.

30 Cf. M. Dean, *The Constitution of Poverty: Toward a Genealogy of Liberal Governance*, London, Routledge, 1991, pp. 216, 190–2; F. Driver, *Power and Pauperism: The Workhouse System, 1834–1884*, Cambridge, Cambridge University Press, 1993, ch. 1.

31 M. Ignatieff, 'State, civil society and total institutions', *Crime and Justice*, 3, 1981, pp. 175–6; M. Ignatieff, *A Just Measure of Pain: The Penitentiary in the Industrial Revolution, 1750–1850*, London, Macmillan, 1978, pp. 44–79.

32 Foucault, *Discipline and Punish*, pp. 218–28.

33 Foucault, 'La poussière et le nuage', p. 35.

34 Ignatieff, 'State, civil society and total institutions', p. 184.

35 Foucault, 'The eye of power', p. 159.

36 P. Corrigan and D. Sayer, *The Great Arch: State Formation as Cultural Revolution*, Oxford, Basil Blackwell, 1985; C. Dandeker, *Surveillance, Power and Modernity*, Cambridge, Polity Press, 1990; Driver, *Power and Pauperism*.

37 G. Burchell, C. Gordon and P. Miller (eds), *The Foucault Effect: Studies in Governmentality*, Hemel Hempstead, Harvester, 1992.

38 M. Foucault, 'Questions of method' in Burchell et al., *The Foucault Effect*, p. 85.

39 See also M. Donnelly, 'On Foucault's uses of the notion ''biopower'' ', in [T. Armstrong (ed.),] *Michel Foucault, Philosopher*, transl. T. Armstrong, New York, Routledge, 1992, pp. 199–203.

40 Foucault, 'Questions of method', p. 74.

41 Recent applications of Foucault's work in the context of colonialism include T. Mitchell, *Colonizing Egypt*, Cambridge, Cambridge University Press, 1989; R. Tolen, 'Colonizing and transforming the criminal tribesman: the Salvation Army in British India', *American Ethnologist*, 18, 1991, pp. 106–25.

42 M. Foucault, *Madness and Civilization: A History of Insanity in the Age of Reason*, New York, Social Science Paperbacks, 1965, pp. 243–55.

Place and Landscape

CRITICAL HUMAN GEOGRAPHY

The Betweenness of Place

Towards a Geography of Modernity

J. Nicholas Entrikin

PLACE AND LANDSCAPE

Placing place and landscape

Place and landscape are often presented, as they are here, as a complementary pair. There is something comforting and familiar about both that make them seem natural allies. Such apparent affinity is made sharper if we compare them with another traditional pairing in human geography, space and location, which in contrast appear cold and somehow inhuman. We analyse space and location, but we experience place and landscape. This ostensible congruity between place and landscape is more apparent than real, however. For even a cursory review of the intellectual history of the two terms reveals that the meaning and function of place and landscape have been quite different, rendering their traditional alignment at best problematic.

Although we have kept the traditional pairing of place and landscape here, we would also like to disrupt it, or at least to raise questions about it (represented by the textual strategy of separating the respective discussions of place and landscape in this introduction). If place and landscape seem a natural pairing, it is only because we have been socialised into a particular view of what is natural – in this case, by a 50-year Anglo-American tradition of regional geography that coupled them, and which was then continued by **humanistic geographers** and also by some within the **new cultural geography**. To denaturalise the naturalness of this pairing is one of our purposes.

Place and landscape do have something in common, however, although it is not their warm and fuzzy nature; it is their role as modes of ordering, in this case, ordering the facts of the geographical world. Rather than thinking of place and landscape as nouns, it is perhaps better to think of them as geographically powerful verbs, as the active process by which geographers assert order in the world. Given the critical role that ordering has played within human geography, it is not surprising that disputes over the meaning of place and landscape have driven some of the most significant debates in the discipline. Those who have most successfully engaged place and landscape have frequently also been those who have moved the discipline. With such high stakes, definitional wranglings over place and landscape have been endemic.

Place

Place is probably the more problematic and contentious of the two. Bound up with issues of **areal differentiation**, uniqueness, and regionalism, place description was the earliest form of geographical inquiry. Systematised by Strabo in his 17-volume *Geographia* (20 BC – 8 AD) as **chorology** – to study parts of the world – geographical descriptions were to be directed toward practical ends and not written for their own sake. That pragmatic bent was often forgotten, however, and subsequent chorological studies of place became only a catalogue of the peculiar and singular characteristics found at some particular corner of the globe.

In this century Richard Hartshorne (1939) in *The nature of geography* argued vigorously for the chorological method, conceiving place as an 'element-complex' (pages 428–31) – an intricate constellation of the various features found at a given location. In the early 1950s Hartshorne's view was contested by Fred Schaefer (1953) in what is still geography's most famous altercation. Arguing that Hartshorne's view was **exceptionalist**, Schaefer proposed that geographers focus not on the particularities of place – an **idiographic** form of enquiry – but generalise those particularities as **morphological laws**. From this view, places gain significance only as examples or instances of broader universal regularities – a **nomothetic** form of enquiry.

Later commentators argued that Hartshorne and Schaefer were in fact much closer intellectually than either thought. Although they differed about the purpose of studying place – uniqueness versus generalisation – they were in accord about the fundamental nature of place as an assemblage of brute facts that could be scientifically collected, represented, and analysed. Both men conceived place from the outside, as a set of externally observable features found at a particular site.

A very different view of place arose in the 1970s with humanistic geography which emphasised not the decentred, externalist perspective of Hartshorne and Schaefer, but started with the inside view of personal experience. Foreshadowed by the writings of J. K. Wright (1947) and William Kirk (1952), humanistic geographers conceived place as an intensely subjective and personal entity. As Ted Relph (1976, page 43) put it, 'The essence of place lies in the largely unselfconscious intentionality that defines places as profound centres of human existence'. This is the other extreme of the Hartshorne/Schaefer view; place as a **subjective** and not as an **objective** phenomenon.

Critiques of this subjective view of place came in the early 1980s from proponents of social theory, who couched the issue in slightly different terms than before. The old dualism of subjectivity–objectivity was transposed into the new one of **agency–**structure (see Section Eight). This requires some explanation. The subjectivist view of place is, in effect, one in which the world is populated by autonomous, imaginative, and creative individuals – agents – who make places in their own image. In contrast, social theory critics argued that such agency needed tempering by an outside structure. People can't make places be whatever they want them to be; individuals live within a set of social relations that constrain their actions and shape their beliefs, including those about place. At the same time, the wider social structure is not all-determining, immutably fixing people and places. The trick is to find a way of mediating between the two, that is, an approach that recognises both agency and structure and their changing interaction with and in place. That mediation was found in various forms of theory that emphasised a **recursive** movement between structure and agency and that mutually constituted both. In this scheme places were defined as the active sites at which such recursivness occurs. Pred (1984), for example, drew upon **structuration theory** (see Section Seven) to represent Stockholm as a 'historically contingent process' where structure, agency, and place continuously act and react with one another; and Massey (1984, ch. 5) implicitly relied on critical **realism** to portray South Wales mining communities as places made from the mutually interactive relation between economic structure and local cultural practices.

The most recent turn in the study of place, albeit still nascent, makes use of various elements of **post-structural** theory, especially those around **identity** (see Section Eight). There are two major claims. The first is that place has no single essence. Rather, it is an open-ended and porous entity, reflecting both the differential experiences of those who live there and the myriad external relations of which it is the locus. The second claim is that the identities of those living in a place, and who contribute to its sense of place, are constructed from geographically dispersed discourses that are themselves forged from spatially diffuse systems of power. There remains in this post-prefixed view the idea of place as a process, but there is the added argument that the most important of those processes occur outside the place itself either as discourses shaping identity or as external flows of information, images, ideas, and commodities. No place is an island unto itself, even islands.

Fixing place

Nicholas Entrikin, whose essay begins this section, is concerned with the classical debates around place, taking as his problematic what he calls 'the betweenness of place'. For Entrikin, places are wedged among a series of related dualisms: objectivism and subjectivism is the most important, but also description and explanation, inside and outside, the symbolic and the literal, ideal and material, the sciences and the humanities. Place, in short, is an awkward thing, not fitting within traditional conceptual polarities. Such intellectual ungainliness, however, is something that should be celebrated, representing as it does both an academic puzzle as well as an experience to be represented on the written page. Entrikin discusses the various forms that 'the betweenness of place' has taken in geography, showing how different people at different times have occupied various positions on, in particular, the objectivist–subjectivist pole. Where he takes the argument in a new direction is in his proposed resolution based upon **emplotment** and **narrative**. Taking his cue from the French philosopher and **hermeneutician**, Paul Ricouer (1983), Entrikin argues that the various dualisms surrounding the representation of place are reconcilable by telling stories or narratives. When we tell good stories we find that we are able to do all the things that we want to do in representing place: integrate subjective and objective, meld explanation and description, and blend literal and symbolic. Like M. Jourdan who spoke prose without knowing it, in telling stories about the earth we unselfconsciously create and represent place.

Doreen Massey in the paper that follows is less concerned with geography's past and more concerned with the here and now, that is, in an epoch of **postmodernity**. Her argument is set against David Harvey's thesis of **time–space compression** (see his article in this volume, Section Five, Chapter 15), and its consequences for place. Harvey argues that because of technological change, itself driven by the imperatives of capitalism, the world has shrunk and that the pace of life has accelerated like never before. In the midst of such change and turmoil, people are clinging to place; it is a bulwark of stability in the choppy seas of (post)modern life. Massey contests Harvey's vision in two ways, and in doing so provides the lineaments of a different view. First, she argues that different people are affected in different ways by time–space compression because of their particular identities. Whereas middle-class male white academics may be able to e-mail around the globe and jet to distant locations, most working class women of colour cannot. This is not to diminish the force of globalisation, but it is to say that its effects are multifaceted and variegated. Second, Harvey essentialises place as a fixed entity from which people take comfort. But for Massey, there are no fixed characteristics, not even fixed spatial boundaries. Places are defined as much by the outside as the inside. Given the multiplicity and changing nature of the relations impinging on place, it is necessarily a site of conflict and contradiction; there is no single identity. This doesn't deny uniqueness of place, but that uniqueness must be analysed as a consequence of the multiple intersection of generalised flows, power structures, discourses, and subjectitivies.

Landscape

Whereas place is classically conceived as bounded and circumscribed, landscape connotes openess and expanse, constituted by multifarious features and forms that stretch as far as the eye can see. The key word here is the eye. For landscape more than anything else, as both Dennis Cosgrove and Gillian Rose make clear in their respective articles in this section, is about a type of seeing. The very existence of landscape requires an observer, a pair of eyes that peer to the horizon.

Although as an idea it dates back to the medieval period, landscape was not formally codified until the late 19th century when a group of German geographers used the term to define geography as a discipline (*Landschaftsgeographie*). In practice this amounted to pigeon-holing various regional landscapes into a set of pre-established morphological boxes. Initially there was some recognition of a separate cultural morphology that interacted with physical forms, but increasingly only the natural landscape – geomorphology – was pursued.

Carl Sauer (1925) in his famous essay, 'The morphology of landscape', attempted to return to the German school's original conception by recognising both the physical and the cultural. As he wrote: 'The **cultural landscape** is fashioned from a natural landscape by a culture group. Culture is the agent, the natural area is the medium, the cultural landscape is the result.' Defined as an empirical and scientific form of enquiry concerned with tangible forms, Sauer and his **Berkeley school** prosecuted a **genetic methodology** that traced the mutually changing interactions on landscape between humans and their environment. The problem with this take on landscape, as Sauer himself came to recognise, was that it presumed some pristine, unbesmirched, natural world with which humans interacted. Nature from the beginning, however, is influenced and constructed by humans. As a result, landscape cannot be the site of two different things interacting – nature and culture – but only the changing history of one of them, human culture.

A more culturally centred view of landscape emerged in the 1950s propounded by two populist and, in many ways, maverick writers, the Englishman W. G. Hoskins and the American J. B. Jackson. Hoskins used the metaphor of the palimpsest – a parchment or tablet reused after earlier writing on it has been partially erased – to reconstruct *The making of the English landscape*, and J. B. Jackson, founder of *Landscape Magazine*, recorded in lyrical detail the virtues of such vernacular landscapes as the Westward moving house and the New Mexican trailer park. In neither case was there a formal method, only fluid prose and especially a sharp eye for landscape minutia. But whether it was discerning feudal furrows in ancient fields, or recognising the tell-tale notches of Western fence posts, landscapes were primarily apprehended by the eye.

The first forays into landscape studies by social theorists in the 1980s combined the historical sensibility of Sauer with the concern for the 'ordinary landscapes' of Hoskins and Jackson. Added was an appreciation of abstract arguments about social and cultural relations, **power**, and **discourse**. Two arguments were made by these new cultural geographers. The first was that it was not enough to focus on just the visible forms of landscape. To do so was to engage in a kind of fetishism, one that obscured the set of social relations that lay behind or beneath, and which provided meaning and explanation. Second, rather than defined by their brute materiality, landscape was better conceived as a system of visual signs and symbols – an **iconography** – that gained meaning from prevailing cultural discourse and the instantiated relations of power. In this new way of thinking, as Cosgrove and Daniels (1988, page 1) wrote, landscape is redefined as 'a cultural image, a pictorial way of representing, structuring or symbolising surroundings'. Although the interpretation of material artefacts remains important, the central task is to step back and interrogate the wider system of social and cultural relations in which the meanings of artefacts are constituted.

Seeing landscapes

Dennis Cosgrove, in the third article of this section, was one of the first to scrutinise landscape from a social-theoretical sensibility. He does so by historicising landscape, setting it within the context of the social relations in which it first emerged and took on meaning. His argument is that terms such as landscape, because of their historical origins, are laden with all kinds of intellectual baggage and therefore should not be appropriated willy-nilly without critical study. This is a problem for humanistic geographers. Landscape, like place, appears an attractive idea for mobilising humanistic sentiments, but its genesis makes it unsuited for that purpose. In making this argument Cosgrove makes two claims. The first is that landscape is a particular way of seeing, one based upon linear perspective, which, in turn, was derived from the application of rationalist geometrical principles. We may think of landscape as a loose and humane notion, but in fact its intellectual ancestry is that of the most exacting and precise sciences – mathematics. The second claim is that the acceptance of the linear perspective view of landscape is bound up with the emerging ideology of capitalism based upon private property and the interests of the bourgeoisie class that it represented. The linear perspective provided an implicit legitimation for the existing social order by representing space as something that could be controlled and dominated 'as an absolute objective entity, [and] transform[ed] into the property of individual or state' (page 325).

Gillian Rose's article picks up Cosgrove's insight of landscape as a way of seeing. But, for her, seeing, or 'the gaze' as she calls it, represents a peculiarly **masculinist** form of acquiring knowledge, implying a detached and disembodied observer who objectivises the things that he sees. Most often objectified by that gaze is woman. This is true not only in the literal sense but also figuratively, whereby objects of the gaze are transcoded as feminine. This is the case for landscape. As Rose (page 350) writes, 'the female figure represents landscape, and landscape a female torso'. From this perspective the male gaze is also bound up with issues of sexual desire, pleasure, and power. Rose thus adds to Cosgrove's argument the recognition that the historical focus on the visual form of landscape was a product not only of the ideology of class but also of the ideology of gender; it was not simply the bourgeoisie, but the male bourgeoisie that made looking at landscape 'an uneasy pleasure of power'.

References

Cosgrove, D., and Daniels, S. (eds) 1988: *The iconography of landscape. Essays on the symbolic representation, design and use of past environments.* Cambridge: Cambridge University Press.

Hartshorne, R. 1939: *The nature of geography: A critical survey of current thought in the light of the past.* Lancaster, PA: Association of American Geographers.

Kirk, W. 1952: Historical geography and the concept of the behavioural environment. *Indian Geographical Journal,* Silver Jubilee Volume, 152–60.

Massey, D. B. 1984: *Spatial divisions of labour: Social structures and the geography of production.* London: Macmillan.

Pred, A. 1984: Place as historically contingent process: structuration and the time geography of becoming places. *Annals of the Association of American Geographers,* **74**; 279–97.

Relph, E. 1976: *Place and placelessness.* London: Pion.

Ricouer, P. 1983: *Time and narrative.* Chicago, IL: Chicago University Press.

Sauer, C. 1925: The morphology of landscape. *University of California Publications in Geography,* **2**(2); 19–54.

Schaefer, F. K. 1953: Exceptionalism in geography: a methodological examination. *Annals of the Association of American Geographers,* **43**; 226–49.

Wright, J. K. 1947: Terrae Incognitae: the place of imagination in geography *Annals of the Association of American Geographers,* **37**; 1–15.

Suggested Reading

Recently there have been a number of useful edited collections around the issue of place. Three very useful ones are: John Agnew and James Duncan (eds), *The power of place: Bringing together geographical and sociological imaginations* (London, Unwin Hyman, 1989); James Duncan and David Ley (eds), *Place/culture/representation* (London, Routledge, 1993); and Michael Keith and Steve Pile (eds), *Place and the politics of identity* (London: Routledge, 1993). Some of the essays in those volumes also examine issues around landscape, but that topic is dealt with more systematically in Denis Cosgrove and Stephen Daniels (1988), and in Trevor Barnes and James Duncan (eds), *Writing worlds: Discourse, text and metaphor in the representation of landscape* (London, Routledge, 1992).

17 J. Nicholas Entrikin
The Betweenness of Place

Excerpts from: J. N. Entrikin, *The betweenness of place: Towards a geography of modernity*, Chapter 2. London, Macmillan (1991)

The existential and naturalistic qualities of place

The geographical concept of place refers to the areal context of events, objects and actions.[1] It is a context that includes natural elements and human constructions, both material and ideal. The French regional geographers captured this sense of place as context in the term *milieu*.[2] Their interests were to describe and understand the natural context associated with particular ways of life, but the concept can be extended to include the symbolic context that we create as agents in the world. These two aspects of our understanding of place reflect differing attitudes that we take toward place. We live our lives in place and have a sense of being part of place, but we also view place as something separate, something external.[3] Our neighborhood is both an area centered on ourselves and our home, as well as an area containing houses, streets and people that we may view from a decentered or an outsider's perspective. Thus place is both a center of meaning and the external context of our actions.

Our frequently noted ability to 'socially construct' places is a particularly modern view that recognizes our freedom to create meaning. Such a view highlights the active agent and the societal constraints on action, but does not overcome the basic tension that exists between the relatively subjective, existential sense of place and the relatively objective, naturalistic conception of place.[4] Reference to intersubjectivity fails to overcome this tension because place as *milieu* extends beyond what we share as subjects to refer to a world that is independent of subjects. The difficulty of combining the existential and naturalistic views reflects the underlying polarity between the subjective and the objective in our understanding of place.[5]

A distinguishing feature of the modern version of this polarity is the increased 'distance' between the subjective and the objective views. This distance is related to the success of the decentered view of the scientist. The scientific theorist strives for an objective, perspectiveless view, a 'view from nowhere'.[6] Place and region tend to 'fragment' into their 'parts' within the analytic and detached view of the theoretician. A large intellectual gap exists between our sense of being actors in the world, of always being in place, and the 'placelessness' that characterizes our attempts to theorize about human actions and events.

This gap has been made to appear even larger as the theoretical view of the scientist has become the model for addressing questions of how we should act. In normative descriptions of scientific practice, the scientist adopts a detached attitude toward the world with the hope of uncovering the reality that exists behind appearances. The rules of scientific method and experimentation are designed to allow the scientist to approximate the ideal of a detached observer in order to construct a view of the world as it exists.[7]

The conflict between the relatively objective, external vision embodied in our theoretical outlook, and the relatively subjective, internal vision that we have of ourselves as individual agents, represents a basic polarity of human consciousness. The tendency has been to reduce one side to the other, by suggesting that all that is real from the subjective view is reducible to the objective or vice-versa.[8] Extreme varieties of positivism and phenomenology represent these two respective forms of reduction.[9]

This polarity is a feature of our understanding of space and is illustrated in the distinction between the idea of existential space and that of geometric space; it is fundamental to our understanding of place.[10] The recognition of this polarity helps to explain why place remains a potentially significant concept in human affairs, despite its peripheral role in scientific discourse. A completely objective view of the world would have no 'here' and 'there', just as it would have no 'past', 'present' and 'future'.[11] In such a view the only meaning of place is that of the location of one object in relation to others. To limit the real to such a view would leave no 'room' for the subject except as another object in the world. It is difficult to imagine the existence of an active subject in a world that contains no 'here'. In order to create room for such a subject we require two irreducible parts to the concept of place: place as the relative location of objects in the world, and place as the meaningful context of human action. As Tuan states: 'Place is not only a fact to be explained in the broader frame of space, but it is also a reality to be clarified and understood from the perspectives of the people who have given it meaning'.[12]

Geography and the study of place

Geographers more than other groups of scholars have considered the concepts of place and region central to their discipline, and thus much of this analysis concerns their arguments. Geography has been described as a science that derives from the naive experience of the similarities and differences among places. Its practitioners have had difficulty balancing this naive sense, however, with the demands associated with the goal of scientific rationality. Although the general problem of the linkage between theory and experience is not unique to geography, geographers face the added complications associated with the spatial character of their concepts and the manner in which these concepts relate various kinds of phenomena. In the synthesis of heterogeneous phenomena according to their relations in space, geographers draw together elements of the world that tend to be analytically separated in the theoretical perspectives of other sciences.[13]

In everyday life we often conflate concepts that would be analytically drawn apart in a more theoretical outlook. One such conflation links objects and events to their locations. Another draws together the relatively objective aspects of a place as an external environment with the relatively subjective aspects of our experience (both direct and indirect) of place.[14] Such conflations are occasionally reflected in the way we speak about the world. For example, our references to 'Jonestown' or to 'Chernobyl' have a 'semantic density' that extends far beyond the geographic locations to include the terrible events that took place there.[15] The reference may also draw together both a descriptive and an emotive sense of those events and their context.

The metonymic quality of our everyday concept of place has parallels in the characterization of place in myth.[16] In mythical thought necessary connections link events and their locations, and the subjective and the objective are weakly differentiated. Places take on the meanings of events and objects that occur there, and their descriptions are fused with human goals, values and intentions. Places and their contents are seen as wholes.

This whole-part quality has been a feature of the geographer's characterization of place, region and landscape. Such a holistic perspective was evident in the works of early modern geographers, for example in the writings of nineteenth-century German geographers Carl Ritter and Alexander von Humboldt, but it has gradually receded in importance in twentieth-century geographic thought. The necessary connections between people and places have been replaced by contingent connections.

Ritter saw the wholeness of a region as a microcosm of a divinely ordered world.[17] This holism was in part an intellectual residual from the pre-Enlightenment era, however, in that the cohesion was other-worldly rather than of this world. According to the literary theorist Mikhail Bakhtin, the apotheosis of such a holistic vision among Enlightenment and post-Enlightenment scholars was found in the writings of J. W. von Goethe.[18] Goethe's literary works illustrated a sense of wholeness that linked past and present, time and space, nature and civilization, individual to humanity, and did so in terms of the concrete. He saw the universal in the concrete. His perspective was the visual, and he was able to see time in space, to see history unfold in parts of the earthspace.

Bakhtin has observed that Goethe's literary writings suggested a necessary linkage between events and places:

> In Goethe's world there are no events, plots, or temporal motifs that are not related in an essential way to the particular spatial place of their occurrence . . .[19]

The modern geographer has occasionally sought to capture this sense of wholeness, totality and necessity in the language of modern science. The historical referent of these geographers generally has been the aesthetic holism of von Humboldt rather than the theistic holism of Ritter. Typically, scientific support has been sought in the biological sciences. For example, in early twentieth-century American geography, Carl Sauer (himself influenced by the combined scientific and aesthetic vision of Goethe) used evolutionary biology and natural history as models for his culture history.[20] The apparent organic quality of areal units was accommodated through the recognition of their usefulness as fictions, a view expressed by the neo-Kantian philosopher Hans Vaihinger and cited by ecologically minded social scientists.[21] Similarly, the early twentieth-century regional studies of the French geographer Paul Vidal de la Blache and his students were grounded in a sense of terrestrial unity and in a naturalistic conception of social science.[22] For the Vidalians, necessary relations were replaced by a 'contingent' necessity based upon ideas of serial causality and chance.[23] In more recent discussions these last remnants of necessity have been replaced by ideas of pure contingency, as in Allan Pred's characterization of place as 'historically contingent process'.[24] The holism of the students of region and landscape has been replaced by the social physics of time geography or by the functional holism of systems analysis.[25]

In our everyday world we are made aware of this contingency through our apparent freedom from place or, somewhat more specifically, from place-based social relations. At the same time, however, we are confident about the objective quality of place and are content to offer explanations that use specific places as part of the cause of actions and events. It would seem, for example, that newsreaders feel that they have received at least a partial explanation for an especially bizarre crime or a peculiar incident when they learn that it took place in Los Angeles. Or, in a different context, it is difficult to imagine a discussion of the presidencies of Lyndon Johnson or Jimmy Carter that did not include reference to being a Texan or a Southerner, respectively, as part of the explanation for some of their actions. We recognize upon reflection that such place references are often based on stereotypes and misconceptions, but they are nonetheless a real and an important part of everyday discourse.

Place [also] serves as an important component of our sense of identity as subjects. The subject's concern for this sense of identity may be no different in kind from that of the geographer, in that the geographer's aim of accurately representing places can also be tied to concerns for social action and cultural identity. We see this link clearly in nineteenth- and early twentieth-century regional studies. For example, one of the themes of the Vidalian tradition was the balance between the diversity of provinces and the unity of the nation that provided the model of the moral and economic order of the Third Republic. Vidal sought to balance unity with diversity in a manner modeled on the liberal ideal of the relation between the individual and community.[26]

Vidal's concern with the moral order of the French nation illustrates the fact that the distinction between a decentered, objective view and a more centered, subjective view is a relative one. I have discussed these differences thus far in terms of the individual subject or agent and the scientist. But the views of place may be thought of as existing along a continuum from the most subjective view of an individual to the more objective view shared by members of a national culture, to the still more objective view of the theoretical scientist. As we move along this continuum we move from relatively centered views to relatively decentered views.

Specificity

Specificity, similar to the related concepts of uniqueness, the concrete and the idiographic, has through varying usages come to refer to a cluster of ideas in geographic thought. It will be helpful to recognize this fact when trying to make sense of the sometimes competing, and occasionally confusing, claims that have been made and continue to be made about the study of place and region. One of the reasons for this confusion may be that it is beyond our intellectual reach to attain a theoretical understanding of place and region that covers the range of phenomena to which these concepts refer. A more modest, but not insignificant, goal is a better understanding of the narrative-like qualities that give structure to our attempts to capture the particular connections between people and places.

Studies of this connection have been at the center of the discipline of geography, but on the periphery of social science. Geographers studying place and region have addressed important issues concerning the relationships between culture and nature and between society and nature. They have done so, however, in a manner that supporters and critics have described as being too subjective to fit comfortably within scientific discourse. Geographers express this ambiguity by their constant reference to what might be described as the 'betweenness' of chorology (or chorography), the study of place and region. For example, chorology has been described as being located on an intellectual continuum *between* science and art, or as offering a form of understanding that is *between* description and explanation.[27] Geographers have not been alone in this search for a middle ground; historians have sought a similar intellectual homeland.[28]

There have been two epistemological responses by geographers to this search. One seeks to broaden the base of scientific epistemologies beyond those that equate scientific explanation with the search for laws. Neo-Kantianism is an example of such a philosophy; a more recent example is transcendental realism. Neo-Kantians have argued that such a goal is only a step toward the ultimate goal of science, which is certain knowledge of the world, both in terms of nomothetic regularities and comprehensive knowledge of the individual case. Transcendental realists have argued that the goal is one of uncovering the basic structures and causal mechanisms of the world, and that these are expressed as lawful relations only in the unusual circumstances associated with closed systems.

The other response distances the study of place and region from social theory and social science and moves it closer to the humanities.[29] This response has been especially evident among historical and cultural geographers. Their emphasis on the concrete relations between culture and the material world has made them particularly sensitive to the complexity of the experiences that cultural groups have in places. They are especially cognizant of the necessary incompleteness of theories, and wary of attempts to generalize from the experience of these groups.

Specificity and humanistic geography

Humanistic geography developed in the 1970s as a mélange of epistemological positions and thematic interests. One of the threads holding together these disparate concerns has been the emphasis given to experience and to meaning. The specificity of place has been associated with the unique experiences of place and the meanings that we associate with these experiences. The emphasis on intentionality connects the observer and the observed in a manner that cannot be drawn apart. Uniqueness thus becomes a function of the quality of experience rather than a description of a world (i.e. a place or a region) that is completely external to the knowing subject.

Denis Cosgrove has described the effect of this emphasis upon meaning in the geographical study of landscape. He has described the differences between the traditional, morphological studies of landscape and the studies of the symbolic dimensions of landscape in a manner that is directly translatable to the study of place and region:

> To regard landscape as both object and subject has important consequences for a discipline seeking to theorise according to determinate rules of scientific procedure the relationships between human beings and their environment as those relationships give rise to characteristically differentiated areas. Morphological analysis, with its concentration on empirically defined forms and their integration, can operate only at a surface level of meaning . . . Below this lie deeper meanings which are culturally and historically specific and which do not necessarily have a direct empirical warranty. Formal morphology remains unconvincing as an account of *landscape* to the extent that it ignores such symbolic dimensions – the symbolic and cultural meaning invested in these forms by those who have produced and sustained them, and that communicated to those who come into contact with them.[30]

The concept of meaning in the social sciences is, of course, a notoriously treacherous one. The literature on the philosophy of the human sciences has most often characterized it in a manner similar to that used by Cosgrove in his reference to the geological metaphor of 'layers' or 'strata'. This has been an especially useful metaphor in the structuralist and phenomenological literature, both of which rely heavily on the image of the underlying structures of meaning and on notions of the 'sedimentation' of meaning.

In the literature of logical empiricism, however, the analogies used have been somewhat different, in that meanings are described as objects, not unlike the objects of the natural sciences.[31] For example, the philosopher of social science May Brodbeck refers to the different types of meaning that are of interest to the human sciences.[32] The first two types refer to the objective meaning of concepts that have space–time instances ($meaning_1$) and are connected to one another in terms of empirical generalizations ($meaning_2$). Both refer to the objective realm of scientific discourse. Other types of meaning described by Brodbeck refer to the unobservable world of mental events, the intent of a thought ($meaning_3$) and the subjective realm of the individual ($meaning_4$). She argues that these latter types of meaning (3 and 4) are of interest to the human scientist as an object of study, but are inappropriate as part of the language of science. Such subjective states require translation into the objective language of science through operational definitions that link them to observable phenomena.

For humanistic geographers this rigid distinction between the objective and the subjective is seen as part of the problem of a naturalistic social science. Humanists criticize both the tendency to treat the human subject as an object, and the failure to acknowledge the subjectivity of the social scientist. They extend their criticisms in an uneven fashion to the attempts by chorologists to create a science of regional studies. In their concern to avoid the objectification of positivist social science, some humanists highlight what Stephen Daniels describes as a private world of 'feelings' that are 'beyond rational scrutiny'.[33] In seeking to capture the holistic quality of the experience of place, humanists seek to understand that experience through the eyes of the 'insider'. Daniels observes that: 'From a humanistic perspective the meaning of a place is inseparable from the consciousness of those who inhabit it'.[34]

The specificity of places is thus a function of the unique experiences that individuals and groups associate with place. Although the method for gaining access to this consciousness is a matter of debate, humanists tend to be less troubled than their more scientifically oriented colleagues by the potential for subjective judgements by the observer in seeking to uncover this meaning. Where the scientific geographer sees a continuum with endpoints labeled subjective and objective, the humanistic geographer sees degrees of intersubjectivity. The blending of the subjective and the objective was a seemingly unintentional consequence of a science of areal differentiation, but it has become one of the goals of more recent attempts to theorize about everyday actions in a manner that roots social theory in space and time.

A 'new' regional geography

The call for a new regional geography has become a familiar refrain in the literature of contemporary geography. Those making such calls have argued for various forms of contextualist social science.[35] Contextualist arguments can be distinguished from those of traditional chorology by their emphasis on the study of space and society rather than nature and society, by their emphasis on theory, and by an explicit concern with meaning. These differences to a certain extent reflect the diverse concerns of contemporary geographers to which the contextualists have sought to give order. They have attempted to mediate the spatial analysts' concern with space, the neo-Marxists' concern with social relations and structures, and the humanists' concern with agency and meaning. Contextualists have also sought to combine a theoretical orientation with the chorologists' interest in specific place and region.[36] Specificity has been described by the contextualists both in terms of uniqueness and in terms of the fusion of place and experience in practical knowledge.

The uniqueness arguments are associated with the view that place and region matter because social processes take place and places vary.[37] Social and economic forces shape places and in turn are shaped by places. General forces produce unique outcomes because they are played out under varying circumstances. For Massey,

> Local uniqueness matters. Capitalist society, it is well-recognised, develops unevenly. The implications are twofold. It is necessary to unearth the common processes, the dynamic of capitalist society, beneath the unevenness, but it is also

> necessary to recognise, analyse and understand the complexity of the unevenness itself. Spatial differentiation, geographical variety, is not just an outcome: it is integral to the reproduction of society and its dominant social relations. The challenge is to hold the two sides together; to understand the general underlying causes while at the same time recognising and appreciating the importance of the specific and the unique.[38]

Massey's argument differs from those of traditional chorologists in its explicit concern with theory and in the type of neo-Marxist theory she employs. However, the recognition of the balance between the general forces and the unique circumstances has been a theme in all discussions of regional studies. The unresolved difficulty concerns the working out of the logical relations that can support such a balance between the general and the specific.

One attempt at reconciliation has been to undergird structurationist theory with a transcendental–realist epistemology. For example, John Agnew argues that such a perspective offers the best alternative to a positivist social science, in that 'structuration theory provides the only base upon which to combine an emphasis on agency with a continued commitment to causation', and that it 'allows for the historical specificity and uniqueness of places while proposing that these "multiple outcomes", if you will, are the product of a "one to many correspondence" between cause and effect'.[39]

The structurationist argument has apparently moved beyond traditional conceptions of chorology by incorporating certain elements of the humanistic geographer's concern with agency and subjectivity. Our holistic experience of place involves, however, both the affective and the cognitive. These affective attachments are part of the motivational context of human action.[40] Motivated action is obscured in structurationist and contextualist accounts by the emphasis on habitual action and the unintended consequences of such actions. The role of the symbolic context of action and of the active agent as an interpreter of that context is mentioned, but rarely developed.

Agnew's transcendental–realist perspective seems to overcome the perennial problem of betweenness in studies of place and region by adopting a strategy somewhat similar to that of the neo-Kantians. Both transcendental realism and neo-Kantianism offer arguments consistent with the traditional conception of science as the search for causes. Each defines causation in a non-Humean fashion, however, by separating causation from the constant conjunction of events. Causes need not be expressed in terms of generic concepts and lawful regularities. Thus each view allows for the possibility of a causal explanation of the specific.

Transcendental realism as it has been applied by geographers appears to involve not only logical commitments concerning the nature of causation, but also theoretical commitments.[41] I have presented in my introductory arguments a realist view that commits me to the belief that a world exists independent of our ideas about it, and that our theories attempt to describe that world. The transcendental realist would accept this simple conception of realism, but would add to it an ontology of structures and mechanisms that gives objects 'causal powers'. Such an ontology appears to be based in part on a theoretical commitment associated with neo-Marxism. Even if we were to assume, as the transcendental realists claim, that physical theories require an ontology of structures and mechanisms, we are given no independent grounds for assuming that social theories must be similarly constructed. The lack of empirical warrant for most social theories and the many differing ontologies that are associated with them would seem to preclude an *a priori* adoption of one over the others.

The question that is raised by the reference to contextualist theory in geography concerns the ability of its proponents to overcome the problems faced by those who have come before them in attempting to develop a theoretical perspective on place and region. What are the criteria of significance of such theories that give sense and importance to aspects of the manifold experience of place and region? What is the theoretical logic that contributes a determinative structure to contextualist arguments? The contextualists' attempts to mediate traditional antinomies in social thought (e.g. the particular–general, micro–macro, agency–structure, subjective–objective, etc.) seem more effective at the meta-theoretical than at the theoretical level.[42] At the level of theory, contextualist arguments have been unable to circumvent disjunctions that arise when we apply the cognitive demands of theorizing to the 'holistic' experience of place. If we accept Tuan's description of experience as a combination of both the affective and the cognitive, then we are led to a conclusion similar to his – that our experience of place is not something that can be fully captured in the objectivist realm of theory.

Our theories tend to disconnect the subject from the world, but the significance of place in the modern world is in part a function of that connection. Contextualist theory reminds us of the fact that the social sciences deal with subjects acting in the world. Contextualists have further recognized that place matters in the world of everyday action. In order to accommodate agency in their theory, however, contextualists limit the individual subject to a habitual action and routinized behaviors. The subject becomes a construction derived from the social rules governing action.

Narrative-like synthesis

A means of describing the world in relation to a subject is through narrative. Narrative understanding has been characterized as a way of 'seeing things together'.[43] It has been described as a distinct form of knowing that derives from the redescription of experience in terms of a synthesis of heterogeneous phenomena.[44] In one of its simplest forms the narrative has two components, the story and the storyteller. Narratives are by definition told from the point of view of a subject or subjects.[45] The type of narrative of greatest interest to the geographer has been the historical narrative. Robert Scholes and Robert Kellogg suggest that the development of the historical narrative was related to the idea of an actual past as opposed to a traditional version of the past and required 'for its development means of accurate measurement in time and space, and concepts of causality referable to human and natural rather than to supernatural agencies'.[46] They distinguish the historical narrative from the mimetic narrative, one that seeks to imitate human action.[47] Humanistic geographers have made this mimetic form a part of the geographer's study of place in the modern world.

In narrative, events are given meaning through their configuration into a whole. Historians and philosophers, like geographers, have debated the nature of such 'wholes'. For example, Ricoeur emphasizes the importance of necessity. He draws upon Aristotle's *Poetics* to make a distinction between episodic plots and simple plots, a distinction that he characterizes as 'one thing after another' versus 'one thing because of another'.[48] The episode implies an accidental quality that is turned into necessity or probabilism in the plot. He argues that: 'To make up a plot is already to make the intelligible spring from the accidental, the universal from the singular, the necessary or the probable from the episodic'.[49]

Ricoeur's discussion of plot relies to a large extent on the writings of the French historian Paul Veyne, who has noted that the idea of the region functions for the geographer in the same way that the plot functions for the historian. One of the implications of the similarity is that regions (and places) are also presented from a point of view:

> For the division of the spatial continuum, the geographer can choose among numberless points of view, and those regions have no objective frontiers and existence. If we undertake, like Ritter, to find the 'true' division into regions, we fall into the insoluble problem of an aggregation of points of view, and into a metaphysics of organic individuality, or into a physiognomy of the landscape (the idea of a geometrical projection being the edulcoration of those superstitions).[50]

Geographers have sought to follow the model of the theoretical scientist, but the perspectival quality of our understanding of place means that those who study place must stop short of the decentered view. As one proceeds toward a decentered, theoretical view, place dissolves into its component parts.

The tension between the relatively subjective and the relatively objective sense of place generally has been overlooked. The tendency to reduce one side to the other has made it easy to ignore. The theoretical reduction of place to location in space could not effectively capture, however, the sense of place as a component of human identity, and the opposing reduction tends to treat place solely as a subjective phenomenon.

Twentieth-century communitarian and regionalist movements have recognized this basic polarity, but their naturalistic models have tended to ignore the individual subject except as an agent of an objective spirit. The subject has been viewed as an agent of the whole, and not as an individual moral agent. Mainstream liberal social thought in the twentieth century separated the agent from this context, but in the process constructed an ideal agent independent of context, outside of place and period. One of the ways in which this dichotomous situation has been brought into view in the late twentieth century is in the understanding of the role of narrative in modern life, both as an element of culture and as a 'meta-code' of human communication.[51]

To link the understanding of place and region to narrative appears on the surface to affirm what many geographers have said throughout the century. Most such references have been made in opposition to a scientific vision of the field in which narrative is seen as a proto-scientific mode.[52] Others have contrasted geographical description with historical narration in maintaining the distinction between the historian's concern with time and the geographer's concern with space.[53] I attempt to set a more positive course by suggesting the significance of place in modern life and the importance of a narrative understanding for capturing this significance.

A key element in this form of understanding is the process of 'emplotment'. According to Ricoeur, 'Plot, in effect, "comprehends" in one intelligible whole, circumstances, goals, interactions, and unintended results'.[54] In doing so it combines both the chorologist's concern for the connection of the objective 'facts' of place into a comprehensible whole and the humanist's concern for the intentional connection between actor and environment. Its goals are similar to the goals of the contextualist – to draw together agents and structures, intentions and circumstances, the general and the particular, and at the same time seek to explain causally.[55] Emplotment differs, however, in that its proponents need not claim equivalent cognitive status with the theoretical view of the scientist. Explicit in narrative is the fact that it is from a point of view. Its relative centeredness is what allows it to incorporate elements of both objective and subjective reality without collapsing this basic polarity between the two views. A difference between the view of the agent in everyday life and the view of the geographer tends to be one of the degree to which the two poles are distinguished. The geographer typically makes a self-conscious attempt to draw these two poles apart and occupy a relatively objective position somewhere between those of the agent and the theoretician. From this vantage point the geographer gains an understanding of place as the context of human actions and events.

The apparent insignificance of place in modern life and the concomitant insignificance of the study of place are related to the confidence that moderns have in the objective view of the theoretical scientist. Our technological control of nature emphasizes the global, the universal and the objective, and the success in the manipulation of nature has led to the application of the same perspective to human society. Such a view is unable to capture the importance of the moral uniqueness of the individual agent and the source of agency in the local, the particular and the subjective. The narratives of place help to redress this imbalance, without camouflaging the underlying tensions between the subjective and the objective and between individual agents and the circumstances within which agents act.

Notes

1 Most of my argument concerning the concept of place may also be applied to the concept of region. Place and region both refer to areal contexts, but may be distinguished in terms of spatial scale.

2 Anne Buttimer, *Society and Milieu in the French Geographic Tradition* (Chicago: Rand McNally and Company for the Association of American Geographers, 1971). A more direct translation of place into French is *lieu*, a concept that suffers from the same variety of usage as does the term place. According to Vincent Berdoulay:

> The term place (*lieu*) is used among French language geographers in an informal sense. As such it is generally not used as a research-inducing concept, as is often the case in Anglo-American geography.

Vincent Berdoulay, 'Place, Meaning, and Discourse in French Language Geography', in John Agnew and James Duncan (eds) *The Power of Place: Bringing Together Geographical and Sociological Imaginations* (Boston: Unwin Hyman, 1989), pp. 124–39, ref. on p. 124.

3 Yi-Fu Tuan, *Space and Place: The Perspective of Experience* (Minneapolis: University of Minnesota Press, 1977).

4 Robert D. Sack, *Conceptions of Space in Social Thought: A Geographic Perspective* (Minneapolis: University of Minnesota Press, 1980); 'The Consumer's World: Place as Context', *Annals of the Association of American Geographers*, vol. 78, 1988, pp. 642–64.

5 Thomas Nagel, *The View from Nowhere* (New York: Oxford University Press, 1986).

6 T. Nagel, *The View from Nowhere.*

7 T. Nagel, *The View from Nowhere*, pp. 13–27.

8 Thomas Nagel, *Mortal Questions* (Cambridge: Cambridge University Press, 1979), pp. 196–213.

9 T. Nagel, *The View from Nowhere.*

10 J. Nicholas Entrikin, 'Geography's Spatial Perspective and the Philosophy of Ernst Cassirer', *The Canadian Geographer*, vol. 21, 1977, pp. 209–22.

11 T. Nagel, *The View from Nowhere*, p. 57 n.1. See also John Urry, 'Social Relations, Space and Time', in Derek Gregory and John Urry (eds) *Social Relations and Spatial Structures* (Basingstoke: Macmillan, 1985), pp. 21–48.

12 Yi-Fu Tuan, 'Space and Place: Humanistic Perspectives', in *Progress in Geography*, vol. 6, 1974, pp. 213–52, ref. on p. 213.

13 Paul Ricoeur has defined narrative as the synthesis of heterogeneous phenomena, and I have borrowed his wording to describe the task of the geographer. Paul Ricoeur, *Time and Narrative* (Chicago: University of Chicago Press, 1983). The concept that underlies this expression is one that has been expressed many times in the history of geographic thought. It has been an important part of the Kantian tradition in geography: for example, Richard Hartshorne, *The Nature of Geography: A Critical Survey of Current Thought in Light of the Past* (Lancaster, Pa.: Association of American Geographers, 1939). A more recent statement from a somewhat different perspective is found in Torsten Hägerstrand's discussion of the contextual approach. Torsten Hägerstrand, 'Presence and Absence: A Look at Conceptual Choices and Bodily Necessities', *Regional Studies*, vol. 18, 1984, pp. 373–80. See also Derek Gregory, 'Suspended Animation: The Stasis of Diffusion Theory', in D. Gregory and J. Urry (eds) *Social Relations and Spatial Structures*, pp. 296–336.

14 R. Sack, *Conceptions of Space in Social Thought.*

15 For a discussion of the semantic density of place names see Roland Barthes, 'Proust and Names', in *New Critical Essays*, translated by Richard Howard (New York: Hill and Wang, 1980), pp. 55–68. In Jonestown, Guyana, a colony named after Jim Jones, the leader of the Peoples Temple in San Francisco, California Congressman Leo Ryan and members of the press were assassinated and over 900 members of the Peoples Temple died from cyanide poisoning in a mass murder/suicide on 18 November 1978. In Chernobyl, a town in the Ukrainian Republic of the USSR, a nuclear power plant experienced a meltdown on 26 April

1986. A radioactive cloud from the accident spread over large areas of the Soviet Union, Eastern Europe and Scandinavia.

16 Ernst Cassier, *The Philosophy of Symbolic Forms: Vol. 2 Mythical Thought* (New Haven: Yale University Press, 1955).

17 R. Hartshorne, *The Nature of Geography*, pp. 40–84; George Tatham, 'Geography in the Nineteenth Century', in Griffith Taylor (ed.) *Geography in the Twentieth Century: The Study of Growth, Fields, Techniques, Aims and Trends* (London: Methuen, 1951), pp. 28–69.

18 M. M. Bakhtin, *Speech Genres and Other Late Essays*, translated by Vern W. McGee and edited by Caryl Emerson and Michael Holquist (Austin: University of Texas Press, 1986), pp. 44–5.

19 M. M. Bakhtin, *Speech Genres*, p. 42; Bakhtin refers to this space–time whole as a 'chronotope', and although his primary concern is with a theory of literary genres, it is a useful term for capturing the sense of space–time 'wholes' that have fascinated both geographers and historians. M. M. Bakhtin, *The Dialogic Imagination*, translated by Caryl Emerson and Michael Holquist (Austin: University of Texas Press, 1981), pp. 84–258.

20 For a discussion of Sauer and natural science see J. Nicholas Entrikin, 'Carl O. Sauer, Philosopher in Spite of Himself', *Geographical Review*, vol. 74, 1984, pp. 387–408. Sauer's historical and evolutionary perspectives are discussed in Michael Williams, ' "The Apple of My Eye": Carl Sauer and Historical Geography', *Journal of Historical Geography*, vol. 9, 1983; pp. 1–28, and in Michael Solot, 'Carl Sauer and Cultural Evolution', *Annals of the Association of American Geographers*, vol. 76, 1986, pp. 508–20. For a discussion of Goethe and Sauer see Martin Kenzer, 'Milieu and the "Intellectual Landscape": Carl O. Sauer's Undergraduate Heritage', *Annals of the Association of American Geographers*, vol. 75, 1985, pp. 258–70; Wiliam Speth discusses the relation of Sauer's ideas to German Romanticism, especially to the arguments of Herder in 'Historicism: The Disciplinary World of Carl O. Sauer', in Martin Kenzer (ed.) *Carl O. Sauer: a Tribute* (Corvallis: Oregon State University Press, 1987), pp. 11–39.

21 Hans Vaihinger, *The Philosophy of the 'As If'*, translated by C. K. Ogden (New York: Harcourt Brace, 1925); J. Nicholas Entrikin, 'Robert Park's Human Ecology and Human Geography', *Annals of the Association of American Geographers*, vol. 70, 1980, pp. 43–58, ref. on p. 48; and 'Carl Sauer, Philosopher in Spite of Himself', p. 388.

22 Vincent Berdoulay, 'The Vidal–Durkheim Debate', in David Ley and Marwyn S. Samuels (eds) *Humanistic Geography: Prospects and Problems* (Chicago: Maaroufa Press, 1978), pp. 77–90, ref. on p. 83; F. Lukermann, 'Geography as a Formal Intellectual Discipline and the Way in which It Contributes to Human Knowledge', *The Canadian Geographer*, vol. 8, 1964, pp. 167–72, ref. on p. 171.

23 F. Lukermann, 'The "Calcul des Probabilities" and the École Française de Géographie', *The Canadian Geographer*, vol. 9, 1965, pp. 128–35.

24 Allan Pred, *Place, Practice and Structure: Social and Spatial Transformation in Southern Sweden: 1750–1850* (Totowa, N.J.: Barnes and Noble, 1986).

25 Torsten Hägerstrand, 'Survival and Arena: on the Life-History of Individuals in Relation to their Geographic Environment', in Tommy Carlstein, Don Parkes and Nigel Thrift (eds) *Human Activity and Time–Geography: Vol. 2, Timing Space and Spacing Time* (London: Edward Arnold, 1978), pp. 122–45; 'Diorama, Path and Project', *Tijdschrift voor Economische en Sociale Geografie*, vol. 73, 1982, pp. 323–39. Functionalist, system-analytic approaches to region are discussed in G. P. Chapman, *Human and Environmental Systems: A Geographer's Appraisal*

(London: Academic Press, 1977); R. J. Bennett and R. J. Chorley, *Environmental Systems: Philosophy, Analysis and Control* (Princeton: Princeton University Press, 1978).

26 Vincent Berdoulay, *La formation de l'école française de géographie (1870–1914)* (Paris: Bibliothèque Nationale, 1981); and *Des mots et des lieux: la dynamique du discours géographique* (Paris: CNRS, 1988), p. 17.

27 John Fraser Hart, 'The Highest Form of the Geographer's Art', *Annals of the Association of American Geographers*, vol. 72, 1982, pp. 1–29.

28 Hayden White, *Topics of Discourse: Essays in Cultural Criticism* (Baltimore: Johns Hopkins University Press, 1978).

29 Cole Harris, 'The Historical Mind and the Practice of Geography', in D. Ley and M. Samuels (eds) *Humanistic Geography: Prospects and Problems*, pp. 123–37; Donald Meinig, 'Geography as an Art', *Transactions of the Institute of British Geographers*, new series 8, 1983, pp. 314–28.

30 Denis Cosgrove, *Social Formation and Symbolic Landscape* (Totowa, N.J.: Barnes and Noble, 1985), pp. 17–18.

31 David Thomas, *Naturalism and Social Science: A Post-Empiricist Philosophy of Social Science* (Cambridge: Cambridge University Press, 1979).

32 May Brodbeck, 'Meaning and Action', in May Brodbeck (ed.) *Readings in the Philosophy of the Social Sciences*, (New York: Macmillan, 1968), pp. 56–78.

33 Stephen Daniels, 'Arguments for a Humanistic Geography', in R. J. Johnston (ed.) *The Future of Geography* (London: Methuen, 1985), pp. 143–58, ref. on p. 151.

34 Stephen Daniels, 'Arguments for a Humanistic Geography', p. 145.

35 For example, see John Agnew, *Place and Politics: The Geographical Mediation of State and Society* (Boston: Allen and Unwin, 1987); Derek Gregory, *Regional Transformation and Industrial Revolution: A Geography of the Yorkshire Woollen Industry* (London: Macmillan, 1982); R. J. Johnston, 'The World Is Our Oyster', *Transactions of the Institute of British Geographers*, new series, vol. 9, 1984, pp. 443–59; Anssi Paasi, 'The Institutionalization of Regions: A Theoretical Framework for Understanding the Emergence of Regions and the Constitution of Regional Identity', *Fennia*, vol. 164, 1986, pp. 105–46; A. Pred, *Place, Practice and Structure*; Nigel Thrift, 'On the Determination of Social Action in Space and Time', *Environment and Planning D: Society and Space*, vol. 1, 1983, pp. 23–57.

36 Anne Gilbert, 'The New Regional Geography in English and French-Speaking Countries', *Progress in Human Geography*, vol. 12, 1988, pp. 208–28; Mary Beth Pudup, 'Arguments within Regional Geography', *Progress in Human Geography*, vol. 12, 1988, pp. 369–90.

37 Doreen Massey and John Allen, *Geography Matters!* (Cambridge: Cambridge University Press, 1984).

38 Doreen Massey, *Spatial Divisions of Labour: Social Structures and the Geography of Production* (New York: Methuen, 1984), pp. 299–300.

39 J. Agnew, *Place and Politics: The Geographical Mediation of State and Society*, p. 42; see also Barney Warf, 'Regional Transformation, Everyday Life, and Pacific Northwest Lumber Production', *Annals of the Assocation of American Geographers*, vol. 78, 1988, pp. 326–46.

40 Jeffrey C. Alexander, *Action and Its Environments: Towards a New Synthesis* (New York: Columbia University Press, 1988), pp. 301–33.

41 Andrew Sayer, 'Explanation in Economic Geography: Abstraction versus Generalization', *Progress in Human Geography*, vol. 6, 1982, pp. 68–88; *Method in Social Science: A Realist Approach* (London: Hutchinson, 1984).

42 Alan Warde, 'Review of *Place, Practice and Structure* by Allan Pred', *Annals of the Association of American Geographers*, vol. 77, 1987, pp. 484–6.
43 Louis O. Mink, 'The Autonomy of Historical Understanding', *History and Theory*, vol. 5, 1966, pp. 24–47, ref. on p. 42.
44 P. Ricoeur, *Time and Narrative*. The models of narrative understanding discussed by Ricoeur and Mink have been offered as models for fields other than history. For example, Donald E. Polkinghorne uses the literature from the philosophy of history and narratology to discuss psychology. He argues that the traditional division in psychology between theory and practice is mediated through narrative knowledge. Donald E. Polkinghorne, *Narrative Knowing and the Human Sciences* (Albany: State University of New York Press, 1988).
45 Robert Scholes and Robert Kellogg, *The Nature of Narrative* (London: Oxford University Press, 1966), p. 4.
46 R. Scholes and R. Kellogg, *The Nature of Narrative*, p. 13. Scholes and Kellogg distinguish between empirical and fictional narratives. They further distinguish empirical narratives as being either 'historical' or 'mimetic'. The latter is seen as tending toward plotlessness. For example, they distinguish the historical tendency of the biography from the mimetic tendency of the autobiography. The modern novel is described as a synthesis of these types.
47 The mimetic function of narrative has been characterized in seemingly contradictory ways. For example, it has been viewed as an essentially conservative force by Roland Barthes. He has argued that mimesis artificially constrains meaning and confirms the already existing order of things. It has also been viewed as a form of invention, a means of creating meaning. For example, Paul Ricoeur has noted these creative aspects of mimesis that are part of the human attempt to know the world through the redescription of experience. The imitation of the relation of humans in the world is fundamental to our sense of geographic description in both everyday and scientific discourse. Both are a form of human practice that seeks to describe the world in familiar and shared symbols. Such descriptions may easily be shaped by ideologies and personal concerns, but as Christopher Prendergast has argued in his discussion of mimesis:

> These [shared and familiar] images and representations may, under the pressure of a certain type of analysis, be shown as embodying a large portion of illusion ('misrecognition'); in the perspectives opened up by the holy trinity, Nietzsche, Freud and Marx, they may be 'fictions' serving particular desires, interests and ideologies. But, in other ways, they are also arguably indispensable to any conceivable social reality or what Wittgenstein calls a 'form of life' . . . ; and perhaps the supreme illusion would be in the assumption that we could live entirely without them, in a euphoric movement of 'unbound' desire and 'infinite' semiosis.

The mimetic activity that is closest to that of the construction of place is similar to the plots that are created in the narration of historical events. Christopher Prendergast, *The Order of Mimesis: Balzac, Stendhal, Nerval, Flaubert* (Cambridge: Cambridge University Press, 1986), p. 7.
48 P. Ricoeur, *Time and Narrative*, p. 41.
49 P. Ricoeur, *Time and Narrative*, p. 41.
50 Veyne states further that:

> Practically, the aggregation of points of view is done in confusion, either by surreptitiously jumping from one point of view to another in the course of the

> account, or by cutting out from the continuum for the sake of a point of view arbitrarily or naively chosen (whether inspired by toponamastics or by administrative geography). In geography and in history, the idea of subjectivity – of liberty and equality of points of view – brings a definitive clarification and tolls the knell of historicism. On the other hand, it does not follow (and Marrou protests against this confusion) that what has happened in time is subjective; just as nothing is more objective than the earth's surface, the object of geography. Geography and history are nominalisms: whence the impossibility of a history à la Toynbee and of a geography à la Ritter, for whom regions or civilizations really exist and are not a question of points of view.

Paul Veyne, *Writing History: Essays on Epistemology* (Middletown, Conn.: Wesleyan University Press, 1984), pp. 296–7, n. 7. I would like to thank Gordon Clark for bringing Veyne's work to my attention.

51 Hayden White, *The Content of the Form: Narrative Discourse and Historical Representation* (Baltimore: Johns Hopkins University Press, 1987), p. 1. See also Fredric Jameson, *The Political Unconscious: Narrative as a Socially Symbolic Act* (Ithaca: Cornell University Press, 1981); and Derek Gregory 'Areal Differentiation and Post-Modern Human Geography', in Derek Gregory and Rex Walford (eds) *New Horizons in Human Geography* (London: Macmillan, 1989), pp. 1–29.

52 Hayden White suggests that:

> Narration is a manner of speaking as universal as language itself, and narrative is a mode of verbal representation so seemingly natural to human consciousness that to suggest that it is a problem might well appear pedantic. But it is precisely because the narrative mode of representation is so natural to human consciousness, so much an aspect of everyday speech and ordinary discourse, that its use in any field of study aspiring to the status of a science must be suspect. For whatever else a science may be, it is also a practice that must be as critical about the way it describes its objects of study as it is about the way it explains their structures and processes. Viewing modern sciences from this perspective, we can trace their development in terms of their progressive demotion of the narrative mode of representation in their descriptions of the phenomena that their specific objects of study comprise . . . . To many of those who would transform historical studies into a science, the continued use by historians of a narrative mode of representation is an index of a failure at once methodological and theoretical.

H. White, *The Content of the Form*, p. 26.

53 See, for example, H. C. Darby, 'The Problem of Historical Description', *Transactions of the Institute of British Geographers*, vol. 30, 1962, pp. 1–13. Derek Gregory suggests that this distinction in Darby's work represents a confusion of the consecutive character of writing with the configurational character of reading. Derek Gregory, personal communication.

54 P. Ricoeur, *Time and Narrative*, p. 142.

55 Some have viewed causality in the historian's descriptions as a narratological element; see H. White, *The Content of the Form*, pp. 142–68; White refers here to Fredric Jameson, *The Political Unconscious*.

18 Doreen Massey

A Global Sense of Place

Reprinted in full from: *Marxism Today,* (June), 24–29 (1991)

This is an era – it is often said – when things are speeding up, and spreading out. Capital is going through a new phase of internationalization, especially in its financial parts. More people travel more frequently and for longer distances. Your clothes have probably been made in a range of countries from Latin America to South-East Asia. Dinner consists of food shipped in from all over the world. And if you have a screen in your office, instead of opening a letter which – care of Her Majesty's Post Office – has taken some days to wend its way across the country, you now get interrupted by e-mail.

This view of the current age is one now frequently found in a wide range of books and journals. Much of what is written about space, place and post-modern times emphasizes a new phase in what Marx once called 'the annihilation of space by time'. The process is argued, or – more usually – asserted, to have gained a new momentum, to have reached a new stage. It is a phenomenon which has been called 'time–space compression'. And the general acceptance that something of the sort is going on is marked by the almost obligatory use in the literature of terms and phrases such as speed-up, global village, overcoming spatial barriers, the disruption of horizons, and so forth.

One of the results of this is an increasing uncertainty about what we mean by 'places' and how we relate to them. How, in the face of all this movement and intermixing, can we retain any sense of a local place and its particularity? An (idealized) notion of an era when places were (supposedly) inhabited by coherent and homogeneous communities is set against the current fragmentation and disruption. The counterposition is anyway dubious, of course; 'place' and 'community' have only rarely been coterminous. But the occasional longing for such coherence is none the less a sign of the geographical fragmentation, the spatial disruption, of our times. And occasionally, too, it has been part of what has given rise to defensive and reactionary responses: certain forms of nationalism, sentimentalized recovering of sanitized 'heritages', and outright antagonism to newcomers and 'outsiders'. One of the effects of such responses is that place itself, the seeking after a sense of place, has come to be seen by some as necessarily reactionary.

But is that necessarily so? Can't we rethink our sense of place? Is it not possible for a sense of place to be progressive; not self-enclosing and defensive, but outward-looking? A sense of place which is adequate to this era of time–space compression? To begin with, there are some questions to be asked about time–space compression itself. Who is it that experiences it, and how? Do we all benefit and suffer from it in the same way?

For instance, to what extent does the currently popular characterization of time–space compression represent very much a Western, colonizer's, view? The sense of dislocation which some feel at the sight of a once well-known local street now lined with a succession of cultural imports – the pizzeria, the kebab house, the branch of the Middle Eastern bank – must have been felt for centuries, though from a very different point of view, by colonized peoples all over the world as they watched the importation, maybe even used the products, of, first, European colonization, maybe British (from new forms of transport to liver salts and custard powder), later US, as they learned to eat wheat instead of rice or corn, to drink Coca-Cola, just as today we try out enchiladas.

Moreover, as well as querying the ethnocentricity of the idea of time–space compression and its current acceleration, we also need to ask about its causes: what is it that determines our degrees of mobility, that influences the sense we have of space and place? Time–space compression refers to movement and communication across space, to the geographical stretching-out of social relations, and to our experience of all this. The usual interpretation is that it results overwhelmingly from the actions of capital, and from its currently increasing internationalization. On this interpretation, then, it is time, space and money which makes the world go round, and us go round (or not) the world. It is capitalism and its developments which are argued to determine our understanding and our experience of space.

But surely this is insufficient. Among the many other things which clearly influence that experience, there are, for instance, race and gender. The degree to which we can move between countries, or walk about the streets at night, or venture out of hotels in foreign cities, is not just influenced by 'capital'. Survey after survey has shown how women's mobility, for instance, is restricted – in a thousand different ways, from physical violence to being ogled at or made to feel quite simply 'out of place' – not by 'capital', but by men. Or, to take a more complicated example, Birkett, reviewing books on women adventurers and travellers in the nineteenth and twentieth centuries, suggests that 'it is far, far more demanding for a woman to wander now than ever before'.[1] The reasons she gives for this argument are a complex mix of colonialism, ex-colonialism, racism, changing gender relations, and relative wealth. A simple resort to explanation in terms of 'money' or 'capital' alone could not begin to get to grips with the issue. The current speed-up may be strongly determined by economic forces, but it is not the economy alone which determines our experience of space and place. In other words, and put simply, there is a lot more determining how we experience space than what 'capital' gets up to.

What is more, of course, that last example indicated that 'time–space compression' has not been happening for everyone in all spheres of activity. Birkett again, this time writing of the Pacific Ocean:

> Jumbos have enabled Korean computer consultants to fly to Silicon Valley as if popping next door, and Singaporean entrepreneurs to reach Seattle in a day. The borders of the world's greatest ocean have been joined as never before. And Boeing has brought these people together. But what about those they fly over, on their islands five miles below? How has the mighty 747 brought them greater communion with those whose shores are washed by the same water? It hasn't, of course. Air travel might enable businessmen to buzz across the ocean, but the concurrent

24

decline in shipping has only increased the isolation of many island communities . . . Pitcairn, like many other Pacific islands, has never felt so far from its neighbours.[2]

In other words, and most broadly, time–space compression needs differentiating socially. This is not just a moral or political point about inequality, although that would be sufficient reason to mention it; it is also a conceptual point.

Imagine for a moment that you are on a satellite, further out and beyond all actual satellites; you can see 'planet Earth' from a distance and, rarely for someone with only peaceful intentions, you are equipped with the kind of technology which allows you to see the colours of people's eyes and the numbers on their numberplates. You can see all the movement and tune in to all the communication that is going on. Furthest out are the satellites, then aeroplanes, the long haul between London and Tokyo and the hop from San Salvador to Guatemala City. Some of this is people moving, some of it is physical trade, some is media broadcasting. There are faxes, e-mail, film distribution networks, financial flows and transactions. Look in closer and there are ships and trains, steam trains slogging laboriously up hills somewhere in Asia. Look in closer still and there are lorries and cars and buses, and on down further, somewhere in sub-Saharan Africa, there's a woman on foot who still spends hours a day collecting water.

Now, I want to make one simple point here, and that is about what one might call the *power geometry* of it all; the power geometry of time–space compression. For different social groups, and different individuals, are placed in very distinct ways in relation to these flows and interconnections. This point concerns not merely the issue of who moves and who doesn't, although that is an important element of it; it is also about power in relation *to* the flows and the movement. Different social groups have distinct relationships to this anyway differentiated mobility: some people are more in charge of it than others; some initiate flows and movement, others don't; some are more on the receiving end of it than others; some are effectively imprisoned by it.

In a sense, at the end of all the spectra are those who are both doing the moving and the communicating, and in some way in a position of control in relation to it: the jet-setters, the ones sending and receiving the faxes and the e-mail, holding the international conference calls, the ones distributing the films, controlling the news, organizing the investments and the international currency transactions. These are the groups who are really in a sense in charge of time–space compression, who can really use it and turn it to advantage, whose power and influence it very definitely increases. On its more prosaic fringes this group probably includes a fair number of Western academics and journalists – those, in other words, who write most about it.

But there are also groups who are also doing a lot of physical moving, but who are not 'in charge' of the process in the same way at all. The refugees from El Salvador or Guatemala and the undocumented migrant workers from Michoacan in Mexico, crowding into Tijuana to make a perhaps fatal dash for it across the border into the USA to grab a chance of a new life. Here the experience of movement, and indeed of a confusing plurality of cultures, is very different. And there are those from India, Pakistan, Bangladesh, the Caribbean, who come half-way round the world only to get held up in an interrogation room at Heathrow.

Or – a different case again – there are those who are simply on the receiving end of time–space compression. The pensioner in a bed-sit in any inner city in the UK, eating British working-class-style fish and chips from a Chinese take-away, watching a US film on a Japanese television; and not daring to go out after dark. And anyway, the public transport's been cut.

Or – one final example to illustrate a different kind of complexity – there are the people who live in the *favelas* of Rio, who know global football like the back of their hand, and have produced some of its players; who have contributed massively to global music, who gave us the samba and produced the lambada that everyone was dancing to last year in the clubs of Paris and London; and who have never, or hardly ever, been to downtown Rio. At one level they have been tremendous contributors to what we call time–space compression; and at another level they are imprisoned in it.

This is, in other words, a highly complex social differentiation. There are differences in the degree of movement and communication, but also in the degree of control and of initiation. The ways in which people are placed within 'time–space compression' are highly complicated and extremely varied.

But this in turn immediately raises questions of politics. If time–space compression can be imagined in that more socially formed, socially evaluative and differentiated way, then there may be here the possibility of developing a politics of mobility and access. For it does seem that mobility and control over mobility both reflects and reinforces power. It is not simply a question of unequal distribution, that some people move more than others, and that some have more control than others. It is that the mobility and control of some groups can actively weaken other people. Differential mobility can weaken the leverage of the already weak. The time–space compression of some groups can undermine the power of others.

This is well established and often noted in the relationship between capital and labour. Capital's ability to roam the world further strengthens it in relation to relatively immobile workers, enables it to play off the plant at Genk against the plant at Dagenham. It also strengthens its hand against struggling local economies the world over as they compete for the favour of some investment. The 747s that fly computer scientists across the Pacific are part of the reason for the greater isolation today of the island of Pitcairn. But also, every time someone uses a car, and thereby increases their personal mobility, they reduce both the social rationale and the financial viability of the public transport system – and thereby also potentially reduce the mobility of those who rely on that system. Every time you drive to that out-of-town shopping centre you contribute to the rising prices, even hasten the demise, of the corner shop. And the 'time–space compression' which is involved in producing and reproducing the daily lives of the comfortably off in First World societies – not just their own travel but the resources they draw on, from all over the world, to feed their lives – may entail environmental consequences, or hit constraints, which will limit the lives of others before their own. We need to ask, in other words, whether our relative mobility and power over mobility and communication entrenches the spatial imprisonment of other groups.

But this way of thinking about time–space compression also returns us to the question of time and a sense of place. How, in the context of all these socially varied time–space changes do we think about 'places'? In an era when, it is argued, 'local communities' seem to be increasingly broken up, when you can go abroad and find the same shops, the same music as at home, or eat your favourite foreign-holiday food at a restaurant down the road – and when everyone has a different experience of all this – how then do we think about 'locality'?

Many of those who write about time–space compression emphasize the insecurity and unsettling impact of its effects, the feeling of vulnerability which it can produce. Some therefore go on from this to argue that, in the middle of all this flux, people desperately need a bit of peace and quiet – and that a strong sense of place, of locality, can form one kind of refuge from the hubbub. So the search after the 'real' meanings of places, the unearthing of heritages and so forth, is interpreted as being, in part, a response to desire for fixity and for security of identity in the middle of all the movement and change. A 'sense of place', of rootedness, can provide – in this form and on this interpretation – stability and a source of unproblematical identity. In that guise, however, place and the spatially local are then rejected by many progressive people as almost necessarily reactionary. They are interpreted as an evasion; as a retreat from the (actually unavoidable) dynamic and change of 'real life', which is what we must seize if we are to change things for the better. On this reading, place and locality are foci for a form of romanticized escapism from the real business of the world. While 'time' is equated with movement and progress, 'space'/'place' is equated with stasis and reaction.

There are some serious inadequacies in this argument. There is the question of why it is assumed that time–space compression will produce insecurity. There is the need to face up to – rather than simply deny – people's need for attachment of some sort, whether through place or anything else. None the less, it is certainly the case that there is indeed at the moment a recrudescence of some very problematical senses of place, from reactionary nationalism, to competitive localisms, to introverted obsessions with 'heritage'. We need, therefore, to think through what might be an adequately progressive sense of place, one which would fit in with the current global–local times and the feelings and relations they give rise to, *and* which would be useful in what are, after all, political struggles often inevitably based on place. The question is how to hold on to that notion of geographical difference, of uniqueness, even of rootedness if people want that, without it being reactionary.

There are a number of distinct ways in which the 'reactionary' notion of place described above is problematical. One is the idea that places have single, essential, identities. Another is the idea that identity of place – the sense of place – is constructed out of an introverted, inward-looking history based on delving into the past for internalized origins, translating the name from the Domesday Book. Thus Wright recounts the construction and appropriation of Stoke Newington and its past by the arriving middle class (the Domesday Book registers the place as 'Newtowne': 'There is land for two ploughs and a half . . . There are four villanes and thirty-seven cottagers with ten acres', pages 227 and 231), and contrasts this version with that of other groups: the white working class and the large number of important minority communities.[3] A particular problem with this conception of place is that it seems to require the drawing of boundaries. Geographers have long been exercised by the problem of defining regions, and this question of 'definition' has almost always been reduced to the issue of drawing lines around a place. I remember some of my most painful times as a geographer have been spent unwillingly struggling to think how one could draw a boundary around somewhere like the 'East Midlands'. But that kind of boundary around an area precisely distinguishes between an inside and an outside. It can so easily be yet another way of constructing a counterposition between 'us' and 'them'.

And yet, if one considers almost any real place, and certainly one not defined primarily by administrative or political boundaries, these supposed characteristics have little real purchase.

Take, for instance, a walk down Kilburn High Road, my local shopping centre. It is a pretty ordinary place, north-west of the centre of London. Under the railway bridge the newspaper stand sells papers from every county of what my neighbours, many of whom come from there, still often call the Irish Free State. The postboxes down the High Road, and many an empty space on a wall, are adorned with the letters IRA. Other available spaces are plastered this week with posters for a special meeting in remembrance: 'Ten Years after the Hunger Strike'. At the local theatre Eamon Morrissey has a one-man show; the National Club has the Wolfe Tones on, and at the Black Lion there's *Finnegans Wake*. In two shops I noted this week's lottery ticket winners: in one the name is Teresa Gleeson, in the other, Chouman Hassan.

Thread your way through the often almost stationary traffic diagonally across the road from the newsstand and there's a shop which as long as I can remember has displayed saris in the window. Four life-sized models of Indian women, and reams of cloth. On the door a notice announces a forthcoming concert at Wembley Arena: Anand Miland presents Rekha, live, with Aamir Khan, Salman Khan, Jahi Chawla and Raveena Tandon. On another ad, for the end of the month, is written 'All Hindus are cordially invited'. In another newsagent's I chat with the man who keeps it, a Muslim unutterably depressed by events in the Gulf, silently chafing at having to sell the *Sun*. Overhead there is always at least one aeroplane; we seem to be on a flight-path to Heathrow and by the time they're over Kilburn you can see them clearly enough to tell the airline and wonder as you struggle with your shopping where they're coming from. Below, the reason the traffic is snarled up (another odd effect of time–space compression!) is in part because this is one of the main entrances to and escape routes from London, the road to Staples Corner and the beginning of the M1 to the North.

This is just the beginnings of a sketch from immediate impressions – but a proper analysis could be done – of the links between Kilburn and the world. And so it could for almost any place.

Kilburn is a place for which I have great affection; I have lived there many years. It certainly has 'a character of its own'. But it is possible to feel all this without subscribing to any of the static and defensive – and in that sense reactionary – notions of 'place' which were referred to above. First, while Kilburn may have a character of its own, it is absolutely not a seamless, coherent identity, a single sense of place which everyone shares. It could hardly be less so. People's routes through the place, their favourite haunts within it, the connections they make (physically, or by phone or post, or in memory and imagination) between here and the rest of the world vary enormously. If it is now recognized that people have multiple identities then the same point can be made in relation to places. Moreover, such multiple identities can either be a source of richness or a source of conflict, or both.

One of the problems here has been a persistent identification of place with 'community'. Yet this is a misidentification. On the one hand communities can exist without being in the same place – from networks of friends with like interests, to major religious, ethnic or political communities. On the other hand, the instances of places housing single 'communities' in the sense of coherent social groups are probably – and, I would argue, have for long been – quite rare. Moreover, even where they do exist this in no way implies a single sense of place. For people occupy different positions within any community. We would counterpose to the chaotic mix of Kilburn the relatively stable and homogeneous community (at least in popular imagery) of a small mining village. Homogeneous? 'Communities' too have internal structures. To take the most obvious example, I'm sure a woman's sense of place in a mining village – the spaces through which she normally moves, the meeting places, the connections outside – are different from a man's. Their 'senses of the place' will be different.

Moreover, not only does 'Kilburn', then, have many identities (or its full identity is a complex mix of all these) it is also, looked at in this way, absolutely *not* introverted. It is (or ought to be) impossible even to begin thinking about Kilburn High Road without bringing into play half the world and a considerable amount of British imperialist history (and this certainly goes for mining villages too). Imagining it this way provokes in you (or at least in me) a really global sense of place.

And finally, in contrasting this way of looking at places with the defensive reactionary view, I certainly could not begin to, nor would I want to, define 'Kilburn' by drawing its enclosing boundaries.

So, at this point in the argument, get back in your mind's eye on a satellite; go right out again and look back at the globe. This time, however, imagine not just all the physical movement, nor even all the often invisible communications, but also and especially all the social relations, all the links between people. Fill it in with all those different experiences of time–space compression. For what is happening is that the geography of social relations is changing. In many cases such relations are increasingly stretched out over space: economic, political and cultural social relations, each full of power and with internal structures of domination and subordination, stretched out over the planet at every different level, from the household to the local area to the international.

It is from this perspective that it is possible to envisage an alternative interpretation of place. In this interpretation, what gives a place its specificity is not some long internalized history but the fact that it is constructed out of a particular constellation of social relations, meeting and weaving together at a particular locus. If one moves in from the satellite towards the globe, holding all those networks of social relations and movements and communications in one's head, then each 'place' can be seen as a particular, unique, point of their intersection. It is, indeed, a *meeting* place. Instead, then, of thinking of places as areas with boundaries around, they can be imagined as articulated movements in networks of social relations and understandings, but where a large proportion of those relations, experiences and understandings are constructed on a far larger scale than what we happen to define for that moment as the place itself, whether that be a street, or a region or even a continent. And this in turn allows a sense of place which is extroverted, which includes a consciousness of its links with the wider world, which integrates in a positive way the global and the local.

This is not a question of making the ritualistic connections to 'the wider system' – the people in the local meeting who bring up international capitalism every time you try to have a discussion about rubbish collection – the point is that there are real relations with real content – economic, political, cultural – between any local place and the wider world in which it is set. In economic geography the argument has long been accepted that it is not possible to understand the 'inner city', for instance its loss of jobs, the decline of manufacturing employment there, by looking only at the inner city. Any adequate explanation has to set the inner city in its wider geographical context. Perhaps it is appropriate to think how that kind of understanding could be extended to the notion of a sense of place.

These arguments, then, highlight a number of ways in which a progressive concept of place might be developed. First of all, it is absolutely not static. If places can be conceptualized in terms of the social interactions which they tie together, then it is also the case that these interactions themselves are not motionless things, frozen in time. They are processes. One of the great one-liners in Marxist exchanges has for long been 'ah, but capital is not a thing, it's a process'. Perhaps this should be said also about places; that places are processes, too.

Second, places do not have to have boundaries in the sense of divisions which frame simple enclosures. 'Boundaries' may of course be necessary, for the purposes of certain types of studies for instance, but they are not necessary for the conceptualization of a place itself. Definition in this sense does not have to be through simple counterposition to the outside; it can come, in part, precisely through the particularity of linkage to that 'outside' which is therefore itself part of what constitutes the place. This helps get away from the common association between penetrability and vulnerability. For it is this kind of association which makes invasion by newcomers so threatening.

Third, clearly places do not have single, unique 'identities'; they are full of internal conflicts. Just think, for instance, about London's Docklands, a place which is at the moment quite clearly *defined* by conflict: a conflict over what its past has been (the nature of its 'heritage'), conflict over what should be its present development, conflict over what could be its future.

Fourth, and finally, none of this denies place nor the importance of the uniqueness of place. The specificity of place is continually reproduced, but it is not a specificity which results from some long, internalized history. There are a number of sources of this specificity – the uniqueness of place.[4] There is the fact that the wider social relations in which places are set are themselves geographically differentiated. Globalization (in the economy, or in culture, or in anything else) does not entail simply homogenization. On the contrary, the globalization of social relations is yet another source of (the reproduction of) geographical uneven development, and thus of the uniqueness of place. There is the specificity of place which derives from the fact that each place is the focus of a distinct *mixture* of wider and more local social relations. There is the fact that this very mixture together in one place may produce effects which would not have happened otherwise. And finally, all these relations interact with and take a further element of specificity from the accumulated history of a place, with that history itself imagined as the product of layer upon layer of different sets of linkages, both local and to the wider world.

In her portrait of Corsica, *Granite island*, Dorothy Carrington travels the island seeking out the roots of its character.[5] All the different layers of peoples and cultures are explored; the long and tumultuous relationship with France, with Genoa and Aragon in the thirteenth, fourteenth and fifteenth centuries, back through the much earlier incorporation into the Byzantine Empire, and before that domination by the Vandals, before that being part of the Roman Empire, before that the colonization and settlements of the Carthaginians and the Greeks . . . until we find that even the megalith builders had come to Corsica from somewhere else.

It is a sense of place, an understanding of 'its character', which can only be constructed by linking that place to places beyond. A progressive sense of place would recognize that, without being threatened by it. What we need, it seems to me, is a global sense of the local, a global sense of place.

Notes

1 D. Birkett, *New Statesman and Society*, 13 June 1990, pp. 41–2.
2 D. Birkett, *New Statesman and Society*, 15 March 1991, p. 38.
3 D. Wright, *On living in an old country*. London: Verso, 1985.
4 D. Massey, *Spatial divisions of labour: Social structures and the geography of production*. Basingstoke: Macmillan, 1984.
5 D. Carrington, *Granite island: A portrait of Corsica*. Harmondsworth: Penguin Books, 1971.

19 Denis Cosgrove

Prospect, Perspective and the Evolution of the Landscape Idea

Excerpts from: *Transactions of the Institute of British Geographers*, 10, 45–62 (1985)

Geographical interest in the landscape concept has seen a revival in recent years. In large measure this is a consequence of the humanist renaissance in geography. Having enjoyed a degree of prominence in the interwar years, landscape fell from favour in the 1950s and 1960s. Its reference to the visible forms of a delimited area to be subjected to morphological study (a usage still current in the German 'landscape indicators' school)[1] appeared subjective and too imprecise for Anglo-Saxon geographers developing a spatial science. The static, descriptive morphology of landscape ill-suited their call for dynamic functional regions to be defined and investigated by geographers contributing to economic and social planning.[2]

Recently, and primarily in North America, geographers have sought to reformulate landscape as a concept whose subjective and artistic resonances are to be actively embraced. They allow for the incorporation of individual, imaginative and creative human experience into studies of the geographical environment, aspects which geographical science is claimed to have devalued at best and at worst, ignored. Marwyn Samuels, for example,[3] refers to landscapes as 'authored', Courtice Rose thinking along similar lines would analyse landscapes as texts,[4] and Edward Relph regards landscape as 'anything I see and sense when I am out of doors – landscape is the necessary context and background both of my daily affairs and of the more exotic circumstances of my life'.[5] American humanist geographers have adopted landscape for the very reasons that their predecessors rejected it. It appears to point towards the experiential, creative and human aspects of our environmental relations, rather than to the objectified, manipulated and mechanical aspects of those relations. It is the latter against which humanism is a protest, which Relph traces to the seventeenth century scientific revolution and its Cartesian division of subject and object. Landscape seems to embody the holism which modern humanists proclaim.

In Britain a revival of landscape is also apparent. Here the humanist critique in geography has been less vocal. Recent landscape study has remained closer to popular usage of the word as an artistic or literary response to the visible scene.[6] Among British geographers interest in landscape was stimulated partly by perception studies, particularly the short-lived excitement over landscape evaluation for planning purposes which surrounded the 1973 reform of local government.[7] This led to various mechanistic theories of landscape aesthetics which, like Jay Appleton's ethologically-founded and influential 'habitat theory' of landscape,[8] had little in common with the humanism proclaimed in North American studies.

Epistemological divergence notwithstanding, landscape is again a focus of geographical interest. With that interest has come a refreshing willingness by geographers to employ landscape representations – in painting, imaginative literature and garden design – as sources for answering geographical questions.[9] The purpose of this paper is to support and promote that initiative while simultaneously entering certain caveats about adopting the landscape idea without subjecting it to critical historical examination as a term which embodies certain assumptions about relations between humans and their environment, or more specifically, society and space. These caveats go beyond landscape as such and touch upon aspects of the whole humanist endeavour within geography.

Landscape first emerged as a term, an idea, or better still, a *way of seeing*[10] the external world, in the fifteenth and early sixteenth centuries. It was, and it remains, a visual term, one that arose initially out of renaissance humanism and its particular concepts and constructs of space. Equally, landscape was, over much of its history, closely bound up with the practical appropriation of space. As we shall see, its connections were with the survey and mapping of newly-acquired, consolidated and 'improved' commercial estates in the hands of an urban bourgeoisie; with the calculation of distance and trajectory for cannon fire and of defensive fortifications against the new weaponry; and with the projection of the globe and its regions onto map graticules by cosmographers and chorographers, those essential set designers for Europe's entry centre-stage of the world's theatre. In painting and garden design landscape achieved visually and ideologically what survey, map making and ordnance charting achieved practically: the control and domination over space as an absolute, objective entity, its transformation into the property of individual or state. And landscape achieved these ends by use of the same techniques as the practical sciences, principally by applying Euclidian geometry as the guarantor of certainty in spatial conception, organization and representation. In the case of landscape the technique was optical, *linear perspective*, but the principles to be learned were identical to those of architecture, survey, map-making and artillery science. The same handbooks taught the practitioners all of these arts.[11]

Landscape, like the practical sciences of the Italian Renaissance, was founded upon scientific theory and knowledge. Its subsequent history can best be understood in conjunction with the history of science. Yet in its contemporary humanist guise within geography, landscape is deployed within a radically anti-scientific programme. Significantly that programme is equally non-visual. Recent programmatic statements of geographical humanism (and critiques of it) in the pages of these *Transactions* are notable for their concentration on verbal, literary and linguistic modes of communication and for their almost complete neglect of the visual and its place in geography.[12] The attack on science is characteristic of much contemporary humanist writing. But the apparent lack of interest in the graphic image is more surprising. Consider the traditions of our discipline, its alignment with cartography and the long-held belief that the results of geographical scholarship are best embodied in the map. Consider too the humanists' proclaimed interest in *images* of place and landscape, and yet their remarkable neglect of the visual.[13] Indeed the clearest statement of the centrality of sight in geography that I know is found in William Bunge's *Theoretical Geography*, a manifesto for spatial science: 'geography is the one predictive science whose inner logic is literally visible':[14] Bunge's book may be closer in spirit to the original humanist authors of the landscape idea than his contemporary humanist critics. The book after all is a celebration of the certainty of geometry as the constructional principle of space.

In fact, the humanist attack on science and its neglect of the visual image in geography are not unconnected. They both result in some measure from the lack of critical reflection on the European humanist tradition, from the conflation of the spatial theme in geography with a positivist epistemology, and from a mystification of art and literature. All three of these aspects will be illustrated in a brief exploration of the landscape idea as a way of seeing in the European visual tradition, emphasizing that tradition's most enduring convention of space representation, linear perspective. In this exploration I shall justify and elaborate the claim that the landscape idea is a visual ideology; an ideology all too easily adopted unknowingly into geography when the landscape idea is transferred as an unexamined concept into our discipline.

Geometry, perspective and renaissance humanism

Traditionally the seven liberal arts of medieval scholarship were grouped into two sets. The trivium was composed of grammar, rhetoric and logic; the quadrivium of arithmetic, geometry, astronomy and music. While in its narrowest definition humanism referred to studies in the trivium (the recovery, secure dating and translation of texts), many early renaissance humanists were equally fascinated by the material of the quadrivium, seeking a unity of knowledge across all the arts.[15] The fifteenth century saw revolutionary advances in both sets of studies, advances which altered their organization, social significance and role in the production and communication of human knowledge of the world and our place within it. In the arena of words, language and written expression the most striking advance was the Gutenberg invention of movable type in the 1440s.[16] In the quadrivium, always more theoretical, the critical advance came from the re-evaluation of Euclid and the elevation of geometry to the keystone of human knowledge, specifically its application to three-dimensional space representation through single-point perspective theory and technique. Perspective, the medieval study of optics, was one of the mathematical arts, studied since the twelfth-century revival of learning, as evidenced for example in Roger Bacon's work. Painters like Cimabue and Giotto had constructed their pictures in new ways to achieve a greater realism (*il vero*) than their predecessors.[17] But the theoretical and practical development of a coherent linear perspective awaited the fifteenth-century Tuscan Renaissance. That movement, despite its emphasis on classical texts, grammar and rhetoric, revolutionized spatial apprehensions in the west. For the plastic and visual arts: painting, sculpture and architecture, and for geography and cosmology, all concerned with space and spatial relations,

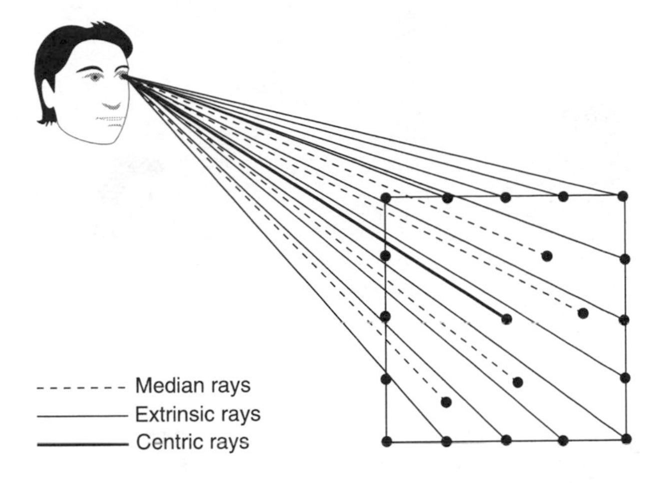

Fig. 19.1 The visual triangle as described by Alberti (from Samuel Y. Edgerton Jr, *The Renaissance rediscovery of linear perspective*, Harper and Row, London, 1975, reproduced with permission)

it was from the quadrivium, from geometry and number theory, that form and structure were determined – even if their content was provided by the trivium.

In 1435 the Florentine humanist and architect Leon Battista Alberti published his *Della Pittura (On painting)*,[18] a work whose authority in artistic theory endured beyond the eighteenth century when Sir Joshua Reynolds, first president of the Royal Academy, used it as the foundation for his lectures on pictorial composition, beauty and the hierarchy of genres. In *Della Pittura* Alberti demonstrates a technique which he had worked out experimentally for constructing a visual triangle which allowed the painter to determine the shape and measurement of a gridded square placed on the ground when viewed along the horizontal axis, and to reproduce in pictorial form its appearance to the eye. The *construzione leggitima* gave the realist illusion of three-dimensional space on a two-dimensional surface. This construction, the foundation of linear perspective, depended upon concepts of the vanishing point, distance point and intersecting plane. Alberti describes it as a triangle of rays extending outwards from the eye and striking the object of vision. There are three kinds of ray (Fig. 19.1).

> The extrinsic rays, thus circling the plane – one touching the other, enclose all the plane like the willow wands of a basket cage, and make . . . the visual pyramid. It is time for me to describe what the pyramid is and how it is constructed by these rays . . . The pyramid is a figure of a body from whose base lines are drawn upward, terminating at a single point. The base of the pyramid is the plane which is seen. The sides of the pyramid are the rays which I have called extrinsic. The cuspid, that is the point of the pyramid, is located within the eye where the angle of the quantity is.[19]

The visual pyramid here described is familiar to every geographer who reads *Area*, although its geographical significance may not always be fully

Fig. 19.2 A seventeenth century 'way of seeing' (familiar to readers of *Area*)

appreciated (Fig. 19.2). We need not concern ourselves here with the details and accuracy of Alberti's construction (except perhaps to note the definition of pyramid, lifted directly from Euclid). But we should observe certain consequences that flow from it. First, form and position in space are shown to be relative rather than absolute. The forms of what we see, of objects in space and of geometrical figures themselves, vary with the angle and distance of vision. They are produced by the sovereign eye, a single eye, for this is not a theory of binocular vision. Secondly, Alberti regards the rays of vision as having origin *in the eye itself*, thus confirming its sovereignty at the centre of the visual world. Thirdly, he creates a technique which became fundamental to the realist representation of space and the external world. The artist, through perspective, establishes the arrangement or composition, and thus the specific time of the events described, determines – in both senses – the 'point of view' to be taken by the observer, and controls through framing the scope of reality revealed. Perspective technique was so effective that the realist conventions which it underlay were not fundamentally challenged until the nineteenth century.[20]

Realist representation of three-dimensional space on a two-dimensional surface through linear perspective directs the external world towards the individual located outside that space. It gives the eye absolute mastery over space. The centric ray moves in a direct line from the eye to the vanishing point, to the depth of the recessional plane. Space is measured and calculated from this line and the rest of what is seen constructed around the vanishing point and within the frame fixed by external rays.

Visually space is rendered the property of the individual detached observer,

Fig. 19.3 Ambrogio Lorenzetti: 'Good Government in the City' detail from Palazzo Pubblico, Siena (ditta O. Böhm)

from whose divine location it is a dependent, appropriated object. A simple movement of the head, closing the eyes or turning away and the composition and spatial form of objects are altered or even negated. Developments from the fifteenth century may have altered the assumed position of the observer, or used perspective analytically rather than synthetically as Alberti and his contemporaries intended,[21] but this visual appropriation of space endured unaltered. Significantly, the adoption of linear perspective as the guarantor of pictorial realism was contemporary with those other realist techniques of painting: oils, framing and production for a market of mobile, small canvases. In this respect perspective may be regarded as one of a number of techniques which allowed for the visual representation of a bourgeois, rationalist conception of the world.

The term bourgeois is appropriate, for linear perspective was an urban invention, employed initially to represent the spaces of the city. It was first demonstrated practically by Alberti's close associate, Filippo Brunelleschi, in a famous experiment of 1425 when he succeeded in throwing an image of the Baptistery at Florence onto a canvas set up in the great portal of the cathedral.[22] If we compare Ambrogio Lorenzetti's well-known frescoes in the Palazzo Pubblico at Siena (Fig. 19.3) which represent good government in the city, painted in the 1340s, with Peitro Perugino's representation of *Christ giving to St Peter the Keys to the Kingdom of Heaven* (Fig. 19.4) painted on the wall of the Sistine Chapel in 1481, the significance of perspective is clear. Lorenzetti shows us the city as an active bustling world of human life wherein people and their environment interact across a space where unity derives from the action on its surface.

> These pre-perspective urban landscapes show not so much what the towns looked like as what it felt like to be in them. We get an impression of the towns not as they might have looked to a detached observer from a fixed vantage point but as they

Fig. 19.4 Pietro Perugino: 'Christ giving to St Peter the Keys to the Kingdom of Heaven' Vatican City, Sistine Chapel (ditta O. Böhm)

might have impressed a pedestrian walking up the streets and seeing the buildings from many different sides.[23]

By contrast, in Perugino's ideal city a formal, monumental order is organized through precise geometry, constructed by the eye around the axis which leads across the chequerboard piazza to the circular temple at its centre. The piazza, geometrical centre of this city, becomes in this genre symbolic of the whole city.[24] The hills and trees beyond reflect the same regimented order as the urban architecture. The people of the city, or rather within it, for they reveal no particular attachment to it, group themselves in dignified and theatrical poses. In the 'ideal townscapes' of the late fifteenth-century Umbrian school of Piero della Francesca humans scarcely appear. They have no need to for the 'measure of man', so neatly captured in Leonardo da Vinci's *Man in a Circle and a Square*, is written into the measured architectural facades and proportioned spaces of the city, an intellectual measure rather than sensuous human life.[25] This alerts us to the fact that perspective and its geometry had a greater significance than merely its employment as a painting technique.

The mathematics and geometry associated with perspective were directly relevant to the economic life of the Italian merchant cities of the Renaissance, to trading and capitalist finance, to agriculture and the land market, to navigation and warfare. Michael Baxandall[26] has shown that merchants attending the *abbaco* or commercial school in their youth undertook a curriculum which provided the key skills of mathematics for application in commerce: accounting, book-keeping, calculation of interest and rates of return, determining proportions in joint risk ventures. One of the most commonly used tests summarizing the various merchant skills was Fra Luca Pacioli's *Summa di Arithmetica, Geometria, Proportione et Proportionalità* (1494).[27] Its author, a close friend of Leonardo, acknowledges Alberti as well as Ptolemy and Vitruvius, and of course Euclid among his sources. While Piero della Francesca had himself written an earlier text, *De Abbaco*, Pacioli's was the first complete manual of practical mathematics to appear in printed book form, following only two years after the first printed geometry and setting the model for a collection of later texts. Pacioli devotes the second book of the volume to geometry and the measurement of distance, surface and volume. He points out the value of such skills for land survey and map making, for warfare and navigation. From a text like this Italian merchants learned to calculate visually or 'gauge' by eye and using π the volume of a barrel, a churn, a haystack or other regular shape, a valuable skill in an age before standard sizes and volumes became the norm. This visual gauging was regarded as a wonderful skill. In the words of Silvio Belli writing of visual survey in 1573: 'certainly it is a wondrous thing to measure with the eye, because to everyone who does not know its rationale it appears completely impossible'.[28] It has been argued that the search for accurate visual techniques of land survey held back Italian innovations in instrumentation for many decades,[29] but the significance accorded to it indicates the importance attached to the power of vision linked to intellect through geometry, and how the principles which underlay perspective theory were the everyday skills of the urban merchant.

Not all land survey was by eye. The astrolabe, quadrant and plane table were in use and discussed in the texts cited. For map makers and navigators these were crucial instruments. But they required geometrical calculation to make their results meaningful. The Italian renaissance was a cartographic as much as an artistic event. Ptolemy whose *Almagest* had always ranked as a key geometrical source became known too for his *Cosmografia*, brought as a Greek text to Florence at the beginning of the fifteenth century. Alberti produced an accurately surveyed map of Rome, Leonardo one of Pavia. These were regarded as revelations of the rational order of created space produced by the application of geometry. Perhaps more closely related to landscape painting was the *pianta prospettiva*, the bird's eye view of cities which became so popular at the turn of the sixteenth century. Among the best known of these is Jacopo de' Barbari's 1500 map of Venice, like so many of its type as much an ideological expression of urban dominion as an accurate rendering of the urban scene.[30] The viewpoint for these maps is, significantly, high above the city, distant, commanding, uninvolved. It is the same perspective that we find in Bruegel's or Titian's landscapes, panoramas over great sweeps of earth space, seas, mountains and promontories.

Linear perspective organizes and controls spatial coordinates, and because it was founded in geometry it was regarded as the discovery of inherent properties of space itself.[31]

Landscape, perspective and realist space

It is known that the first artist references to specific paintings as 'landscape' (*paesaggio*) come from early sixteenth-century Italy. One of the most often quoted is that from 1521 referring to Giorgione's *Tempesta*.[32] Both Kenneth Clark and J. B. Jackson, in discussions of landscape in this period, sense a relationship between the new genre and notions of authority and control. Noting the appearance of 'realist' landscape in upper Italy and Flanders, the second mercantile core of early modern Europe, Clark claims that it reflected 'some change in the action of the human mind which demanded a new nexus of unity, enclosed space', and suggests that this was conditioned by a new, scientific way of thinking about the world and an 'increased control of nature by man'.[33] Jackson refers to a widespread belief that the relationship between a social group and its landscape could be so expertly controlled as to make appropriate a comparison between environmental bonds and family bonds,[34] thereby allowing landscape to become a means of moral commentary. Perspective was the central technique which allowed this control to be achieved in the new paintings of landscape. In Leonardo's writings the importance of perspective is in no doubt: 'for Leonardo, as for Alberti, painting is a science because of its foundation on mathematical perspective *and on the study of nature*'.[35] Leonardo himself wrote that

> Among all the studies of natural causes and reasons light chiefly delights the beholder – and among the great features of mathematics the certainty of its demonstrations is what pre-eminently tends to elevate the mind of the investigator. Perspective must therefore be preferred to all the discourses and systems of human learning.[36]

Geometry is the source of the painter's creative power, perspective its technical expression. For Leonardo, perspective 'transforms the mind of the painter into the likeness of the divine mind, for with a free hand he can produce different beings, animals, plants, fruits, landscapes, open fields, abysses and fearful places'.[37] Linear perspective provides the certainty of our reproductions of nature in art and underlies the power and authority, the divine creativity of the artist.

Landscape, prospect and visual ideology

While it is not suggested that perspective stands alone as the basis for realism and landscape painting – the demand for *il vero* in Renaissance art was a complex social and cultural product[38] – it is argued that the realist illusion of space which was revolutionized more by perspective than any other technique was, through perspective, aligned to the physical appropriation of space as property, or territory. Surveyors' charts which located and measured individual estates, for example in England after the dissolution of monasteries; cartographers' maps which used the graticule to apportion global space, for example the line defined by Pope Alexander VI dividing the new world between Portugal and Spain; engineers' plans for fortresses and cannon trajectories to conquer or defend national territory, as for example Vauban's French work or Sorte's for the Venetian defences against Austria; all of these are examples of the application of geometry to the production of real property.[39] They presuppose a different concept of space ownership than the contingent concept of a feudal society where land is locked into a web of interdependent lordships based on fief and fealty. The new chorographies which decorated the walls of sixteenth-century council halls and signorial places,[40] and the new taste for accurate renderings of the external world which gradually moved from background to main subject matter, were both organized by perspective geometry and achieve aesthetically what maps, surveys and ordnance charts achieve practically. Landscape is thus a way of seeing, a composition and structuring of the world so that it may be appropriated by a detached, individual spectator to whom an illusion of order and control is offered through the composition of space according to the certainties of geometry. That illusion very frequently complemented a very real power and control over fields and farms on the part of patrons and owners of landscape paintings.[41] Landscape distances us from the world in critical ways, defining a particular relationship with nature and those who appear in nature, and offers us the illusion of a world in which we may participate subjectively by entering the picture frame along the perspectival axis. But this is an aesthetic entrance not an active engagement with a nature or space that has its own life. Implicit in the landscape idea is a visual ideology which was extended from painting to our relationship with the real world whose 'frame and compass' Elizabethans so admired and which Georgian English gentlemen would only approach through the language of landscape painting or the optical distortion of their Claude Glass.

The Italian word for perspective is *prospettiva*. It combines senses which in modern English are distinct: 'perspective' and 'prospect'. Perspective itself has a number of meanings in English, but as the projection of a spatial image onto a plane it first appears in the later decades of the sixteenth century. This usage is found for example in John Dee's Preface to the first English translation of *Euclid* (1570). Dee, the Elizabethan mathematician, navigational instrument maker and magician, links this use of perspective to painting in a classically renaissance way:

> great skill of Geometric, Arithmetik, Perspective and Anthropographie with many other particular arts hath the Zographer need of for his perfection . . . This mechanical Zographer (commonly called the Painter) is marvelous in his skil, and seemeth to have a divine power.[42]

Dee is writing at the opening of a decade which will see Saxton's county maps published and when a new 'image of the country' was being produced as an aspect of Elizabethan patriotism, using maps and landscape representations as instruments of Tudor power and nationalist ideology.[43]

By 1605 we can find reference to perspective as a form of insight, a point of view, as in the phrase 'getting something into perspective', or seeing it in its *true* light, its correct relationship with other things. Many of the early references quoted in the Oxford English Dictionary to support the definition of perspective as a drawing contrived to represent true space and distance relations refer to landscape and garden layout.[44] The visual ideology of perspective and of landscape as ways of seeing nature, indeed a true way of seeing, is certainly current in the English Renaissance. When we turn to the word *prospect* we find it used to denote a view outward, a looking forward in time as well as space. By the end of the sixteenth century prospect carried the sense of 'an extensive or *commanding* sight or view, a view of the landscape as affected by one's position'.[45] This neatly reflects a period when command over land was being established on new commercially-run estates by Tudor enclosers and the new landowners of measured monastic properties. That command was established with the help of the surveyors' 'malicious craft', the geometry which wrote new perspectives across real landscapes.[46]

By the mid-seventeenth century 'prospect' had become a substitute for landscape. The command that it implied was as much social and political as spatial. Commanding views are the theme of country house painting, poetry and landscaping throughout the seventeenth and eighteenth centuries . . . , and a number of recent studies have revealed the degree to which landscape was a vehicle for social and moral debate during this period.[47] The prospects designed for men like the Duke of Marlborough at Blenheim who had made their fortunes from war had an appropriately military character in their blocks of woodland set against shaven lawns. This no doubt reinforced the image of power and authority, at least for those who wielded it. The survey skills which calculated and laid out these landscapes produced fortification plans, ordnance charts and campaign maps as well as serving the requirements of the parliamentary enclosers. It is not surprising that in his critique of emparkment and landscaping Oliver Goldsmith in *The Deserted Village* should describe the park that has replaced Sweet Auburn in military metaphors: 'its vistas strike, its palaces surprise'. In those great English landscape parks prospect also signified the future. Control was as much temporal as spatial. Their clumps of oak and beech would not be seen in full maturity by those who had them planted, but security of property ensured for later scions of the family tree the prospect on inheritance of commanding a fine view. The prospect of the eye was equally commercial, such woodland in the landscape was an economic investment. It represented prospecting in wood, as those who scoured the landscape in the following century seeking gold would be described.[48]

Landscape and the humanist tradition in geography

Landscape comes into English language geography primarily from the German *landschaft*. Much has been written about the fact that the German word means area, without any particularly aesthetic or artistic, or even visual connotations.[49] My own knowledge of German usage is too meagre to contest this claim, but some comment is warranted. In Humboldt's *Kosmos*, regarded by many as one of the two pillars upon which German geography was erected, a whole section is devoted to the history of the love of landscape and nature up to the time of Goethe whom Humboldt greatly revered and who was a major visual theorist.[50] English geographers could have taken their landscape concept from John Ruskin and discovered a usage not very different from Humboldt's.[51] More directly, *Landschaft* in the work of Hettner and Passarge, the main sources for English language geographers like Carl Sauer and R. E.

Dickinson of the landscape concept, was confined to the study of *visible* forms, it was the eye which determined their selection and inclusion. Moreover, *Landschaft*, as Sauer's classic paper – 'Morphology of Landscape' – makes clear,[52] was to be studied by the chorological method and its results transmitted descriptively in prose and above all by the map. Given what we know of the traditional links between cartography, chorography and landscape painting it is difficult to accept the argument that *Landschaft* sustained in German geography the entirely neutral sense of *area* or *region* as its English and American devotees of the inter-war period claimed. Certainly there is a thread of interest in German geography for *Gestaltende Geografie*, study of aesthetic holism in landscape, that runs from Humboldt through Ewald Banse to Gerhart Hard.[53]

Anglo-Saxon geographers introducing landscape as an areal concept were not unaware of the problems caused by its common usage as a painters' term. But in the interests of a scientific geography they were keen to distance their concept of landscape from that of painters or literary writers; poets and novelists. Thus the links between landscape, perspective and the control of space as property – the visual ideology common to landscape painting and cartography – have gone unrecorded and unexplored by geographers. This is particularly surprising today when we are far clearer about the role that geography has played in the evolution of the bourgeois concept of individual and national space.[54] Landscape remains part of our unexamined discourse, to be embraced by humanist geographers as a concept which appears to fulfil their desire for a contextual and anti-positivist geography. Whereas in the past landscape geographers actively distanced their concept from that of common usage, today writers like Samuels, Meinig, Wreford Watson and Pocock take the opposite position.[55] In both periods of its popularity in geography landscape as an artistic concept is given the role of potential or actual challenger to geographical science. Marwyn Mikesell's claim (with its interesting reference to perspective) is an example of this view:

> the perspective of the geographer is not that of the individual observer located at a particular point on the ground. The geographer's work entails map interpretation as well as direct observation, and he makes no distinction between foreground and background. The landscape of the geographer is thus very different from that of the painter, poet or novelist. By means of sampling, survey or detailed inventory, he achieves the comprehensive but synthetic perspective of the helicopter pilot or balloonist armed with maps, photographs and a pair of binoculars.[56]

The distinction seems spurious, it is drawn at the level of technique rather than aims and objectives. Given what we know of Leonardo's detailed notes on how light falls upon different rock formations, or of Constable's inventories of cloud formations and atmospheric conditions, of Turner's strapping himself to a ship's mast the better to observe the movement of the storm, or of Ruskin's instructions to painters to rival the geologist, botanist and meteorologist in their knowledge of topography, geology, vegetation and skies, it is likely that had they had access to the battery of techniques with which Mikesell would arm his geographer they would all have made good use of them. Certainly Christoforo Sorte would have revelled in their use to improve his 'chorographic art', and both Bruegel and Titian produced landscapes that have a perspective far above the ground and are as comprehensive and synthetic as Mikesell could wish for. Above all the geometry which underlay perspective, the constructional principle of landscapes, and which gave certainty to their realism, is the same geometry which determines the graticule of Mikesell's maps and delimits the boundaries or locates the elements of his geographical landscapes.

Beyond the issue of specific techniques there are also methodological similarities between landscape in painting and in geography, similarities which have allowed geographers to adopt unconsciously something of the visual ideology integral to the landscape idea. Like other area concepts in geography, region or *pays*, landscape has been closely associated in geography with the morphological method.[57] Morphology is the study of constituent forms, their isolation, analysis and recomposition into a synthetic whole. When applied to the visible forms of a delimited area of land this is termed *chorology*.[58] The result of a landscape chorology is a static pattern or picture whose internal relations and constituent forms are understood, but which lacks process or change. Indeed, one of the criticisms of chorology in the post-war years was precisely that it failed to explain the processes giving rise to the forms and spatial relations it described. The idea of change, or process, is very difficult to incorporate into landscape painting, although there are certain conventions like the *memento mori* or the ruined building which occasionally do so. But one of the consistent purposes of landscape painting has been to present an image of order and proportioned control, to suppress evidence of tension and conflict between social groups and within human relations in the environment. This is true for the villa landscapes painted by Paolo Veronese in the strife-ridden Venetian countryside of the later sixteenth century, it was equally true for the arcadian image of English landscape parks in the Georgian period of rural conflict and transformation. In this sense the alignment of geographical landscape with morphology serves to reproduce a central dimension of the ideology of the landscape idea as it was developed in the arts.

Despite appearances the situation is little different in much of contemporary geographical use of landscape. Too often geographical humanists make the mistake of assuming that art and within it, landscape, are to do with the subjective, somehow standing against science and its proclaimed objective certainties.[59] The subjectivism of art is a recent and by no means fully accepted thesis, a product above all of the artistic self-image generated in the Romantic movement. Originally, as we have seen, landscape was composed and constructed by techniques which were considered to ensure the certainty of reproducing the real world. Equally, again as we have seen, there is an inherent conservatism in the landscape idea, in its celebration of property and of an unchanging status quo, in its suppression of tension between groups *in* the landscape. When we take over landscape into geography, and particularly into public policy we inevitably import in large measure the realist, visual values with which it has been loaded: its connections with a way of seeing, its distancing of subject and object and its conservatism in presenting an image of natural and social harmony. John Punter has pinpointed the place of these social and visual values in contemporary discussions of landscape and the conservation and planning of areas defined as having 'landscape value'.[60] A vast field awaits research into contemporary visual and social values in landscape.[61]

To return, however, to the opening point of this paper. Humanist geographers have spent a great deal of time and energy challenging the orthodoxy of positivism, they have opened up a debate on the language of geography – the constraints and opportunities of language. Some have even begun to explore the ideological assumptions inherent in our concepts of space itself.[62] All of these are important matters. But the ideology of vision, the way of seeing implicit in much of our geography still awaits detailed examination. At the most obvious level, we warn students of the pitfalls of accepting the authority of numbers, of the dangers of misused statistics, but virtually never those of accepting the cartographic, still less the landscape, image. Less obviously, but more significantly for geographical scholarship, geography and the arts, or geography as art, is frequently presented as a refuge from tendentious social and political debates within the discipline, and the 'soul' of geography a resort in which we can express our 'passions' in the neutral and refined area of subjectivity and humane discourse, expressing ourselves in those reverential tones that serve purely to sustain mystification. Geography and the arts are too important for this. Both bear directly upon our world, both can challenge as well as support the ways we structure, modify and see that world.

In *Theoretical Geography* Bunge came closer than any other recent geographical writer to acknowledging the significance of the graphic image in geography. His later, brilliant use of cartography as a subversive art bears testimony to his insight.[63] Bunge was equally clear that geometry was the language of space, the guarantor of certainty in geographical science, visually and logically. As shown, the relationship between geometry, optics and the study of geographic space is very strong in European intellectual history since the Renaissance.[64] In Bunge's thesis spatial geometry was aligned to a powerful claim for geography as a generalizing positivist science, a very different conception of science from that understood by the founders of modern geometry and perspective, many of whom still recalled the magic of Pythagoras and regarded metaphysics as being as much a branch of science as empirical study,[65] and for whom the trivium and quadrivium were equal contributors to the seven liberal arts. In rejecting science *tout court*, humanist geographers have severed links with spatial geometry, concentrated on the material of the trivium and failed, among other things, to develop a proper critique of landscape.

Such a division was not true of Renaissance humanist geographers. John Dee was as close to Ortelius and Mercator as he was to Sir Philip Sidney, admired the magician Cornelius Agrippa's work as much as he did that of Copernicus. Cusanus' closest friend, the executor of his will, was Piero dal Pozzo Toscanelli. Toscanelli, from a Florentine merchant family, was a doctor, student of optics and the foremost geographer of his day. As a member of the Greek Academy at Florence, he studied one of its greatest intellectual trophies, Ptolemy's *Cosmografia* brought from Constantinople in the early years of the fifteenth century. In this work Ptolemy describes a projection for the world map which uses the same geometrical construction as the Florentine humanists employed to develop linear perspective.[66] With the aid of this study Toscanelli produced a map which he sent with a letter to Christopher Columbus encouraging the Genoese navigator's exploration west on the grounds that the distance from Europe to China was shorter than was then commonly believed by cartographers. The geographical consequences of this collaboration of art, science and practical skill need not be spelled out here. But the example of this geographical colleague of the great humanists Alberti and Brunelleschi may remind contemporary humanists in geography to pay equal attention to the Albertian revolution as to that of Gutenberg.

Notes

1 Geipel, R. (1978) 'The landscape indicators school in German geography', in Ley, D. and Samuels, M. (eds) *Humanistic geography prospects and problems* (London) pp. 155–72.
2 See for example the comments on landscape in Harvey, D. (1969) *Explanation in geography* (London) pp. 114–15.
3 Samuels, M. (1979) 'The biography of landscape', in Meinig, D. (ed.) *The interpretation of ordinary landscapes* (Oxford) pp. 51–88.
4 Rose, C. (1981) 'William Dilthey's philosophy of historical understanding: a neglected heritage of contemporary humanistic geography', in Stoddard, D. R. (ed.) *Geography, ideology and social concern* (Oxford) pp. 99–133.
5 Relph, E. (1981) *Rational landscapes and humanistic geography* (London) p. 22. This sense of landscape as an all inclusive, quotidian phenomenon owes a great deal in North American geography to the work of J. B. Jackson. See for example the most recent collection of Jackson's landscape essays (1980), '*The necessity for ruins and other topics*' (Amherst).
6 See the discussion by Punter, J. V. (1982) 'Landscape aesthetics: a synthesis and critique', in Gold, J. and Burgess, J. (eds) *Valued environments* (London) pp. 100–23.
7 Penning-Rowsell, E. C. (1974) 'Landscape evaluation for development plans', *J. R. Tn Plann. Inst.*, 60: 930–4.
8 Appleton, J. (1975) *The experience of landscape* (London).
9 Pocock, D. C. D. (ed.) (1981) *Humanistic geography and literature: essays in the experience of place* (London); Daniels, S. J. (1981) 'Landscaping for a manufacturer: Humphrey Repton's commission for Benjamin Gott at Armley in 1809–10', *J. Hist. Geog.*, 7: 379–96; Cosgrove, D. (ed.) (1982) 'Geography and the Humanities', *Loughborough Univ. of Techn., Occ. Pap.*, No. 5.
10 This phrase is taken from Berger, J. (1972) *Ways of seeing* (London), where some of the social implications of visual conventions are challengingly explored
11 Examples are numerous. One of the earliest is Francesco Feliciano (1518) *Libro d'aritmetica e geometria speculativa, e practicale*, more commonly *Scala & Grimaldelli* (Venice). One of the most comprehensive was Cosimo Bartoli (1564) *Del modo di misurare le distantie . . .* (Venice).
12 Meinig, D. (1983) 'Geography as Art' *Trans. Inst. Br. Geogr.* NS. 8: 314–28;

Wreford-Watson, J. (1983) 'The soul of geography', *Trans Inst. Br. Geogr.* NS. 8: 385–99; Billinge, M. (1983) 'The Mandarin dialect', *Trans. Inst. Br. Geogr.* NS. 8: 400–20; Pocock, D. C. D. (1983) 'The paradox of humanistic geography', *Area*, 15: 355–58.

13 As always, there are exceptions, although to my mind none have examined the visual in relation to geographical study as such: Pocock, D. C. D. (1981) 'Sight and Knowledge', *Trans. Inst. Br. Geogr.* NS. 6: 385–93; Tuan, Yi-Fu (1979) 'The eye and the mind's eye', in Meinig, *The interpretation of ordinary landscape* (Note 3) pp. 89–102.

14 Bunge, W. (1966) *Theoretical geography* (2nd ed. Lund), p. xiv.

15 Yates, F. A. (1964) *Giordano Bruno and the Hermetic Tradition* (London) pp. 160–1 discusses the relations of quadrivium and trivium in Renaissance humanism, arguing that 'the two traditions appeal to entirely different interests. The humanist's bent is in the direction of literature and history; he sets an immense value on rhetoric and good literary style. The bent of the other tradition is towards philosophy, theology, and also science (at the stage of magic)'. This argument depends on a very restricted definition of humanism (see her fn. 3, p. 160), ignores the visual arts which combined literary reference (ut pictura poesis) with 'scientific' skill, and fails to account for the large number of Renaissance scholars equally at home in philosophy and science as they were concerned with grammar, rhetoric and classical texts, for example Giangiorgio Trissino and Daniele Barbaro in sixteenth-century Venice.

16 Eisenstein, E. L. (1979) *The printing press as an agent of change* (Cambridge).

17 Martines, L. (1980) *Power and imagination: City-States in Renaissance Italy* (London).

18 Alberti, L. B. (1966) *On painting* (trans. J. R. Spencer, London).

19 Ibid pp. 47, 48.

20 Even photography was constricted by conventions of perspective realism, landscape painting having far more influence on early photography than vice-versa. See Galassi, P. (1981) *Before photography: painting and the invention of photography* (New York).

21 Ibid. pp. 16–17.

22 For a detailed discussion of Brunelleschi's experiment see Edgerton, S. J. Jr. (1975) *The Renaissance rediscovery of linear perspective* (London) pp. 145–52.

23 Rees, R. (1980) 'Historical links between geography and art', *Geogr. Rev.* 70: 66

24 This group of paintings, produced before the centrally planned church became architecturally popular, includes Raphael's *Spozalizio* and Carpaccio's *Reception of the English Ambassadors* in the St Ursula cycle. The sacred significance of the circle and centre is an enormous topic with cross-cultural implications. See Tuan, Yi-Fu (1974) *Topophilia: a study of environmental perception attitudes and beliefs* (London).

25 The distinction between mind, or intellect, and sense was central to much Renaissance thought, and is discussed in Yates, *Giordano Bruno* (note 15) p. 193. Geometry is of course an intellectual activity. Nicolo Tartaglia calls it 'the pure food of intellectual life' (il puro cibo della vita intellettuale) *Euclide Magarense, philosopho* (Venezia, 1543) p. F11, in the first translation of Euclid into Italian. None the less, one of the reasons why humanists like Alberti accepted the significance of numbers and proportions was that the same proportions which pleased the intellect also seemed to please our eyes and ears. This is a cornerstone of Renaissance aesthetics.

26 Baxandall, M. (1972) *Painting and experience in fifteenth-century Italy* (London).

27 Fra Luca Pacioli (1494) *Summa di arithmetica, geometria, proportione et proportionalità* (Venice). See the reference to the significance of this work in Braudel, F. (1982) *Civilization and capitalism, 15th–18th Century. Vol. II: The Wheels of Commerce* (London) p. 573.

28 Silvio Belli (1565) *Libro del misurar con la vista . . .* (Venezia) preface, pp. 1–2 ('certamente è cosi meravigliosa il misurar con la vista, poi che ogni uno, che non sa la ragione par del tutto impossible').

29 Rossi, F. (1877) *Groma e squadra, ovvero storia dell' agrimensura italiana dai tempi antichi al secolo XVII* (Torino).

30 Schulz, J. (1978) 'Jacopo de 'Barbari's view of Venice: map making, city views, and moralized geography before the year 1500', *The Art Bull.*, LX: 425–74; Mazzi, G. (1980) 'La repubblica e uno strumento per il dominio', in Puppi, L. (ed.) *Architettura e utopia nella Venezia del cinquecento* (Milano) p. 59–62. It has been pointed out that, like contemporary ideal townscapes, the Barbari map lacks all human presence.

31 Renaissance writers never tire of emphasizing that geometry provides certainty, eg. Pacioli, *Summa di arithmetica . . .* (note 27) p. 2r 'e in la sua Metaphysica afferma (Euclid) le scientie mathematiche, essere nel primo grado de certezza'

32 Gombrich, E. (1971) 'The renaissance theory of art and the rise of landscape', in Gombrich, E. *Norm and Form: studies in the art of the renaissance* (London) 109.

33 Clark, K. (1956) *Landscape into art* (Harmondsworth).

34 Significantly, the title of the essay by Jackson, J. B. (1979) 'Landscape as theatre' in *Landscape*, 23: 3; and reprinted in Jackson, *The necessity for ruins* (note 5).

35 Blunt, A. (1962) *Artistic theory in Italy, 1450–1600* (Oxford) p. 26 Italics added

36 Quoted in Ibid. p. 50.

37 Leonardo was a master not merely of linear perspective but also of that other and distinct form of perspective, *aerial* perspective, which plays a complementary role in creating the illusion of space through the manipulation of tone, light and shade and colour intensity. While based on optical theory and experiment, aerial perspective is not geometrically founded. Leonardo's work with colour and chiaroscuro allowed him to convey the 'mood' of space, and he saw the superiority of painting over other arts to lie in its ability to employ aerial perspective.

38 Martines, *Power and imagination* (note 17); Baxandall, *Painting and experience* (note 26).

39 A point that has not gone entirely unnoticed by historical geographers. See for example Ian Adam's work on the role of land surveyors in eighteenth-century Scottish agrarian change. Adams, I. H. (1980) 'The agents of agrarian change', in Parry, M. L. and Slater, T. R. (eds) *The making of the Scottish countryside* (London) pp. 155–75, esp. pp. 167–70.

40 For example the great gallery of maps painted by Ignazio Dante in the Vatican (1580–83) or the similar commissions to Christoforo Sorte to paint walls in the Ducal Palace at Venice (1578 and 1586).

41 Cosgrove, D. (1982) 'Agrarian change, villa building and landscape: the Godi estates in Vicenza 1500–1600', in Ferro, G. (ed.) *Symposium on historical changes in spatial organisation and its experience in the Mediterranean world* (Genova) pp. 135–56; Daniels, D. J. (1982) 'Humphrey Repton and the morality of landscape', in Gold, J. and Burgess, J. (eds) *Valued environments* (note 6) pp. 124–44.

42 Quoted in McLean, *Humanism and the rise of science . . .* (note 32) p. 138. The translation of *Euclid* was by Billingsley. For Dee's importance for geography and cartography see Taylor, E. G. R. (1954) *The mathematical practitioners of Tudor*

and Stuart England (London) pp. 26–48. For Dee and magic see Yates, *Girodano Bruno* (note 15) pp. 148–50.

43 Morgan, V. (1979) 'The cartographic image of the country in early modern England', *Trans. R. Hist. Soc.* 29: 129–54.

44 The whole issue of garden design along circular and orthogonal lines is too large to discuss here but is obviously very closely related to the geometry under discussion, to spatial theory and those of microcosm, marcrocosm and medicinal concepts. The first such garden was designed in Padua in the late sixteenth century by Daniele Barbaro, translater of Vitruvius and commentator on *Euclid.* See Jackson, J. B. (1980) 'Nearer than Eden' and 'Gardens to Decipher' in *The necessity for ruins* (note 5) pp. 19–35 and 37–53.

45 Oxford English Dictionary (OED), italics added.

46 Thompson, F. M. L. (1968) *Chartered surveyors; the growth of a profession* (London); Harvey, P. D. A. (1980) *The history of topographic maps; symbols, pictures and surveys* (London). The idea that surveying was a malicious and magical art was founded in part on the negative consequences for traditional land rights of new concepts of private property enshrined in the legal document that the surveyor produced, in part on the recognition of connections between the geometry of survey techniques and that of hermetic magicians. In the book burnings under Edward VI books containing geometrical figures were particularly at risk.

47 Turner, J. (1979) *The politics of landscape: rural scenery and society in English poetry 1630–1690* (Oxford); Adams, J. (1979) *The artist and the country house. A history of country house and garden view painting in Britain 1540–1870* (London); Barrell, J. (1980) *The dark side of the landscape: the rural poor in English painting 1631–1741* (Cambridge); Rosenthal, M. (1982) *British landscape painting* (London).

48 The OED notes that the verb 'to prospect' emerged in the nineteenth century referring to the particularly capitalist activities of speculative gold mining and playing the stock exchange. It is interesting to note how 'speculation' has itself roots in visual terminology.

49 Mikesell, M. (1968) 'Landscape', in *International encyclopaedia of the social sciences* (New York) p. 577–79. Dickinson, R. E. (1939) 'Landscape and Society', *Scott. geogr. Mag.* 55: 1–15; Hartshorne, R. (1939) *The nature of geography. A survey of current thought in the light of the past* (Lancaster, Pa.).

50 Humboldt, A. Von (1849–52) *Cosmos: a sketch of a physical description of the Universe* (London), Vol. II. The relationship between the landscape idea and attitudes to nature in the nineteenth century is of course enormously complex. On Goethe and geography see Seamon, D. (1978) 'Goethe's approach to the natural world: implications for environmental theory and education', in Ley and Samuels, *Humanistic geography* (note 1) pp. 238–50.

51 Cosgrove, D. (1979) 'John Ruskin and the geographical imagination' *Geog. Rev.* 69: 43–62.

52 Sauer, C. O. (1926) 'The morphology of landscape', reprinted in LEIGHLY, J. (ed.) (1963) *Land and life: selections from the writings of Carl Ortwin Sauer* (Berkeley and Los Angeles).

53 Banse, E. (1924) Die Seele der Geographie (Brunswick); HARD, G. (1965) 'Arkadien in Deutchland', *Die Erde*, 96: 31–4.

54 Harvey, D. (1974) 'What kind of geography for what kind of public policy', *Trans. Inst. Br. Geogr.*; Harvey, D. (1984) 'On the history and present condition of geography: an historical materialist manifesto', *Prof. Geogr.* 35: 1–10.

55 Notes 3 and 12.
56 Mikesell, 'Landscape' (note 49) p. 578.
57 Explicitly so by Sauer, 'Morphology of Landscape' (note 52), and equally in physical geography where landscape in the title suggests a morphological study of landforms.
58 Van Paasen, C. (1957) *The classical tradition of geography* (Groningen).
59 See for example the diagram which serves as the foundation for the discussion of spatial concepts in Sack, R. D. (1980) *Conceptions of space in social thought: a geographial perspective* (Minneapolis) p. 25.
60 Punter, J. 'Landscape aesthetics . . . ' (note 6).
61 Some of the essays in Gold, and Burgess, *Valued environments* (note 6) begin to broach this field, as have papers presented in recent IBG sessions of 'Geography and the Media'.
62 Sack, *Conceptions of Space . . .* (note 59).
63 Bunge, W. (1973) 'The geography of human survival', *Ann. Ass. Am. Geogr.* 63: 275–95.
64 This is distinct from the relations of Greek geometry which apparently were derived from a tactile–muscular apprehension of space, an apprehension which was non-visual. Ivins, *Art and geometry* (New York) pp. 79–80.
65 Yates, *Giordano Bruno* (note 15) pp. 144–56.
66 Edgerton, *The Renaissance rediscovery . . .* (note 22).

20 Gillian Rose

Looking at Landscape: The Uneasy Pleasures of Power

Excerpts from: G. Rose, *Feminism and geography,* Chapter 5. Cambridge; Polity Press (1993)

Landscape is a central term in geographical studies because it refers to one of the discipline's most enduring interests: the relation between the natural environment and human society, or, to rephrase, between Nature and Culture. Landscape is a term especially associated with cultural geography, and although 'literally [the landscape] is the scene within the range of the observer's vision',[1] its conceptualization has changed through history. By the interwar period, for its leading exponents, such as Otto Schlüter in Germany, Jean Brunhes in France and Carl Sauer in the USA, the term 'landscape' was increasingly interpreted as a formulation of the dynamic relations between a society or culture and its environment: '*the process of human activity in time and area*'.[2] The interpretation of these processes depended in particular on fieldwork, and fieldwork is all about looking: 'the good geographers have first been to see, then they have stopped to think and to study the conclusions of others before finally recording their findings for us in maps and print'.[3] Just as fieldwork is central not only to cultural geography but also the discipline as a whole, however, so too the visual is central to claims to geographical knowledge:[4] a president of the Association of American Geographers has argued that 'good regional geography, and I suspect most good geography of any stripe, begins by looking'.[5] The absence of knowledge, which is the condition for continuing to seek to know, is often metaphorically indicated in geographical discourse by an absence of insight, by mystery or by myopia; conversely, the desire for full knowledge is indicated by transparency, visibility and perception. Seeing and knowing are often conflated.

More recent work on landscape has begun to question the visuality of traditional cultural geography, however, as part of a wider critique of the latter's neglect of the power relations within which landscapes are embedded.[6] Some cultural geographers suggest that the discipline's visuality is not simple observation but, rather, is a sophisticated ideological device that enacts systematic erasures. They have begun to problematize the term 'landscape' as a reference to relations between society and the environment through contextual studies of the concept as it emerged and developed historically, and they have argued that it refers not only to the relationships between different objects caught in the fieldworker's gaze, but that it also implies a specific way of looking. They interpret landscape not as a material consequence of interactions between a society and an environment, observable in the field by the more-or-less objective gaze of the geographer, but rather as a gaze which itself helps to make sense of a particular relationship between society and land. They have stressed the importance of the look to the idea of landscape and have argued that landscape is a way of seeing which we learn; as a consequence, they argue that the gaze of the fieldworker is part of the problematic, not a tool of analysis. Indeed, they name this gaze at landscape a 'visual ideology', because it uncritically shows only the relationship of the powerful to their environment. This is an important critique of the unequal social relations implicit in one element of geographical epistemology . . . .

Questions of gender and sexuality have not been raised by this newer work, however. This seems an important omission . . . . A consequence has been that, historically, in geographical discourse, landscapes are often seen in terms of the female body and the beauty of Nature. Here, for example, is . . . [a] quotation from . . . [Younghusband that] highlight[s] the parallels . . . between [a] 'live, supple, sensitive, and active' Nature and a female body:

> It is [in] the face and features of Mother-Earth that we geographers are mainly interested. We must know something of the general principles of geology, as painters have to know something of the anatomy of the human or animal body . . . the characteristic of the face and features of the Earth most worth learning about, knowing and understanding is their beauty.[7]

Stoddart's celebration of geography's exploration and fieldwork tradition similarly conflates the exploration of Nature with the body of Woman; for example, his frontispiece is an eighteenth-century engraving representing Europe, Africa and America as three naked women.[8] This feminization of what is looked at does matter, because it is one half of what Berger characterizes as the dominant visual regime of white heterosexual masculinism: 'women appear', he says' but 'men act'.[9] This particular masculine position is to look actively, possessively, sexually and pleasurably, at women as objects. Now, Berger's comments refer to the female nude in Western art; but I will suggest in this chapter that the feminization of landscape in geography allows many of the arguments made about the masculinity of the gaze at the nude to work in the context of geography's landscape too, particularly in the context of geography's pleasure in landscape. [I] . . . suggest that geography's look at landscape draws on not only a complex discursive transcoding between Woman and Nature . . . but also on a specific masculine way of seeing: the men acting in the context of geography are the fieldworkers, and the Woman appearing is the landscape. This compelling figure of Woman both haunts a masculinist spectator of landscape and constitutes him.

Landscape as visual ideology

Recent critiques of the landscape in geography insist that landscape is a form of representation and not an empirical object. As Daniels and Cosgrove remark, 'a landscape is a cultural image, a pictorial way of representing, structuring or symbolising surroundings'.[10] Whether written or painted, grown or built, a landscape's meanings draw on the cultural codes of the society for which it was made. These codes are embedded in social power structures, and theorization of the relationship between culture and society by these new cultural geographers has so far drawn on the humanist marxist tradition of Antonio Gramsci, Raymond Williams, E. P. Thompson and John Berger. All of these authors see the material and symbolic dimensions of the production and reproduction of society as inextricably intertwined.[11] Cosgrove, one of the most prominent theorists of the new critique of the landscape idea, defines culture as:

> . . . symbolisation, grounded in the material world as symbolically appropriated and produced. In class societies, where surplus production is appropriated by the dominant group, symbolic production is likewise seized as hegemonic class culture to be imposed on all classes.[12]

In his work, landscape becomes a part of that hegemonic culture, a concept which helps to order society into hierarchical class relations.

Cosgrove points out that landscape first emerged as a term in fifteenth- and early sixteenth-century Italy, and he argues that it was bound up with both Renaissance theories of space and with the practical appropriation of space. Euclidean geometry was 'the guarantor of certainty in spatial conception, organisation and representation',[13] and its recovery paved the way for Alberti's explication of the technique of three-dimensional perspective in 1435. Other geometrical skills were being developed contemporaneously, especially by the urban merchant class, and these too involved the accurate representation of space: calculating the volume and thus the value of packaged commodities; map-making to guide the search for goods and markets; and surveying techniques to plot the estates that the bourgeoisie were buying in the countryside. All of these spatial techniques were implicated in relations of power and ownership. Cosgrove is particularly interested in Alberti because, using his manual, artists could render depth realistically, and so establish a particular viewpoint for the spectator in their painting – a single, fixed point of the bourgeois individual. (Cosgrove does remark that this individual was male, but does not develop the point.[14]) From this position, the spectator controlled the spatial organization of a composition, and Cosgrove argues that this was central to landscape images. Merchants often commissioned paintings of their newly acquired properties, and in these canvases, through perspective, they enjoyed perspectival as well as material control over their land. Cosgrove concludes that the idea of landscape is patrician because it is seen and understood from the social and visual position of the landowner. Other writers agree and emphasize the erasure of the waged labour relation in landscape painting. In the context of eighteenth-century English landscape painting, for example, Barrell notes that the labourers in these images are denied full humanity, and Bryson argues that the fine brush-work technique favoured in Western art until the late nineteenth century effaces the mark of the artist as waged worker.[15] It is argued then that landscape is meaningful as a 'way of seeing' bound into class relations, and Cosgrove describes landscape as a 'visual ideology' in the sense that it represents only a partial world view.[16]

This is an extremely important critique of the ideologies implicit in geographical discourse. Its strengths are evident in the interpretation shared by cultural geographers of the mid-eighteenth-century double portrait of Mr and Mrs Andrews, by the English artist Thomas Gainsborough (Fig. 20.1).[17] In their discussions of this image, geographers concur that pleasure in the right-hand side of the canvas – those intense green fields, the heaviness of the sheaves of corn, the English sky threatening rain – is made problematic by the two figures on the left, Mr and Mrs Andrews. Berger, whose discussion of this painting geographers follow, insists that the fact that this couple owned the fields and trees about them is central to its creation and therefore to its meaning: 'they are landowners and their proprietary attitude towards what surrounds them is visible in their stance and their expressions'.[18] Their ownership of land is celebrated in the substantiality of the oil paints used to represent it, and in the vista opening up beyond them, which echoes in visual form the freedom to move over property which only landowners could enjoy. The absence in the painting's content of the people who work the fields, and the absence in its form of the signs of its production by an artist working for a fee on a commission, can be used to support Cosgrove's claim that landscape painting is a form of visual ideology: it denies the social relations of waged labour under capitalism. *Mr and Mrs Andrews*, then, is an image on which geographers are agreed: it is a symptom of the capitalist property relations that legitimate and are sanctioned by the visual sweep of a landscape prospect.

However, the painting of Mr and Mrs Andrews can also be read in other ways. In particular, it is possible to prise the couple – 'the landowners' – apart, and to differentiate between them. Although both figures are relaxed and share the sense of partnership so often found in eighteenth-century portraits of husband and wife, their unity is not entire: they are given rather different relationships to the land around them. Mr Andrews stands, gun on

Fig. 20.1 *Mr and Mrs Andrews*, by Thomas Gainsborough. Reproduced by courtesy of the Trustees, The National Gallery, London

arm, ready to leave his pose and go shooting again; his hunting dog is at his feet, already urging him away. Meanwhile, Mrs Andrews sits impassively, rooted to her seat with its wrought iron branches and tendrils, her upright stance echoing that of the tree directly behind her. If Mr Andrews seems at any moment able to stride off into the vista, Mrs Andrews looks planted to the spot. This helps me to remember that, *contra* Berger, these two people are *not* both landowners – only Mr Andrews owns the land. His potential for activity, his free movement over his property, is in stark contrast not only to the harsh penalties awaiting poachers daring the same freedom of movement over his land (as Berger notes), but also to the frozen stillness of Mrs Andrews. Moreover, the shadow of the oak tree over her refers to the family tree she was expected to propagate and nurture; like the fields she sits beside, her role was to reproduce, and this role is itself naturalized by the references to trees and fields.[19] [In addition,] . . . this period saw the consolidation of an argument that women were more 'natural' than men. Medical, scientific, legal and political discourses concurred, and contextualize the image of Mr and Mrs Andrews in terms of a gendered difference in which the relationship to the land is a key signifier. Landscape painting then involves not only class relations, but also gender relations. Mr Andrews is represented as the owner of the land, while Mrs Andrews is painted almost as a part of that still and exquisite landscape: the tree and its roots bracketing her on one side, and the metal branches of her seat on the other.

Many feminist art historians have argued that heterosexual masculinism structures images of femininity: following that claim, my interpretation of the figure of Mrs Andrews stresses her representation as a natural mother. Obviously, her representation also draws on discourses of class and even nation. I emphasize her femininity, however, because there are feminist arguments which offer a critique not just of the discourses that pin Mrs Andrews to her seat, but also of the gaze that renders her as immobile, as natural, as productive and as decorative as the land. Such arguments consider the dynamics of a masculine gaze and its pleasures. The next subsection introduces their claim that more is involved in looking at landscape than property relations.

Woman, landscape and Nature

This subsection begins to examine the gaze which sees landscapes, and it focuses on the construction of the landscape as feminine. I concentrate mainly on feminist interpretations of nineteenth-century landscape paintings in Europe and North America. The massive social, economic and political upheavals in those places during that period – upheavals which included the colonial explorations through which geography developed as a discipline – meant that many of the schema previously used by artists to represent the world seemed increasingly outmoded, and new iconographies were sought to articulate the changes producing and reproducing the lives of art's audience, the bourgeoisie. By the mid-nineteenth century, the emergence of this new public for paintings was fuelling a vigorous debate about the role of art: art was drawn into debates about social, political and moral standards which might structure the emerging modern world and, as feminists have remarked, central to these wider issues was the figure of Woman – fallen, pure, decadent, spiritual.[20] Parker and Pollock suggest that the very importance attached to Art in the realm of Culture reasserted the association of women with the natural:

> . . . woman is body, is nature opposed to culture, which, in turn, is represented by the very act of transforming nature, that is, the female model or motif, into the ordered forms and colour of a cultural artefact, a *work* of art.[21]

Woman becomes Nature, and Nature Woman, and both can thus be burdened with men's meaning and invite interpretation by masculinist discourse . . . . It should be emphasized that the 'naturalization' of some women is asserted more directly than that of others: allegorical figures especially, but also, in bourgeois and racist society, working-class and black women.[22] Thus the visual encoding of nineteenth-century Western hegemonic masculinist constructions of femininity, sexuality, nature and property are at their most overtly intertwined in the landscapes with figures set in the colonies of Europe and America. To take an example relevant to one of geography's heroic self-images, Theweleit has suggested that the image of the South Sea maiden 'began to construct the body that would constitute a mysterious goal for men whose desires were armed for an imminent voyage, a body that was more enticing than all the world put together',[23] and perhaps the most well-known paintings which fuse beautiful, sexual, fertile, silent and mysterious Woman with a gorgeous, generous, lush Nature are Gauguin's paintings of Tahitian women. In perfect stillness, they offer the produce of their island to him in the same gesture as they offer themselves, their breasts painted like fruits and flowers.[24] The first French encounter with Tahiti is described by Stoddart as one of the founding moments of scientific geography, and the encounter that he chooses to elaborate is a sexual one. Tahitian women represent the enticing and inviting land to be explored, mapped, penetrated and known.[25] This subsection concentrates on the representation of female figures in landscapes, then, in order to examine one moment of the complex transcoding of femininity and Nature in the field of vision. I suggest that, as well as contextualizing stories of geography's beginnings, the conflation of Woman and Nature can also say something about contemporary cultural geography's visual pleasure in landscape.

Lynda Nead has demonstrated the complexity of the social relations which were mediated in images of the landscapes at the heart of Empire, and she stresses the importance of gender relations to the representation of both class and nation. Nead suggests that, in the face of the transformations of the Victorian era, 'confirmation and reassurance . . . were two of the most important functions of nineteenth-century cultural discourse',[26] and one of the most resonant symbols in England was that of the village in the countryside. The social stability associated with the village – people and land in traditional harmony – was so strong that by the 1840s landscape painting was for many art critics a contender as the truly national art genre of England.[27] A contrast between the town and the country has a long tradition in English culture, of course, but by the mid-nineteenth century, despite the continuing arguments for the urban as the centre of civilization and progress, images of the country-side showed a rural idyll which gained much of its impact in opposition to representations of the city as polluted and depraved.[28] The fields and villages of England were painted as embodying all the virtues that the towns had lost – stability, morality and tranquility – and social harmony was fundamental to this discursive construction. The rural idyll was envisioned as a village community. Everyone knew their place, and the harmony of such a community was centrally represented through 'natural' gender differences. Ideas about natural order were epitomized in the 'natural' difference between men and women, with women naturally natural mothers. Nochlin stresses the importance of the rural working mother figure to the rhetoric of Nature and the natural in her discussion of nineteenth-century French paintings of peasant life: 'The peasant woman, as an elemental, untutored – hence eminently ''natural'' female – is the ideal signifier for the notion of beneficient maternity'.[29] And Nochlin describes how the stress on the naturalness of this role led to peasant women being equated directly with the land and animals they tend in many of these genre scenes – both were shown as essentially reproductive.

The supposed closeness of women to Nature was also explicit in other painting genres of the period, particularly those in which classical, fantastical or allegorical women appear surrounded by wild Nature. Dijkstra has catalogued these imaginary scenes in European and American nineteenth-century art.[30] Often nude, in England these images of women required a classical gloss to withstand the puritanism of some critics, although bourgeois patrons adored them.[31] Elsewhere, in Europe and America, less excuse was needed to paint nudes: sleep was a popular allegory allowing scenes of women in unself-conscious abandon, oblivious to the spectator's gaze. In the eyes of nineteenth-century morality, such sexual potential brought these women excitingly close to Nature, and they are found in fields and woods throughout late nineteenth-century bourgeois art: 'Passive but fertile, they personify what had come to be a standard conception of woman as the infinitely receptive, seed-sheltering womb of a sweltering earth'.[32] As nymphs and dryads they entwined themselves in trees, or lay on the leaf-covered earth, languid and passive, so that, according to Dijkstra's somewhat over-empathetic account, 'we can almost hear them call to us like animals waiting to be fed'.[33] In a final iconographic twist, women became allegories of nature itself; for the seasons, for weather, for the time of day, for flowers.[34] In making such a parallel between Woman and Nature these paintings offered the possibility that women could be used as Nature was: 'did not the earth, nature herself, meekly permit her body to be plowed, seeded, stripped, and abused by man?'[35] Nature and Woman were equally vulnerable.

This equivalence between Woman and Nature leads Armstrong to compare the female nude in Western art directly to a landscape:

> The female nude, when free of narrative situations, is most often constituted frontally and horizontally – as a kind of landscape, its significant part the torso, its limbs merely elongations of the line created by the supine, stretched-out torso.[36]

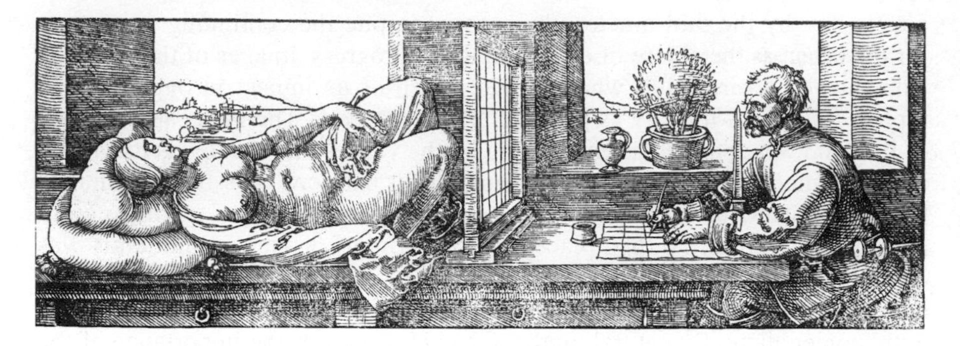

Fig. 20.2 *A Draughtsman Drawing a Nude*, by Albrecht Dürer

The female figure represents landscape, and landscape a female torso, visually in part through their pose: paintings of Woman and Nature often share the same topography of passivity and stillness. The comparison is also made through the association of both land and Woman with reproduction, fertility and sexuality, free from the constraints of Culture. Incorporating all of these associations, both Woman and Nature are vulnerable to the desires of men. Armstrong examines this vulnerability by arguing that if Art and the spectator constitute both Woman and Nature as what they work on and interpret, they do so especially by looking at both in a similar manner. Both are made to invite the same kind of observation. Rarely do the women in landscape images look out from the canvas at the viewer as an equal. Their gaze is often elsewhere: oblivious to their exposure, they offer no resistance to the regard of the spectator. Perhaps they will be looking in a mirror, allowing the viewer to enjoy them as they apparently enjoy themselves. If they acknowledge the spectator/artist, they do so with a look of invitation. The viewer's eye can move over the canvas at will, just as it can wander across a landscape painting, with the same kind of sensual pleasure. Here is another parallel between Woman and landscape: the techniques of perspective used to record landscapes were also used to map female nudes, and the art genre of naked women emerged in the same period as did landscape painting (Fig. 20.2).

One of the earliest discussions of this kind of visual power over the representation of women was Berger's.[37] Like his reading of *Mr and Mrs Andrews*, his arguments focus on the question of ownership. Speaking of the woman in a nude painting, he says that 'this nakedness is not, however, an expression of her own feelings; it is a sign of her submission to the owner's feelings or demands'.[38] Just as he argues that the painting of a landscape in oils was a sensuous celebration of land ownership, so he claims that the representation of a woman in oils turns her too into a commodity, passive and prostrate, able only to welcome the gaze of the owner of the canvas. Being an owner gives material and visual power over property, whether that be land or the image of a woman.

Feminist art historians have acknowledged the force of Berger's account, but they suggest that not only the commodification of art and sex (and land) is involved in 'the landscape of the reclining torso';[39] so too are the (hetero)-sexual fantasies of both artist and spectator. It is the imagined and desired sexuality of the female nude that is offered to the (implicitly masculine) spectator. Nochlin was one of the first feminists to argue that the sexuality of the Western female nude was represented only through masculine desires:

> As far as one knows, there simply exists no art, and certainly no high art, in the nineteenth century based upon women's erotic needs, wishes, or fantasies. Whether the erotic object be breasts or buttocks, shoes or corsets, a matter of pose or of prototype, the imagery of sexual delight or provocation has always been created *about* women for men's enjoyment, by men.[40]

This means that the sensual topography of land and skin is mapped by a gaze which is eroticized as masculine and heterosexual. This masculine gaze sees a feminine body which requires interpreting by the cultured knowledgeable look; something to own, and something to give pleasure. The same sense of visual power as well as pleasure is at work as the eye traverses both field and flesh: the masculine gaze is of knowledge and desire.

This discussion of the visual representation of women and landscape has concentrated on the complex construction of images of 'natural' Woman as the objects of male desire. I have argued that Nature and Woman are represented through masculinist fantasies, and that makes looking pleasurable. Women are seen as closer to Nature than men because of the desirable sexuality given to them in these images and other discourses. In a rare and welcome discussion of pleasure in landscape images, Daniels reveals this desire at work.[41] Noting Berger's claim that painting has an energy which pulls the viewer further from the visible *status quo* than they could manage alone, he suggests that images of the countryside evoke deep and pleasurable emotional responses which can empower; and this pleasure is described in Berger's words, as 'a going further than he could have achieved alone, towards a prey, a Madonna, a sexual pleasure, a landscape, a face, a different world'.[42] This conflation of hunting, a virgin and the single male orgasm stands as a summary of the pleasure of landscape. Pleasure in landscape, it appears, is for straight men's eyes only.

Notes

1 R. E. Dickinson, 'Landscape and society', *Scottish Geographical Magazine*, 55 (1939), pp. 1–14, p. 1.
2 Dickinson, 'Landscape and society', p. 6.
3 P. A. Jones, *Field Work in Geography* (Longmans, Green, London, 1968), p. 1.
4 On the importance of the visual to the contemporary discipline, see D. Cosgrove, 'Prospect, perspective and the evolution of the landscape idea', *Transactions of the Institute of British Geographers*, 10 (1985), pp. 45–62, p. 46, p. 58. For a baroque example of the visual as a metaphor of knowledge, see E. W. Soja, 'The spatiality of social life', in *Social Relatioins and Spatial Structures*, eds D. Gregory and J. Urry (Macmillan, London, 1985), pp. 90–127.
5 J. Fraser Hart, 'The highest form of the geographer's art', *Annals of the Association of American Geographers*, 72 (1982), pp. 1–29, p. 24.

6 See, for example, J. S. Duncan, 'The superorganic in American cultural geography', *Annals of the Association of American Geographers*, 70 (1980), pp. 181–98; P. Jackson, *Maps of Meaning: an Introduction to Cultural Geography* (Unwin Hyman, London, 1989). For a discussion of the exclusion of women from landscape studies, see J. Monk, 'Approaches to the study of women and landscape', *Environmental Review*, 8 (1984), pp. 23–33.

7 F. Younghusband, 'Natural beauty and geographical science', *The Geographical Journal*, 56 (1920), pp. 1–13, p. 3.

8 D. R. Stoddart, *On Geography and its History* (Blackwell, Oxford, 1986).

9 J. Berger, *Ways of Seeing* (British Broadcasting Corporation, London 1972), p. 47.

10 S. Daniels and D. Cosgrove, 'Introduction: the iconography of landscape', in *The Iconography of Landscape: Essays on the Symbolic Representation, Design and Use of Past Environments*, eds D. Cosgrove and S. Daniels (Cambridge University Press, Cambridge, 1988), pp. 1–10, p. 1.

11 For an excellent discussion of this tradition, see S. Daniels, 'Marxism, culture and the duplicity of landscape', in *New Models in Geography, Volume 2: The Political Economy Perspective*, eds R. Peet and N. Thrift (Unwin Hyman, London, 1989), pp. 196–220.

12 D. Cosgrove, 'Towards a radical cultural geography: problems of theory', *Antipode*, 15 (1983), pp. 1–11, p. 5.

13 Cosgrove, 'Prospect, perspective and the evolution of the landscape idea', p. 46.

14 D. Cosgrove, 'Historical considerations on humanism, historical materialism and geography', in *Remaking Human Geography*, eds A. Kobayashi and S. Mackenzie (Unwin Hyman, London, 1989), pp. 189–226, p. 190.

15 J. Barrell, *The Dark Side of the Landscape: The Rural Poor in English Painting, 1730–1840* (Cambridge University Press, Cambridge, 1980); N. Bryson, *Vision and Painting: The Logic of the Gaze* (Macmillan, London, 1983), esp. pp. 89–92.

16 Cosgrove, 'Prospect, perspective and the evolution of the landscape idea', p. 47. The term 'ways of seeing' is after Berger, *Ways of Seeing*.

17 Daniels, 'Marxism, culture and the duplicity of landscape', p. 213; Daniels and Cosgrove, 'Introduction: the iconography of landscape'; M. Gold, 'A history of nature', in *Geography Matters! A Reader*, eds D. Massey and J. Allen (Cambridge University Press, Cambridge, 1984), pp. 12–33, pp. 20–3; J. R. Short, *Imagined Country: Society, Culture and Environment* (Routledge, London, 1991), p. 170.

18 Berger, *Ways of Seeing*, p. 107.

19 A. Bermingham, *Landscape and Ideology: the English Rustic Tradition 1740–1860* (Thames & Hudson, London, 1987), pp. 14–16; S. Daniels, 'The political iconography of woodland in later Georgian England', in *The Iconography of Landscape: Essays on the Symbolic Representation, Design and Use of Past Environments*, eds D. Cosgrove and S. Daniels (Cambridge University Press, Cambridge, 1988), pp. 43–82.

20 L. Nead, *Myths of Sexuality: Representations of Women in Victorian Britain* (Blackwell, Oxford, 1988), pp. 165–8.

21 R. Parker and G. Pollock, *Old Mistresses: Women, Art and Ideology* (Routledge & Kegan Paul, London, 1981), p. 119. See also Duncan's discussion of the disappearance of the image of woman in postwar American art; C. Duncan, 'The MOMA's hot mammas', *The Art Journal*, 48 (1989), pp. 171–8.

22 Many black feminists have commented that black women are constructed in racist discourses as even more 'natural' than white women; see P. Hill Collins, *Black Feminist Thought: Knowledge, Consciousness, and Empowerment* (Harper Collins, London, 1990), p. 170; S. L. Gilman, 'Black bodies, white bodies:

towards an iconography of black female sexuality in late nineteenth-century art, medicine and literature', *Critical Inquiry*, 12 (1985), pp. 204–42; A. Hurtado, 'Relating to privilege: seduction and rejection in the subordination of white women and women of color', *Signs*, 14 (1989), pp. 833–55, p. 847; V. Smith, 'Black feminist theory and the representation of the "Other" ', in *Changing Our Own Words: Essays on Criticism, Theory, and Writing by Black Women*, ed. C. A. Wall (Routledge, London, 1990), pp. 38–57, esp. pp. 45–6.

23 K. Theweleit, *Male Fantasies. Volume One: Women, Floods, Bodies, History* (Polity Press, Cambridge, 1987), p. 296. On Tahiti, see also M. Bloch and J. H. Bloch, 'Women and the dialectics of nature in eighteenth-century French thought', in *Nature, Culture, Gender,* eds C. P. MacCormack and M. Strathern (Cambridge University Press, Cambridge, 1980), pp. 25–41, p. 31.

24 L. Nochlin, *Women, Art, Power and Other Essays* (Thames & Hudson, London, 1989), pp. 139–41.

25 Stoddart, *On Geography and its History*, p. 35.

26 Nead, *Myths of Sexuality*, p. 128.

27 Nead, *Myths of Sexuality*, p. 40.

28 Nead, *Myths of Sexuality*, p. 39. For a contrasting interpretation of the Parisian interest in the countryside, see N. Green, *The Spectacle of Nature: Landscape and Bourgeois Art in Nineteenth Century France* (Manchester University Press, Manchester, 1990).

29 L. Nochlin, 'The *Cribleuses de Blé:* Courbet, Millet, Breton, Kollwitz and the image of the working woman', in *Malerei und Theorie: Das Courbet-Colloquium 1979* (Stäatische Galerie im Städelschen Kunstinstitut, Frankfurt, 1980), pp. 49–74, p. 52.

30 B. Dijkstra, *Idols of Perversity: Fantasies of Feminine Evil in Fin-de-Siécle Culture* (Oxford University Press, New York, 1986). See also L. Davidoff, J. L'Esperance and H. Newby, 'Landscape with figures: home and community in English society', in *The Rights and Wrongs of Women*, eds J. Mitchell and A. Oakley (Penguin, Harmondsworth, 1976), pp. 139–75.

31 J. Maas, *Victorian Painters* (Cresset Press, London, 1969), p. 164.

32 Dijkstra, *Idols of Perversity*, p. 82.

33 Dijkstra, *Idols of Perversity*, p. 99.

34 Parker and Pollock, *Old Mistresses*, p. 13. In fact, flower painting was the one genre in which women were well-represented as artists in this period. Women were accepted as artists in this area because there was thought to be some kind of reciprocity between artist and subject; women were often described as flowers by Victorian gallants, and 'the flower analogy places both women and their work in the sphere of nature'. The encoding of nature as feminine not only gave rise to a series of visual representations of women as passive and fertile as nature itself then; it also limited the possibilities for women as artists; Parker and Pollock, *Old Mistresses*, p. 54. For an earlier account of the relation between Woman and Nature, see C. Fabricant, 'Binding and dressing Nature's loose tresses: the ideology of Augustan landscape design', *Studies in Eighteenth Century Culture*, 8 (1979), pp. 109–35.

35 Dijkstra, *Idols of Perversity*, p. 83.

36 C. M. Armstrong, 'Edgar Degas and the representation of the female body', in *The Female Body in Western Culture: Contemporary Perspectives*, ed. S. R. Suleiman (Harvard University Press, Cambridge, Mass., 1986), pp. 223–42, p. 237. For literary examples, see J. D. Porteous, 'Bodyscape: the body-landscape metaphor', *Canadian Geographer*, 30 (1986), pp. 2–12.

37 Berger, *Ways of Seeing*, pp. 45–64.
38 Berger, *Ways of Seeing*, p. 52.
39 Armstrong, 'Edgar Degas and the representation of the female body', p. 237.
40 Nochlin, *Women, Art, Power, and Other Essays*, p. 138.
41 Daniels, 'Marxism, culture and the duplicity of landscape'.
42 Berger, quoted in Daniels, 'Marxism, culture and the duplicity of landscape', pp. 203 and 215.

Agents, Subjects, and Human Geography

AGENTS, SUBJECTS AND HUMAN GEOGRAPHY

Questions of agency

The question of human agency – of the powers and capabilities of human beings – has recurred throughout the history of geography. Many modern scholars situated geography so firmly within the tradition of natural history and (later) the natural sciences that the conceptual space for human agency was severely restricted. This was most obviously so in the formulations of **environmental determinism**, which claimed that human activities are decisively controlled by the physical environment in which they take place. But even critics of this extremist position could nonetheless represent geography as a natural not a social science (as did Paul Vidal de la Blache, for example), and even those who were prepared to endorse geography's proximity to cultural history could nonetheless insist that it was a study not at all concerned with individuals (as did Carl Sauer, for example).

Humanistic geographies

These views have been challenged by various post-positivist geographies. **Humanistic geography** emerged in the 1970s as a critique of the 'geometric determinism' of **spatial science**; its proponents took exception to the physicalist metaphors and abstract spatial logics of location theory and locational analysis (for example, using the 'friction of distance' to describe and account for human interaction), and they insisted on filling these empty landscapes with the creativity and clamour of 'real' human beings. The model of the natural sciences provided an inadequate foundation for human geography, so these critics argued, because human geography – like the other humanities and social sciences – inquired into a world that was meaningful to its inhabitants *and whose meanings entered directly into the constitution of that world*. For this reason, a properly human geography would have to place at its centre the purposive actions of meaning-endowing human subjects.

Humanistic geography drew on two main intellectual pools. The first was the humanities, which historical geographer Donald Meinig (1983) characterised as 'that special body of knowledge, reflection and substance about human experience and human expression, about what it means to be a human being on this earth'. He had most prominently in mind the study of literature and history, and made much of the aesthetic and interpretative sensibilities of scholars in these disciplines. As Meinig indicated, many of the working practices of this style of humanistic geography avoided explicit formalisation and were concerned to elucidate the particularity of place through an attentiveness to human *experience* (Tuan, 1977). In Yi-Fu Tuan's more idiosyncratic essays, the argument is conducted through an imaginary (or 'ideal') conversation between the author and the reader, in which Tuan offers a series of 'ironic observations on familiar and exotic forms of geographical knowledge and experience' based on his contemplation of the world around him (Tuan, 1976; 1989). Tuan's writings are hardly typical, but he shares with many other humanistic geographers whose concerns also circle around the moral, the aesthetic, and the experiential a reticence towards the conjunction of formal theory and analytical style that characterises both spatial science and mainstream social science. Writers like these worried that such formalisation would get in the way of their attentiveness to the world, that it would both limit their openness and diminish their responsiveness to the human condition. Many humanistic geographers endorsed the instructions of the great socialist-humanist historian E. P. Thompson. He recommended that the apprentice historian should always be 'listening': opening oneself to the archive, to the world, in such a way that, through the common humanity linking people in the past to people in the present, interpretations could be discovered without the intervening (and, by implication, interfering) mediation of formal theoretical constructs (Thompson, 1978). Thompson had **structural Marxism** squarely in his sights when he wrote that, and it would be wrong to equate all formal theory with that particular project. In any event, structural Marxism had made its presence felt in the critical study of both literature and history (which is why Thompson, one of the most literary of historians, was so exercised by it), and its effects were by no means wholly negative. Indeed, it was not long before other humanistic geographers began to appreciate the theoretical force of what were in practice some of the most advanced studies in literary criticism and art history in particular and started to incorporate these and other insights into theoretically informed readings of **cultural landscapes** as **texts** and as images (see Chapter 19).

The second source for humanistic geography was provided by the social sciences, where theoretical self-consciousness was much more apparent. To be sure, many writers insisted that there was a world of difference between the high-level abstractions of spatial science and some of its successor projects – so-called '**grand theory**' – and the more modest, 'grounded', theories that they believed to be more appropriate to humanistic inquiry (see Chapter 5). Some of the studies under this sign were directed towards an explication of the meanings that are embedded within the day-to-day conduct of social life – the interpretative geography of **lifeworlds** – whereas others were directed towards an elucidation of the structures that shape human actions in particular places (see, for example, Ley, 1977; 1989). This bifurcation, if that is what it was, set the stage for an exploration of the intersections between the two. For if, as Marx had argued, 'people make history' (an affirmation of the powers and knowledges that inhere in human **agency**) but they do so 'not just as they please nor under circumstances of their own choosing' (a recognition of the conditions and constraints that are imposed by social structures), then it becomes necessary to reconcile studies of human agency with studies of social structure. This middle-ground was mapped most suggestively by those within the **structuration** school. Following the lead of the British sociologist Anthony Giddens, they argued that it was a mistake to treat agency and structure as opposing poles – most usually by thinking of 'structure' as a series of constraints enclosing a space within which (but not beyond which) 'agency' could be freely exercised – and that it made most sense to treat them instead as mutually dependent. Thus in the day-to-day conduct of social life, agents draw upon the rules and resources made available by structures, which are in turn reproduced or transformed by the actions in which they are implicated. When these ideas were translated into human geography, it was argued that the analysis of the production of *social life* ought to proceed in concert with the analysis of the production of *social space*. This was not a matter of disciplinary convention; it rested, instead, on the theoretical claim that the meshing of 'agency' and 'structure' could be disclosed most effectively through close attention to the time–space settings and sequences of human activity (Pred, 1981; 1984; Thrift, 1983).

The cultural turn and the displacement of humanism

These two streams from the humanities and the social sciences soon braided into one another, both inside human geography – where a revivified cultural geography made the interpretation of meaning a much more theoretically informed affair – and within the academy at large, where the emergence of an interdisciplinary 'cultural studies' added new dimensions to the analysis of cultural practices and social identities. This cultural turn has not been without its critics, not least those who fear that it attaches such importance to tracing the aesthetic and the textual that the more starkly material concerns of political economy have been marginalised, even erased. Whatever one makes of this charge (and we think it substantially mistaken), these developments involve a sustained critique of the core assumptions of humanism and have made possible the construction of a *post-humanist* geography.

There have been two main axes of critique. The first has challenged humanism's concept of the human subject as an autonomous and sovereign individual. This critique is sometimes called the 'decentring of the subject' and has drawn on both **structuralism** and **post-structuralism**. The arguments are as contentious as they are complicated, but in outline they direct attention towards the ways in which human subjects are constituted at the intersection of multiple, often competing, **discourses** and towards the ways in which, in consequence, they are marked by ethnicity, class, gender, and sexuality (see Chapter 7). In the course of these inquiries critics have claimed that the subject of humanism, hidden behind its screen of supposed universality, was in fact white, bourgeois, masculine, and straight: 'the' human condition was defined in a remarkably narrow way, and other worlds of human experience were marginalised (see, for example, Rose, 1993). To understand the different ways in which subjectivities are constituted and identities are constructed much greater attention is being paid to the particular spaces in which and through which these processes take place: hence many geographers have become drawn to a project of 'mapping the subject' (Pile and Thrift, 1995).

This bleeds into a second line of attack. Humanistic geography drew much of its intellectual strength from its critique of the fictional subject at the core of mainstream spatial science – 'rational economic man' – and based much of its work on the claim that the creativity and diversity of human agency could not be restricted to the working out of a narrowly *instrumental* rationality. For humanistic geographers, the purposes and meanings embedded in human action were not confined to a peculiarly economic, means–ends calculus of utility maximisation. In substituting a richer range of motivations and satisfactions, however, humanistic geography typically retained a focus on intentions; even when conceptual space was made for the *un*intended consequences of action, as it was in structuration theory, it was still tacitly assumed that the origins of human action lay in human consciousness. In short, humanistic geography, like human geography more generally, drew back from an engagement with the *unconscious*. Perhaps this reticence was the product of the discipline's earlier forays into psychology and the construction of behavioural geographies that seemed to differ little from the mechanistic models of spatial science (see Ley, 1981). In any event, this second critique of humanistic geography draws not on behavioural psychology but on psychoanalytic theory and seeks to illuminate the intricate ways in which desire animates human action and interaction. This belated interest by geographers in the writings of Freud and especially Lacan has overlapped with the rise of a feminist geography and has been given further impetus through engagements with **post-colonial** theory. Hence there have been attempts at 'mapping desire' – tracing the connections between bodies, sexualities, and identities in space (Bell and Valentine, 1995) – and at disclosing the ways in which geographies of colonial dispossession and imperial domination are marked by an erotics of power and destabilised by ambiguity, fear, and paranoia. Running throughout this developing stream of work is a deep concern with the ways in which **subjectivity** is constituted through the production of **spatiality** (Kirby, 1996; Rose, 1995; Pile 1996).

The politics of humanistic and post-humanistic geography

Humanistic geography was never a purely academic project. It had its conservative protagonists, who often espoused a more or less elitist view of culture, but it was also characterised by a distinctively radical spirit. This progressive temper informed an interest in the daily lives of ordinary people, the production of ordinary landscapes, and geographies of popular culture, and through its emphasis on **reflexivity** humanistic geography elaborated both an ethic of social concern and a politics of collective human agency. For these reasons it is scarcely surprising that so many geographers became interested in exploring the relations between humanistic geography and **historical materialism** (Kobayashi and Mackenzie, 1989). As we have indicated, much of this work was, frankly, ethnocentric: although humanistic geography took as one of its central interpretative concerns the constitution of social meaning, and although it often learned much from cultural anthropology (particularly about **ethnography**), most of its studies remained tied resolutely to the West and were contained by its horizons of meaning. Equally, the version of historical materialism that appealed to most humanistic geographers was a distinctively **Western Marxism**, which offered its own ideology of universalism (see Chapter 2): although it was much more sensitive to questions of culture than was the political economy of classical Marxism, it continued to focus almost unwaveringly on the politics of class.

The new avenues that we have identified have been directed towards forms of politics that are by no means limited to class, however, most particularly towards the connections between place and **identity politics** (Keith and Pile, 1993) and between place and **radical democracy** (Massey, 1995). In doing so they have set in motion explorations of the complex connections between the politics of subjectivity and the politics of spatiality, and they promise a much more radical engagement with the problematic of difference (see our introduction to section Eight). It is too early to determine how this will all turn out, but at least this much is clear: a post-humanist geography, far from being brutally indifferent to the lives of ordinary men and women, as some of its humanist critics fear, represents a serious attempt to render more adequately the complexity of their meanings and actions and hence to contribute to a more inclusive political practice.

Geographies of human agency and human subjectivity

Our first selection is an important essay by Stephen Daniels. He writes as a cultural-historical geographer and hence shares many of Meinig's sensitivities about the power of the humanities and the importance of historicity to human geography. But he radicalises Meinig's critique by claiming that humanistic geographers too often have only a superficial understanding of these intellectual filiations. He shares their objections to the positivist philosophy that underwrites spatial science, but he argues that too much attention has been paid to philosophies such as phenomenology and existentialism that needlessly limit the *analytical* power of humanistic inquiry. These philosophies have typically licensed what Daniels sees as a conservative, even reactionary, geography that purports to recover an 'authentic' relation between people and places through a more or less intuitive understanding of an inner world of subjectivity. He argues that these humanistic geographies obscure the changing relations of power that are involved in the historical construction of 'authenticity'; that their attempts to derive 'transcendent truths' from readings of imaginative literature founder on an inattention to context and convention; and that their determined focus on the cultural has involved a dangerous distancing from the economic. Daniels's counter-argument turns on the critical force of narrative as a way of combining the analytical and the evaluative; narrative has the ability both to 'conserve theoretical categories in solution' and, as it explicates 'meaning through context', to capture the fluid movements between agency and structure.

It is those movements that concern Nigel Thrift in the second essay. This is part of a project to establish a fully *contextual* social theory, by which Thrift means one that is rooted in the continuous flow of social action in time and space. Consistent with this aim, he begins by locating the twin problematics of 'human agency' and 'social structure' in the ways in which European social theory sought to come to terms with – and to find the terms for – the changed circumstances of post-Enlightenment capitalist society. Thus, in outline, a focus on the ordering of populations and the anatomy of society sustained 'structural' modes of inquiry, whereas a focus on individuality and identification sustained inquiries into 'agency'. These were rarely critical interventions, however, and much mainstream social theory was narrowly instrumental. Thrift's purpose is different: he wants to reconcile these two problematics in a *non*-functionalist manner. In other words, he seeks to show the mutual dependence of structure and agency in such a way that they don't endlessly reproduce each other: that 'newness' is able to enter the world. If this is to be a genuinely critical intervention, however, Thrift believes that the insights of what he calls the 'structurationist school' need to be wired to a revitalised historical materialism. This would provide a more rigorous sense of contradiction and conflict than the structurationists offer, and it would allow Marxism to deal more effectively with the constitution of society and, in particular, the constitution of subjectivities. Indeed, Thrift (1995) has remarked that it was the attempt at 'mapping the subject' that was, for him, the crucial message of the essay.

Stephen Pile's essay signals the turn towards psychoanalytic theory by its focus on the ways in which 'the self' has been conceptualised in post-positivist human geography. Tracing through humanistic geography's encounters with phenomenology, symbolic interactionism, and structuration theory, Pile concludes that – with the partial exception of Thrift's essay included here – they have all failed to problematise the human subject to which they all supposedly pay such close attention. In Pile's view, it would be more productive to work with psychoanalytic theory to 'reinterpret' agency and structure in such a way one 'steps beyond' their analysis by 'reveal[ing] the intricate inter-relationships between the personal and the social' (page 422). Psychoanalytic theory is hardly a unified field, of course, and Pile does not present it as such: rather, he offers a series of suggestions for further reflection and exploration based, for the most part, on Lacan. His suggestions are tantalising, and even in this abbreviated form they reveal the spatialised vocabulary with which Lacan worked (cf. our introduction to section Five) and indicate some of the complexities that need to be incorporated into the fabrication and mapping of 'peopled places and placed people.'

References

Bell, D. and Valentine, G. 1995: *Mapping desire: geographies of sexualities.* London: Routledge.

Keith, M. and Pile, S. (eds) 1993: *Place and the politics of identity.* London: Routledge.

Kirby, K. 1996: *Indifferent boundaries: spatial concepts of human subjectivity.* New York: Guilford.

Kobayashi, A. and Mackenzie, S. (eds) 1989: *Remaking human geography.* London: Unwin Hyman.

Ley, D. 1977: Social geography and the taken-for-granted world. *Transactions of the Institute of British Geographers,* **2**; 498–512.

Ley, D. 1981: Behavioral geography and the philosophies of meaning. In Cox, K. R. and Golledge, R. (eds), *Behavioral problems in geography revisited.* London: Methuen, 209–30.

Ley, D. 1989: Modernism, post-modernism and the struggle for place. In Agnew, J. and Duncan, J. (eds), *The power of place: bringing together geographical and sociological imaginations.* Boston, MA: Unwin Hyman, 44–65.

Massey, D. 1995: Thinking radical democracy spatially. *Environment and Planning D: Society and Space,* **13**; 283–8.

Meinig, D. 1983: Geography as an art. *Transactions of the Institute of British Geographers,* **8**; 314–28.

Pile, S. 1996: *The body and the city: Psychoanalysis, space and subjectivity.* London: Routledge.

Pile, S. and Thrift, N. (eds) 1995: *Mapping the subject: Geographies of cultural transformation.* London: Routledge.

Pred, A. 1981: Social reproduction and the time-geography of everyday life. *Geografiska Annaler,* **63B**; 5–22.

Pred, A. 1984: Place as historically-contingent process: structuration theory and the time-geography of becoming places. *Annals of the Association of American Geographers,* **74**; 279–97.

Rose, G. 1993: No place for women. In *Feminism and geography: The limits of geographical knowledge.* Cambridge: Polity Press; Minneapolis, MN: University of Minnesota Press, 41–61.

Rose, G. 1995: Distance, surface, elsewhere: a feminist critique of the space of phallocentric self/knowledge. *Environment and Planning D: Society and Space,* **13**; 761–81.

Thompson, E. P. 1978: *The poverty of theory and other essays*. London: Merlin.
Thrift, N. 1983: On the determination of social action in space and time. *Environment and Planning D: Society and Space*, **1**; 23–57.
Thrift, N. 1995: Author's response. *Progress in Human Geography*, **19**; 528–30.
Tuan, Y-F. 1976: Humanistic geography. *Annals of the Association of American Geographers*, **66**; 266–76.
Tuan, Y-F. 1977: *Space and place: The perspective of experience*. London: Arnold.
Tuan, Y-F. 1989: *Morality and imagination: Paradoxes of progress*. Madison, WI: University of Wisconsin Press.

Suggested reading

For a philosophical discussion of humanism and its critique see Kate Soper, *Humanism and anti-humanism* (London, Hutchinson, 1983). The range and vitality of humanistic geography is shown by the essays contained in David Ley and Marwyn Samuels (eds), *Humanistic geography: Prospects and problems* (London, Croom Helm, 1978) and in Kobayashi and Mackenzie (1989). There is also an effective secondary survey of humanistic geography in chapter 3 of Paul Cloke, Chris Philo, and David Sadler, *Approaching human geography: An introduction to contemporary theoretical debates* (London: Paul Chapman; New York: Guilford Press, 1991).

For the ways in which human geography drew on social theory to illuminate the intersections between 'agency' and 'structure', see chapter 4 of Cloke *et al.* (1991). Much of this has centred on structuration theory and its (constructive) critique of historical materialism, which is best approached through volume 1 of Anthony Giddens, *A contemporary critique of historical materialism* (London, Macmillan, 1981) and his *The constitution of society* (Cambridge, Polity Press, 1984). Critical assessments include David Held and John Thompson (eds), *Social theory of modern societies: Anthony Giddens and his critics* (Cambridge, Cambridge University Press, 1989) and Christopher Bryant and David Jary (eds), *Giddens' theory of structuration: A critical appreciation* (London, Routledge, 1990). Both include assessments of Giddens's concepts of space and spatiality.

Two important collections which have much to say about geographies of identity and subjectivity are Keith and Pile (1993) and Pile and Thrift (1995), and there is a particularly lucid and provocative discussion in Kirby (1996). For introductions to *Lacan*, see Malcolm Bowie, *Lacan* (London, Fontana, 1991) and Elizabeth Grosz, *Jacques Lacan, A feminist introduction* (London, Routledge, 1990), and for a brilliantly creative use of psychoanalytic theory in cultural geography see Pile (1996).

21 Stephen Daniels
Arguments For A Humanistic Geography

Reprinted in full from: R. J. Johnston (ed.), *The future of geography*. London: Methuen (1985)

Since its first use by Yi-Fu Tuan (1976) the term 'humanistic geography' has become a keyword in geographic thought. Perhaps inevitably its meaning has dispersed with its use. For Tuan humanistic geography was a perspective that disclosed the complexity and ambiguity of relations between people and place, qualities eclipsed by the positivist perspective of much human geography. In Tuan's writings a humanist perspective is not a methodology nor even a philosophy but a pattern of ironic observations on familiar and exotic forms of geographic knowledge and experience. Subsequently humanistic geography has been formulated as a substantial programme with distinctive questions and ways of answering them. The term has also been used in weaker senses, sometimes merely to label any human geography which emphasizes human agency and awareness. In this essay I will first review the main arguments of self-consciously humanistic geographers from a perspective that emphasizes the historical dimensions of geographic knowledge. I will then suggest a prospect for humanist geography that, in terms of ideas and methods, is grounded in historical understanding.

Humanism and positivism

What unites humanistic geographers is their disenchantment with the writings of positivist human geographers. Criticisms that positivist assumptions and procedures do not adequately explain human issues are intellectual and moral, sometimes inextricably so. For David Ley (1980) spatial analysis is so abstracted from the world it claims to explain as to lose sight entirely of its subject, a 'geography without man' that is at once intellectually deficient and morally blind. All reasoning implies abstraction of some kind and degree. Humanist geographers propose that reasoning in humanistic geography should conserve contact with the world of everyday experience and recognize, if not celebrate, the human potential for creativity. The scope of this brief is broad in terms of both subject matter and approach. It includes the populist essays on social geography in *Humanist Geography: Prospects and Problems* (Ley and Samuels 1978) and the more elitist essays on novelists and poets in *Humanist Geography and Literature* (Pocock 1981b).

It is customary to describe the positivist reformulation of human geography as a 'revolution'; Ley and Samuels (1978) describe their humanistic reformulation as a 'reawakening', a sign of their moral evangelism and their conservative claim to restore human geography to its rightful role, 'the study of the earth as the home of man' (Ley 1980, 3). Despite or because of their disaffiliation with much post-war geography, positivist revolutionaries were intent on constructing an ancestry made up of scholars from many sciences, to legitimate their undertaking (Taylor 1976, 36). Humanist geographers have been no less intent on creating a tradition to disinherit positivist geography and legitimate theirs. Because of their interests in landscape and culture eminent geographers like Paul Vidal de la Blache and Carl Sauer have been retrospectively converted to humanistic geography. If alive, they would probably protest that their scientific persuasions had been overlooked. Humanist social geographers appeal to a non-positivist, sometimes explicitly humanist, tradition in sociology. They have in the process disentombed the non-positivist persuasions of scholars like Robert Park and Alfred Weber and reclaimed them from the ancestry of spatial analysis (Smith 1981). Tracing a tradition beyond modern academic disciplines is both easier and less meaningful because the term humanist has been used to describe almost any thinking which addresses human issues undogmatically (Ley and Samuels 1978, 4–10).

Moral criticisms of positivist geography are usually situated in and draw authority from a more general questioning of ways of reasoning and acting in modern western culture, particularly the powerful combination of science, technology and capitalism which conventionally defines rationality. Relph (1976) sees logical connections between various narrowly conceived, efficient and reiterative constructions; intellectual constructions like Central Place Theory, material constructions like suburban housing and perceptual constructions like tourist guides. Relph envisages more 'authentic', more wholesomely human ways of understanding, creating and perceiving the world and finds examples in the lives of various pre- or non-industrial peoples. Relph is not alone in his conspiratorial view of modern western culture nor in his romanticism of other cultures but it is worth examining his argument more closely because it reveals the historical myopia that afflicts much humanist geography. As an example of an authentic place, an expression of an 'I–thou relationship between man and god', Relph (1976, 75) includes a photograph of Tintern Abbey, or more precisely its ruins. The problem here is the historical complexity of both the subject and its representation. To be sure, Cistercians were pious but they were also unromantically mindful of their worldly interests as the peasants who were cleared from their estates and the monarch who ruined their abbeys realized. Relph (1976, 67) maintains his photographs are intended to 'demonstrate' not just 'illustrate' his argument but his pictorial interpretation of Tintern Abbey actually conflicts with his written interpretation. The view of Tintern Abbey Relph photographed has a history no less than the abbey itself. The view was first constructed by eighteenth-century tourists and artists according to a conventional picturesque formula that still informs tourist perceptions of England as a series of beauty spots and the photographs they take of them. Relph thus reproduces a tourist's way of seeing that he is at pains a few pages later in his writing to deplore: 'this is inauthenticity at its most explicit; the guided tour to see works of art and architecture that someone else has decided are worth seeing . . . such guides stress the picturesque and the monumental' (Relph 1976, 85).

Concepts

Much humanistic geography concentrates on developing a set of concepts to articulate issues of value and meaning ignored or misconstrued within positivist human geography. These concepts have, or are given, a social, sometimes psychological reference as well as a spatial one. Concepts like 'landscape' and 'region' have been retrieved from the periphery or from beyond the pale of positivist geography. In the process they have been infused with a greater sense of perception and experience than during the heyday of regional and landscape geography (Tuan 1979). Humanists reject the reduction of space and place to geometrical concepts of surface and point; humanist conceptions of space and place are thick with human meanings and values (Entrikin 1976, 623–5).

Place is a key concept in humanistic geography; much humanistic writing is devoted to illustrating and clarifying it. From a humanistic perspective the meaning of a place is inseparable from the consciousness of those who inhabit it. The scope of place as a concept varies according to the extension of the thoughts, feelings and experiences that make up the consciousness of inhabitants. For any one person a favourite armchair may be a place, also a room, street, city, nation, perhaps the whole earth (Relph 1976, 141; Tuan 1977, 149). From a humanistic perspective place is not so much a location as a setting, less a thing than a relationship.

Most humanistic geographers acknowledge the implication of process in place – a sense of place is something that develops through time – but few explore or explain it (Cosgrove 1978, 70). This is not just because some see place as 'an essentially static concept' (Tuan 1977, 179) but because the form of writing they use to explicate place cannot adequately account for issues of process and development. Much humanistic commentary on place is cast in a categorical form and conducted in the present tense. Categories and subcategories of place and place-consciousness are usually illustrated by a variety of examples from various historical periods but it is seldom clear how particular meanings of place develop in particular settings. Meanings are noted, seldom narrated.

The concept of place itself has a complex history. Like 'landscape', 'nature' and 'community' it resonates with ideological implications but these are scarcely explored. Place is a positive word. To be displaced, out of place or placeless is a negative, even 'unnatural', condition. Common to many uses of the word place, including those by humanistic geographers, are connotations of stability, belonging and propriety. What is seldom noted is how these connotations are informed by a strict, even oppressive, sense of social order and control, explicit in the expectation that the poor should know their place. Place is a predominantly conservative notion, one that can perhaps help us understand, even resist, destructive changes but one that threatens to oppose change itself. As Buttimer (1981) recognizes, intellectual appreciations of a sense of place among the underprivileged can amount to a tacit condoning of conditions like poverty and injustice.

In developing its concepts humanistic geography assumes its philosophical tone. This is less an echo of the analytical tradition that prevails in the philosophy departments of Britain and North America than the more speculative philosophical tradition of continental Europe. The style and subject matter of existentialism and phenomenology have proved alluring to humanistic geographers. Analytical philosophers are more reticent about dramatic thoughts or experiences and more modest about what philosophy can do, often arguing it can at best reduce obscurity that is largely linguistic in origin and can do so best by adopting a restrained style. The claims and language of existentialists and phenomenologists are characteristically less restrained. This is not to say their writings are irrational but the humanistic geographers who invoke their authority have done remarkably little reasoning themselves. Philosophy in humanistic geography is less reasoning than reporting. It often amounts to quoting passages with a geographical flavour from this or that existentialist or phenomenologist or just classifying incidents in life or literature in terms of technical concepts like 'lifeworld' or 'insideness' (Buttimer 1976; Samuels 1978; Seamon 1981a). Such concepts are occasionally modified in translation to English and to geography but are seldom themselves the subject of philosophical scrutiny. In a study of 'insideness' and 'outsideness' literature is seen as a 'testing ground to confirm and amplify existing phenomenological claims' (Seamon 1981a, 85). If we accept that reason should be at least potentially critical this seems not just a rejection of positivist criteria, of hypothesis testing, but a rejection of reason itself.

Methods

There is less reticence about how humanistic geography should be conceived than about how it should be done. Ley and Samuels (1978, 121) suggest that conceptual clarification is itself a method, a way of achieving a clearer or deeper understanding of an issue. There is some agreement that humanistic methods cannot be defined as a set of formal procedures or techniques. One informal procedure appealed to is phenomenological reduction, the suspending of conventional ways of knowing, such as ordinary language or academic expertise, to achieve a primordial intuition of the world (Seamon 1981a). It is not hard to see why this procedure has not been satisfactorily used or explained. It rests on the fallacy that a convention like language is an impediment to understanding. Some languages or terminologies may occlude certain qualities. For example, an urban geography constructed around concepts like 'housing' and 'land use' may well not capture the values and experiences of people living in particular 'homes' and 'neighbourhoods' (Buttimer 1981). But it is only by failing to discriminate some things that a language can discriminate at all. We may argue about the kind of understanding particular languages offer but language itself is a convention that makes understanding possible. Seamon (1981b) speculates that phenomenological reduction might disclose essential 'time–space routines' to neighbourly places that underlie or transcend the particularities of history, culture and personality. His calling such essences 'place ballets' shows that if they do exist language, or some other form of communication, is essential to their recognition. In Seamon's account such essences could only be identified by someone who knew what a 'ballet' was.

Most humanistic geographers forgo a phenomenological search for essences for more existential accounts of experiences of particular people in particular places. This they argue implies reducing the detachment and authority of the geographer. 'An existentially aware geographer', announces Buttimer (1974, 24), 'is less interested in establishing intellectual control over man through preconceived analytical models than he is in encountering people and situations in an open, intersubjective manner'. Methods of participant observation developed in sociology and anthropology provide an example for humanistic social geographers in how to account for how those under study understand their world (Smith 1981). Some, less social, humanistic geographers (Rowles 1978) notionally blur the distinction between investigator and investigated, a distinction that is inevitably sharpened when an investigator writes up his research for academic purposes or even conceives of what he or she does as research in the first place.

The incapacity of standard statistical techniques to describe the fluency of human life and in some cases their capacity to distort it out of all recognition has provoked a cautious if not hostile attitude among humanists to quantification. While some settle for a conversational language to capture the nuances of everyday experience others adopt a more self-consciously literary style to express more profound experiences or to explicate the profound in the apparently everyday. Olsson (1978, 110) sees quantitative geography and humanistic geography as essentially opposed, the former aspiring to 'formal reasoning', the latter to 'creative writing'. Such a creative humanistic geography should, Olsson maintains, celebrate ambiguity and allusiveness. What is striking about humanistic aspirations to creativity is the equation of creativity with experiment. In his essay in *Humanistic Geography and Literature* Olsson (1981, 126) asks,

> What does it *now* mean to yearn for home? Jokingly serious, seriously joking?
> Joker trumps the trumps! Follow suit!
> To yearn for home is to experience double bind. It is to be torn between irreconcilable identities, sometimes enjoying the ~~illusory~~ freedom of singing with the wind, sometimes missing the ~~real~~ subjugation of being tied to the ground.

It is not difficult to see what Olsson is getting at here but hard not to suggest that it could have been expressed more straightforwardly, if less playfully, without the various conceits. You don't have to be a linguistic puritan to conclude that the literary styles of humanistic geographers are no less gratuitous than the quantitative or algebraic flourishes of some positivist geographers (Billinge 1983).

Humanistic geography and the humanities

Some of the more puzzling humanistic writings are those about literature. This is not so much because of their style but because few of their authors have been reading literature at all. In discussions of D. H. Lawrence (Cook 1981),

George Eliot (Middleton 1981), Doris Lessing (Seamon 1981a), Mary Webb (Paterson and Paterson 1981) and various English novelists of 'place' (Pocock 1981a) there is little or no recognition of the literary conventions these novelists employ, for example their methods of narration or description. It is no defence to say (Pocock 1981b, 9–10) that these are matters for the literary critic not the geographer. The issue of place and consciousness of place cannot be isolated from the issue of how these are, can or perhaps cannot be represented (Barrell 1983). When analysis of how authors communicate is evaded it is easy as well as convenient to substitute phrases like 'heightened perception' and 'mysterious intuition' (Pocock 1981b, 11–12). The mystery here is of the geographer's making. Isolated from the issue of literary convention the arguments of humanistic geographers, and through them the writings of novelists, slide into soft focus. For Pocock (1981b, 15) 'the starting point' of a geographical reading of literature is not language but 'the artist's perceptive insight'; 'literature simply is perception', and 'in the last analysis . . . it is perhaps not so much what the poet, novelist or playwright says, but what he *does* to us that matters'. For a humanist to abstract the perceptive insight of an artist from the language in which it is embodied is paradoxically to diminish an artist's humanity. From this viewpoint authors appear not to work, as they do and very humanly, with the possibilities and restraints of artistic form and language, or, in a larger context, with those of the society in which they work. They become instead vehicles for transcendent truths. Literature becomes 'literary revelation' (Pocock 1981b, 9).

While a number of human geographers have used literary evidence few have used what seems a more obvious source for a highly visual discipline: the visual arts. Rees (1973) has noted the potential of landscape painting for studying questions like national consciousness and attitudes to nature. In the quest for a tradition in which the sciences and humanities were not so strictly demarcated he has suggested parallels between the ways geographers and painters know and represent the world. It is useful to regard the visual arts as ways of knowing as well as ways of seeing. But the potential of paintings can only be realized if geographers analyse more carefully, as they do in map interpretation, the form this knowledge takes; how paintings represent the world through various conventions such as medium, perspective, framing, symbolism. We might then appreciate for example that many eighteenth-century English landscape paintings are not so much tender-hearted renderings of an open countryside as tough-minded assertions of landed property. Oil paint and perspective combine to assemble a visible world for private consumption in pictures that were originally hung in the country houses of their owners, sometimes next to maps of their property. This is not to reduce their meaning to ideology or to deny their beauty, only to emphasize that art is not invariably the humane thing some humanistic geographers assume it to be (Daniels 1982b). Relations between beauty and morality are complex, sometimes contradictory (Daniels 1982a; Cosgrove and Thornes 1981).

The naive approach of most humanistic geographers to the arts, especially their neglect of artistic form, reveals their poor working knowledge of the humanities, especially literary and art criticism and history. Scholars in the humanities are presently looking to the social sciences for a more adequate account of meaning and value, one that places the work of an author or artist in a wider social and economic context. Those humanistic geographers who turn their backs on social science yet are blind to the humanities may, or may not, sense their isolation.

Objectivity and subjectivity

A dualism that has disabled much geographic thought is that between objective and subjective knowledge. In their readiness to renounce positivist notions of objectivity such as the detachment and neutrality of the observer, some humanists have settled for, even welcomed, positivist notions of subjectivity, especially the reduction of values and perceptions to private feelings and experiences. This limited notion of subjectivity has influenced conceptions of subject matter and ways of studying it. A discussion of 'experiential field work' in the study of the perceptions of old people concludes 'if I am able to come to know Marie [an old person] well, I can gain a special sensitivity to her geographical experience. Together we may develop a shared awareness' (Rowles 1978, 175). The point of such studies as this seems less to make verifiable claims about values than to report feelings and experiences. When humanistic geographers do on occasion address the critical issue of how the claims of a humanistic study to be true are to be assessed their accounts are scarcely persuasive. For Pocock (1983, 356) verification in humanistic geography is a form of intuition: 'Does the description ring true with my own and others' experience?' By ruling out the pre-conditions of rational argument Pocock logically cannot demonstrate the truth of his own assertion. Argument is here reduced to swopping experiences and verification to matching them.

The reduction of subjectivity to feelings which are, for better or worse, beyond rational scrutiny is a general historical development. It dates from the nineteenth century (Williams 1976, 259–64) and a bourgeois conception of life in which a subjective, private world of intimate feelings, symbolized by the suburban home, was separated from and compensated for an objective, impersonal world of hard facts symbolized by the workplace. Despite, or because of, their disenchantment with commercial values, this distinction has been reproduced by humanistic geographers. While sensitive to the more personal qualities of relations between people and between people and place they have proved reluctant to address the economic dimension of these relations (Sayer 1979, 33). There is, to my knowledge, no humanistic geography of work relations or the workplace and little emphasis on the economic implications of the home or private life. This is curious because to neglect the economic is to neglect a dimension of life that is *experienced* forcefully, especially when money or resources are scarce. It is as if some humanistic geographers are not so much criticizing positivist geography as compensating for it, accepting a division of labour which delegates some to go out and analyse the objective world and others stay at home and cultivate the subjective.

Humanistic geographers with a stronger sense of the social dimension of human life and knowledge (Gibson 1978; Ley 1978) have recognized that values are not a matter of inner feelings nor are all facts as apparently clean as facts like the population or latitude of a town. It is a fact that every day in English towns people pay for goods. This is a more informative and explanatory description of these actions than stating that 'every day people exchange pieces of paper or metal for various items' because its meaning derives from and implies certain commercial values that underpin modern English urban society. There are other possible descriptions of those actions, or a sub-set of them, for example 'keeping up with the Joneses' or 'sustaining the capitalist mode of production' and each of these implies a particular system of values, in the latter case a system perhaps not shared by most of the participants. Such descriptions are more controversial. These controversies, which abound in social science, are not settled by intuition but by making rational inferences from factual evidence. It is the same with disputes explicitly about values such as justice. To move from facts to values is not necessarily to slip into solipsism. On this account objective knowledge is not value-free, but a blend of evaluative and non-evaluative description (Graham 1982).

Historical understanding

How may objective knowledge which is evaluative as well as explanatory be acquired and expressed? Cole Harris (1978) describes a procedure that characterizes study in history, a discipline which has for long, and unselfconsciously, dealt with issues of value and meaning that humanistic geographers, generally with little grounding in history, have rediscovered. This historical procedure is 'hardly an explicit methodology, it is better thought of as a habit of mind'. The 'historical mind', as Harris calls it, is 'contextual not law finding', more a matter of incremental learning and judgement than technique. The product of immersion in the sources for a period, it 'is open, eclectic and curiously undefinable'. Harris suspects it is not solely relevant to the study of the past: 'as a form of explanation it is part of everyday experience'. In the remainder of this essay I want to clarify and amplify this procedure, or at least what I see as its main form. In the process I hope to show that it is as much a method as a mentality. It is a method that, despite the censure of some positivists, is still used in geography and deserves affirming: narrative.

Consider this extract from a well-known text on the human geography of North America (Paterson 1979, 175).

> The semi-feudal conditions of land tenure in some of the early settlements, and the pressure on the land of an increasing immigrant population soon produced a drift westward toward the mountains. Here in the rougher terrain of the upper Piedmont and later of the Appalachian valleys, independent farmers carved out their holdings, accepting the handicaps of infertility and remoteness in exchange for liberty of action. To the eighteenth-century farmer the exchange seemed a reasonable one; his twentieth century descendant, occupying the same hill farm, suffers the handicaps without the same compensation.

This is the kind of account which positivists regarded at best as a loose, weakly explanatory combination of factual description and implicit law-like generalization. In positivist terms it might be reformulated as an explanation sketch of pioneer settlement, the significance of some facts about eighteenth-century pioneers indicated by the law-like first sentence about migration. Such a sketch would require filling out by specifying the necessary and sufficient conditions of the law, or at least their statistical probability, and, in greater and stricter detail, the facts which that law explained (Harvey 1969, 421–2). Such arguments have persuaded some human geographers to strive for a positive form, or at least style, of temporal explanation. It is not unusual for developments to be described in the language of systems theory or the symbolism of flow diagrams, as processes with discrete stages connected by lines of causal linkage. If we set aside positivist criteria we can see a different explanatory form to the above passage. Without pretending that it is full or complete we may appreciate that its explanatory force is stronger and more rigorous than positivist criteria suggest.

The explanation is primarily contextual not causal. The point of the passage is not to demonstrate the necessity of pioneer settlement but to establish what westward pioneering meant in eighteenth-century North America. This is done by situating it in an evaluative context, that of the liberty of the pioneers. This context is constructed of a set of overlapping descriptions. Some are more or less straight ('rough terrain', 'infertility'); some more emotive ('suffers the handicaps'); some more technical ('the semi-feudal conditions of land tenure'). One description of pioneering suggestive of its hardship, 'carving out holdings', is directly overlapped by another, 'accepting the handicaps of infertility and remoteness in exchange for liberty of action', which enlarges its meaning by showing how the pioneers themselves conceived their action. The passage shows how narratives can negotiate the perspective of those who lived through the past and that of the modern observer. The advantages of hindsight are exploited. The key concept 'semi-feudal' was unavailable to an eighteenth-century pioneer, or at least not understood in the way it is by modern readers. The final description of his twentieth-century descendant, trapped in a region that once promised opportunity, amplifies through irony the values of the 'drift westward towards the mountains'. The opening causal sentence is a key part of the explanation but subordinate to an overall contexual account that incorporates descriptions of varying evaluative force. The account is both particular and thematic. It is also objective. It is open to amplification or to question by further empirical research which might indicate a more appropriate context for events, one perhaps that establishes a less ideological, more economic meaning.

If I have burdened this brief extract with interpretation it is to show why narrative might appeal to humanistic geographers. It explicates meaning through context and in the process mediates, or should mediate, the views of the 'outsider', the narrator, and those of the 'insider', the participants in the history the narrator constructs. The language of even the most scholarly narratives is mostly conversational, suggesting that it is less abstracted than many academic forms of explanation from the meaning of everyday life.

Narratives conserve a more seamless sense of the fluency of relations between people and between people and place than do systems or structural modes of temporal explanation. But they do not reproduce the flow of lived experience. Narratives are not lived but told. Narrative is an essentially retrospective mode of understanding but it is not confined to studying the dead nor is it necessarily scholarly. For a variety of continuing processes, for example urbanization, farming, race relations, a life, a career, it can explicate what seemed to participants and observers at the time of occurence insignificant or meaningless. A historically minded scholar might make sense of his or her own life in a way similar to the way they explain the lives of those they study. In other words historical understanding is not essentially an empirical notion, one restricted to those pursuits we call 'history', but a philosopical one.

In narrative the essential relationship between events is not antecedent to consequent, but part to whole. Having followed, and thus understood, a narrative, the element of time diminishes as a complex array of happenings – sudden events, steady pressures, long campaigns, impulsive actions – are seen together as parts of a single, if often complex, configuration: 'a landscape of events' (Mink 1969–70, 549). Humanistic geography is sometimes characterized as a broader or clearer perception of the world. An appreciation of narrative form takes some of the mystery out of this description and gives it some methodological muscle.

Although philosophical in tone humanistic geography is rarely if ever theoretical, indeed is often anti-theoretical. This is understandable given the tendency of both positivist and radical geographers to reify theoretical concepts and over-privilege their explanatory status. But to be on principle anti-theoretical is to relinquish the potential of explanatory power. Narrative can conserve theoretical concepts as it were in solution, preventing them precipitating into discrete structural categories. It is perhaps not surprising that Marxist narratives, informed, often implicitly, by a historical theory, provide the clearest examples. An example of a Marxist geographical narrative is Harvey's (1979) study of the Basilica of Sacre Coeur in Paris.

Harvey narrates the conception and building of the Basilica (a process that was neither linear nor smooth) at varying levels, some theoretical, some concrete. He recounts episodes of often violent conflicts and experiences in Paris during and after the Franco–Prussian War, many on the hill of Montmartre. He then recapitulates these episodes in terms of broader conceptual relations: town and country, church and state, bourgeoisie and proletariat, economy and society. As I read it the point of his narrative is not to test or exemplify these concepts in an ahistorical positivist manner. Nor is it merely to chronicle certain events in Paris in a historically empty antiquarian manner. Rather it is to explicate through descriptions of varying scope, from the shooting of Communards to the development of a religious cult, the meaning of a landmark that was intended to mythologize the history it commemorates and which remains one of the most striking and poorly understood in Paris.

> The building hides its secrets in sepulchral silence. Only the living cognizant of this history . . . can truly disinter the mysteries that lie entombed there and thereby rescue that rich experience from the deathly silence of the tomb and transform it into the noisy beginning of the cradle (Harvey 1979, 381).

It is not necessary to make revolutionary inferences from historical analysis to see how narrating landscapes can revitalize them. Narrative as a form of historical interrogation can recover the conflicts and hardships which so often constitute the making of landscapes and which the conventional idea of landscape, with its implications of harmony and peace, seems to deny.

My emphasis on the formal characteristics of narrative should not give the impression that narrative is a formula which can be learned independently of particular subject matter and applied to it. Narratives are explicated from the evidence of specific times and places. This process, a dialectic of discovery and construction, can involve complex and delicate adjudications: between the perspectives of participants and observers, between particular incidents and general themes, between competing theories. Interpretation and judgement are ingredients to narrative, not operations performed before or after the evidence has been collected and processed. Pointing out the formal characteristics of narrative does not make the writing of good narratives any easier than it ever was. But it may help show why the effort is worthwhile.

Conclusion

If humanistic geography is to be more than a criticism of positivist geography it needs a more thoroughly reasoned philosophical base, a closer understanding of the conventions through which human meanings are expressed, a more adequate account of what humanistic methods are or might be, and above all a greater historical understanding. Humanists might extend their claims into areas such as economic geography presently dominated by positivists and radicals. In the process they might be forced to modify their claims through negotiation with other geographical perspectives. In consequence humanistic geography might lose a sense of separate identity. If human geography as a whole assimilated its more persuasive arguments there might be no need for a group of human geographers to proclaim their study emphatically human. Many humanistic geographers would surely welcome this.

References

Barrell, J. (1983) Review of Pocock, D. C. D. (1981) *Humanistic Geography and Literature*, in *Journal of Historical Geography*, 9 (1), 95–7.

Billinge, M. (1983) 'The mandarin dialect: an essay on style in contemporary geographic writing', *Transactions, Institute of British Geographers*, NS 8, 400–20.

Buttimer, A. (1974) *Values in Geography*, Washington, DC: Association of American Geographers Research Paper No. 24.

Buttimer, A. (1976) 'Grasping the dynamism of the lifeworld', *Annals, Association of American Geographers*, 66, 277–92.

Buttimer, A. (1981) 'Home, reach and the sense of place', in A. Buttimer and D.

Seamon (eds), *The Human Experience of Space and Place*, London, Croom Helm, 166–87.

Cook, I. G. (1981) 'Consciousness and the novel: fact or fiction in the works of D. H. Lawrence', in D. C. D. Pocock (ed.), *Humanistic Geography and Literature*, London, Croom Helm, 66–84.

Cosgrove, D. (1978) 'Place, landscape and the dialectics of cultural geography', *Canadian Geographer*, 22, 66–71.

Cosgrove, D. and Thornes, J. E. (1981) 'Of truth and clouds: John Ruskin and the moral order in landscape', in D. C. D. Pocock (ed.), *Humanistic Geography and Literature*, London, Croom Helm, 20–46.

Daniels, S. (1982a) 'Humphry Repton and the morality of landscape', in J. R. Gold and J. Burgess (eds), *Valued Environments*, London, Allen and Unwin, 124–44.

Daniels, S. (1982b) 'Ideology and English landscape art', in D. E. Cosgrove (ed.), *Geography and the Humanities*, Loughborough University, Department of Geography Occasional Paper No. 5, 6–13.

Entrikin, J. N. (1976) 'Contemporary humanism in geography', *Annals, Association of American Geographers*, 66, 615–32.

Gibson, E. (1978) 'Understanding the subjective meaning of places', in D. Ley and M. Samuels (eds), *Humanistic Geography: Prospects and Problems*, London, Croom Helm, 138–54.

Graham, E. (1982) 'Objectivity, values and bias in human geography', paper read at annual conference of the Institute of British Geographers, Edinburgh.

Harris, C. (1978) 'The historical mind and the practice of geography', in D. Ley and M. Samuels (eds), *Humanistic Geography: Prospects and Problems*, London, Croom Helm, 123–37.

Harvey, D. (1969) *Explanation in Geography*, London, Edward Arnold.

Harvey, D. (1979) 'Monument and myth', *Annals, Association of American Geographers*, 69, 362–81.

Ley, D. (1978) 'Social geography and social action', in D. Ley and M. Samuels (eds), *Humanistic Geography: Prospects and Problems*, London, Croom Helm, 41–57.

Ley, D. (1980) *Geography without Man: A Humanistic Critique*, University of Oxford, School of Geography Research Paper No. 24.

Ley, D. and Samuels, M. (1978) 'Introduction: contexts of modern humanism in geography', in D. Ley and M. Samuels (eds), *Humanistic Geography: Prospects and Problems*, London, Croom Helm, 1–17.

Middleton, C. A. (1981) 'Roots and rootlessness in the life and novels of George Eliot', in D. C. D. Pocock (ed.), *Humanistic Geography and Literature*, London, Croom Helm, 101–20.

Mink, L. O. (1969–70) 'History and fiction as modes of comprehension', *New Literary History*, 1, 541–58.

Olsson, G. (1978) 'Of ambiguity or far cries from a memorialising mamafesta', in D. Ley and M. Samuels (eds), *Humanistic Geography: Prospects and Problems*, London, Croom Helm, 109–22.

Olsson, G. (1981) 'On yearning for home: an epistemological view of ontological transformations', in D. C. D. Pocock (ed.), *Humanistic Geography and Literature*, London, Croom Helm, 121–9.

Paterson, J. H. (1979) *North America*, 6th edn, New York, Oxford University Press.

Paterson, J. H. and Paterson, E. (1981) 'Shropshire: reality and symbol in the work of Mary Webb', in D. C. D. Pocock (ed.), *Humanistic Geography and Literature*, London, Croom Helm, 209–20.

Pocock, D. C. D. (1981a) 'Place and the novelist', *Transactions, Institute of British Geographers*, NS 6 (3), 337–47.
Pocock, D. C. D. (1981b) 'Introduction: imaginative literature and the geographer', in D. C. D. Pocock (ed.), *Humanistic Geography and Literature*, London, Croom Helm, 9–19.
Pocock, D. C. D. (1983) 'The paradox of humanistic geography', *Area*, 15 (4), 355–8.
Rees, R. (1973) 'Geography and landscape painting: an introduction to a neglected field', *Scottish Geographical Magazine*, 89, 148–57.
Relph, E. (1976) *Place and Placelessness*, London, Pion.
Rowles, G. D. (1978) 'Reflections on experiential fieldwork', in D. Ley and M. Samuels (eds), *Humanistic Geography: Prospects and Problems*, London, Croom Helm, 173–93.
Samuels, M. (1978) 'Existentialism and human geography', in D. Ley and M. Samuels (eds), *Humanistic Geography: Prospects and Problems*, London, Croom Helm, 22–40.
Sayer, A. (1979) 'Epistemology and conceptions of people and nature in geography', *Geoforum*, 10, 19–43.
Seamon, D. (1981a) 'Newcomers, existential outsiders and insiders: their portrayal in two books by Doris Lessing', in D. C. D. Pocock (ed.), *Humanistic Geography and Literature*, London, Croom Helm, 85–100.
Seamon, D. (1981b) 'Body, subject, time–space routines and place ballets', in A. Buttimer and D. Seamon (eds), *The Human Experience of Space and Place*, London, Croom Helm, 148–65.
Smith, S. J. (1981) 'Humanistic method in contemporary social geography', *Area*, 13 (4), 293–8.
Taylor, P. J. (1976) 'An interpretation of the quantitative debate in human geography', *Transactions, Institute of British Geographers*, NS 1, 129–42.
Tuan, Yi-Fu (1976) 'Humanistic geography', *Annals, Association of American Geographers*, 66, 266–76.
Tuan, Yi-Fu (1977) *Space and Place: the Perspective of Experience*, London, Edward Arnold.
Tuan, Yi-Fu (1979) *Landscapes of Fear*, Oxford, Basil Blackwell.
Williams, R. (1976) *Keywords: A Vocabulary of Culture and Society*, London, Fontana.

22 Nigel Thrift

On the Determination of Social Action in Space and Time

Excerpts from: *Environment and Planning D: Society and Space* 1, 23–57 (1983)

Introduction: the problem of translation

[Since the early 1970s] human geographers have become more and more involved with social theory. This involvement has ranged from the extreme determinism of some structural Marxist approaches, which hope to read off the specifics of places through the general laws or tendencies of capitalism, through to the extreme voluntarism of most 'humanistic' geography, which hopes to capture the general features of place through the specifics of human interaction. But I think it is true to say that, because of the nature of human-geographical subject matter, extant social thoery has proved very difficult for human geographers to handle. The reason for this is quite simple. It is very difficult to relate what are usually very abstract generalizations about social phenomena to the features of a particular place at a particular time and to the actions of 'individuals' (as discussed later, this is a difficult and problematic term to use) within that place. Of course, this is not a problem peculiar to human geography. Social historians have been having something of the same problem as the focus of their subject has moved from 'the circumstances surrounding man to man in circumstance' (Stone, 1979, page 23); in particular, as this subject has moved to the use of selective examples of 'individuals' to illustrate the 'thinking in–acting out' of *mentalité*.

How is this problem of *translation* to be overcome; indeed, can it be overcome? Conventionally, the problem is now represented in human geography as a polarization between social structure and human agency, a polarization that is also known in the guise of the debates on the relative importance to be given to economy and culture or to determinism and free will. But in fact, there are at least four major strands to this outwardly homogenous reaction to the problem of the relation of structure to agency.

At the limit, the implication is that altogether too much ground has been ceded to structural social theory.[1] This is a proposition that appeals, naturally enough, to empiricists, who deal only in the given as it gives itself and who continually mistake a minute description of some regularity for theory. But it also finds allies amongst humanists, who pine for an anthropological philosophy with the category 'man' at its heart.

A second response, one favoured by certain jumbo Marxists, is that social theory was never meant to be applied to the small scale and the unique. An either/or situation is assumed to exist. The choice is either general theory or unique description. No doubt, this reaction is partly due to a reductionism that characterizes some Marxist analyses, both of the concrete to the merely abstract and of social science to philosophy. But perhaps it is also partly linked to a view of social theory that is still wedded to a conception of (social) science and scientific statements as merely being generalizations about social phenomena. [This paper, therefore, is 'realist' in intent (compare Sayer, 1981).] It is an unintended consequence of this view that 'consideration of the possibility of coming to terms with the unique aspects of situations (as against their common characteristics) within a generalizing frame of reference is neglected because of the seeming refractoriness of the problem to the conventional canons of scientific analysis' (Layder, 1981, page 49). The significance of small-scale human interaction is therefore bound to be minimized.

A third view is that a major shift is required in the theoretical centre of gravity of social science that will lead towards a new 'structurationist' problematic based upon a theory *of* social action (or practice) that complements theories *about* social action (Dawe, 1979). This is a view that has gained considerable support in human geography (for example, see Carlstein, 1981; Gregory, 1981; Pred, 1981b; Thrift and Pred, 1981) and is now being put forward as the touchstone of recent developments in other subjects, for instance in historical sociology (Abrams, 1980; 1982) and in administrative science (Ranson et al, 1980).

A fourth view, the one to which I shall subscribe in this paper, is that it is possible to produce general knowledge about unique events, but that this is best achieved through the interpenetration of these structurationist concerns with existing, specifically Marxist, social theory, because Marxism, for all its very definite sins and omissions, has a strong notion of *determination*. No doubt such an extension will prove anathema to many Marxists, but, to foreshadow the argument,

> it is unclear why a preoccupation with the material practices of everyday life – or for that matter the structure of popular belief – is either Utopian or undesirable from a Marxist point of view. Nor is there any reason to counterpose the personal and familial with global and overall views (Samuel, 1981, page xxi).

This does not mean, however, that I am in favour of the 'rambling impressionism' (Abrams, 1982, page 328) which it is possible to find in certain texts that have been concerned with, for example, daily life in the past. Rather, I am looking for a theoretically structured approach to the 'real world of real human beings', which is not 'held at a safe distance by [the] extreme forms of idealist abstraction' (Selbourne, 1980, page 158) that are so characteristic of a substantial proportion of the Marxist tradition.

This paper is therefore arranged as follows. A general synoptic overview of modern social theory leads to a consideration of the four major concerns of what I shall call the structurationist 'school'. I will want to argue that these concerns are crucial to any nonfunctionalist Marxist social theory which must take into account 'contextual' as well as 'compositional' determinations. I will then sketch, in the final section of the paper, an outline of what such a social theory might begin to look like when extended to the smaller scale and to the consideration of unique events. A new kind of regional geography/sociology would be integral to this project.

The two responses to the problem of structure and agency: determinism and voluntarism

Most social theory is not reflexive. It does not consider its own origin in the theoretical and practical thought of a period, as this is determined by the prevailing social and economic conditions. In particular, social theory born under capitalism must reflect many of the features of capitalism and, in particular, the basic contradictions of this system.

Since the Enlightenment, and the intellectual vacuum caused by the gradual dwindling away of notions of cosmic order, humankind has been beset by two tendencies, partly causes of and partly the result of capitalism. The first has been the tendency towards the seemingly ever-greater scale and extent at which the production and reproduction of society takes place; a tendency marked off by such indicators as the continuing concentration and centralization of capital (Marx), the dramatic increase in and concentration of population (Malthus), the growth of the State and the penetration of its bureaucracy into every corner of our lives (Weber, the Frankfurt School), rapid time–space convergence, and the formation of all those 'masses' – mass audiences, mass consumption, mass culture, and so on (Arnold, Veblen, the Frankfurt School). The experience of these phenomena is now commonplace in our daily lives. But for the nineteenth-century middle and upper classes, from which so many significant social theorists were drawn, these were *new* experiences that were conceived as real and immediate problems. To give but one example, domestic and foreign visitors to Manchester in the early nineteenth century constantly remarked upon the vast and noisy working-class crowds, whether these crowds were celebrating at a fair or a wake, demonstrating at a Chartist rally, shopping at night, or even travelling to and from work. There was a general sigh of relief as the workers filed into the factories and workshops where they could be neatly closeted away for the day. It is therefore not surprising that for the nineteenth-century middle class the problem of scale was above all perceived as centering around words like 'order' and 'control' and around the question of how social order could come about or was possible in such rapidly changing circumstances.

To deal with the changed situation, new practical and theoretical categories are formulated and old categories are revised. The idea of a population to be counted out and analyzed, the idea of a moral topography, and the whole semantic field of political economy (containing such terminology as 'manufacture', 'industry', 'factory', 'class', 'capital', 'labor' and, finally, 'industrial revolution') gradually all come into being . . . . Above all, the medical metaphor of the anatomy of society leads to a concern with 'structural' explanation . . . .

This tendency was reinforced by the one that grew up beside it. The conception of the individual had gradually and decisively changed – to its precise antithesis. In medieval times 'individual' meant inseparable or indivisible . . . . A 'single individual' would have had no meaning. The relatively closed and static medieval community was one founded on a minimal division of labour and the sharing of many tasks. The system was based on the control of social interaction in order to limit interactions to members of the known social world. Everyone therefore knew everyone else and there was no need for a particularly developed system of control and surveillance. It could be local . . . . This system gradually gave way, for a variety of interrelated reasons that eventually lead to capitalism – the increasing division of labour, the rise of wage labour, the rise of a calculative rationality, the increase in urban populations, and so on – to a strong *individuality* based upon the concept of human rather than divine will, on the idea that a human being had a choice over the form of self that might be sought, and to self-control (Weintraub, 1978). The connection between individuality and the problem of social control is once more a strong one. The idea of the individual acquires its modern secularized meaning in relation to the State (rather than the Church) (Foucault, 1977; Elias, 1978; 1982). Through all number of new institutions like the civil service and the police, the State gradually builds up a 'grid of intelligibility'. Everyday life is brought into scientific discourse . . . . Individuality is linked with identification; as a concept, individuality now pertains to specific social groups and to the idea of an individual as a set of developing attributes describing a 'career' over time (Ginzburg, 1979). The individual is made an operational concept, the object of scientific knowledge (Foucault, 1972). The underlying epistemology is again medical, but it is now based on the diagnosis of *symptoms*. It leads directly to the deciphering of signs, as in semiotics, and to Freud.

In capitalism these two tendencies come together as a major contradiction between socialized production and private appropriation. Capitalist societies are both collectivist and individualistic. On the one hand, each individual lives in a highly socialized world; on the other, each individual lives in a privatized world . . . . The position is uneasy and ambiguous. This tension, I submit, is as obvious in social theory as it is in everyday life (Dawe, 1979). These two contradictory tendencies, pulling either one way or the other, are therefore found in most nineteenth- and twentieth-century social theory. Thus each particular social theory tends to stray, in varying degrees, towards either the determinism of capitalist society or to the voluntarism of capitalist individuality. Thus there are 'two' sociologies (Dawe, 1970; 1979), 'two' anthropologies (Sahlins, 1976), 'two' Marxisms (Albrow, 1974; Veltmeyer, 1978; Gouldner, 1980; Hall, 1980), and so on and so forth. In reality, of course, such a distinction is a simplification of the highest order. Most social theories, and variants thereof, put forward by particular 'individuals', are an admixture of both dimensions, and are best represented as points on a continuum between the two polarities. This is what Fig. 22.1 tries to do, in a *very* crude fashion, for a number of recent debates in the social sciences. The theoretical positions of particular individuals on the axes in Fig. 22.1 are made to appear more simple than is actually the case because of the reduction in dimensionality that is necessary to present such a diagram.

In a related and important way, social theory tends towards either the *compositional* or the *contextual* (Hägerstrand, 1974). [Simpson (1963) and Kennedy (1979) make a similar distinction between *immanent* and *configurational* approaches.] In the compositional approach, which reaches its apogee in the 'structural–genetic' method of Marx . . . , human activity is split up into a set of broad structural categories founded on the property of 'alikeness' and derived via a formal–logical method based on the tool of abstraction. These categories are then recombined as an explanation of society or, at least, of parts of it. In the contextual approach, *elements* of which can be found in Schutz's phenomenology, in Berger and Luckmann's phenomenological–dialectical approach, in Goffman's frame analysis, in Harré's architectonics, and in Hägerstrand's time–geography, human activity is treated as a social event in its immediate spatial and temporal setting and the categories so derived are based on a property of 'togetherness' that must not be split asunder. Too often, of course, such attempts to construct a contextual explanation have ended in nominalism or monism or, paradoxically, have only pointed yet more forcefully to the power of social structure . . . . But that such attempts have

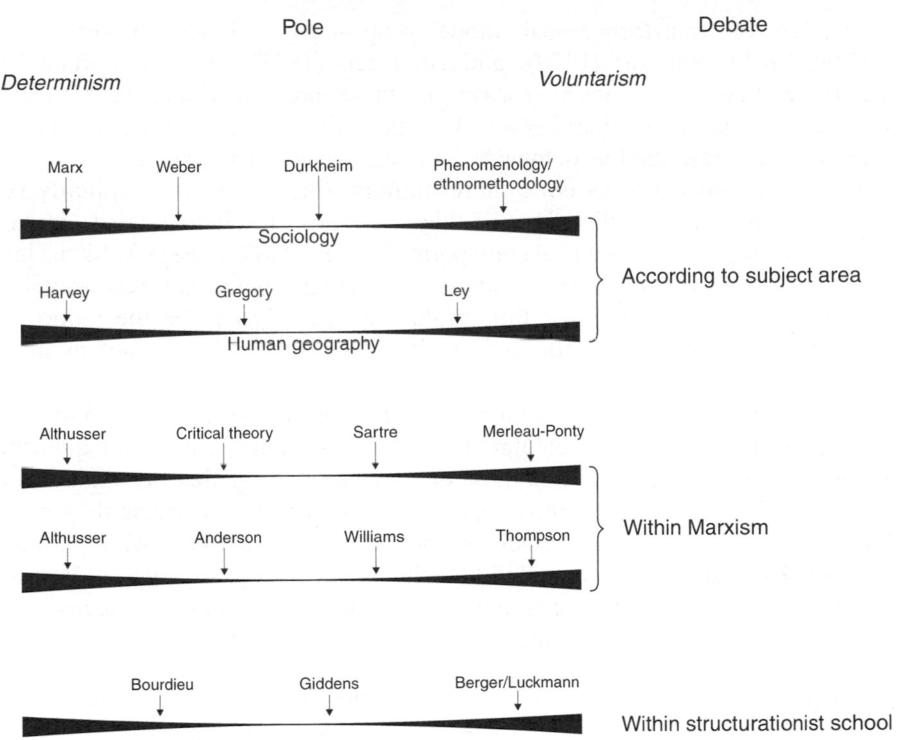

Fig. 22.1 Determinism and voluntarism in various social theorists and subject areas – an approximation

ended in failure does not necessarily alter the terms of their critique or lead to the conclusion that the critique has no force. In contrast to the compositional approach, and in many cases in reaction to it, the contextual approach is an attempt to recapture the *flow of human agency* as a series of situated events in space and time. Once again this distinction between compositional and contextual approaches is a simplification – most social theories can be located somewhere on a continuum between these two polarities – but it has important implications for the argument that follows.

The structurationist school: towards a nonfunctionalist social science

In [the previous section], I tried to show . . . that a basic duality, one that can partly be traced to capitalism, lies at the heart of modern social theory between determinist and voluntarist approaches and between their versions of human agency, respectively, 'plastic' and 'autonomous' man (Hollis, 1977). Can this duality be recast in such a way that it dialectically recombines social structure and human agency? Certainly this is the hypothesis of a number of authors who espouse, in one form or another, a theory of '*structuration*', elements of which were first put forward by Berger and Luckmann (1966), but which has now reached maturity with the more sophisticated 'recursive' or 'transformational' model proposed by Giddens (1976; 1977; 1979a; 1981), Bhaskar (1979), and Bourdieu (1977).[2] It is important to emphasize that the theories put forward by these three individuals have strong differences, but the similarities are, I think, still sufficiently great for these authors to be described as belonging to a structurationist 'school'.

Four common concerns unite these authors. First, they are (explicitly or implicitly) antifunctionalist. That is, they recognise that functionalist 'explanation' is simply an evasion.[3] At one point, Giddens (1979a, page 7) likens his project to '*show*[*ing*] what a non-functionalist social science actually involves' (my emphasis), and this might well be taken to be the motto of the structurationist school, for each of the other three shared concerns flow into and out of this node.

The second concern is a common message, that neither a structural–determinist (objectivist) nor a voluntarist (subjectivist) approach is satisfactory, joined to a common goal, to link these two approaches together in a dialectical synthesis. Structural–determinist approaches are criticized because they treat human practices as being mechanical and devoid of creativity, what Castoriadis (1975) and others have called 'alterity',[4] the quality of newness. Voluntarist approaches, on the other hand, are equally problematic because, in concentrating on interaction, they become blind to the fact that

> interpersonal relations are never, except in appearance, individual-to-individual relationships . . . the truth of the interaction is never entirely contained in the interaction. This is what social psychology and interactionism or ethnomethodology forget when reducing the objective structure of the relationship between the individuals to the conjunctural structure of their interaction in a particular situation or group (Bourdieu, 1977, page 81).

No, social structures are characterized by their *duality*. They are both constituted by human practices, and yet at the same time they are the very medium of this constitution. Through the processes of socialization, the extant physical environment, and so on, individuals draw upon social structure. But at each moment they do this they must also reconstitute that structure through the production or the reproduction of the conditions of production and reproduction. They therefore have the possibility, as, in some sense, capable and knowing agents, of reconstituting or even *transforming* that structure. Hence, the 'transformational' model (Bhaskar, 1979). Social life is therefore fundamentally *recursive* (Giddens, 1979a) and expresses the mutual dependence of structure and agency. Social structure cannot exist independently of motivated (but not necessarily reasoned) activity, but neither is it simply the product of such activity.

However, by far the more important problem is how to forge a nonfunctionalist link between structure and agency. Here individual members of the structurationist school differ in their approach (see Fig. 22.2), but no doubt each would concur with Bhaskar (1979, page 51) that

> we need a system of mediating concepts . . . designating the 'slots', as it were, in the social structure into which active subjects must slip in order to reproduce it; that

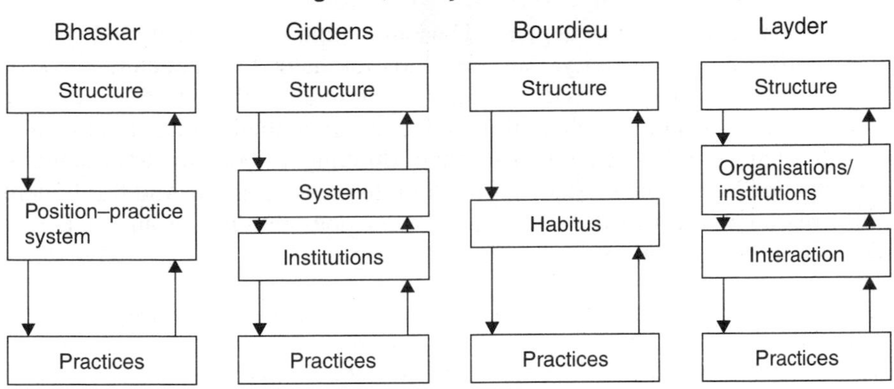

Fig. 22.2 Mediating concepts in the schemas of various members of the structurationist school

> is, a system of concepts designating the 'point contact' between human agency and social structures.

What these Sartrean *mediating concepts* are differs from author to author. Bourdieu's . . . answer is to insert a third 'dialectical' level between social structure and human practices, a 'semistructure' called 'habitus' consisting of cognitive, motivating ('reason-giving') structures that confer certain objective conditions and predefined dispositions on actors based on the objective life-chances which are incorporated into the strategies involved in particular interactions, these interactions being improvisations regulated by the habitus. Thus each class, for instance, has a particular habitus that results from a common set of material conditions and, therefore, expectations . . . . Each mode of production has, so to speak, its own modes of perception. Giddens . . . , by contrast, refines existing concepts. For example, structure is limited in its meaning to rules and resources, to particular structural properties. An essentially new concept of 'system' is then added as reproduced and regular social practices. Institutions are the basic building blocks of these social systems. Finally Bhaskar (1979, page 51) offers the rather more general concept of a position–practice system; he argues that

> it is clear that the mediating system we need is that of *positions* (places, functions, rules, tasks, duties, rights, etc.) occupied (filled, assumed, enacted, etc.) by individuals, and of the *practices* (activities, etc.) in which in virtue of their occupancy of these positions (and vice versa) they engage.

A third major tenet of the structurationist position is the lack of a theory of (practical) action (Giddens), of the acting subject, of a theory of practice (Bourdieu) in most of social theory. There is, therefore, a need for an explicit theory of practical reason and practical consciousness that can account for human intentionality and motivation. This project is vital for three reasons. First, as already pointed out, there is no direct connection between human action and social structure in social theory. Indeed the properties possessed by social forms are often very different from those possessed by the individuals upon whose activity they depend. Thus motivation and intentionality may characterize human activity, but they do not have to characterize social structure or transformations in it. Second, it is also obvious that ordinary people do not spend their whole time reflecting upon their social situation and how to change it. Indeed much of their thinking is bent towards actually carrying out preassigned tasks with no definite goals in mind. And third, there is no reason to believe that when an actress does rationalize an act it will necessarily be the real reason that she is doing it (Wright Mills, 1959). Thus

> . . . people in their conscious activity, for the most part unconsciously reproduce (and occasionally transform) the structures governing their substantive activities of production. Thus peole do not marry to reproduce the nuclear family or work to sustain the capitalist economy. Yet it is nevertheless the unintended consequence (and inexorable result) of, as it is also a necessary condition for, their activity. Moreover, when social forms change, the explanation will not normally lie in the desires of agents to change them that way, though as a very important theoretical and political limit it *may* do so (Bhaskar, 1979, page 44).

It is therefore important to make a distinction between *practical* and *reflexive* (or discursive) consciousness and reason. Practical activity is not theorized reflexively. Practical consciousness is 'tacit knowledge that is skilfully applied in the enactment of courses of conduct, but which the actor is not able to formulate discursively' (Giddens, 1979a, page 57). The reasons an actor gives for an action are not necessarily the real ones – which *may* operate outside the understanding of an agent. They are part of a set of 'reason-givings', which must, of course, relate to the real reasons, but which can be refracted or even reflected back as their inverse.[5]

The fourth major component of the structurationist case, one that is necessarily related to the third, is that time and space are central to the construction of all social interaction and, therefore, to the constitution of social theory. This does not just mean that social theory must be historically and geographically specific. More importantly, social theory must be about the time–space constitution of social structure *right from the start*. Thus, for Giddens (1979a, page 54), 'social theory must acknowledge, as it has not done previously, time–space intersections as essentially involved in all social existence'. Such a viewpoint has a number of important consequences. Social structure, for example, cannot be divorced from spatial and temporal structure. The two have to be theorized conjointly (see Giddens, 1981; Gregory, 1982), not as the impact of one upon the other. Further, human agency must be seen for what it is, a continuous flow of conduct in time and space constantly interpellating social structure. Such a view of human agency is necessarily *contextual*. As Bourdieu (1977, page 9) puts it: 'practices are defined by the fact that their temporal structure, direction and rhythm are constitutive of their meaning'. The point is that human action takes place in time as a continual time- (and space-) budgeting process and as an irreversible sequence of actions. There is no doubt that social theory, which has a time that is only partly that of practice, has tended to forget this. Yet it is the ad hoc improvisatory strategy imposed on people's practices by the fact that they have a limited time in which to carry out particular activities (and a limited time in which to decide to do them) that is a crucial part of practice, practical consciousness, and practical meaning. Practice, therefore, is always situated in time and space. This is one link to structure for the structurationists, since the places at which activity is situated are the result of institutions which themselves reflect structure – home, work, school, and so on. These institutions form nodes in time and space around which human activity is concentrated. As Giddens (1979a) points out, this is an area that human geographers through the study of 'time-geography', have been particularly involved in for the last ten years or so (Hägerstrand, 1970; 1973; Thrift, 1977; Thrift and Pred, 1981).

These, then, are the four common elements of the structurationist school. In the next section I will argue that they provide an important means of progressing towards a nonfunctionalist Marxism. However, in order to justify this claim it is important, at this juncture, to understand what I take to be at the core of Marx's work and what, in effect, I consider to be 'historical materialism'. For me, Marx's work has a threefold significance. First, it subscribes to the simple transhistorical fact that at the core of all human social structure is 'the production of material life itself'. Second, and at the next level of abstraction, social structure is constructed around particular organizations of production which characterize particular times and places. One aim of 'historical materialism' is to uncover the 'mechanisms' which underlie these organizations. Third, 'historical materialism' tries to reconstruct the 'concrete in thought' (and *not* the empirical, which is inevitably contingent) as the result of the successive addition of determinations from different levels of abstraction. When placed together these determinations form a hierarchical set of dynamic or process categories (compare Sayer, 1979; 1981; Zeleny, 1980). I consider other intermittently important features of Marx's work to be more period and place specific. As Giddens (1979a; 1981) notes, some of these features must be seen as the excess baggage of nineteenth-century thought. They are therefore transitory. These features include functionalism, evolutionism, and essentialism (see Keat and Urry, 1982). Yet other features can now be seen to exist at a different level of historical specificity to the level at which Marx considered them. Thus, in the light of recent research, the overarching transhistorical statement – the history of all hitherto existing society is the history of class struggles – is more properly considered at the more historically specific mode-of-production level of abstraction (compare Giddens, 1981).

The implications of the structurationist school for Marxist social theory

The interpenetration of the insights of the structurationist school with Marxism to form a nonfunctionalist Marxism will be no easy task. The problem is well put by Giddens (1981, page 16):

> In diverging from functionalism . . . we need to be able to recognize *both* what might be called the theorem of 'knowledgeability' – that we are all purposeful,

> knowledgeable agents who have reasons for what we do – *and* that social processes at the same time work 'behind our backs', affecting what we do in ways of which we are unaware. Marx summed this up in the famous aphorism, 'Men make history, but not in circumstances of their own choosing'. However, working out the implications of this unobjectionable statement is difficult.

What seems certain is that the structurationist approach cannot simply be laid alongside Marxism . . . . The implications of the approach are more far-reaching than this and demand a thorough reworking of Marxism. But neither is the approach so far removed from at least some interpretations of Marx as to make this an impossibility.

This is not to suggest, however, that the marriage will be an easy one. In particular, the writings of members of the structurationist school suffer from a number of failings. First, they provide no clear notion of *determination*. Marxism can be criticized for many things, but at least the forces and relations of production on which it has concentrated give it a strong notion and scheme of determination. Although the structurationist school acknowledges the strength of the forces and relations of production (see especially Giddens, 1981), the complexity of determination that is provided in their schemes is obviously much greater (Gregory, 1982). Partly, of course, this is the result of a concentration on other things; for example, the constitution and reproduction of individual subjectivities (Bhaskar), cultural capital (Bourdieu), or authoritative resources (Giddens). But partly it is also the result of an unresolved ambiguity about the relative importance of economic determination which sometimes allows structurationist models to appear far more individualist and voluntarist than is the intention . . . . Second, *conflict* is dealt with only in the most axiomatic and general way in the writings of the structurationist school. Third, and closely related to the second point, although Giddens, for example, proclaims that history is indistinguishable from social science, his notion of history still tends to the systematic and general. The use by the structurationist school of history remains firmly *compositional* rather than contextual, and this has important implications. For the structurationists, Anderson's (1980, pages 21–22) judgement on E. P. Thompson is no less valid:

> A *historical*, as opposed to an axiomatic, approach to the problem would seek to trace the *curve* of (deliberate) enterprises, which has risen sharply, in terms of mass-participation and scale of objective, in the last two centuries, from previously low levels. Even so, however, it is important to recall that there are huge areas of existence which remain largely outside *any* form of concerted agency at all . . . .
> . . . The area of self-determination has been widening in the past 150 years. But it is still very much less than its opposite. The whole purpose of historical materialism, after all, has precisely been to give men and women the means with which to exercise *a real popular self-determination for the first time in history*.

The point of historical materialism is to allow conscious, directed, and reflexive human agency to become the social structure. It is ironic, then, that the whole point of Marx's work, to identify in an objective and rational fashion the bounds to human agency so that social structure might be refashioned, sometimes seems to be seen by the structurationist school as deterministic, although actually it is (or, at least, tries to be) a realistic assessment of the prevailing conditions under which agency must work or has had to work.

These reservations apart, the benefits that would flow from the interpenetration of Marxism and the structurationist school are many. Certainly, there is no all-embracing panacea to be found in Marx for the problems that beset Marxism now (Giddens, 1979b). Most obviously, the problem of functionalism has to be solved. This will require a change in the orientation of Marxism towards a number of new areas of study which are not just additions to the domain of Marxism but are integral to the revision of Marxist theory.

The reproduction of labour-power

Nearly all the new areas of study are to be found grouped under the problem of an adequate account of the *reproduction of labour-power* within the realm of practices, of the constitution and reproduction of individual subjects, which Urry (1981a, page 39) calls 'civil society':

> The problem is that labour-power, unlike all other commodities, is not produced by capitalists for profit. It is produced, or reproduced, in the sphere of civil society. So while all other commodities are produced within capitalist production, labour-power is produced elsewhere, outside capitalist relations of production.

In *Capital* labour-power is relevant only insofar as it enters, active and refreshed, into the sphere of production. The reproduction of this labour-power and the constitution of labour-power as a conscious subject are things Marx dealt with barely, if at all. Concrete individuals, worker and capitalist alike, are depicted only to the extent that they are 'personifications' or 'functionaries' of economic categories . . . ; that is, as instances of the economic existences of men and women undifferentiated by other determinations.

Some Marxists, of course, regard all efforts to deal with individuals as somehow irredeemably humanist. There is, however, nothing in Marxist literature that leads to this conclusion; not even, indeed, in the work of that most antihumanist of all Marxists, Althusser. As Molina (1979, page 239) points out, Marx:

> explicitly formulates a clear distinction between a theoretical treatment of individuals as *personifications* (as bearers of economic categories) and a treatment of individuals as individuals (according to individual differences which do not arise from the economic relation itself). The question and its answer correspond clearly to a difference between what is the problem of the states of the category of 'individual' in Marx's theory and what would be the problem of a theory of individuality as such, the latter problem not being present in Marx.

Thus the problem of individuals and individuality is not closed off in Marx; rather individuals appear shorn of every determination except those coming from economic relations. On this skeleton, however, can be hung the flesh of other, more contingent determinations. Marx, then, provides a set of theoretical principles with which to think both the problem of conceptualizing the different historical laws of the existence of individuality and the most basic determinations of that individuality . . . The reproduction of labour-power . . . is of crucial importance, in particular as it informs and makes concrete the study of three closely interrelated areas of Marxist analysis: class conflict, ideology and hegemony, and personality.

Human labour-power is unlike other commodities. Because it has creative capacities it is the source of expanded capital and profit. But it is also a source of resistance. Such labor-power has conscious will, it is possessed of *agency* (Willis, 1977). As Yeo and Yeo (1982, page 147) put it:

> The act of selling labour-power is not the same in meaning whenever and however it is done . . . . What does the labour-power actually consist of? It is abstract only for the purposes of commensuration. How is it seen and valued by the sellers and buyers? What is its meaning in relation to the rest of life? What is obtained, denied, dreamed of in return for the sale? All these questions are historical and material, not abstract and transcendant and not to be answered by any single theorized version of a constant, still less a hierarchical relationship between work and leisure. The wage relation is more than an economic one: producing things entails relations: relations are *social*, between whole beings who exist, including capitalists, in many dimensions.

Labour-power, its sale, its purchase, its existence, this is the subject of conflict. And class conflict is meant to occupy a central place in the Marxian framework. Yet all too often, in Marx and in subsequent Marxist analyses, class conflict occupies an interstitial, functional position sandwiched between capital and labor and acting more like a coefficient of friction between the two warring blocs than as a central dynamic. Tribe's (1981, page 32) criticism of the Brenner/Dobb position in the 'transition from feudalism to capitalism debate' is just as relevant elsewhere:

> The gap was however tentatively closed through the invocation of politics. The Marxist precept, 'the class struggle is the motor of history' is inserted between the shafts of the feudal wagon and is encouraged to drag the immobile economic order of feudalism into capitalism through periodic dialectical interventions. But this 'politics' on closer inspection turns out to be constituted by a projection of economic relations onto a space which they already occupy: politics is used to explain the contradictory progress of the economic order, but it is itself no more than an expression of this order. Notwithstanding such difficulties, 'politics' does the job: the gap between feudalism and capitalism is closed and an orderly genesis is re-established.

It is little wonder, then, that Marxism has no adequate theory of 'politics' and that there is still little idea, either of how the process of class structuration . . . in the realm of practices takes place [that is, as the formation of *social* classes within civil society (Urry, 1981a)] or of how class awareness and class consciousness are constituted, notwithstanding the explanations by Thompson (1963) and Foster (1974). Similar problems beset Marxist notions of how the dominant order is maintained, especially outside work. Categories like ideology and hegemony have been, of course, the subject of notoriously functionalist usage (see Abercrombie et al, 1980); further, they are often wielded in such a general and abstract fashion that it is difficult to use them to capture the diversity of practices in civil society (see Urry, 1981a). Finally, Marxism has no adequately developed theory of concrete human personality. Indeed, as presently constituted, it cannot have.

The filling out of each and every one of these three areas of analysis, with their emphasis on everyday life and the formation of individuality, demands the development of a *contextual* dimension to Marxist theory, a dimension which can be constructed only by building upon the major elements of structurationist writing: the importance of nonfunctionalist links between structure and agency, practical reason, and time and space intersections. The investigation of subjects like class conflict, ideology and hegemony, and the formation of personality all demand not only a measure of theoretical generality but also a theory that can deal with specificity as a *theoretical* as well as an empirical object. For example, part of the reason why some people do certain things at certain places in certain periods while other people at other places in other or the same periods do not, or why different forms of protest and patterns of organization arise in apparently similar conditions . . . , is quite specific to particular contexts, to particular *places*. In brief, there is a need to develop a *contextual science*, a sociology of detail or, as others have called it, 'sociopsychology' (Bhaskar, 1979), the 'science of the singular' (Sève, 1975) or the 'science of the specific' (Layder, 1981), in which determinations can be traced out as they occur in particular individuals and particular groups of individuals in a particular locality. . . .

A nonfunctionist theory of social action in space and time: a research agenda

In this final section, I want to sketch in some of the components of what an historically specific, nonfragmented, contextual theory of human action, which at the same time has clear lines of communication to a compositional theory about human action, might look like. I will want to be able to incorporate the four structurationist concerns into this outline without reducing it, however, to just them. Thus the outline stresses the integral importance, for example, of practical reason and action and of concrete interaction in time and space.

What follows at this juncture is, of necessity, programmatic and skeletal. And little of it is novel. Many parts of the outline are already being filled in by various workers, most especially in social history, historical sociology, and historical anthropology, but also now in human geography and regional sociology. My aim here is to stress two things; first, the essential *unity* of the concerns of workers in many subject areas, and second, the very great importance of *empirical* investigations as a necessary moment in any social theory. Along the way, I also hope to show that there is a place for a *reconstructed regional geography* (a regional geography that builds upon the strengths of traditional regional geography, for example, the feel for context, but that is bent towards theoretical and emancipatory aims) (Baker, 1979; Gregory, 1981), not only as the focus of all these diverse concerns, but also as the subject and object of a theory of social action. The 'region' (I expand upon this term below) can be seen, in this conception, as the 'actively passive' (I have borrowed this term from Sartre) meeting place of social structure and human agency, substantive enough to be the generator and conductor of structure, but still intimate enough to ensure that the 'creature-like aspects' (Heller, 1982, page 21) of human beings are not lost.

It is important to note at the outset, however, the compartmentalized nature of this account. This is, at one level, an admission of failure; social activity in any region takes place as a continuous *discourse*, rooted in a staggered series of shared material-situations that constantly arise out of one another in a dialectically linked distribution of opportunity and constraint, presence and absence. A region is lived *through*, not in. The term 'discourse' is intended to convey something more than the fact that human communities exist socially through the medium of language, important though this aspect is (Pred, 1981b). It is also intended to convey the wider sense of a lived world of material practices implicit in Williams's (1977, page 110) notion . . . of *hegemony* as a silent but deafening contextual field:

> a wide body of practices and expectations, over the whole of living: our senses and assignments of energy, our shaping perceptions of ourselves and the world. It is a lived system of meanings – constitutive and constituting – which as they are experienced as practices appear reciprocally confirming. It thus constitutes a sense of reality for most people in the society, a sense of absolute because experienced reality beyond which it is difficult for most members of the society to move in most areas of their lives. It is, that is to say, in the strongest sense a culture, but a culture which has also to be seen as the lived dominance and subordination of particular classes.

Further, 'discourse' must be seen as an *active* category. The inhabitants of the culture or cultures of a region must not be seen as simply the passive recipients of class or any other social relations, as cultural dupes for whom socialization is simply conditioning. This would be to fall into the trap of the functionalism that underlies all social control explanations . . . . Rather, each culture has *limitations* and an ability to *penetrate* the existing order, the one often being linked to the other . . . . For example, Willis (1978, page 6) has shown that even commodities, which are, after all, the hallmark of capitalism, have uses other than blind consumption:

> Though the whole commodity form provides powerful implications for the manner of its consumption, it by no means enforces them. Commodities can be taken out of context, clarified in a particular way, developed and repossessed to express something deeply and thereby to change somehow the feelings which are their product. And all this can happen under the very nose of the dominant class – and with their products. We might even say that the characteristic of a certain kind of creative cultural development is the exploitation of qualities, capacities and potentials in those profane things which the dominant society has thrown aside, produced as 'business', or left undeveloped for cultural meaning.

No doubt, any reconstituted regional geography can start conventionally enough with a compositional account of 'the regional setting'. This involves, first of all, all those geographical determinations that can be grouped under the general heading of topography; such things as geology, hydrology, and climatic conditions, which have likely been changed already by the impacts of societies over the years. An account of the organization of production in a region is then needed, which involves ascertaining the level of the productive forces and the form of the productive relations, concentrating on the labour process (Dunford, 1981).[6] This emphasis on production leads, under capitalism in particular, to an outline of the class structure of a region and to the history of class formation, themes which are inevitably crosscut by other divides; the prevailing sexual division of labour, of course, but also ethnic, racial, and religious divisions. Finally, the local form of the state must be taken into account. Already written into such a compositional analysis is the *possibility* of quite dramatic interactions along the boundaries of one or all these divides. Wilson (1981, page 68), for example, demonstrates how feuding and the institution of the vendetta in Corsica were explicit in an extremely complex socioeconomic system, which:

> included transitional pastoralism alongside horticulture and itinerant cereal culture on land away from centres of population, collective land use alongside private property, work in family units on land of both kinds alongside share cropping and wage labour, and a dependence on mutual aid within and among families to carry out basic tasks.

A compositional account of a region is obviously a difficult enough task in itself, both theoretically and empirically. Perhaps it is for this reason that most analyses stop at this point. But, since the concern here is with developing a (contextual) theory of social action that can interpenetrate and ultimately change the conventional (compositional) theoretical account about social action, it is necessary to go one step further. The process of discourse in which human agency is produced and spun out cannot be reduced to a compositional account without jettisoning the very qualities that make it a discourse; the intricacies of interaction, the specificity of particular times and spaces, the sense of living as meeting, the context.

The locale

The region, initially at least, must not be seen as a *place*; that is a matter for investigation. Rather, it must be seen as made up of a number of different but connected *settings for interaction*. Giddens (1979a; 1981) uses the word *locale* to carry this meaning. It is a useful term and I will adopt it here.

Any region provides the opportunity for action and the constraints upon action; that is, the base for what is known about the world and the material with which to do (or not to do) something about it. In any region the *life paths* of particular individuals *can* interact, simply because they are collateral, near to one another in time and space. Whether they *will* interact, however, depends on the particular pattern of production (and, in a related way, consumption), and that, in turn, results in (and stems from) a particular pattern of locales, that punctuate the landscape. Each life path is, effectively, an *allocation of time* between these different locales. In any particular organization of production, certain of these locales will be *dominant*; that is, timc *must* be allocated to them. They are economic (and state) imperatives. Under capitalism, for example, home (reproduction), work (production), and, later, school (reproduction) are determinant. (Under other modes of production other nodes might be determinant. In Bali, for example, religion takes up more time than even work.) These dominant locales provide the most direct link between the interaction structure of a region and objective social structure, because they are the main sites of class production and reproduction. Such locales have five main effects. First, they structure people's life paths in space and time. They provide the main nodes through which a person's life path *must* flow and, since these locales are class structured and/or class differentiated, they will structure people's life paths in ways that are class specific (Fig. 22.3). As Therborn (1980, page 23) puts it:

> being 'in the world' is both *inclusive* (being a member of a meaningful world) and *positional* (having a particular place in the world in relation to other members of it, having a particular gender and age, occupation, ethnicity, and so on).

Second, these institutions can have effects on *other* people's life paths through the constraints they place on a person's ability to interact with other people

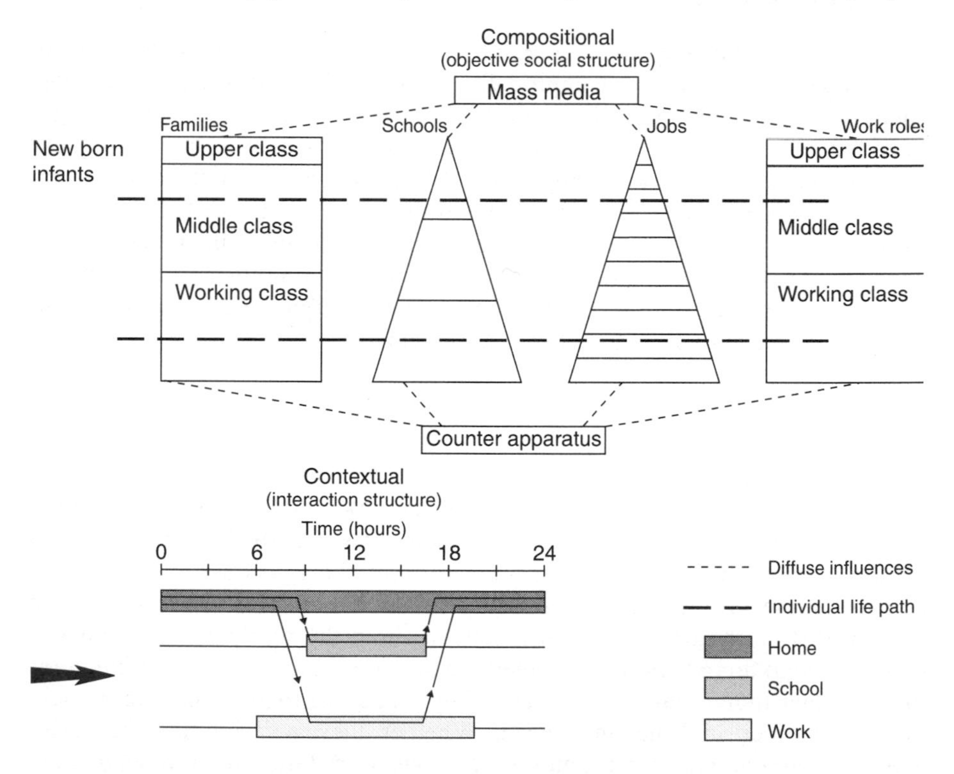

Fig. 22.3 The life path seen as a compositional ordering and a contextual field (adapted from Therborn, 1980, page 87)

engaged in activities within them. Third, they provide the main arenas (in time terms) within which interaction with other people takes place; thus they are sites of class and other conflicts, the medium and the source of most practical and reflexive reason, and, generally, the major context in which knowledge–experience about the world is gathered and common awareness engendered. Fourth, they provide the activity structure of the day-to-day *routines* that characterize most parts of people's lives. [The importance of routine as a 'second order' or 'officializing' strategy (Bourdieu, 1977) cannot be overestimated, since routine can quickly make pieces of behaviour appear not only natural but also disinterested.] And fifth, they are the major sites of the processes of *socialization* (seen in the active sense) that take place from birth to death, within which collective modes of behaviour are constantly being negotiated and renegotiated, and rules are learned but also created (Ricouer, 1981). It is also important to consider the particular internal *organizations* of time and space *within* each type of dominant locale [which Giddens (1981) calls 'regionalization']; that is, to look both at the routines that characterize different homes, different workplaces, and so on, and at the content of each locale in terms of accumulation of knowledge–experience. In the workplace, for instance, the spatial and temporal organization of the particular labour process, and how this varies from industry to industry, will be crucial . . . .

To these dominant locales must, of course, be appended the consideration of the position and internal organization of interrelated institutional locales (in particular, locales at which consumption activities take place) which will, within a particular organization of production, be sometimes of importance and sometimes not. To say that these institutions are dependent upon the dominant locales for their existence and importance is not to denigrate them, it is simply to situate them within a particular historical pattern of determination. For instance, under capitalism the allocation of time to 'leisure' (a specifically capitalist term) is determined, in the first instance, by hours of work. Thus, in many of the industrial areas of early nineteenth-century England, the pub was the only important extradomestic leisure institution, because hours of work (combined with income) made any alternative problematic. Finally, it is important to consider the *counterinstitutions* that challenge the ruling orthodoxy (Therborn, 1980). Such institutions often tend to be coincident with other sites, like those of work and school; indeed they often grow up in reaction to them. Their importance (especially as focuses of human agency) varies with history and location. The reasons for such variation are addressed below.

The study of the effects of changes in the functioning of such institutional locales on people's life paths is now becoming quite common. For instance, Hareven (1975; 1982) and Pred (1981a) have considered how changes in work hours and the organization of work each rebounded on the organization of family life and 'leisure' time in the nineteenth-century United States. Similarly, Stedman Jones (1974) and Thrift (1981; 1984) have documented how changes in work hours and the rise of compulsory primary education led to corresponding changes in the nature of the family, the social definition of home, and the use of 'leisure' time in nineteenth-century England. Joyce (1980) has considered, in some detail and in specific regional settings, how the factory becomes the dominant focal point of lived experience and popular culture in nineteenth-century English factory towns and, importantly, how the flow of factory life spilt over into the (few) hours outside work, into the form that home, school, and leisure took.

It is important to point out that, increasingly, a locale does not have to be local. The phenomenon of time–space convergence has meant that the setting for interaction common to certain social groupings has become differentially more extensive. For example, the middle class typically acts out its presence in a more extensive spatial setting than the working class. Interaction is more often at a distance. Factors such as commuting and the increased mobility associated with many white-collar jobs have ensured that this is so. Further, as the locales of social groupings have become more differentiated by their spatial extent, so they have become more fragmented by a process of homogenization that is the result of factors like the influence of a relatively free housing market and the housing estate, the advent of mass media and educational systems, and so on. Space is increasingly created as a series of commodified enclaves (mobility is crucial in differentiating these spaces), within which at least parts of everyday life can be carried on by one social group in isolation from other social groups. Giddens (1981) calls this a tendency to 'sequestration' and relates it to the rise of so-called 'total' or 'greedy' institutions like asylums and hospitals as well as to commodification in the strict sense.

But these tendencies do not mean that the *region*, seen as a particular intersection of these locales, has therefore lost its coherence or its distinctiveness. Common experiences engendered by the mass media, common school curricula, or the homogenization of residential space are still mediated by distinctively local references. Moreover, although the circumference of most parts of everyday life may have increased with practices like commuting, most individuals still follow well-worn paths during most parts of their daily existence. Even more importantly, as a result of distinctive and cumulative historical patterns of class and other forms of social differentiation, the social complexity of regions has probably, on the whole, increased rather than decreased, bringing with it richer possibilities for interaction. This observation is enhanced by Urry's (1981b, page 464) argument that variations in local class structures have not only been ignored but are also of increasing importance: 'Important changes in contemporary capitalism are at present heightening the economic, social and political significance of each locality'. And, of course, the complexity of locality can be appropriated as a resource. Crossick (1977), for example, has argued that in the late nineteenth and early twentieth centuries the English middle class seized upon local politics as a means of exerting influence, an observation that might be extended to much of the current literature on urban social movements.

Four aspects of social action

To capture the sense of social action as discourse through and in a region is no easy task, and it is true to say that, as yet, this goal has not been achieved, although it is often enough hinted at. Programmes like those of the structurationist school remain provisional and all too often are linked to rather general statements backed up by three or four references to particular cases. A concentrated programme of theoretically informed empirical research is needed if this theoretical elaboration is not to occur in a vacuum . . . . In this final subsection I will consider only four particular pretheoretical aspects of this programme which seem to me to be amongst the most pressing. These may be considered as an affirmation of the importance of studying human agency and social structure as a duality, but more importantly as an historical and geographical inquiry into the respective variations in this duality, as they are fought through in conflict, in particular contexts.

Personality and socialization: the historical geography of life-path development Personality can be seen as the total constellation of an individual's psychological characteristics and it therefore subsumes the different aspects of personality, such as conception of *self* . . ., *identity* . . ., and *individuality*. Personality necessarily involves the three sets of relations that Giddens (1979a) considers necessary for a theory of the subject – the unconscious, practical consciousness, and discursive or reflexive consciousness. Seen in the contextual sense, personality is a constant *process* of 'internalization' or 'interiorization' of social relations along the course of a life path set within civil society. 'Social relations penetrate the body, structure its psychic ''central fire''; psychic structure is social structure alive in the heart of the body' (Bertaux and Bertaux-Wiaume, 1981, page 174). Personality, therefore, can never be universal. Rather it is a continuously negotiated and renegotiated expression of social and economic relations that vary, in other than their most basic form, according to locale and region.

Sève (1972; 1975; 1978) has provided the most complete interpretation of what a materialist and contextual theory of human personality might look like as a 'general theory of the concrete individual'. In this conception every adult personality appears as a series of sedimented activity-experiences carried forward in time at the tip of the life path. These activity-experiences have taught us 'how to do' and therefore 'how to be', messages which are constantly reaffirmed by the routine of everyday life, but which may also be reinterpreted in the light of new activity-experience – memory is an active not a passive process. Sève argues that at the core of personality is an historically (and, by implication, geographically) specific structure of allocation of time to particular activity-categories.[7] Personality, therefore, is an expression of the *objective social logic* of this variable set of activities into categories; in other words, you are what you do and you do what you are. [This representation, although now situated in the context of particular individuals, is not so very different from Bourdieu's (1977; 1980) notion of habitus.]

Quite obviously, a conception of personality like this could, if taken to its logical conclusion, come dangerously close to the idea of individuals as cultural dupes and as merely an encapsulated microcosm of social relations. This makes it even more imperative to stress the transformative side of the individual, which, in turn, must be seen as an aspect of the penetrative possibilities of the society and region in which she is situated. An individual is a product of social interaction rather than of individual action. Therefore the study of personality involves, necessarily and integrally, the study of *socialization* as a process of domination *and* resistance, the one dialectically linked to the other. Willis (1977), for example, shows how some working-class schoolchildren build up a resistance to mental work out of their resistance to the authority of school. Manual work is invested with seemingly opposite qualities to mental work – aggressiveness, solidarity, sharpness of wit, masculinity – and becomes a positive affirmation of freedom. Here resistance has the effect that working-class schoolchildren accept working-class jobs through 'free' choice. They willingly embrace their own repression. . . .

[T]he concept of personality as a continuously socialized trajectory varying with time and space in time and space [emphasizes that] . . . divisions in society do not occur just because of grand compositional relations such as class, sex, or race. Perhaps one of the most important . . . forces is the population cohort into which an individual is born,[8] since this influences an individual's life chances, subsequent history of socialization, and 'structure of feeling' (Williams, 1977; 1979; Thrift and Pred, 1981).[9] Particular cohorts, sorted by social groups, have different collective experiences of period and place, and, compared with other cohorts, they experience the impact of particular events differentially. These factors mark out their personality from those of other cohorts.

Penetration and the availability of knowledge The study of ideology and hegemony is tainted, seemingly inevitably, by functionalism. In the three most extreme cases, ideology becomes a level or instance which acts as a receiver of the impulses of the economy, or ideology is passed down to the working class by capital and uncritically accepted, or everything is reduced to a constitutent part of an overwhelming 'ideological state apparatus'.[10] In the contextual sense, neither ideology or hegemony can be seen in these ways. Rather, within the overall structure of discourse, certain social practices may have 'ideological' or 'hegemonic' effects which are not, however, necessarily functional to the continuation of the existing dominant order and which may, indeed, not have had their origins in that order (Urry, 1981a). And what effect these practices have on particular individual subjects is highly contingent and depends on the balance of other social practices. Certainly, it is doubtful that a general notion of ideology or hegemony can contribute much to the actual analysis of particular societies. Rather, it is more appropriate to consider, as has been pointed out above, the degree of understanding of or penetration into the existing social order that particular social groupings (and, therefore, the individuals within them) can achieve. Such an understanding will always be contingent and will always be linked to various limitations on that understanding, so that no social grouping will ever achieve more than a partial penetration into the conditions of its existence. Seen in this way – as a creative process of limitation and a limiting process of creation – a number of factors can be considered. An important one is variability in the availability of knowledge, which will depend upon the particular setting of a region and will be mediated by class and other group memberships as they are present in that region. At least five types of interrelated *unknowing* can exist in a locale or region at any one time (Thrift, 1979):[11]

(a) *unknown*, and not possible to know, in terms of being totally unknown at a particular time, either to a society, a locale, or a region, or to members of them;
(b) *not understood*, in terms of not being within the frame of meaning of a society, a locale, or a region, or not being within the frame of meaning of certain members of a society, locale, or region;
(c) *hidden*, in terms of being hidden from certain members of a society, locale, or region;
(d) *undiscussed*, in terms of being taken for granted as 'true' or 'natural', either by a society, a locale, or a region, or certain members of them; and
(e) *distorted*, in terms of being known only in a distorted fashion by a society, a locale, or a region, or certain members of them.

Studies on the spatially variable acquisition of *written*[12] knowledge and ideas have become quite common in social history, although usually at the large scale (for example, Burke, 1978; Martin, 1978; Darnton, 1979; Eisenstein, 1979). However, studies are beginning to appear on the process of acquisition of reading material with application to more local areas (for example, Judt, 1979; Ginzburg, 1980; Spufford, 1979; 1981; compare Williams, 1980). These studies show the strong effects of isolation on what a person knows. Similarly, studies of the specific effects of mass media are becoming more common (see Williams, 1980). Such studies must combine the diffusion of knowledge with information about the institutions through which this knowledge has diffused. Such institutions usually have had particular degrees of class allegiance and bias, which structure the information that is disseminated . . . An interesting but neglected part of this new field of study concerns the impact of certain kinds of knowledge on people's thought processes as they switch from practical to 'reflexive' or 'discursive' reason. Ginzburg (1980) provides one particularly useful example of this impact in his study of a sixteenth-century Italian miller. Having catalogued the written knowledge available to the miller, Ginzburg shows how it was systematically misappropriated because the miller had been socialized into a predominantly oral culture. Abstract words like 'matter', 'nature', 'unity', 'elements', and 'substance' were all related quite literally to cues from the particular region in which the miller lived. Even metaphors were taken in a quite literal sense; '*like* cheese' becomes cheese, for instance. And God *is* a father, God *is* a lord. These processes of misappropriation are crucial because they are political. They have strong parallels now. Yet we know so very little about them. When do people pass from practical reason to reflexive reason? What makes them leave off from what they are doing to think reflexively about it? The fact that the evidence of these inner thought processes must be indirect . . . does not mean that such processes cannot be studied.

The example of Ginzburg's (1980) miller also points to the crucial importance of language as a creative element of discourse in terms of a particular group's penetration of social practice within a particular region, for language is still primarily a practical tool that gains its meaning from doing as doing gains its meaning from language. Language is therefore always in a state of becoming (Pred, 1981b). It is a semantic field that shifts as the practices and projects of the material world alter, setting new limits as old ones are overtaken, inventing new meanings for old words, or bringing new words and meanings into existence. The limitations of language are intimately related to social practices carried out within a regional context . . . . Hélias (1978, page 334), for example, lists a whole web of words that have fallen out of use in the Breton language, their demise reflecting the shattering of one particular discourse:

> Nothing is left of my early civilisation but wreckage. There are still some trees, but no more forests. To speak only of its objects, as soon as they were dispersed, they lost almost all their meaning. Museums have been built for them; and sometimes, with touching care, the large room of some farmhouse has been reconstructed down to the last detail. But that room doesn't live anymore, doesn't work anymore.

Sociability and community The ability of any social grouping, set within a region, to penetrate the conditions of its existence must depend upon its social institutions and whether these are distinctive or are combined in a distinctive way. Yet we know very little, in any systematic fashion, about how particular *organizations* of social institutions allow, promote, or inhibit understanding. Thus:

> not all complex and lively communities are politically radical . . . there are good grounds for supposing that above a certain size, towns develop political patterns owing little or nothing to the presence or absence of active political debate among the populace. Other factors intervene, notably the predominant local occupations and the way in which *they* are organized, and the ideas and identity they help impose upon those who work in them. Sociability, the tendency to active public life rather than an isolated private existence, appears to have been most significant in small communities where the population was, if not economically homogenous, then at least economically interdependent (Judt, 1979, page 155).

There are some factors that are obviously local. For people living in certain areas of dispersed dwelling, the difficulty of meeting with one another is a very simple example. Factors like this have become less important over time, however, What is perhaps more important is the overall institutional context of sociability. Judt (1979), for instance, shows how in Provence, in the period from 1881 to 1914, communal festivals and church attendance declined in importance and were replaced or augmented by union meetings, political clubs, and cafés based on the same sense of community but bent to more consciously political ends . . . The problem of sociability is obviously linked to the interrelated problems of *sense of community* and *sense of place*, yet even the scale at which such factors work is problematic. What areas actually have a self-sufficient character? The answer may be a town, a village, a neighbourhood, or even a single street (compare Clark, 1973; Macfarlane, 1977; Calhoun, 1978; Joyce, 1980; Neale, 1981). Sense of community has presumably varied geographically and historically. Thus sense of community is assumed to have been greater in the classical European working-class communities prior to the Second World War than it is now (Joyce, 1980; MacIntyre, 1980; Therborn, 1980). And it is regarded as a stronger factor in multiethnic societies and in countries with certain communal institutions like the cafe, the street, and the market. Such a sense is:

> a matter of long term co-operation. Many of the results of this co-operation are not conscious goals in the minds of participants. More exactly, many actions may fit these 'goals' without being explicitly instrumental. At particular junctures people may decide to pursue one or another task of societal develoment; practices they may consider as instrumental are later taken for granted. At the simplest level, we all need to limit the range of possibilities which we take into consideration when choosing an action. Habit is by no means the least important way in which this is done; cultural rules are another; social constraints on the availability of information add to the limitation. The efficiency of habit and culture clearly depends on the familiarity of situations and events. Community both depends on this familiarity and helps to produce it. Being able to predict the behaviour of those with whom one must deal is one of the social advantages of community membership. This ability comes not only from long observation of particular persons *but* from the systematicity of the communal relationships. The former provides for collective definitions of relationships and the obligations they entail and expectations they justify. The latter increases people's investment in particular relationships, and causes them to be much more influenced by the wishes of others (Calhoun, 1980, pages 126–127).

But a sense of community and a sense of place can, of course, be a two-edged sword. Once again, it is a matter both of penetration *and* of limitation. Thus, Joyce (1980, page 116) links working-class deference in nineteenth-century factory towns precisely with sense of community and place:

> The *milieus* of ordinary life were impregnated with the authority and influence of an élite, so that the habit of neighbourhood community, expressed in the terminology of sociology as 'communal sociability', was the source of subordination as well as of class selfhood.

Once again, penetration and limitation are inextricably linked.

Conflict and capacity Each of the three aforegoing aspects of social action come together as part of a closely interrelated inquiry into the nature of confict and the *capacity* (Wright, 1978) of particular social groups living through particular regions to carry on class conflict and other forms of conflict. Capacity can be thought of here as the ability of different social groupings, themselves organized in particularly historically and geographically specific ways, to organize and then to carry on various historically and geographically specific kinds of opposition to other social groupings. In

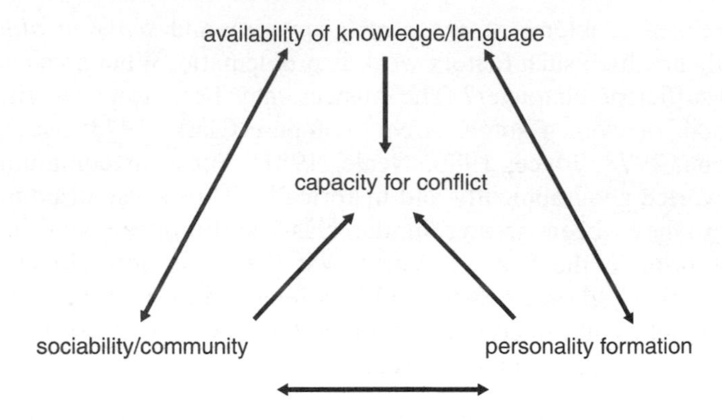

Fig. 22.4 Elements of conflict in context

particular, capacity can be measured by the ability of a group to penetrate the conditions of its existence and to express solidarity with that critique through the insertion of new social practices into its discourse. Marx's likening of the French peasantry to a 'sack of potatoes' is one (notoriously incorrect) example of a judgement on a social group's capacity. Capacity is clearly a function of the ability of a social grouping to produce transformative individuals (personality formation), particular forms of knowledge, and particular forms of sociability, each of these being inseparable from the others (see Fig. 22.4). None of these functions can be torn apart from the discourse of the region through which they are lived as social practices, for their very nature is tied up with the *particularity* of the interrelation of those practices, which together constitute a particular 'structure of choice' (Przeworski, 1977) with a unique form of *social logic* (Offe and Wiesenthal, 1980). [The classic example of such a logic is, of course, the rationality of practices implicit in Thompson's (1971) idea of a moral economy (see also Scott, 1976; Popkin, 1980; Bechhofer and Elliot, 1981).] Transgressions of this logic may result in what seem to be 'spontaneous' protests, but these are really only the visible part of an offended organization of practice in a region. As Groh (1979, page 278) points out:

> in the case of strikes which broke out apparently without warning and, in traditional terms, 'spontaneously', we have been able to show that the stereotypical formulation of a trade union or party press – that an industrial struggle had 'broken out with the elemental force of an act of nature – was usually just a rationalization of their own inability to identify the aims and causes of the action. This thesis is supported by the interesting fact that the police authorities observing labour conditions and organizations were often able to predict the location of strikes and their participants. The same applies to 'social protest' and 'collective violence'. Here, too, the outbreaks and development of the incidents were not spontaneous . . . to put it in extreme terms, in our field spontaneity is at worst a bourgeois myth and at best bad psychology. Until we have found a better word, we prefer to speak of non-organized rather than spontaneous actions, directing our attention to their specific logic.

Conclusions

In this paper, I have tried to show . . . that it is possible to conjoin a theory about social action, that of Marx, to the structurationist analysis of social action, utilizing the richness and the importance of what Marx called the 'active life process' as it *must* take place in space and time, but retaining the crucial element of determination. I have no doubt that it is the lack of a theory of social action that has disabled Marxism, both practically and theoretically, and that has led to it becoming, in practice, a force for emancipation and a force for great oppression (Dawe, 1979; Giddens, 1981). A more *human* (not humanist) Marxism is needed, one that can give *theoretical* respectability to the practical 'problems of motivating people to behave in altruistic and considerate, dignified, and conscientious ways without transcendental goals. This is not a matter of "ideals" or "morals" but of a daily practical mechanism of conduct, keyed-in to practices and institutions' (Hirst and Woolley, 1982, page 139).

The position I have outlined in this paper is very much a first approximation, open to revision as further work is carried out; comparatively little is yet known about so many of the issues that have been raised. But, whatever the problems, there is little doubt that it is becoming increasingly rare to find social action in space and time treated by social theorists as simply an afterthought or as the mere imprint of social structure or as belonging, in some way, to an autonomous realm of existence. Space and time are always and everywhere social. Society is always and everywhere spatial and temporal. Easy enough concepts, perhaps, but the implications are only now being thought through.

Notes

1 There is a strong distinction, not usually appreciated, between structuralism and the structural–genetic approach of Marx (compare Hall, 1980; Zeleny, 1980). Marx was *not* a structuralist.

2 I have chosen to concentrate on these three authors only. A number of other authors have what may at first appear as partially overlapping concerns, for instance, with the problem of nonfunctionalist theories of structure and agency (for example, Castoriadis, 1975; Dawe, 1979; Harré, 1978; 1979; Kosik, 1976; Sartre, 1964; 1976; Touraine, 1977; 1981; Urry, 1981a) or with the history of ego development (for example, Elias, 1978; 1982; Habermas, 1979). But these authors usually take an overwhelmingly compositional approach to these problems, generally, but not always, ignoring the way that they are embedded in space and time.

3 'Functionalism' . . . is a term that can be applied to a number of first-order mistakes that characterize too much of current social science. These include: (i) the attributing of 'needs' to social systems; (ii) the assumption that social systems are functionally ordered and cohesive; (iii) the imputing of a teleology to social systems; (iv) the characterization of effects as causes; and (v) the setting up of empirically unverifiable propositions via tautological statements. Parts of Marx's work were undeniably functionalist. Excellent critiques of functionalism (and its partner, reductionism) in Marxism can be found in Giddens (1979a; 1981) and Urry (1981a).

4 The French word *alterité* has recently been borrowed from nineteenth-century French hermeneutics and the even older form of the word in poetics. The first problem with using it here to describe a quality of newness is that it can be used only in the sense of the newness that results from intrinsically *social* action (interaction) which produces new forms of social action. The second problem is that Sartre (1976) has already brought the term into English in a somewhat different usage. 'Alterity' is still therefore, in part, a translator's device, an uncritical transposition into English which works only because English does not have such a word.

5 Sartre (1976) makes a useful distinction here between *comprehension*, the understanding of human activity in terms of the purposes of its agents, and *intellection*, the understanding of human activity which is not necessarily in terms of the purposes of its agents. In a similar vein, Bourdieu (1977) points out that we need not only a theory of practice but also a theory of theory and a theory of the relation of theory to practice (and vice versa).

6 The labour process is probably the main link between the compositional and the contextual. First, it is the site of an important series of mediating concepts to do with the management of time. Second, 'the labour process (properly conceived) is the very locus of structuration since, in Marx's words, "labour is, first of all, a process between man and nature" through which "he acts upon external nature and changes it, and in this way he simultaneously changes his own nature"' (Gregory, 1982, page 214). Of course, a lot of weight in this sentence is borne by 'properly conceived'. Third, although the labour process is a pivotal concept it remains true that we have yet to learn how to handle it contextually when more than one labour process is found in a region.

7 I disagree with the crude analogies and the humanist emphasis in Sève's work.

8 Obviously, the interaction of the various models of the development of ego-identity with this factor are important here, although I will not go into them in this paper (for example, see Erikson, 1959; 1963; 1975; Piaget, 1965; Kohlberg, 1971; Habermas, 1979).

9 Williams (1977; 1979) is trying to capture with the concept of 'structure of feeling' the quality of *presence* of a particular society, social group, or region; that is, how the society, social group, or region appears to those who live through it at the point that they live it. This has . . . connections with other phenomena, for example, the commodification of memory currently taking place (see Davis, 1979).

10 It is ironic that Althusser, who was able to see ideology as a matrix of social practices, displaced nearly all such practices into state apparatuses. Clearly, many of the 'apparatuses' that Althusser considered to be in the domain of the state, for example, the family, exist in civil society (see Larrain, 1979; Therborn, 1980; Urry, 1981a).

11 This schema bears some resemblance to that proposed by Habermas (1982, page 264) in a different (compositional) context.

12 There are, of course, many other forms of knowledge than just the written. I take the written as one example of what is needed.

Selected references

Abercrombie N, Hill S, Turner B S, 1980 *The Dominant Ideology Thesis* (George Allen and Unwin, Hemel Hempstead, Herts)

Abrams, P, 1980, 'History, sociology and historical sociology' *Past and Present* number 87, 3–16

Abrams, P, 1982 *Historical Sociology* (Open Books, Shepton Mallet, Somerset)

Albrow, M, 1974, 'Dialectical and categorical paradigms of a science of society' *Sociological Review* **22** 183–201

Anderson, P, 1980 *Arguments within English Marxism* (New Left Books, London)

Baker, A R H, 1979, 'Historical geography: a new beginning?' *Progress in Historical Geography* **3** 560–570

Bechhofer F, Elliott B, 1981, 'Petty property: the survival of a moral economy' in *The Petite Bourgeoisie. Comparative Studies of the Uneasy Stratum* Eds F Bechhofer, B Elliot (Macmillan, London) pp 182–199

Berger P, Luckmann T, 1966 *The Social Construction of Reality. A Treatise in the Sociology of Knowledge* (Penguin Books, Harmondsworth, Middx)

Bhaskar R, 1979 *The Possibility of Naturalism. A Philosophical Critique of the Contemporary Human Sciences* (Harvester Press, Hassocks, Sussex)

Bourdieu P, 1977 *Outline of a Theory of Practice* (Cambridge University Press, Cambridge)

Bourdieu P, 1980 *Le Sens Pratique* (Editions de Minuit, Paris)

Burke P, 1978 *Popular Culture in Early Modern Europe* (Maurice Temple Smith, London)

Calhoun C J, 1978, 'History, anthropology and the study of communities' *Social History* **3** 363–373

Calhoun C J, 1980, 'Community: towards a variable conceptualization for comparative research' *Social History* **5** 105–129

Carlstein T, 1981, 'The sociology of structuration in time and space: a time–geographic assessment of Giddens's theory' *Svensk Geografisk Arsbok* **57** 41–57

Castoriadis C, 1975 *L'Institution Imaginaire de la Société* third edition (Editions du Seuil, Paris)

Clark D B, 1973, 'The concept of community: a re-examination' *Sociological Review* **21** 63–82.

Crossick G, 1977, 'The emergence of the lower middle class in Britain: a discussion' in *The Lower Middle Class in Britain 1870–1914* Ed. G. Crossick (Croom Helm, London) pp 11–60

Darnton R, 1979 *The Business of the Enlightenment. A Publishing History of the Encyclopédie 1775–1800* (Harvard University Press, Cambridge, MA)

Davis F, 1979 *Yearning for Yesterday. A Sociology of Nostalgia* (Free Press, New York)

Dawe A, 1970, 'The two sociologies' *British Journal of Sociology* **21** 207–218

Dawe A, 1979, 'Theories of social action' in *A History of Sociological Analysis* Eds T. Bottomore, R. Nisbet (Heinemann Educational Books, London) pp 362–417

Dunford M F, 1981, 'Historical materialism and geography' research paper 4, Department of Geography, University of Sussex, Brighton, England

Eisenstein E L, 1979 *The Printing Press as an Agent of Change* 2 volumes (Cambridge University Press, Cambridge)

Elias N, 1978 *The Civilizing Process. The History of Manners* (Basil Blackwell, Oxford)

Elias N, 1982 *The Civilizing Process. State Formation and Civilization* (Basil Blackwell, Oxford)

Erikson E H, 1959 *Identity and the Life Cycle* (W W Norton, New York)

Erikson E H, 1963 *Childhood and Society* second edition (W W Norton, New York)

Erikson E H, 1975 *Life History and the Historical Moment* (W W Norton, New York)

Ferrarotti F, 1981, 'On the autonomy of the biographical method' in *Biography and Society. The Life History Approach in the Social Sciences* Ed. D. Bertaux (Sage, Beverley Hills, CA) pp 19–27

Foster J, 1974 *Class Struggle and the Industrial Revolution* (Weidenfeld and Nicholson, London)

Foucault M, 1972 *The Archaeology of Knowledge* (Tavistock Publications, Andover, Hants)

Foucault M, 1977 *Discipline and Punish* (Allen Lane, London)

Giddens A, 1976 *New Rules of Sociological Method* (Hutchinson, London)

Giddens A, 1977 *Studies in Social and Political Theory* (Macmillan, London)

Giddens A, 1979a *Central Problems in Social Theory, Action, Structure and Contradiction in Social Analyses* (University of California Press, Berkeley, CA)

Giddens A, 1979b 'Postscript 1979' in *The Class Structure of the Advanced Societies* second edition (Hutchinson, London)

Giddens A, 1981 *A Contemporary Critique of Historical Materialism* (Macmillan, London)

Ginzburg C, 1979, 'Clues. Roots of a scientific paradigm' *Theory and Society* **7** 273–288

Ginzburg C, 1980 *The Cheese and the Worms. The Cosmos of a Sixteenth Century Miller* (Routledge and Kegan Paul, Henley-on-Thames, Oxon)

Gouldner A W, 1980 *The Two Marxisms. Contradictions and Anomalies in the Development of Theory* (Macmillan, London)

Gregory D, 1981, 'Human agency and human geography' *Transactions of the Institute of British Geographers* new series **6** 1–18

Gregory D, 1982, 'Solid geometry: notes on the recovery of spatial structure' in *A Search for Common Ground* Eds. P Gould, G Olsson (Pion, London) pp 187–219

Groh D, 1979 'Base-processes and the problem of organization: outline of a social history research project' *Social History* **4** 265–283

Habermas J, 1979 *Communication and the Evolution of Society* (Heinemann Educational Books, London)

Habermas J, 1982, 'A reply to my critics' in *Habermas. Critical Debates* Eds J B Thompson, D Held (Macmillan, London) pp 219–283

Hägerstrand T, 1970, 'What about people in regional science?' *Papers of the Regional Science Association* **24** 7–21

Hägerstrand T, 1973, 'The domain of human geography' in *Directions in Geography* Ed. R J Chorley (Methuen, Andover, Hants) pp 67–87

Hägerstrand T, 1974, 'Tidgeografisk beskrivning – syfte och postulat' *Svensk Geografisk Arsbok* **50** 86–94

Hall S, 1980, 'Cultural studies: two paradigms' *Media, Culture and Society* **2** 57–72

Hareven T K, 1975, 'Family time and industrial time. Family and work in a planned corporation town, 1900–1924' *Journal of Urban History* **1** 365–389

Hareven T K, 1982 *Family Time and Industrial Time. The Relationship between the Family and Work in a New England Industrial Community* (Cambridge University Press, Cambridge)

Harré R, 1978, 'Architectonic man: on the structuring of lived experience' in *Structure, Consciousness and History* Eds R H Brown, S M Lyman (Cambridge University Press, Cambridge) pp 139–172

Harré R, 1979 *Social Being. A Theory for Social Psychology* (Basil Blackwell, Oxford)

Hélias P, 1978 *The Horse of Pride. Life in a Breton Village* (Yale University Press, New Haven, CT)

Heller A, 1982, 'Habermas and Marxism' in *Habermas. Critical Debates* Eds J B Thompson, D Held (Macmillan, London) pp 21–41

Hirst P, Woolley P, 1982 *Social Relations and Human Attributes* (Tavistock Publications, Andover, Hants)

Hollis M, 1977 *Models of Man* (Cambridge University Press, Cambridge)

Joyce P, 1980 *Work, Society and Politics. The Culture of the Factory in Later Victorian England* (Harvester Press, Hassocks, Sussex)

Judt T, 1979 *Socialism in Provence 1871–1914. A Study in the Origins of the Modern French Left* (Cambridge University Press, Cambridge)

Keat R, Urry J, 1982 *Social Theory as Science* second edition (Routledge and Kegan Paul, Henley-on-Thames, Oxon)

Kennedy B A, 1979, 'A naughty world' *Transactions of the Institute of British Geographers* **4** 550–558

Kohlberg L, 1971, 'From is to ought' in *Cognitive Development and Epsitemology* Ed. T. Mischel (Academic Press, New York) pp 153–176

Kosik K, 1976 *The Dialectics of the Concrete* (D Reidel, Dordrecht, The Netherlands)

Lakoff G, Johnson M, 1980 *Metaphors We Live By* (Chicago University Press, Chicago, IL)

Larrain J, 1979 *The Concept of Ideology* (Hutchinson, London)

Layder D, 1981 *Structure, Interaction and Social Theory* (Routledge and Kegan Paul, Henley-on-Thames, Oxon)

Macfarlane A, 1977, 'Historical anthropology and the study of communities' *Social History* **5** 631–652

MacIntyre S, 1980 *Little Moscows. Communism and Working Class Militancy in Inter-War Britain* (Croom Helm, London)

Martin H J, 1978, 'The bibliotheque bleue. Literature for the masses in the Ancien Regime' *Publishing History* **3** 70–102

Molina V, 1979, 'Notes on Marx and the theory of individuality' in *On Ideology* Centre for Contemporary Cultural Studies (Hutchinson, London) pp 230–258

Neale R S, 1981 *Bath. A Social History 1680–1850* (Routledge and Kegan Paul, Henley-on-Thames, Oxon)

Offe C, Wiesenthal H, 1980, 'Two logics of collective action: theoretical notes on social class and organisation form' in *Political Power and Social Theory* volume 1, Ed. M Zeitlin (JAI Press, Greenwich, CT) pp 67–115

Piaget J, 1965 *The Moral Development of the Child* (Free Press, New York)

Popkin S, 1980, 'The rational peasant. The political economy of peasant society' *Theory and Society* **9** 411–471

Pred A, 1981a, 'Production, family and "free-time" projects: a time–geographic perspective on the individual and societal change in nineteenth-century US cities' *Journal of Historical Geography* **7** 3–36

Pred A, 1981b, 'Societal reproduction and the time–geography of everyday life' *Geografisker Annaler* **63** series B, 5–22

Przeworski A, 1977, 'Proletariat into a class: the process of class formation from Karl Kautsky's *The Class Struggle* to recent controversies' *Politics and Society* **4** 342–401

Ranson S, Hinings B, Greenwood R, 1980, 'The structuring of organisational structures' *Administrative Science Quarterly* **25** 1–17

Ricouer P, 1981 *Hermeneutics and the Social Sciences. Essays on Language, Action and Interpretation* (Cambridge University Press, Cambridge)

Sahlins M, 1976 *Culture and Practical Reason* (Chicago University Press, Chicago, IL)

Samuel R, 1981, 'People's history' in *People's History and Socialist Theory* Ed. R Samuel (Routledge and Kegan Paul, Henley-on-Thames, Oxon) pp xiv–xxxvii
Sartre J-P, 1964 *The Problem of Method* (Methuen, Andover, Hants)
Sartre J-P, 1976 *Critique of Dialectical Reason. I. Theory of Practical Ensembles* (New Left Books, London)
Sayer A, 1981, 'Abstraction: a realist interpretation' *Radical Philosophy* number 28, 6–16
Sayer D, 1979 *Marx's Method. Ideology, Science and Critique in Capital* (Harvester Press, Hassocks, Sussex)
Scott J, 1976 *The Moral Economy of the Peasant* (Yale University Press, New Haven, CT)
Selbourne D, 1980, 'On the methods of *History Workshop*' *History Workshop* number 9, 150–161
Sève L, 1972 *Marxisme et Théorie de la Personnalité* second edition (Editions Sociale, Paris)
Séve L, 1975 *Marxism and the Theory of Human Personality* (Lawrence and Wishart, London)
Sève L, 1978 *Man in Marxist Theory and the Psychology of Personality* (Harvester Press, Hassocks, Sussex)
Simpson G G, 1963, 'Historical science' in *The Fabric of Geology* Ed. C C Albritton (Stanford University Press, Stanford, CA) pp 24–48
Spufford M, 1979, 'First steps in literacy: the reading and writing experiences of the humblest seventeenth century autobiographers' *Social History* **4** 407–435
Spufford M, 1981 *Small Books and Pleasant Histories. Popular Fiction and Its Readership in Seventeenth-Century England* (Methuen, Andover, Hants)
Stedman Jones G, 1974, 'Working-class culture and working class politics in London, 1870–1900. Notes on the remaking of a working class' *Journal of Social History* **4** 460–508
Stone L, 1979, 'The revival of narrative' *Past and Present* number 85, 3–24
Therborn G, 1980 *The Ideology of Power and the Power of Ideology* (New Left Books, London)
Thompson E P, 1963 *The Making of the English Working Class* (Weidenfeld and Nicolson, London)
Thompson E P, 1971, 'The moral-economy of the English crowd in the eighteenth century' *Past and Present* **50**, 76–136
Thrift N J, 1977 *An Introduction to Time–Geography. Concepts and Techniques in Modern Geography 13* (Geo-Abstracts, Norwich)
Thrift N J, 1979, 'Limits to knowledge in social theory: towards a theory of human practice' available as a mimeograph from Department of Human Geography, Australian National University, Canberra, Australia
Thrift N J, 1981, '"Owners' time and own time": the making of a capitalist time-consciousness. 1300–1880' in *Space and Time In Geography* Ed. A R Pred (C W K Gleerup, Stockholm) pp 56–84
Thrift N J, 1984 *Spontaneity or Order? Time Consciousness in Medieval and Early Modern England* (Cambridge University Press, Cambridge) forthcoming
Thrift N J, Pred A R, 1981, 'Time–geography: a new beginning' *Progress in Human Geography* **5** 277–286
Touraine A, 1977 *The Self-Production of Society* (Chicago University Press, Chicago, IL)
Touraine A, 1981 *The Voice and the Eye. An Analysis of Social Movements* (Cambridge University Press, Cambridge)

Tribe K, 1981 *Genealogies of Capitalism* (Routledge and Kegan Paul, Henley-on-Thames, Oxon)
Urry J, 1981a *The Anatomy of Capitalist Societies. The Economy, Civil Society and the State* (Macmillan, London)
Urry J, 1981b, 'Localities, regions and social class' *International Journal of Urban and Regional Research* **5** 455–474
Veltmeyer H, 1978, 'Marx's two methods of sociological analysis' *Sociological Inquiry* **48** 101–112
Weintraub R R, 1978 *The Value of the Individual. Self and Circumstances in Autobiography* (Chicago University Press, Chicago, IL)
Williams R, 1977 *Marxism and Literature* (Oxford University Press, London)
Williams R, 1979 *Politics and Letters* (New Lcft Books, London)
Williams R, 1980 *Culture* (Fontana Books, London).
Willis P E, 1977 *Learning to Labour. How Working-Class Kids get Working-Class Jobs* (Saxon House, Farnborough, Hants)
Willis P E, 1978 *Profane Culture* (Routledge and Kegan Paul, Henley-on-Thames, Oxon)
Wilson S, 1981, 'Conflict and its causes in Southern Corsica, 1800–1835' *Social History* **6** 33–69
Wright E O, 1978 *Class, Crisis and the State* (New Left Books, London)
Wright Mills C, 1959 *The Sociological Imagination* (Oxford University Press, London)
Yeo E, Yeo S, 'Ways of seeing: control and leisure versus class and struggle' in *Popular Culture and Class Conflict 1590–1914. Explorations in the History of Labour and Leisure* Eds E Yeo, S Yeo (Harvester Press, Hassocks, Sussex) pp 128–154
Zeleny J, 1980 *The Logic of Marx* (Basil Blackwell, Oxford)

23 Stephen Pile

Human Agency and Human Geography Revisited: A Critique of "New Models" of the Self

Excerpts from: *Transactions of the Institute of British Geographers: New Series* 18, 122–139 (1993)

Introduction

In this essay, I explore the ways in which 'the self' has been conceptualized in recent human geography, and how the relationships between individuals and their social situation have been understood.

There are of course as many approaches to geography as geographers, but here post-positivist social geography is taken to have two dominant perspectives: humanism . . . and historical materialism . . . . I will argue that both these approaches have produced one-sided accounts of 'the social'; and that this is because of their conceptual splitting of the social into context and intentionality (by the humanists) and into structure and agency (by the historical materialists). Even the developing *rapprochement* of the late 1980s between humanistic and historical materialist approaches has failed to alter the basic dislocation of structure and agency. A search needs to be instigated into alternative models of the self, as a means of understanding the position of the person within the social. For, as Sibley (1991, page 33) argues, 'an analysis of the self can be seen as integral to theoretical accounts of social space'. In this paper, I offer an alternative. I argue that it is inconceivable that subjectivity can be understood, and therefore that a truly *human* geography can be imagined, without drawing on the insights of psychoanalysis.

The problem is this: after the division of 'the social' into structure and agency (or into context and intentionality), structure (context) is seen as external while agency (intentionality) is seen as internal. The effect of externalizing structure is to make it taken-for-granted (not yet known) and impersonal (denying the personal in the social). The effect of internalizing agency is to make it individual, uniquely ours (so there can be no irremovable barriers to self-awareness), and to relegate social influences to role playing or the internalization of social norms. All the while the unconsciousness is systematically erased from these positions. This is no accident, but necessary if structure and agency are to be held conceptually separate. Psychoanalytic theory, in its theories of the unconscious, describes how the social enters, constitutes and positions the individual. Similarly, by showing that desire, fantasy and meaning are a (real) part of everyday life, it shows how the social is entered, constituted and positioned by individuals. One problem, for example, that the 'new models' share is that they cannot specify psychic structures, or say how or where or why particular psychic structures form where/when they do, or say how people are different, or where personality comes from, or how people experience personality, or precisely how people become people, or how it is that they can set out to make history.

This essay begins with a discussion of the search for the self conducted by humanistic geographers in the mid 1970s; two key moments are identified in this search, namely phenomenology (especially Tuan and Buttimer) and symbolic interactionism (especially Ley). Nevertheless, by the end of the 1970s, both the agenda and practice of humanistic geography were coming under attack; most decisive was Derek Gregory's (1981) intervention, and I therefore deal with Gregory's article in some detail in the second section. This has led to the marginalization of humanistic geography; where debates about the constitution of the human subject have been stifled by Gregory's introduction of the rational (and one-sided) category of human agency, though against his own intentions. In the third section, I look at the subsequent development of humanistic and historical materialist geography; I show that each has developed its own variant of the same model of the self. I argue that more has been lost than gained by the (attempted) murder of humanistic geography, and that both humanistic and historical materialist geography share a fatal splitting of structure from agency. In two final sections, I establish that psychoanalytic theory offers a different theory of the self, one which neither denies nor relies on the structure–agency dichotomy; I demonstrate its deeper understanding of the self by returning to the themes already raised by geographers, but concentrating on two aspects of the fragmented self: the 'unknowing' subject and the 'placed' subject.

Throughout this essay, I have selected authors with whom I have a great deal of sympathy; even so, these are also authors whom I take to be symptomatic of the condition of the geographical imagination, though I would not consider them to be either typical or representative. Moreover, in a paper such as this, it is impossible to do more than illustrate something of the nature of psychoanalysis and current views of its significance in cultural theory (see Frosh, 1987; Brennan, 1989; Donald, 1991). And, in making this argument, I make a number of generalizations and skim over certain complexities in the geographical literature; I hope that I will be forgiven these. My aim in producing this critical assessment is to recover a debate about the relationship between subjectivity, space and society, which I believe has been consistently marginalized for the last decade.

In search of the self

The turn to spatial analysis in the 1960s provoked two fundamental forms of critique; one was based in historical materialism, but it was the humanistic geographers who objected to positivism's implicit models of human subjectivity. In contrast to positivism, it was their intention to adopt a model of 'man', in which 'he' was centred (as producer and product), self-reflexive and self-conscious, intentional and active.[1]

> The purpose of the humanist campaign was to put man, in all his reflective capacities, back into the center of things as both a producer and a product of his social world and also to augment the human experience by a more intensive, hence self-conscious reflection, upon the meaning of being human (Ley and Samuels, 1978, page 7).

The question of meaning and value was placed at the heart of humanistic concern; in contrast to the spatial sciences, which were grounded in naturalism, Ley (1981b, page 214) warned.

> the facts of human geography cannot be viewed independently of a subject whose concerns confer their meaning, a meaning that directs subsequent action. Unlike the natural sciences, then, the social sciences cannot escape the task of interpreting the domain of consciousness and subjectivity.

Although most humanistic geographers shared this distaste for passive models of man, they were divided in their assumptions about the nature of the relationship between man and his lived world. For example, on the one hand, Anne Buttimer (1976), drawing on phenomenology, saw the self as essentially personal, though the 'essential' self, living in shared life-world horizons, is able to relate to others through intersubjectivity. On the other hand, David Ley believes that the 'experiential' self is not uniquely personal, the distinctiveness of the individual's inner life derives from the symbolic interactions between individuals: 'life-world is an inter-subjective one of shared meanings, of fellow men with whom he engages in face-to-face relationships' (1977, 505). The question these writers address relates to the origins, nature and interpretation of the inter-subjective meaning; the resolution relates to the model of consciousness and social context adopted.

This early debate about social being was developed using two key philosophies (see Gregory, 1981, page 10 and Gregory, 1989a, page 360): phenomenology (e.g. Buttimer, 1976) and symbolic interaction[ism] (e.g. Ley, 1977, 1981b). In the next two subsections, I will look at how these geographers built their models of man from these sources.

Phenomenology and the lifeworld

For geographers drawing on phenomenology, it is 'man's . . . special capacity for thought and reflection' that is held to be distinctive (Tuan, 1976, page 267). This perspective also raised special questions about the relationship between people, territory and nature: for example, they 'are able to hold territory as a concept, envisage its shape in the mind's eye, including those parts they cannot currently perceive' and 'the quality of human emotion and thought gives place a range of human meaning inconceivable in the animal world' (Tuan, 1976, page 269). Phenomenology provided a people-centred form of knowledge based in human awareness, experience and understanding; which were seen to be necessary to the study of, and conscious reflection on, the meaning of being human, of being located in time and space.

> The phenomenological notion of intentionality suggests that each individual is the focus of his own world, yet he may be oblivious of himself as the creative center of that world . . . Each knower should recognize himself as an intentional subject, i.e., as a knower who uses words – intended meanings – to render his intuitions objective and communicable (Buttimer, 1976, page 279).

It is argued, each knower is situated in a pre-existing social world, or 'lifeworld'. The lifeworld is characterized by sets of unquestioned meanings and routines, which determine behaviour. The lifeworld may be likened to an ocean in which individuals float like icebergs, and where the depths and dynamics of the sea can be only vaguely sensed. The demand of phenomenology is that the knower should return to the self, in order to expose the lifeworld, by bringing it to consciousness.

> A core concern of 'pure phenomenology' was the analysis and interpretation of consciousness, particularly the conscious cognition of direct experience. One endeavours to peel off successive layers of a priori judgement and to transcend all preconceptions in order to arrive at a consciousness of pure essences (Buttimer, 1976, page 279).

The ability to know is placed at the centre of the phenomenological endeavour, but meaning is not found entirely within the knower's consciousness, nor can it be found exclusively in direct experience. Meanings are not simply revealed by reflection, interpretations also have to be shared with others. Indeed, the notion of intersubjectivity is essential because it allows the geographer to understand personal and collective dimensions of behaviour. Nevertheless, Buttimer (1976, page 288) identifies a problem: 'we need a language and a set of categories which will enable us to probe lifeworld experience and to communicate about it'.

This is not just a peripheral question of using the right words, but a central problem because it stems from the presumed relationship between mind, essence and behaviour. If each knower is to successively strip away a priori judgement, then this will mean the progressive suspension of a priori language; phenomenological knowledge is incommunicable because it must deny the social conventions of communication. The problem stems from the dichotomy between inner mind and outer behaviour; the true, inner self is covered by an untrue, outer social world. The phenomenologist acts as archaeologist, digging ever deeper into the inner depths; but the archaeo-phenomenologist can never be sure if the true self has been found, especially because it is the already-buried (false) self which does the digging, using the tools provided by the lifeworld. Because it wants to disinter 'essential meanings', phenomenology forecloses on an investigation of, not only the opaqueness of communication or why it is opaque at all, but also the role of communication in the (re)production of the subjective and experiential self in the contextual and material world. Phenomenology has radical intentions, to question the unquestioned, but suggesting a (true) inner world and a (false) outer world denies that there is an internal relationship between them; one merely covers the other, though this cannot be explained.

Symbolic interactionism and social context

Symbolic interactionism is mainly concerned with the symbolic life of social worlds; it emphasizes the links between symbols of all kinds, and the way in which individuals construct and subsequently maintain their self-images (see Rock, 1979). These self-images are both taken to be symbolic expressions of the individual self (subjectivity), and also shared according to the type of interaction (intersubjectivity), but within the boundaries of the social setting (contextuality). This model provided a people-centred alternative for those humanistic geographers who were concerned about the excessive idealism of phenomenology. Thus, Ley (1978, page 45) argued that

> . . . the environment is not in the head. Consciousness cannot break loose from a concrete time–space context, from the realities of everyday living; notions of pure consciousness are as much an abstraction from human experience as any isotropic plain.

The realities of everyday life are negotiated by people in concrete contexts, and so, Ley argues, place is socially constructed (in contrast to both phenomenology and positivism). The concept of place is pivotal because it is the contact zone between brute reality (i.e. context and environment) and symbolic meaning (i.e. shared meanings and self-image). This argument relies on assumptions about the relationship between the self, symbolic meaning and the realities of everyday living. These might be teased out through Ley's discussion of place: the home.

> The home is the most articulate landscape of expression of the self and can reinforce either a positive self-image or, in the case of dreary public housing in an unwanted location, it may sustain an identity of a peripheral and low status member of society with little ability to mould his [sic] environment. There is, then, a reciprocal symbolic interchange between people and places . . . Like other commodities, space is engaged not only as a brute fact, but also as a product with symbolic meaning (Ley, 1981b, page 220).

At the centre of Ley's analysis is not the home, but the identity of the home dweller. The home is a symbolic landscape for an individual. The home is set within a context of social inequality, which sets limits on the ability of the home dweller to change their home. It is through this discussion of social inequality that Ley seeks (a) to avoid a fixation upon consciousness and (b) to deal with context, by suggesting the preconditions and consequences of thought and action (see Ley, 1981a, page 252). Elsewhere, Ley describes the interchange between place and identity as 'dialectical'; arguing that the social world is 'the product of human creativity', but the social world also has 'a certain autonomy', though 'their autonomy is always contingent' (1978, page 52).

Though using concepts like 'reciprocal' and 'dialectical', Ley offers a dislocated analysis of the internal relationship between people and place; for example, the concept 'reciprocal' does not imply that 'home' would simply be an expression of the self, or merely reinforce a self-image. This displacement is the result of certain presuppositions: (1) there is a brutal world of a priori neutral meaning, which can therefore sustain either negative or positive self-images; but (2) people's experience is social, involving relationships with others; and, (3) the self is social, arising and developing from symbolic interactions. The brutal world (of causes) and the symbolic world (of reasons) are not each other's dialectical or reciprocal opposite, because different things are going on in each of these worlds.

Ley's sense of reciprocity and dialectics enables two important insights: first, that there is an internal relationship between people and place; and second, that there is a certain symmetry here. On the other hand, these ideas cannot differentiate between different kinds of internal relationship, and they cannot describe asymmetrical relationships. For example, people encounter place as both brute reality and symbolic product, but it remains unclear how brute reality can be experienced except as a symbolic product. When dealing with personhood, symbolic interactionism denies the possibility that there may be a place in the mind where 'brute facts' do not apply; it denies landscapes and environments that are purely mental, and which conform to no rules but their own.

To recapitulate: the humanistic project is a sustained inquiry into the ways in which human subjects internalize – at a variety of 'levels', from the body itself through levels of being, experience and reflection – senses of space, place, environment and landscape that 'constitute' these subjects as selves 'in the world' (Philo, personal communication). On the other hand, phenomenology and, to a lesser extent, symbolic interactionism seemed to some to be unable to provide convincing accounts either of personal experience or of social context (see Smith, 1979).

The attempted murder of humanistic geography[2]

By the end of the 1970s, the humanistic geographers' agenda for understanding the lived world, set out most clearly in Ley and Samuels' 1978 edited collection, was becoming questioned (for example, by Smith, 1979). Eventually, Gregory tried to terminate the project:

> . . . when its protocols are translated into substantive work its neo-romantic vision of what critique entails often seems to invite a withdrawal *from* (or at least a contemplation, even a decoration *of*) the world, rather than any sustained engagement *with* it. The signs are not difficult to see: the casual ransacking of fictional writing as a ready means of recovering the most obvious images of intentionality, prised away from the material structures which help give them their effectivity, and divorced from any serious recognition of what Steiner calls 'the sociology of the text, and of our relations to the text', the exasperating interrogation of the mundane and the transparently trivial, devoid of any attempt to locate social actions in wider sequences of social reproduction and transformation, and blinded by a belief in the self-legitimating power of 'experience' to the need for any sort of theoretical effort or examination; and the mannered preoccupation with stylistic form and literary experimentation, which annihilates communication by dressing up the utterly familiar shape of its arguments in the tattered and violently coloured weeds of a latterday Dada or the beautifully stitched and carefully draped silks of the Emperor's tailor (Gregory, 1981, page 2).

Between violence and beauty, Gregory had stitched up humanistic geography. But Gregory is working with a particular conception of knowledge, and critique; his reading itself a stylized piece of rhetoric. The question is whether more was lost than gained. We must return to this excerpt and look more closely at Gregory's argument. Gregory is complaining: on the one hand, about the humanistic geographer's concern with intentionality, human action, and the lived world in its tiniest details; and, on the other, about their lack of consideration of the material context within which people live their lives. In humanism, the lifeworld is under-bounded and one-sided; paradoxically, covering too much and too little.

Thus, Gregory surmises that the problems of humanistic geography stem from its inability to understand the relationship between structure and agency (Gregory, 1981, page 2). The terms of the debate are now set: Gregory proceeds to discuss the work of Vidal de la Blache, E. P. Thompson and Anthony Giddens. These writers are very different, but each he believes can contribute something to a fuller conceptualization of the structure–agency duality. They offer a way of understanding the 'boundedness' and contingency of practical life (page 5). From these writers, Gregory learns that the real problem is

> to find a real model for the social process which allows an autonomy to social consciousness within a context which, in the final analysis, has always been determined by social being. (Thompson, 1978, p. 81; cited by Gregory, 1981, page 7)

Thompson, having identified the problem, is then found wanting in terms of an answer. Instead, Giddens's version is preferred, because his framework demonstrates the essential 'recursiveness' of social life (Gregory, 1981, page 8). Recursiveness describes the way in which actors reproduce the systems of communication, power and sanction by routinely drawing on existing structures of signification, domination and legitimation. Thus, in drawing on these structures

> the actors involved are displaying some degree of 'penetration' of practical life, whether they are able to verbalize their knowledge or not, and that consequently *structure is not a constraint on or a barrier to action but is instead essentially involved in its reproduction* (Gregory, 1981, page 10, emphasis in original).

People are now actors, who to some degree penetrate their lifeworld; but, whether or not they can speak its name, it is structure that is the medium for the reproduction of action. But, as Anthony Giddens (1976, page 127) argues, 'structures only exist as the reproduced conduct of situated actors with definite intentions and interests'. This insight is lost in Gregory's anti-humanistic rhetoric; this oversight is only partly rectified later when he argues that 'society does not exist independently of human activity' (Gregory, 1982, page 255). For Gregory, subjectivity must be grounded; actors' understanding of their world is 'inescapably embedded in *but irreducible to coincident sets of structural relations*' (page 256). Humanists may have been 'incapable of overcoming the limitations of their mechanistic determinism and voluntaristic idealism' (page 10), but then so is Gregory's schema. Because the point of contact between human agency and social structure is understood '*at the level of social relations*' (page 256), structures have a mechanical autonomy and agency has an irreducible idealism. This view is an inversion of Ley's image of the dialectic between consciousness and context, but where Ley's point of contact is at the level of human experience.

The evisceration[3] of the person from the social, ingrains the problem: 'social systems are both the medium and the outcome of the practices that constitute them: the two are recursively separated and recombined' (Gregory, 1982, page 11). Having separated practices (the situated actions of individuals) from social systems (the situating sructures of society), the difficulty for Gregory (as for humanistic geography) is to identify the mechanisms of recombination. He again opts for a dualism: on the one hand, Gregory identifies a cultural sphere which bears the meaning of men's actions; on the other, there is an economic sphere which is independent of man's will. Culture is the realm of discourse, and the economic, of silence; there can never be a dialogue between discourse and silence, and so the economic always remains determinant in the last instance.

The structure–agency split has led Gregory to valorize structure at the expense of people's actions, even while he claims that they are 'simultaneous equations which cannot be solved separately' (page 16). Because questions of human agency become questions of *determinate effects* or *determination* (pages 13 and 15 respectively), and of the (materialist) properties and (a priori) constitution of structural elements (page 15): namely, the economy and culture.

Even the term human agency serves to convert the individual into a rational (and rationally recoverable) category. Thus, the voluntarist idealism of the individual and the mechanistic determinism of the economy are thus *only* mediated by culture; while this formulation avoids one pitfall of reductionism, it simultaneously reduces people to the one-sided structural category of 'agency'. It remains in doubt as to whether this can be seen as 'a major advance' (page 10) on humanistic geography; which simultaneously avoided reductionism (a charge never levelled at humanistic geography) and placed man at the centre of things.

The humanist response to Gregory's article was unsurprisingly dismissive (see Ley, 1982); instead the debate was carried forward elsewhere, in papers by Thrift (1983) and Cosgrove (1983).

Thrift's (1983) 'On the determination of social action in space and time' is still one of the most comprehensive treatments of personality, socialization and geography available. And this in itself should be surprising. The paper is set within a structurationist perspective . . . and draws on writers such as Anthony Giddens, Pierre Bourdieu, Raymond Williams and Roy Bhaskar; yet barely resembles Gregory's model. Deploying a series of debates in social theory, Thrift (un)covers a number of novel topics; most of which have remained uniquely Thrift's concerns. The last section of the paper identifies two major areas for research . . . : first, the locale; and, second, social action, which is taken to have four interrelated aspects: personality and socialization, hegemony and knowledge, sociability and community, and conflict and capacity.

Here, I will concentrate on Thrift's conception of personality and socialization. Basic hypotheses concerning the nature of man are taken from Marx: the human being is an animal which can individuate itself only in the midst of society . . . , and people make history but not in conditions of their own choosing . . . ; but these ideas present formidable problems which Thrift is concerned to explore . . . . He identified two immediate areas of difficulty: *first*, the process of class formation, class awareness and class consciousness and its relationship to class conflict, ideology and hegemony; *second*, the process of personality formation and the interrelationship of various 'segments' of lived experience, such as housing, sexuality, the family, and work. Both of these areas might be considered to be different sides of the same coin, namely social being; but Thrift is not in a position to recognize this because of his (inherited) dismemberment of social process from social action. In other words, the self and personality formation are relegated to one aspect of social action, which is actually understood in terms of social process (i.e. socialization).

Thrift, ever mindful to relate social theory and geography, considers personality and socialization to be the historical geography of life-path development. For Thrift the term 'personality' encompasses all aspects of psychology, including subjectivity, identity and individuality . . . . Following Giddens (1979), he argues that there are three necessary features of a theory of the person: the unconscious, practical consciousness and discursive consciousness.

> Seen in the contextual sense, personality is a constant process of 'internalization' or 'interiorization' of social relations along the course of a life path set within civil society (Thrift, 1983, page 43).

For Thrift, personality is never transhistorical or transgeographical, but 'a continuously negotiated and renegotiated expression of social and economic relations that vary, in other than their most basic form, according to locale and region' and 'a series of sedimented activity-experiences carried forward at the tip of the life path' (Thrift, 1983, page 43). This conception of the personality is easily integrated into structuration: where personality is both structured by objective social logic, and structuring through resistance to domination. This is a dialectical process, which is termed 'socialization' and contextualized through the notion of the life-path . . . . Such an approach is able to recognize the influences of various social forces on any individual; not only race, class and gender, but also time and space. Curiously, however, for Thrift the most appropriate level of analysis of these *individual* circumstances is the population cohort. Thrift ends his discussion by noting that people are different at different times and in different places.

There is much to learn from Thrift's discussion of personality and socialization: he does raise the issue of the unconscious, and even locates people in bodies (still an innovative idea). However, Giddens' criteria for a theory of personality are not met by Thrift's own 'socialization' thesis; that is, Thrift erases the unconscious and elides practical and reflexive consciousness. Thrift notes that the prime danger of a socialization thesis is that people become merely dupes to the system. However, his idea that personality is the result of constant 'internalization' offers no possibility that people can set out a place of their own from which to (re-)negotiate social forces and make history (because history makes people). It must be noted that humanism's stress on the internalization of meaning in the constitution of the human subject also invokes this problem.

In the same year as Thrift's article, Denis Cosgrove's 'Towards a radical cultural geography' appeared. Although drawing on similar writers such as Raymond Williams and E. P. Thompson, Cosgrove was attempting to develop a different position to the heavily structuration and time geography influenced writings of Gregory and Thrift. Cosgrove is less concerned with systems, and more with the way that 'human beings experience and transform the natural world as a human world through their direct engagement as reflective beings with its sensuous, material reality' (Cosgrove, 1983, page 1). This formulation at once reveals its Marxist origins and its similarities with Gregory and Thrift's notions of human agency, but it also implies a difference. By introducing the category of experience, Cosgrove is able to suggest that material production and reproduction is 'sustained through codes of communication' (page 1) and that 'if all human production is symbolically constituted we may restate modes of production as modes of symbolic production' (page 8). Unfortunately, these insights are not developed, and they are sidelined elsewhere in Cosgrove's discussion. Having recognized the importance of the symbolic in shaping the nature of the individual's self, Cosgrove then proceeds to see it almost as an effect of the material, but certainly limited by it. Thus, Cosgrove's statement that '. . . the material world is constituted culturally yet remains itself the condition of culture' (page 9) remains curiously ambiguous and imprecise. It does not 'bridge the gap' between the material and the cultural, or between social formation and meaning (also page 9). Indeed, it would be more accurate to say that Cosgrove cannot bridge the gap because his formulation relies on the distance between them.

Although Gregory, Thrift and Cosgrove were working separately, their shared assumption of dichotomies almost killed off the debate on the self. Such questions were no longer important: first, because the new category of human agency valorized action and relegated everything personal to triviality; and, second, because the material side of both 'structure' and 'agency' was problematized in the attempt to uncover the 'realities' of space, place and landscape. So that, by the . . . late 1980s, the 'problem' of human agency, let alone the 'problem' of people, barely gets a look in.

New models' of structure and agency

Structure and agency: the humanistic version

It is symptomatic of the (attempted) killing of humanistic geography that it should not have renewed its models. Thus Ley reaffirms their commitment to putting people at the centre, where

> a model of personhood was adopted which upheld the dimension of meaning; human values and experience were integral to a study of people and place. Issues of human agency were central; society consisted of intentional, acting people, and the concepts of subjectivity and intersubjectivity led to the widespread perspective of the social construction of reality . . . . But action, and particularly the culture-binding routines of everyday life, are none the less commonly taken for granted and opaque to actors (Ley, 1989, pp. 228–9).

Ley still feels it necessary to affirm that this humanism is not the humanism that Gregory savaged, and that this mutilation still scars humanistic writing. Thus, Ley (page 230) wants to negate the 'treatment of subjectivity and experience as a virtual fetish, separated from context and material life'. So it is not surprising to find Ley's article in a collection designed to explore the convergence of humanism and materialism. He begins by suggesting that humanism is oriented towards subjectivity and experience, and materialism towards context and material life. Ley argues that this, and the disavowal of the poverty and power of grand theory, allows possibilities for the integration of humanism and materialism, by restoring the relationship between the social milieu and the hermeneutics of everyday communication.

> There are tantalizing possibilities for a renewed human geography, one which is more representational (or empirical) than the recent past, more contextual in its regional, cultural and historical specificity, which seeks to integrate facts and meanings, both the objective and subjective (Ley, 1989, page 244).

There are problems with this synthesis; partly because it actually institutes a series of equivalent dichotomies, while claiming to overcome or transcend them. Most notable are the dichotomies of meaning and fact, experience and material, subjectivity and context, and subject and object. While the use of dichotomies enables the description of the content of these categories, it disables an understanding of the ruptures and intersections between them. Though Ley shows that humanism does (and did) take account of the culture-building routines that take place behind people's backs, this is where the lifeworld, and contextuality, remain – simultaneously opaque and taken-for-granted. Re-stating old aspirations has not brought humanism closer to an understanding of 'the making, remaking and appropriation of place' (page 227). Thus, everyday life can still be described as

> the pattern and meanings of, and reasons for, human actions are structured into and by the societies into which we are born. We both create and are created by society and these processes are played out within the context of everyday life (Eyles, 1989, page 103).

But these processes cannot be differentiated. Eyles cannot say how it is that people create society, and he cannot say how it is that people are created by society. For humanistic geography the examination of 'the social construction of place, landscape, or region, *as* the interplay between people and context which they both inherit and help redefine' (Ley, 1989, page 229, emphasis added) must remain a tantalizing possibility; always just out of reach, like the taken-for-granted world: always unable to locate and specify the interplay of people and context, because there is a gap between them.

For humanist geographers to acknowledge and repeat that people create society – though not in contexts, or lifeworlds, of their own choosing – it is unobjectionable; ironically, this is the same problematic that the historical materialists have been struggling with.

Structure and agency: the historical materialist version

While the humanists have always acknowledged the presence of the material (through concepts like lifeworld and context), some historical materialists such as Steve Daniels (1985, 1989) and Denis Gosgrove (1989) have injected 'humanity' into their geographies; both by exploring ways in which 'culture' structures the social and by moving away from the economic determinism of structural Marxism. Cosgrove's and Daniels' writing in the mid-1980s constituted a rich vein of ideas (Cosgrove, 1982, 1983; Daniels, 1985). With Williams' notion of the structure of feeling close to the centre of their projects, it might be expected that an understanding of the social (and material) construction of the self would play an increasing role in their analyses of (in their case) the iconography of landscapes. However, their work has developed in other directions: Cosgrove (1989) concentrates on the features of Venetian culture; and, Daniels (1989) on Marxism's colonization of art theory.

So, in this section, I take a closer look at the recent writing of Gregory (1989a, 1989b), who explicitly raises once again the question of agency;

though this issue is embedded in the context of contemporary debates about modernity, knowledge and geography (I have commented on these debates elsewhere, see Pile and Rose, 1992). Gregory detects a crisis in modernity (1989a), but he sees this juncture both historically and positively. He weaves a new web of social theory and spatial relations; new postmodern ideas about the decentred self are used to articulate a new geography concerned with areal differentiation (1989b). But, I argue, this vision and associated models (Sayer, 1989; Thrift, 1989a, 1989b) remain nevertheless one-sided; turning a centred self into a decentred self does not alter the presumed point of contact between structure and agency, the level of social relations.

In the 1980s, Gregory senses that a series of questions are being asked about the 'will to power' which lies behind all regimes of truth, including geography. In addressing these questions seriously, Gregory wants to clarify what precisely is being asked of geography, and how geography is implicated in the transformation of the intellectual landscape. He presents a model of the making and unmaking of modern human geography, which suggests an inexorable movement towards a postmodern human geography. The collapse of the old certainties of modernist social science means that

> instead of presupposing that societies are totalities with clear-cut boundaries, it becomes necessary to show how social relations are stretched across varying spans of time and space (Gregory, 1989a, page 354).

Giddens' work remains pivotal because the theory of structuration is a sustained attempt 'to dismantle the dualism between agency and structure and replace it with a duality' (Gregory, 1989a, page 354). Unfortunately Gregory does not spell out the difference between dualism and duality, but it seems safe to argue that structure and agency are still considered separate in some sense, even if only analytically. For Gregory, however, Giddens' understanding of agency is inadequate:

> while he evidently does not conceive of the human subject as somehow 'preformed', he undoubtedly offers an account of the constitution of human subjects which is, at bottom, ahistorical. He treats subjectivity in strictly developmental terms, drawing on the ideas of Erikson, Freud, Lacan and others to establish the transformation of the body into an instrument of acting-in-the-world, and subsequently using Goffman, Hägerstrand, and others to emphasize the significance of time–space routinization for sustaining the stratification of personality (Gregory, 1989a, page 377).

Leaving aside the incompatibility of the ideas of Erikson, Freud and Lacan for the moment (see below), what Gregory really dislikes is the emphasis on time–space routinization, which seems to ignore the effect of time–space distanciation. Habermas's version is preferred instead, because

> his theory of social evolution, whatever its demerits, is at least designed to draw attention to the different ways in which human subjects are constituted in different types of society. It is this insight which is missing from structuration theory (Gregory, 1989a, page 377).

Gregory argues that Habermas's conception of the (centred?) human subject is nevertheless insufficient because of his denial of the spatiality of social life.

Thus, neither Habermas nor Giddens offer a complete solution to the problem of the constitution of the human subject, but together they might. Gregory offers a 'historical geography of the person' which is strategically complete because it would take account of both time–space routinization and time–space distanciation in the time–space constitution of the human subject. Gregory then proceeds to list some examples which show that 'a purely abstract account of human subjectivity is inadequate for any critical theory with practical intent' (1989a, page 379). I can only agree with this statement, but it worries me that Gregory argues, in the course of a critique of humanism's centring of the human subject, that

> concepts of 'the person', for example, differ widely over space and through time and so, paradoxically, it is their very importance which ensures that they cannot provide a constant foundation for the human sciences (1989b, page 70).

Gregory wishes to put people at the centre of historical geography's concerns, partly because concepts of 'the person' are so variable in time and space. Even so, people, despite their importance, cannot provide a constant foundation for the human sciences; the question is where this leaves concepts of 'the person' in relation to critical theory. If 'constant' requires no naughty people vandalizing space–time theory by behaving badly, it also signifies Gregory's belief in the need for an ahistorical and aspatial foundation for critical knowledge. Subjectivity, on the verge of being essential, is, to follow the implications of the argument, to be marginalized as nonuniversal (i.e. historical and geographical!). Hence the partial recovery of Habermas's universalist theory of social evolution, rather than a return to the psychoanalytical sources of Giddens' theory of the self (i.e. Erikson, Freud and Lacan). All this leaves the properties of structures as the proper foundation for critical social theory. Only these properties can provide the terms in which everything else, including 'the person', is to be understood. Thus, Gregory believes people are constituted in the dimensions of time and space, routinization and distanciation, production and consumption, social integration and system integration, modernity and postmodernity, signification and subjectivity, amongst others; these are the properties of social relations. The carving of structure from agency remains.

So often in his writing, Gregory takes us to the edge of our knowledge and shows us where the gaps lie. Then, frustratingly, he turns away from those problems and, instead, proceeds to tell us how we got here. In this case, he tells us that the structure–agency dualism is not simply unhelpful, but also counterproductive. Then, he suggests that subjectivity should be at the centre of any critical human geography. He then retreats from this position, back into the structure–agency dichotomy from which he was trying to escape.

The remaking and remodelling of human geography implies an agenda; albeit an unsettled one. Some are after a mutual accommodation between humanism and historical materialism (following Ley), while others would like the tension between humanism and historical materialism to be a creative one (following Cosgrove). However provisional this agenda, I believe it already contains the seeds of its own downfall. Neither humanism nor historical materialism can show how the human subject is (re)created, with(in) forms of power because they each rely on 'directly facing the dualism of structure and human agency' (Kobayashi and Mackenzie, 1989, page 10). Resolutions of the dialectic/reciprocity between structure (lifeworld, context) and agency (intentionality, subjectivity) are offered either at the level of social relations or at the level of direct experience. But none of this is the point: the structure–agency dichotomy is itself the problem.

Structure, agency and the self

At its most sophisticated, what geographers have learnt about the relationship between the self and society has been aptly summed up by Andrew Sayer (1989, pages 211, 213): 'what you are depends not just on what you have, together with how you conceive yourself, but on how others relate to you, on what they understand you to be and themselves to be . . .' and that '. . . to a considerable extent people have to adopt meanings, roles and identities which pre-exist them'. I might add that time and space are constitutive of these relations, as Gregory suggests; and, following the humanistic geographers, that people both create and are created by society; and, following the Marxists, that people make history but not under circumstances of their own choosing. Nevertheless, I believe that the re-statement of 'common sense' is no longer enough to provide an understanding of the relationship between structure, agency and people.

The work reviewed so far shares some assumptions, a dualistic architecture, and a problem. The mind is assumed to consist only of consciousness; what is not known is the consequence of a lack of sufficient reflection. It is also assumed that rational procedures of thought can uncover the forgotten or hidden or the deep or the structural. 'Even the prereflective qualities of the "taken for granted" world simply habitualize certain experiences with the clear understanding that these unarticulated feelings can be brought back into consciousness as and when the actor so desires' (Burgess, 1982, page 18). The self is (potentially) transparent and recoverable. These assumptions are created by, and maintain a dualistic splitting of structure and agency, even where they are acknowledged as being two sides of the same coin: for example, dialectical or reciprocal or recursive.

By starting from a different theoretical space, psychoanalysis offers insights which we can no longer ignore. As Thrift (1989a, 1989b) argues, a number of questions begin to arise once the processes of becoming self aware (the 'knowing' subject) and of subject formation (the 'placed' subject) are problematized.

> All manner of problems present themselves – the explanation of practical reason, how people fashion accounts, the nature of the self. These problems are intensely geographical. People are socialized in localized contexts (although the institutions of socialization are now rarely local) and the exigencies of these contexts produce different people with different capacities to think, to co-operate, to dominate, and to resist (Thrift, 1989a, pages 152–3).

For me, this suggests that the constitution of everyday life, or of society, or of structure, can only be understood if the whole nexus of subjectivity and personality formation is brought out of the shadows of our geographical imagination (see Thrift, 1989b, page 263; and Thrift, 1985, 1986).

There is a joke that goes something like this. A motorist is lost in a strange city. The driver stops to ask a passer-by the way to 'Z'. The pedestrian replies: 'Well, I wouldn't start here'. There will never be a crossing between the two-way streets either of humanism and historical materialism, or of agency and structure, because this is not the place to go from. Instead, we must start somewhere else: with psychoanalysis, with its conception of the fragmented self and the implications for social geography.

It is not so much that psychoanalysis provides the crossing points between agency and structure, although it does contain many theories which would be helpful. Nor is it that psychoanalysis dissolves the distinction between agency and structure, although these terms are rendered redundant. It is more that it can act as an interlocutor (like Buttimer's new language), re-interpreting both 'agency' and 'structure' in terms of each other; it provides a set of resources which show what is possible (and what is not). It is more that it can reveal the intricate inter-relationships between the personal and the social, which enter into the depths of the person and provide the axes around which the self is organized, at the point of contact between the individual and the collective; it steps beyond the analysis of structure–agency.

The subject of psychoanalysis

This of course is the crunch section; the (already tired) reader is expecting some quick answers, a neat exposition of psychoanalysis and how it can be mapped onto geography and vice versa. A reasonable wish, but also a desire that cannot be fulfilled here; not least because theory

> does not make anything happen, or get things done, or make certain things matter more than others. The more complete a theory is, the more powerless it becomes. The theory-maker can easily fall victim to his own success: having been animated at the start by an array of precise local issues, he can end up as the hapless proprietor of a gratuitous verbal universe. He can lose all sense of there being causes to defend, follies to expose, sufferings to alleviate and pleasures to pursue. The noise of theory can be as forlorn as the rustle of dried leaves on a dull day. Against this background, speech itself becomes glorious: the subject is in question, and *jouissance* breaks cover even as death threads its way invisibly from one syllable to the next. In speech, everything is still to play for (Bowie, 1991, pp. 159–60).

Partly, I have repeated Bowie's remarks because I do not wish to repeat a series of technical debates, which might reify (psychoanalytic) theory to the point where it becomes as forlorn as the rustle of dried leaves. Partly I have introduced this statement to say that psychoanalysis is not a complete theory, it is still alive and vital (despite drawing primarily on the work of two dead white males!). For example, over time, Freud developed three models of the mind; and, Lacan changed his concept of 'the name of the father' to 'The-Name-of-the-Father'. Partly, I am suggesting that psychoanalysis is a form of knowledge which can only be grounded in theoretically-informed empirical work (as the entire corpus of Freud's work demonstrates). For example, Habermas (1971) sees Freudian psychoanalysis as the epitome of a critical-dialectical method, because it deploys both theory and practice, in a systematic method of interpretation (see Pile, 1990). Partly, because they reveal a further critique of the 'new models' of the self – their seeming completeness. It may also be inferred that in the new models, *jouissance* does not break cover; instead they bear the mark of death, a victim of their own success.

Now I want to change the mood of the paper, to introduce an air of hesitancy, to say that everything is still to play for. Up to here I have been certain of my criticisms, but now my point is not that psychoanalysis is theoretically consistent or complete (these would only be consolations), but that it may help us ask the right questions in the right way, and it may help us accept that 'right' is contingent, arbitrary and strategic. Psychoanalysis is a place from which *jouissance* can break cover, and help us

> to find a model for the social process which allows an autonomy to social consciousness within a context which, in the final analysis, has always been determined by social being (Thompson, 1978, page 81).

But if psychoanalysis is a starting place, what kind of place might this be? There is not of course one answer, and this surely is a good thing, but here I will concentrate on the fragmented self.[4] Two aspects will be discussed: *first*, the 'unknowing' subject, who is neither whole (i.e. unfragmented) nor wholly consciousness (or indeed capable of this); and, *second*, subject formation. In order to clarify these themes, I will develop, first, Freud's discussion of the unconscious, and, second, an example drawn from Lacan's discussion of the mirror stage of the child's psychic development.

Psychoanalysis, geography and the 'unknowing' subject

I have demonstrated that 'new models' of human geography rely on the implicit idea of the (partly, partially) knowning subject; in contrast psychoanalysis reveals the 'unknowing' subject. Freud created a dynamic psychology (in contrast to the models adopted by behavioural geography), which was concerned with the source, aim and content of human motivation and interaction. This is a concern which humanistic and historical materialist geography shares. The foundation of Freud's project, however, relies on the notions of repression and the unconscious. Freud's conception of the unconscious means that consciousness cannot form the basis for understanding human behaviour and experience. I will spend a little time describing Freud's understanding of the unconscious, if only to show what the arguments between psychoanalysts are about. At the end of this subsection, I will raise some of the questions that this understanding poses for the geographical imagination.

For Freud, 'the essence of society is repression of the individual and the essence of the individual is repression of himself [sic]' (Brown, 1959, page 3).

As the social world impinges on the child, it continually tries to transform its desires, or frustrations, into acceptable expressions. The infant uses a number of means to deny/deflect dangerous impulses, but the most important defence is repression. The child protects itself by keeping painful material away from consciousness (Freud, 1915; 1926). So, when the motivations of behaviour are dangerous or unacceptable, they become hidden: a state of reason is imposed by society which must be unreasonable for the individual. There are two points to make: first, this account describes and explains the burying of parts of the self which phenomenology can only assume; second, the notion of repression is simultaneously personal and social, and therefore avoids the Scylla and Charybdis of the structure–agency dichotomy.

The effect of repression is to produce an internal splitting of the mind into a conscious and an unconscious: 'the repressed is the prototype of the unconscious for us' (Freud, 1923, page 353). Repressed ideas are dynamic: both because they are forced into the unconscious, and because they have a motivating effect on human actions. The unconscious consists of wishful impulses where there is no negation of desires, no repression (Freud, 1915, page 190). Moreover, the unconscious has no conception of time or place, contradiction or reality: it is the opposite of the world of common sense, rationality and order, and has its own inner life. And, because the unconscious lies outside the consciousness, it cannot be easily controlled. There is a continuous struggle, at all levels of the mind, to maintain repression; but sometimes this battle is lost, and the repressed returns to compel people to express their desires and frustrations, though in disguised ways (e.g. in dreams, parapraxes, neuroses, etc.). Because of repression, we are alienated from parts of ourselves and we can only explain ourselves with reference to an unconscious place, hidden deep inside ourselves. This means that people's behaviour is motivated and constrained by forces, from the inside out and the outside in, which lie outside their control or easy access.

There has been a great deal of controversy about Freud's account of the unconscious and the structure of the mind. This debate has focused on two interrelated points: first, the constitution of the unconscious and its effects on behaviour; and, second, the structure of the psyche and its origin. The first issue relates to the extent to which biological instincts can be used to explain repression and the existence of an unconscious. The second question relates to the original self, asking whether human beings are coherent unities which then become split, or whether they are inherently split, contradictory and fragmented. As a result of these arguments, four variants of psychoanalysis have developed since Freud's death – the ego psychologists (e.g. Erikson), the object relations school (e.g. Winnicott), and the Lacanian and Kleinian schools (see Frosh, 1987). Although they may share certain ideas, these positions are mutually incompatible; for example, Lacan believed that the child is born without a self, while Erikson argued that the infant is born with a complete (though not fully formed) self, while Freud thought the unconscious forms as a result of the encounter between a conscious self and its maturational environment (compare with Giddens' appropriation of Erikson, Freud and Lacan above).

This story only touches on a small part of psychoanalytical theory; for example, Freud (1915, 1923) used another set of concepts to understand the fragmented self. He suggested that the self was also split into the Id, Ego and Super-ego. One set of ideas did not replace the other, but instead describe different aspects of the self. The mind becomes so conflictual and so dynamic that certainties about 'the self' themselves become a fantasy.

> Far from adopting a unified view of selfhood from classical psychology, Freud found the human mind to be all too plainly self-divided and disputatious. His mental models, far from being clear-contoured experimental hypotheses, easily came apart into riddles and paradoxes. The notion of an integrated ego, buoyantly pursuing its goals and deflecting antagonists, has the force not of an observable fact or of a logical necessity but of a wish, a hope, a recommendation. (Bowie, 1991, pages 20–21)

Indeed, it may be that the concept of a knowing subject is itself mere consolation; and if this is so then we must think very carefully about the basic premises that underpin our attitude to what constitutes knowledge. The 'new models' remain grounded in the certainties of rational knowledge of the integrated ego, which in turn is based on the commensurate, and necessary, abjection of irrationality. Psychoanalysis undermines such certainties by recognizing the rationality of irrationality (and irrationality of rationality). Without a 'conscious', rational subject on which to centre knowledge, psychoanalysis constantly transgresses the boundaries between knowledge and unknowledge. In sum, psychoanalysis offers a two-fold critique of the 'new models' of the self: *first*, that the unknowing subject is at least as important as the knowing subject, and probably more so; and, *second*, that knowledge based on particular notions of rationality must disintegrate under psychoanalytic scrutiny. Moreover, by ignoring the personal and individual unconscious, there is no place from which people can set out to resist social forces and make history. Psychoanalysis not only offers such a place – the unconscious (where there is no repression, no negation of desire) – it also explains why it is so difficult to change ourselves and our world (however provisional 'ourselves' and 'our world' are rendered by the insights of critical theories).

There is perhaps one further point to make here, and this relates to methodology. For example, we may now suspect the vogue for legitimizing empirical enquiry on the grounds of the self-reflection, and this is no longer confined to humanistic geography. Pyschoanalysis questions whether the researcher can separate himself from himself in order to look back on himself. Here self-reflection can be seen as being a) about (self) distancing, a typically masculine position, b) fundamentally narcissistic, and c) impossible. Instead, psychoanalysis suggests another model for research; if psychoanalysis is the talking cure, then its primary mode is listening and interpreting. And it is learning 'listening' that could help us go beyond the tired, robotic models which are currently being used to describe the economic, the political, the social and the cultural. What is needed to interpret both meaning and practice is a level of interpretation which has been called 'depth hermeneutics' (Habermas, 1971; Ingelby, 1981; Pile, 1990). These issues are beginning to be broached by geographers, so I will not develop them here (see Burgess *et al.*, 1988a, 1988b; Pile, 1991).

Psychoanalysis, geography and the 'placed' subject

In the previous subsection, I suggested that psychoanalysis may help geography's concern with the constitution of the human subject. In this section, [I] will argue that it will also enhance our understanding of the human experience of space and place.

If psychoanalysis were taken to be a kind of humanism, then there are many directly psychological 'objects' that we may like to historicize and 'geographicize'. If psychoses and neuroses are taken to lie at the intersection of personal and social structures, then there are certain consequences. It is at least common sense that our moods affect what we do and who we are. Humanistic geography has already begun to describe our feelings towards landscapes coining terms such as topophilia (pleasure, love), topophobia (fear) and topocide (aggression, loss). Psychoanalysis offers other human experiences, such as schizophrenia, paranoia, guilt, disgust, narcissism and so on. Though the psychoanalytic literature could itself become the object of study (it has a very particular historical geography), the suggestion here is that psychoanalytic principles would help humanistic geography re-interpret the circumstances of various 'mind-sets', across time and space. Yet, humanistic geographers have neither 'mapped' these mind-sets, nor used them in the geographies they already produce.

If phychoses and neuroses are worth looking at, then we must be aware that they do have a historical and geographical context. They (particular psychic structures, particular social formations, particular concepts of the person) are not to be found always and everywhere, or even at all; as Gregory suggests. Nevertheless, I am wary of the suggestion that we might simply map moods or psychological problems or fantasies or love, nor do I wish to simply describe ambivalence, paranoia, guilt, or disgust towards particular landscapes. Such cartographies could easily remain superficial descriptions; replicating phenomenological discussions of human emotionality.

If psychoanalysis were to be taken as a form of historical materialism, then it may be used to describe social practices: for example, sexuality (about versions of femininity and masculinity, about sexual practices and taboos) and aggression (about abusive language, rape, serial killing, warfare, and so on); fantasy and desire; authority, discipline (including the law, punishment, religion and education) and forms of (in)sanity; childhood, infant care and public welfare (as shaped by patriarchal capitalism), including the conditions and meaning of mothering, feeding practices, post-natal care and health visiting; and gender relations, group interactions and 'Otherness'. Although these are extremely important, I believe that psychoanalysis has more to offer than different accounts of socio-spatial relations and how these become internalized in ways that dominate, or at least guide, individual human actions.

More importantly, psychoanalysis offers other models of the lived world, of space, of geography; none of which have been explored, nor even commented upon by geographers. As an example, in the following discussion, I will draw on the world of Lacan to suggest that the world is lived through three kinds of space: real, imaginary, and symbolic. Each is always present, but individuals are located differently in relation to each of these spaces; we are placed, but we are irrevocably lost.[5] Lacan, in attempting to describe this relation, suggested that these realms were like a conjuror's trick, in which three separate hoops become a solid whole, but the hoops never intersect. A difficult idea at its easiest; easy answers are forbidden in this world because these are also stages. At each we are forced to place ourselves in relation to a different authority, and individuals are differentiated along axes of domination, such as gender, class, 'race', able-bodiedness, and so on. Here, I will draw out the implications of spatial relations in a discussion of the mirror stage (see Rose, 1975, 1986; Macey, 1978, 1988).

To cut a long story short, for Lacan (1973, 1977), the self is formed in relation to otherness; so one's self can only be experience as an alienated 'I'. (This notion changes the emphasis of much humanistic geography; for example, instead of looking at whether a particular landscape, like the city, is alienating or not, the question would be how and why it is alien.) The child becomes fragmented in two basic phases: the mirror stage and the imago stage. These stages move the child through three orders: from the real (via the mirror stage) to the imaginary, and thence (via the imago stage) to the symbolic order.[6] Here I will develop Lacan's discussion of the mirror stage, because it is here that the self is first seen to be conflictual. Any discussion of the mirror stage must begin with the term itself: 'stade du miroir'. As usual with Lacan, there is a pun at work: *the mirror stage* is not simply a temporal phase in the child's development, but also 'a stadium (*stade*) in which the battle of the human subject is permanently being waged' (Bowie, 1991, p. 21).

Between six and eighteen months old, the child sees itself reflected in a mirror (which can also be taken as a metaphor for the reaction of others to the behaviour of the child); now, here there is something for the child to celebrate. The infant can now surmise that 'That is me'. But the child cannot rejoice in its discovery of itself; the infant remains dependent upon adults to provide its needs. If the unreality of the 'image' is recognized, the child must accept its own lack of mastery and autonomy. Instead the child remains perversely and laughably fascinated with its own specular body. This fascination operates by and through two spatial relationships: first, the distance between the child's body and its mirrored body; and, second, the place of the child's body and its setting within the mirror image. The map of the body, setting and mirror both captivates and consoles the child, but it is an illusion, a trap, a decoy; geography is the medium of deception, it offers 'ground truth' but cannot be trusted.

This phase/battlefield makes the future development of the self possible, by providing the child with a minimal means of ordering the world; and this is profoundly spatial. So far, using Lacan, I have only demonstrated that the psyche is 'spatialized'; in part ordered through the geometries of the social body and its encounters with others, encounters which are of course socially structured. The argument may be returned to geography as an empirical and theoretical discipline; fieldwork can be specified as a way of looking at the world, a world which can now be recognized as partly belonging to the imaginary (see [note] 6). Thus, the geography of the imaginary may be cast as the search for 'the radiance of the surface world', which is now accepted as a real illusion, a decoy, a consolation; this is the world which phenomenology actively seeks to deny, and in which symbolic interactionism is embedded.

Briefly, then, *the imago stage* regulates the child's entry into language, which places the child in relation to (power-ridden) systems of cultural categories. According to Lacan, we are finally forced to take our place in society, in language, in a host of symbolic relations which are not ours. Language is central to this splitting: it insists on difference and provides the symbolic structures in which relationships can take place. During maturation, the child is forced to take up a position in relation to the cultural categories it experiences. Along different lines, children are forced to occupy particular, exclusive and bounded symbolic space(s). Thus, men's and women's (symbolic) spaces are different, incommensurable and this notion can be extended to the other dimensions through which individuals are forced to adopt orientations, such as 'race', class, able-bodiedness, and so on; these ideas remain alien to historical materialism which seeks to ground its critical theories in universal structures (see Deutsche, 1991).

This brief example allows me finally to suggest three inter-related ways in which psychoanalysis may enhance our understanding of 'the meaning of being human' (from Ley), 'the historical geography of the person' (from Gregory) and 'the cultural constitution of the material world' (from Cosgrove).

Firstly, if we only have access to the world through a symbolic (which is not ours), then more attention has to be paid by geographers to 'language' (i.e. exchanges – of all kinds – in general). The whole field of communication must be problematized, interpreted and set in its place, including such geographical icons as place, territory, landscape; for, not only do these symbols take on meaning only in relation to a given order, but they also highlight particular meanings while simultaneously hiding others. Geographers can make a primary contribution by exploring the spatial dimensions of the symbolic: for example, by uncovering the use of spatial metaphors in everyday speech – how they affect the way in which we understand the world, and how we act in the world as a result of that understanding (e.g. Kolodny, 1973; Livingstone and Harrison, 1981; Pred, 1990; Pile, 1990; Pile and Rose, 1992). This would allow geographers to (dare I say) deconstruct the hegemonic ideas through which we understand our world(s): for example, through the analyses of the mass media (e.g. Burgess *et al.*, 1991) or of the semiotics of everyday life (e.g. Keith, 1988).

Secondly, if taken as a 'new model' for human geography, then the (real, imaginary, symbolic) world becomes multi-layered and multi-dimensional; riddled with paradox, ambiguity and ambivalence; internally fragmented, incoherent, inconstant and contradictory (and there is a strong link here to Marxist dialectics, but deepened, broadened and enhanced); and about both process (from Marx) and disjuncture (from Foucault). And all this I would suggest might be covered by the concept of 'articulation'; jointed, separate, connected, particular, spoken, written, learnt; where the term would also disclose the intersection of the different dimensions of oppression. And allow us to say something of the real, imaginary and symbolic structuring of the human experience of geography.

Finally, it is important to note that psychoanalysis does not reduce the world to a text, nor does it suggest an epistemology based solely on deconstruction, iconography or discourse analysis, nor does it reduce society to encounters between the (fragmented) self and the other(s). For, as Lacan notes, the real is always glued to the heel of a purely mental or textual world. Moreover, in Lacan's work, the symbolic bears the law (which gives experience meaning) and may take many forms: such as, economic laws of exchange, the judiciary, the law-of-the-father, incest taboos, and so on. Psychoanalysis neither ignores the dimensions of power, social structures and the material world; nor reduces explanation to childhood sexuality. Instead, it recomposes the relationship between the person, social action and social structure; it forces us to think again about the models we have of the material, social, cultural world.

Ideology is not a fashionable term in contemporary social theory (because it invokes the spectre of *false* consciousness), yet psychoanalysis offers a model of ideology based on systematically distorted communication; ideology becomes the *symbolic* representation of the *imaginary* relationship to the *real* conditions of existence (following Althusser following Lacan; see Jameson, 1991, pages 51–4). Psychoanalysis can describe and explain how power infects language and how language places people in relation to power. The question is whether geography has been able to do the same, and whether it will be able to offer a sustained critique of ideology without psychoanalysis. Perhaps the real pay-off for geography, then, is that it will be able to contribute to contemporary debates on the politics of identity, able to help articulate a politics of movement (and not merely position), able to respond to the demand for a politics of desire.

Conclusion

I have argued that we need to seek new models, with new concepts, if we are to understand the relationship between people, space and society. There is no longer any point in remaining in the cul-de-sac of the structure–agency dichotomy, as the 'new models' do. I have offered another road to go down, psychoanalysis; but this is no easy choice. There are many varieties of psychoanalysis, and each is contested and incomplete; there is no single psychoanalytic concept which is not the subject of fierce debate. Varieties of psychoanalysis have different assumptions and problems; but, taken together, and rereading in the light of criticism and defence, there are positions to be taken up which go beyond the limitations of any one group. In this paper, I have drawn out an alternative model of the self and social context, drawing on Freud's notion of the unconscious and Lacan's distinction between the real, the imaginary and the symbolic. There are many other resources to be drawn on; I am particularly interested in the trajectory created by feminist's clash with Lacan and psychoanalysis (Mitchell, 1974; Grosz, 1989; Gallop, 1982; Brennan, 1989). There are other possibilities. There are also other models of the self (see, for example, Philo, 1991).

Though Lacan allows us to transcend the nature/culture debate and understand the positioning of the self (and geography has really got to grips with neither), and though Freud allows us to understand specific psychic structures (id, ego, ego-ideal; conscious, pre-conscious, unconscious) in a masculine-oriented world, neither can explain the multiple dimensions of identity because their aetiology is restricted to singular traumas (or phases) of maturation/socialization. And this is where geographers can contribute. Stated simply, the symbolic is not simply phallic, the penis is not the centre of everything, and the psychic structure is itself splintered; and we as geographers can help specify these multiplicities, as we consider the relationships in peopled places and placed people.

My hope, in writing this essay, is that psychoanalytic controversies will be discussed by geographers, and that they in turn will inform those debates. Psychoanalysis may not have all the answers, but it does provide a way of knowing, and a set of propositions, which we can no longer ignore. Without psychoanalysis, geography will not be in a position from which to develop the agenda that Cosgrove set out a decade ago.

> As revolutionary practice, cultural geography can not only reveal the symbolic contribution of human agency in producing and sustaining landscapes, and the degree to which those landscapes themselves structure and maintain symbolic production, but it can examine critically emergent forms of spatial organisation and landscape (Cosgrove, 1983, page 10).

Psychoanalysis will extend this problematic by re-conceptualizing the relationship between the social (whether the symbolic, spatial organization, landscape, place, territory, and so on) and the self (conscious, unconscious, self-image, identity, intentional, unintentional, and so on); only this time the points of contact will be the intricate, dynamic and power-ridden relations between them. Psychoanalysis will help us understand why the apparently unobjectionable, and therefore taken-for-granted, Marxist formula 'people make history, but not in circumstances of their own choosing' is just the beginning.

Notes

1 I have repeated (here and subsequently) the use of the term 'man' in order to show that this is not a gender neutral way of ordering the world (Lloyd, 1984).
2 The figure of death (especially by bodily catastrophe) stalks the rest of this essay, partly because of Lacan's suggestion that the symbolic is threaded through with morbidity (also Bowie, 1991, p. 159–60; cited in main text).
3 I have used the verb eviscerate to evoke the idea of a body being in parts (Lacan's *l'homelette*); I will continute to use such terms for this reason. Moreover, the idea of wholeness of the body is taken to be as much an unfulfillable wish as is the idea of wholeness of the mind.

4 There are many other ways of suggesting what is fundamental about psychoanalysis for geographers. For example, it would be possible to use Lacan's (1973) four fundamental concepts: the unconscious and repetition; the gaze as *objet petit a*; the transference and the drive of the Other. Or to use the features common to all psychoanalyses: the unconscious, repression, autobiographical reconstruction, childhood sexuality, and so on.
5 It is not a coincidence that the real is like the unconscious; there, but inaccessible.
6 Lacan's triadic structure of the real, imaginary and symbolic remains useful in part because the definitions remained stable throughout his work. The real represents 'that which always returns to the same place'; by this Lacan means a condition of existence (whether material or fantasized). Before the real, the imaginary falters, and over it, the symbolic stumbles; nevertheless, the real is impossible. The imaginary may be defined as the realm of images, and is entered in the mirror stage (see main text). The symbolic is not simply the realm of symbols, but about the structure of meanings, which acquire value only in relation to a given order. This, for Lacan, is the order which constitutes the adult (imago) subject. Reality may be said to be constituted by the real, the imaginary and the symbolic; none of which may be said to be directly in opposition to another. Hence, the allusion to the conjuror's illusion (see main text). With Lacan's contribution, reality may be recognized as doubly spatial: both in terms of a spatialized epistemology and a spatialized ontology.

Selected references

References to Freud in either the *Pelican Freud Library* (Pelican, Harmondsworth) (PFL) or the *Standard Edition of the Complete Works of Sigmund Freud* (Hogarth Press, London) (SE).

Bowie, M. (1991) *Lacan* (Fontana, London)

Brennan, T. (ed.) (1989) *Between feminism and psychoanalysis* (Routledge, London)

Brown, N. O. (1959) *Life against death: The psychoanalytic meaning of history* (Wesleyan University Press, Middleton, Conn.)

Burgess, J. (1982) 'Emotions, meaning and motives: a critique of geographical interpretations of self and social context'; unpublished paper

Burgess, J., Harrison, C. and Maiteney, P. (1991) 'Contested meanings: the consumption of news about nature conservation', *Media, Culture and Society* 13: 499–519

Burgess, J., Limb, M. and Harrison, C. M. (1988a) 'Exploring environmental values through the medium of small groups. Part One: theory and practice', *Environ. Plann. A* 20: 309–26

Burgess, J., Limb, M. and Harrison, C. M. (1988b) 'Exploring environmental values through the medium of small groups. Part Two: illustrations of a group at work', *Environ. Plann. A* 20: 457–76

Buttimer, A. (1976) 'Grasping the dynamism of lifeworld', *Ann. Ass. Am. Geogr.* 66: 277–92

Cosgrove, D. (1982) 'The myth and the stories of Venice: an historical geography of a symbolic landscape', *J. His. Geogr.* 8: 145–69

Cosgrove, D. (1983) 'Towards a radical cultural geography: problems of theory', *Antipode* 15: 1–11

Cosgrove, D. (1989) 'Historical considerations on humanism, historical materialism and geography', in Kobayashi, A. and Mackenzie, S. (eds) *Remaking human geography* (Unwin Hyman, London) pp. 189–205

Daniels, S. (1985) 'Arguments for a humanistic geography', in Johnston, R. J. (ed.) *The future of geography* (Methuen, London) pp. 143–58

Daniels, S. (1989) 'Marxism, culture, and the duplicity of landscape', in Peet, R. and Thrift, N. (eds) *New models in geography: Volume 2* (Unwin Hyman, London) pp. 196–220

Deutsche, R. (1991) 'Boys town', *Environ. Plann. D: Soc. Space* 9: 5–30

Donald, J. (ed.) (1991) *Psychoanalysis and cultural theory: thresholds* (Macmillan, London)

Eyles, J. (1981) 'Why geography cannot be Marxist: towards an understanding of lived experience', *Environ. Plann. A* 13: 1371–1400

Eyles, J. (1989) 'The geography of everyday life', in Gregory, D. and Walford, R. (eds) *Horizons in human geography* (London, Macmillan) pp. 102–117

Freud, S. (1915) 'Papers on metapsychology' (1984, PFL 11) pp. 99–222

Freud, S. (1923) 'The ego and the id' (1984, PFL 11) pp. 350–407

Freud, S. (1926) 'Inhibitions, symptoms, anxiety' (1979, PFL 10) pp. 237–333

Frosh, S. (1987) *The politics of psychoanalysis: an introduction to Freudian and post-Freudian theory* (Macmillan, London)

Gallop, J. (1982) *Feminism and psychoanalysis: the daughter's seduction* (Macmillan, London)

Gallop, J. (1988) *Thinking through the body* (Columbia University Press, New York)

Giddens, A. (1976) *New rules of sociological method: positive critique of interpretative sociologies* (Hutchinson, London)

Giddens, A. (1979) *Central problems in social theory: action, structure and contradiction in social analyses* (Macmillan, London)

Gregory, D. (1981), 'Human agency and human geography', *Trans. Inst. Br. Geogr.* 6: 1–18

Gregory, D. (1982) 'A realist construction of the social', *Trans. Inst. Br. Geogr.* 7: 254–56

Gregory, D. (1989a) 'The crisis of modernity? Human geography and critical social theory', in Peet, R. and Thrift, N. (eds) *New models in geography: Volume 2* (Unwin Hyman, London) pp. 348–85

Gregory, D. (1989b) 'Areal differentiation and post-modern human geography', in Gregory, D. and Walford, R. (eds) *Horizons in human geography* (Macmillan, London) pp. 67–96

Grosz, E. (1989) *Sexual subversions: three French feminists* (Allen and Unwin, Sydney)

Habermas, J. (1971) *Knowledge and human interests* (Heinemann, London)

Ingelby, D. (ed.) (1981) *Critical psychiatry: the politics of mental health* (Penguin, Harmondsworth)

Jameson, F. (1991) *Postmodernism: or, the cultural logic of late capitalism* (Verso, London)

Keith, M. (1988) 'Racial conflict and the "no-go areas" of London', in Eyles, J. and Smith, D. M. (eds) *Qualitative methods in human geography* (Polity Press, Cambridge) pp. 39–48

Kobayashi, A. and Mackenzie, S. (1989) 'Introduction', in Kobayashi, A. and Mackenzie, S. (eds) *Remaking human geography* (Unwin Hyman, London) pp. 1–14

Kolodny, A. (1973) *The lay of the land: metaphor as experience and history in American life and letters* (University of North Carolina Press, Chapel Hill)

Lacan, J. (1973) *The four fundamental concepts of psychoanalysis* (1977, Peregrine, Harmondsworth)

Lacan, J. (1977) *Ecrits: a selection* (Tavistock Publications, London)

Ley, D. and Samuels, M. S. (1978) *Humanistic geography: prospects and problems* (Croom Helm, London)
Ley, D. (1977) 'Social geography and the taken for granted world', *Trans. Inst. Br. Geogr.* 2: 498–512
Ley, D. (1978) 'Social geography and social action', in Ley, D. and Samuels, M. S. (eds) *Humanistic geography: prospects and problems* (Croom Helm, London) pp. 41–57
Ley, D. (1981a) 'Cultural/humanistic geography', *Prog. Hum. Geogr.* 5: 249–57
Ley, D. (1981b) 'Behavioural geography and the philosophies of meaning' in Cox, G. R. and Golledge, R. G. (eds) *Behavioural problems in geography revisited* (London, Methuen) pp. 209–30
Ley, D. (1982) 'Rediscovering man's place', *Trans. Inst. Br. Geogr.* 7: 248–53
Ley, D. (1989) 'Fragmentation, coherence, and the limits to theory in human geography', in Kobayashi, A. and Mackenzie, S. (eds) *Remaking human geography* (Unwin Hyman, London) pp. 227–44
Livingstone, D. N. and Harrison, R. T. (1981) 'Meaning through metaphor: analogy as epistemology', *Ann. Ass. Am. Geogr.* 71: 95–107
Lloyd, G. (1984) *The man of reason: male and female in Western philosophy* (Methuen, London)
Macey, D. (1978) 'Jacques Lacan', *Ideology and Consciousness* 4: 113–29
Macey, D. (1988) *Lacan in contexts* (Verso, London)
Mitchell, J. (1974) *Psychoanalysis and feminism: a radical reassessment of Freudian psychoanalysis* (Penguin, Harmondsworth)
Philo, C. (compiler) (1991) *New words, new worlds: reconceptualising social and cultural geography* (Department of Geography, St David's University College, Lampeter)
Pile, S. (1990) 'Depth hermeneutics and critical human geography', *Environ. Plann. D: Soc. Space* 8: 211–32
Pile, S. (1991) 'Practising interpretative geography', *Trans. Inst. Br. Geogr.* 16: 458–69
Pile, S. and Rose, G. (1992) 'All or nothing? Politics and critique in the modernism–postmodernism debate', *Environ. Plann. D: Soc. Space* 10: 123–36
Pred, A. (1990) 'In other wor(l)ds: fragmented and integrated observations on gendered languages, gendered spaces and local transformation', *Antipode* 22: 33–52
Rock, P. (1979) *The making of symbolic interactionism* (Macmillan, London)
Rose, J. (1975) 'The imaginary', in Maccabe, C. (ed) *The talking cure: essays in psychoanalysis and language* (1981, Macmillan, London) pp. 132–61
Rose, J. (1986) *Sexuality in the field of vision* (Verso, London)
Sayer, A. (1989) 'On the dialogue between humanism and historical materialism in geography', in Kobayashi, A. and Mackenzie, S. (eds) *Remaking human geography* (Unwin Hyman, London) pp. 206–26
Sibley, D. (1991) 'The boundaries of the self', in Philo, C. (compiler) *New words, new worlds: reconceptualising social and cultural geography* (Department of Geography, St. David's University College, Lampeter) pp. 33–5
Smith, N. (1979) 'Geography, science and post-positivist modes of explanation', *Prog. Hum. Geog.* 3: 356–83
Thompson, E. P. (1978) *The poverty of theory and other essays* (Merlin Press, London)
Thrift, N. (1983) 'On the determination of social action in space and time', *Environ. Plann. D: Soc. Space* 1: 23–57
Thrift, N. (1985) 'Flies and germs: a geography of knowledge', in Gregory, D. and

Urry, J. (eds) *Social relations and spatial structures* (Macmillan, London) pp. 366–403

Thrift, N. (1986) 'Little games and big stories: accounting for the practice of personality and politics in the 1945 general election', in Hoggart, K. and Kofman, E. (eds) *Politics, geography and social stratification* (Croom Helm, Beckenham) pp. 86–143

Thrift, N. (1989a) 'Introduction' to new models of civil society, in Peet, R. and Thrift, N. (eds) *New models in geography: Volume 2* (Unwin Hyman, London) pp. 149–56

Thrift, N. (1989b) 'Introduction' to new models of social theory, in Peet, R. and Thrift, N. (eds) *New models in geography: Volume 2* (Unwin Hyman, London) pp. 255–66

Tuan, Yi-Fu, (1976) 'Humanistic geography', *Ann. Ass. Am. Geogr.* 66: 266–76

Geography and Difference

GEOGRAPHY AND DIFFERENCE

A discipline in difference

The discipline of geography is founded on two kinds of difference that slide over and into one another. Human geography has always had a keen interest in the diversity of the world's populations, and its interest in cultural *difference* has sustained a series of sometimes close connections with social and cultural anthropology. In the closing decades of the 19th century, for example, Friedrich Ratzel described an 'anthropogeography' that related cultural differences to biophysical diversity, and parallel projects were pursued into the 20th century by Paul Vidal de la Blache, H. J. Fleure, Carl Sauer, and many others.

All of these writers were equally interested in how these relationships varied from place to place, and subsequent attempts to codify these co-variations also depended on *spatial difference*. This was developed in different ways and with different inflections, but, in general, those geographers who were most interested in the particularity of place – in regional geography, or what Richard Hartshorne (1939) called areal differentiation – *absolutised* difference by describing and mapping the uniqueness of place, whereas those geographers who were most interested in **spatial science** *relativised* difference by searching for patterns that revealed a systematic order (or 'spatial structure') to the distribution of differences over space.

Questions of cultural and spatial difference have continued to be raised by most post-positivist geographies, but they have been significantly rephrased. In the first place, systems of cultural difference have been wired to grids of power and to a much more complicated understanding of the politics of identity. Central to this project has been the attempt to expose and subvert the construction of *hierarchies* of cultural difference. These are systems of social meaning that privilege some constructions (e.g. 'white', 'straight') and marginalise others (e.g. 'black', 'lesbian'). These systems have been so deeply sedimented over time that the powers and rights that have been accorded to the privileged terms have been taken for granted by those who have received them. Audrey Kobayashi and Linda Peake (this section) have described the attempt to subvert these congealed hierarchies of cultural difference as the 'denaturalisation' or (more technically) the 'deconstruction' of discourse, a critical practice intended to call into question 'not just the explicit statements of belief about what is "natural" but also those values that are so naturalized that they have not previously come under question'. In the second place, and closely connected to these proposals, the intersections between domination and difference – between grids of power and the politics of identity – have been grounded in a much more sophisticated understanding of the production of place, space, and 'nature'. In particular, human geographers have become interested in the ways in which cultural differences are *inscribed in and constituted through* systems of spatial difference, which has in turn raised a new series of questions about the relations between social practices and representations of space.

We will consider each of these developments in turn, though it should be clear from these preliminary comments that much of our discussion will necessarily revisit issues raised in our previous introductory essays.

Denaturalising discourse

Since the early 1980s a revitalised history of geography has revealed the discipline's long-standing complicity in naturalising and legitimising dominant systems of cultural difference. The rise of **historico-geographical materialism** did much to elucidate the class inflections of the modern discipline of geography: its bourgeois origins, its service as a sort of spatial technology for capital, and its marginalisation of working-class interests and struggles (Harvey, 1984). More recently, geography's involvement in the ideologies and privileges of European colonialism and imperialism has also attracted considerable critical scrutiny (Driver, 1992; Godlewska and Smith, 1993), and its intricate entanglement with various 'scientific' racisms has been exposed (Livingstone, 1992). These complicities are not matters of the remote past, however, and versions of a supposedly critical human geography have also been indicted for their unreflective **masculinism** (Rose, 1993), their latent homophobia (Bell, 1991), and their extraordinary, **ethnocentric**, reluctance to 'learn from other regions' (see Chapter 2).

These critiques are embedded in the formation of avowedly feminist, queer, and post-colonial geographies, but they have also been drawn into transdisciplinary systems of thought that have attracted attention throughout the Anglo-American academy: in particular, **postmodernism, post-structuralism**, and **post-colonialism**. As their very names suggest, all of these discourses seek to establish a different relation from those variously troubled pasts – hence the 'post' – by denaturalising intellectual and cultural traditions. They have paid special attention to the ways in which those traditions have transmitted a series of claims about hierarchies of cultural difference that have come to seem perfectly natural and entirely normal. These various 'posts' have used their critical genealogies of dominant conjunctions of power and knowledge to try to recover marginalised voices and to valorise other cultural differences. Indeed, taken together, there is no doubt that postmodernism, post-structuralism, and post-colonialism have made 'difference' one of the keywords in the vocabularies of the contemporary humanities and social sciences: and human geography is no exception (see Gregory, 1994).

Projects of this sort raise difficult issues of interpretation – how are we to understand others whose situations are different from our own? – and geographical inquiry has thus established far deeper affinities with cultural anthropology through a common interest in **ethnography**. Particular attention has been paid to the relations of power that inhere in fieldwork and to polyphonic experiments with 'multiple voices' in written accounts (Crang, 1992; Katz, 1992). This in turn has prompted a renewed concern over the politics of representation and authorisation: most starkly, how are we to convey otherness without reducing other people to variants of ourselves still positioned at the centre and as the norm (cf. Chapter 3)? As Linda McDowell observes in Chapter 7, we need to rethink our notions of difference in order to recognise, in fact to navigate, the inescapable tension 'between challenging the oppositional construction of others and continuing to make political demands on their behalf'.

On the one side, then, it is important to establish commonality and connection, and to recognise our responsibilities for what Michael Ignatieff calls 'distant strangers', by affirming those ways in which we share in a common humanity. The incautious celebration of difference can otherwise all too easily license the grotesque politics of purity and 'ethnic cleansing' that have sustained recent conflicts in Eastern Europe and Africa or, closer to home, the neoconservative hostility to virtually all forms of international aid in the name of fiscal rectitude and ethnocentric claims about 'charity beginning at home' (see Corbridge, 1993). But these cautions do not carry the implication that we are all so many different versions of the same. On the other side, therefore, as Hélène Cixous puts it,

> What is the 'Other'? If it is truly the 'other', there is nothing to say; it cannot be theorized. The 'other' escapes me. It is elsewhere, outside: absolutely other. It doesn't settle down. But in History, of course, what is called 'other' is an alterity [an otherness] that does settle down, that falls into the dialectical circle. It is the other in a hierarchically organized relationship in which the same is what rules, names, defines, and assigns 'its' other (Cixous, 1986, page 70).

This is a difficult passage, but Cixous is drawing attention to one of the most insidiously general ways in which 'otherness' has been constructed within the canonical formulations of that imperial 'History-with-a-capital-H' which has played such a prominent part in the installation of what the French philosopher Jacques Derrida called the West's 'white mythologies'. We can unpack Cixous's argument through the example provided by Edward Said's (1978) critique of the discourse of *Orientalism*. Said describes Orientalism as an institutionalised system of power and knowledge by means of which the West constructed and then projected a set of bipolar identities onto the East. Thus this fictionalised 'Orient' was figured as 'irrational', 'unchanging', and 'feminine', so that the 'Occident' emerged in contrastive clarity as the home of reason, of a restless modernity, and a militant masculinity. Through the discourse of Orientalism, then, the West does indeed 'rule, name, define, and assign "its" other', as

Cixous suggests, and that other, the exotic East, is imaginatively possessed by the West as 'its' other and indelibly marked as everything the West is not: as a lack. Orientalism is not a synonym for colonial discourse, and there are important distinctions to be drawn between the imaginative geographies constructed by Western writers to represent (for example) Africa, India, and South America, but the general discursive strategy is undoubtedly repeated in many other cultural registers. So, for example, a parallel operation installs a fundamental **phallocentrism** within Western metaphysics, whose operation 'is bound up with an opposition between "man" and "woman" in which the former is dominant and proclaims its own privileged singularity in opposition to the inferior, plural "otherness" of the latter' (Bondi and Domosh, 1992) – again defined as a lack which, within Lacanian psychoanalytic theory, becomes the lack of a phallus.

These arguments have been developed with most force against the West's writing of 'History', but their point is not confined to a single discipline. The oppositions that they identify confound disciplines in the humanities, such as philosophy and literary criticism (Young, 1990), *and* disciplines in the social sciences, such as anthropology and sociology (Fabian, 1983; McGrane, 1989). And although they characterise 'Geography-with-a-capital-G' too – through the complicities we listed earlier – they also have implications for critical human geographies conducted under the sign of postmodernism in particular.

In an extensive review essay, Edward Soja and Barbara Hooper (1993) draw attention to various radical attempts to overturn entrenched systems of difference and domination by reversing the binary distinctions on which they rely. Soja and Hooper argue that these distinctions are characteristic of what they call a distinctively *modernist* **identity politics**, in which particular political struggles are organised around oppositional hierarchies such as capital–labour, coloniser–colonised, white–black, man–woman, and so on. In each case, struggles over 'radical subjectivity' are directed towards overturning the privilege that is typically accorded to the first term over and against the second term. Soja and Hooper claim that

> Modernist identity politics characteristically projects its particular radical subjectivity, defined within its own oppressive binary structure, as overarchingly (and often universally) significant.

Exactly that manoeuvre was performed by David Harvey (1989) in the closing pages of his critique of *The condition of postmodernity*. There he called for 'the treatment of difference and "otherness" not as something to be added on to more fundamental Marxist categories' – and remember that he was writing from within historico-geographical materialism and its characteristic predeliction for class-based politics – but as 'something that should be omni-present from the beginning'. So far so good. But Harvey then went on to suggest that those other differences – of 'race', gender, and sexuality, for example – should be 'recuperated' *within 'the overall frame of historical materialist enquiry (with its emphasis upon the power of money and capital circulation) and class politics (with its emphasis upon the unity of the emancipatory struggle)'* (emphasis added). Not surprisingly, this was read by feminist and postmodernist critics as an attempt to condense politics around the class struggle – in Soja and Hooper's terms, as an attempt to project a particular radical subjectivity as 'overarchingly significant' – and hence as an attempt to privilege class over other axes of difference and discrimination. It sparked a turbulent debate (Deutsche, 1991; Massey, 1991) in the course of which Harvey (1992) tried to insist that 'the dialectic of commonality and difference' was in fact at the very centre of his project.

But this misses the point made by Soja and Hooper. They call for a 'disordering of difference from its persistent [modernist] binary structuring' – *a project that displaces the dialectic altogether* – and vigorously uphold the various attempts of a radical postmodernism, post-structuralism, and post-colonialism to interrupt and call into question the dichotomies of a conventional identity politics. Within what we identified in our introduction to Section Seven as a post-humanist geography, many writers now urge a recognition of **hybridity**, of the sheer complexity of the ways in which, as Michael Watts (Chapter 26) once put it, identities are 'cobbled together' as *multiple* not singular subjectivities. In an important reflection on 'the challenge of difference', Craig Calhoun (1995) observes that this idea of multiplicity subverts the very notion of the 'individual', with its standard implication 'that the person cannot be internally divided' because each one of us 'inhabits a single self-consistent **lifeworld**'. Against this, Calhoun argues that we are not entirely 'self-same': that we routinely come to terms with a variety of different identities and inhabit a plurality of different lifeworlds. Indeed, in his latest writings Said (1993) has himself set on one side that claim repeated 'throughout the exchange between Europeans and their "others" that began systematically half a millenium ago', namely that 'there is an "us" and a "them", each quite settled, clear, unassailably self-evident'. Instead, he insists that 'all cultures are involved in one another; none is single and pure, all are hybrid, heterogeneous, extraordinarily differentiated'.

Placement, displacement, and the production of differences

Said's work is instructive for another reason: it is distinguished by a remarkable geographical sensitivity (Gregory, 1995). Said shows that the conventional binary oppositions between 'the same' and 'the other' are put in place through the production of **imaginative geographies** whose boundaries install partitions between the two. This process takes place within cultures and between cultures, and its distinctions often have real material force in the production and reproduction of (in)human geographies of exclusion (Sibley, 1995; Radcliffe, 1996). For this reason, particular attention has been paid to the ways in which places and spaces become racialised, gendered, and sexualised, and to the implication of these processes in the fabrication of identities (see, for example, Keith and Pile, 1993; Bell and Valentine, 1995).

Much of this work has a double edge. On the one side, it clearly reactivates a long-standing concern within both spatial science and critical human geography over the unequal operation and outcome of processes of segregation and discrimination; on the other side, it holds out the promise of a more affirmative politics of location that is not mapped within the conventional metrics of centre and margin. This radically alternative geography has been variously described as a 'paradoxical space' attentive to the 'geometrics of difference' (Rose, 1993) and as a heterodox and differential 'third space' that literally displaces the traditional binary orderings of difference (Soja and Hooper, 1993). Tantalising as these last suggestions are, we think it important to retain and sharpen the double edge. Although identities are undoubtedly more fluid than traditional constructions make out, and although identities are constituted in different ways in different places, Geraldine Pratt and Susan Hanson (1994) sound an important note of caution against overvaluing the fluidity of subject positions. They insist that, for most people, social axes of power, position, and privilege cut in such a way that identities and differences are created and fixed in place, so that the ideal of displacement – of mobility in subject position, of 'paradoxical spaces' and 'third spaces' – must be held in tension with geographies of placement.

This intersects with another concern. Liz Bondi and Mona Domosh (1992) note that the invocation of travel, movement, and mobility as **metaphors** – loosed from their material moorings – can reduce postmodern geographies to a kind of 'tourism' in which the voices and experiences of others are articulated and appropriated by those in positions of power (see also Chapter 10). But they also note, with Donna Haraway (1992), that there is

> a very fine line between the appropriation of another's (never innocent) experience and the delicate construction of the just-barely-possible affinities, the just-barely-possible connections that might actually make a difference in local and global histories.

These are the affinities and mutualities, the circles of connection and conversation, that we outlined in our introduction to Section One. And so we return to where we began this book: to worlding geography, which can now be seen as a project of attending to difference in ways that enable human geography to *make* a difference.

Geography and difference

In the first essay, Audrey Kobayashi and Linda Peake suggest that racism and sexism are deeply embedded in geography's history and that this can be traced back to the European '**Enlightenment project**' at the end of the 18th century. This provided a series of binary oppositions – culture–nature, man–woman, white–black, and so on – that proved to be remarkably durable. They constituted an ideological order, a geographical imaginary, whose oppositions and hierarchies came to seem thoroughly 'natural'; hence Kobayashi and Peake want to call its privileges into question through the conduct of what they call an 'unnatural discourse'. What is striking about their critique, we think, is its determination to consider the connections between these pervasive dualisms. They do so by focussing on 'race' and gender and by showing how these social constructions are caught up in the policing of distinctions between 'nature and culture'. Like David Demeritt in Chapter 13, they insist that an important part of this project is a close attention to language as a social practice.

To be sure, nobody is suggesting that racism and sexism will be overcome through changing linguistic practices alone. But there is nonetheless a very important sense in which language enters decisively into the poetics and politics of geographical inquiry. Language is a system predicated on difference, and it is also (and for that very reason) sometimes a powerful means of silencing, suppressing, and even erasing difference. In the second essay, Michael Brown provides a particularly vigorous illustration of what we have in mind. Through an incisive critique of what have become conventional geographies of AIDS (acquired immune deficiency syndrome), Brown shows that those spatial representations that focus on vectors of disease transmission in space – the standard operating procedure for diffusion studies conducted under the sign of spatial science – have effectively closeted gay men: their 'viral focus reduces the already marginalized gay body to a mere vector for illness'. He also considers the ways in which these reductions, marginalisations, and silencings reinforce oppressive social distinctions: the 'prevention' of AIDS is typically presented in such a way that it sets up what Brown identifies as 'an us–them dichotomy socially founded on fear and homophobia', in which 'gay bodies' threaten 'the body public', in which 'they' threaten 'us'. Brown's question – which is Kobayashi and Peake's question too – is: who is this 'we'? He argues that analyses of this kind tacitly close their audience around the straight community, and so reproduce and underwrite the enforced normalisations of a repressive society: those who are excluded are marginalised socially and spatially. This is not an argument for ignoring difference and distance, however, and in the closing sections of his essay Brown reflects on the much more constructive part what he calls 'the ironies of distance' can play in the disclosure of what would otherwise remain 'hidden geographies'.

The final essay, by Michael Watts, is located at a political and intellectual crossroads: its conjuncture is formed by the inauguration of the 'New World Order', with its post-Cold-War geopolitical realignments, its market triumphalism, and its aggravated polarisation of the

North–South divide, and by the emergent conversations between political economy and cultural studies. But Watts rapidly draws geography into this mix. He notes the intensified internationalisation of late-20th-century capitalism – the complex interpenetrations of the local and the global – and considers some of the ways in which **time–space compression** (see Chapter 15) speaks to geographies of difference. Recalling both Haraway's and Slater's remarks about learning from the margins, he turns to Nigeria – one of the sites of his own remarkable research programme – to confront what we might call, as a sort of short-hand, the *articulation* of difference: in other words, how differences are constituted at the multiple hinges of the local and the global *and* how they are given voice – how they are represented – on a precarious and shifting 'common ground'. When Watts invokes the work of Walter Benjamin in order to think about the Maitatsine movement in West Africa, he also tacitly invites us to consider the ways in which our own, often peculiarly Western, notions can still illuminate other places and, in the same moment, how much we still have to learn not only about but also from those other places.

References

Bell, D. 1991: Insignificant others: lesbian and gay geographies. *Area*, **23**; 323–9.

Bell, D. and Valentine, G. (eds) 1995: *Mapping desire: Geographies of sexualities*. London: Routledge.

Bondi, L. and Domosh, M. 1992: Other figures in other places: on feminism, postmodernism and geography. *Environment and Planning D: Society and Space*, **10**; 199–214.

Calhoun, C. 1995: *Critical social theory: Culture, history and the challenge of difference*. Oxford, UK, and Cambridge, MA: Blackwell.

Cixous, H. and Clément, C. 1986: *The newly born woman*. (Trans.) Manchester: Manchester University Press.

Corbridge, S. 1993: Marxisms, modernities, and moralities: development praxis and the claims of distant strangers'. *Environment and Planning D: Society and Space*, **11**; 449–72.

Crang, P. 1992: The politics of polyphony: reconfigurations in geographical authority. *Environment and Planning D: Society and Space*, **10**; 527–50.

Deutsche, R. 1991: Boys town. *Environment and Planning D: Society and Space*, **9**; 5–30.

Driver, F. 1992: Geography's empire: histories of geographical knowledge. *Environment and Planning D: Society and Space*, **10**; 23–40.

Fabian, J. 1983: *Time and the other: How anthropolgy makes its object*. New York: Columbia University Press.

Godlewska, A. and Smith, N. (eds) 1994: *Geography and empire*. Oxford, UK, and Cambridge, MA: Blackwell.

Gregory, D. 1994: *Geographical imaginations*. Oxford, UK, and Cambridge, MA: Blackwell.

Gregory, D. 1995: Imaginative geographies. *Progress in Human Geography*, **19**; 447–85.

Haraway, D. 1992: *Simians, cyborgs and women: the reinvention of nature.* New York: Routledge. 113.

Hartshorne, R. 1939: *The nature of geography.* Washington, DC: Association of American Geographers.

Harvey, D. 1984: On the history and present condition of geography. *Professional Geographer,* **36**; 1–11.

Harvey, D. 1989: *The condition of postmodernity: An enquiry into the origins of cultural change.* Oxford, UK, and Cambridge, MA: Blackwell.

Harvey, D. 1992: Postmodern morality plays, *Antipode,* **24**; 300–26.

Katz, C. 1992: All the world is staged: intellectuals and the projects of ethnography. *Environment and Planning D: Society and Space,* **10**; 495–510.

Keith, M. and Pile, S. (eds) 1993: *Place and the politics of identity.* (London: Routledge).

Livingstone, D. 1992: 'A sternly practical pursuit': geography race and empire. In *The geographical tradition: Essays in the history of a contested enterprise.* Oxford, UK, and Cambridge, MA: Blackwell.

Massey, D. 1991: Flexible sexism, *Environment and Planning D: Society and Space,* **9**; 31–57.

McGrane, B. 1989: *Beyond anthropology: Society and the Other.* New York: Columbia University Press.

Pratt, G. and Hanson, S. 1994: Geography and the construction of difference. *Gender, place and culture,* **1**; 5–29.

Radcliffe, S. 1996: Imaginative geographies, postcolonialism and national identities: contemporary discourses of the nation in Ecuador. *Ecumene,* **3**; 23–42.

Rose, G. 1993: *Feminism and geography.* Cambridge: Polity Press; Minneapolis, MN: University of Minnesota Press.

Said, E. 1978: *Orientalism.* New York: Pantheon.

Said, E. 1993: *Culture and imperialism.* New York: Alfred Knopf.

Sibley, D. 1995: *Geographies of exclusion: Society and difference in the West.* London: Routledge.

Soja, E. and Hooper, B. 1993: The spaces that difference makes: some notes on the geographical margins of the new cultural politics. In Keith, M. and Pile, S. (eds), *Place and the politics of identity.* London: Routledge, 183–205.

Young, R. 1990: *White mythologies: Writing History and the West.* London: Routledge.

Suggested reading

On the most dominant form of ethnocentrism, see Ella Shohat and Robert Stam, *Unthinking Eurocentrism: Multiculturalism and the media* (London: Routledge, 1994); for a much more advanced – and correspondingly more difficult – discussion, see Young (1992).

On feminism and geography, see Bondi and Domosh (1992) and Rose (1993); the various reviews of Rose's book also give a good sense of the political and intellectual vitality of feminist geography. Work on geography and sexuality has yet to achieve the theoretical sophistication of work on geography and gender, but see the brilliantly provocative collections of essays gathered together in Bell and Valentine (1995) and in Beatriz Colomina (ed.) *Sexuality and space* (Princeton, NJ: Princeton Architectural Press, 1992).

24 Audrey Kobayashi and Linda Peake

Unnatural Discourse. 'Race' and Gender in Geography

Excerpts from: *Gender, Place and Culture* 1, 225–43 (1994)

Throughout its development, Western geography has been involved in the *construction* of (*inter alia*) 'races' and genders. Since its earliest involvement in exploration and scientific classification of the world, it has had a racist role, in that it has (first and foremost) supported the establishment of Eurocentric/Western domination both politically and intellectually. Geographers have literally and metaphorically mapped those boundaries, including physical boundaries, that separate and exclude the world of privilege from the world of the 'other' along racial lines. Indeed, geographers were major players in the development of scientific racism during the nineteenth century, at a time when the political and economic context of the West provided a welcome venue for deterministic biologisms (Livingstone, 1992, Ch. 7). For at least the past century, a body of geography textbooks and geographical travel accounts have passed on a legacy of naturalistic assumptions, as Europeans have re-invented colonial societies through their 'imperial eyes' (Pratt, 1992), thus imposing ideological regimes that facilitate their domination. Geographers have recently attempted, in both theoretical and empirical work, to overcome this past (see Domosh, 1991; Driver, 1991, 1992; Livingstone, 1992) but, regrettably, naturalistic assumptions – if not overtly racist prejudices – are still common within the discipline. Although their role is now somewhat muted, some purveyors of cultural geography continue to map the world along lines that perpetuate notions of racial difference and, implicitly or explicitly, racial hierarchy (but see Jackson, 1989).

Geography also has a sexist legacy, although its boundaries here are somewhat more obscure. Women remained invisible throughout most of the history of the discipline (Monk, 1983; Mayer, 1989), but where they have been represented, it is primarily in subordinate domestic roles. The world of work and commerce has been the world of men. Geography has thus until relatively recently supported the notion of separate domestic and public spheres, based on an ideological divide that has both provided a rationale for limiting access by women to the public sphere, and theoretically obscured our understanding of gender relations as complex relations of power. Moreover, feminist geographers have tended to address gender primarily in relation to class relations; witness the '*Antipode* debate' of the mid-1980s over the relationship between patriarchy and capitalism (Foord & Gregson, 1986; McDowell, 1986; Gier & Walton, 1987; Knopp & Lauria, 1987; Johnson, 1987; Gregson & Foord, 1987) which, while productive, ignored questions of racism almost entirely, and illustrated the limitations of treating a complex set of interrelationships in binary terms. Such lapses, however, are an indication of the extent to which inherited paradigms occlude new insights.

Our major concern here is not with the obviously sexist or racist tendencies within the history of geography, although these do need to be mapped across our intellectual legacy and a major study is long overdue.[1] Rather, we are concerned with the ways in which contemporary society is permeated by vestiges of thought categories, particularly those associated with the 'Enlightenment', that impose logic and rationality in nominalist and dualistic terms (Lloyd, 1984).[2] Despite the general rejection by contemporary scholars of absolutism and nominalism, and despite the current trend of debunking the Enlightenment, it was during this period that some of the most enduring aspects of dominant discourse were developed. The conceptual tools now at our disposal to challenge Enlightenment thinking were supplied within that very context. We shall continue to have to engage this intellectual heritage, therefore, for some time to come.

The social project of the Enlightenment was expressed through a series of interwoven discourses that naturalized racist and sexist categories. Racialization provided a rationale for nationhood exercised through the negotiation of territorial power:

> The popular Enlightenment concern with national characteristics often explicitly identified these characteristics racially. That this should have seemed so 'natural' a conflation is attested to by the intersection of nation with native: those properly of the nation are native to it, born and bred at its breast; Natives, by contrast, are those natural in racial kind to foreign, hostile, dominated lands. The latter are *naïve*, simple, lacking art, culture, and the capacity for rational self-determination [Williams, 1983, pages 213–216]. They are, in short, to be kept in their place, politically or geographically. (Goldberg, 1992, page 557)

Similar logic established the notion of 'natural women, cultured men' (Sydie, 1988), creating a basis for a sexual division of power according to putative biological endowments. Unlike the tendency with 'natives', however, women usually have been constructed as pleasant, docile creatures, the perfect complement to Man's authority:

> In *The Philosophy of Right*, Hegel contrasted the 'happy ideas, taste and elegance' characteristic of female consciousness with male attainments which demand a 'universal faculty'. 'Women are educated – who knows how? – as it were by breathing in ideas, by living rather than by acquiring knowledge. The status of manhood, on the other hand, is attained only by the stress of thought and much technical exertion.' [Hegel, 1952, pages 263–264] . . . In western thought, maleness has been seen as itself an achievement, attained by breaking away from the more 'natural' condition of women. Attitudes to Reason and its bearing on the rest of life have played a major part in this; a development that occurred in the seventeenth century has been particularly crucial. (Lloyd, 1984, page 38)

Dualistic thinking organizes the world according to oppositional categories – man/nature; culture/nature; man/woman; state/civil society; theory/practice; black/white – that order our existence. Such categories are more than a means of imposing intellectual order; they also exert and maintain political power and they almost always involve the privileging of one over the other (Collins, 1990, page 225). The metaphors of Enlightenment dualism have proved amazingly durable and powerful in influencing subsequent thought and in the construction of racist and sexist social relations that characterise modernity. Indeed, these dualisms have become so naturalized in everyday thought, language and actions that it is difficult to transform their metaphysical heritage in changing human relations.

From a geographical perspective, the intellectual heritage of the Enlightenment is one that involves *placing* the world within an ideological order, creating environments according to socially constructed and naturalized categories.

The remainder of this paper is devoted to a critical analysis of social constructions of gender and 'race.' We focus on the naturalization of these categories, as both academic and more broadly social constructs, and posit what we believe is a potentially constructive direction for geographical studies of 'race' and gender.

Social construction and naturalization

The incorporation of work on 'race' and gender in geography cannot be understood without reference to the parallel incorporation of values as a significant dimension of our research. One major shift in the discipline over the past two decades has been away from positivist philosophies towards a wider acceptance of critical theory. This has involved an initial affirmation of values as *part* of scholarship, and a subsequent transformation of concepts of values, incorporating developments in Marxist geography and the 'phenomenological moment' of the early 1970s, becoming increasingly critical with the contemporary debate over post-modernism . . . and the recognition that all forms of knowledge, including our analytical categories, are value-laden.[3]

Post-modernism, as one strand developing from critical theory, draws upon linguistic theories in its emphasis on the analysis of discourse as a means of understanding regimes of power, and on the methodology of deconstruction as a means of challenging and resisting dominant discourses. We agree with Calhoun (1993) that post-modern 'theory' is untenable, as 'pseudohistory', but recognize its importance as a *concerted* effort to break loose from the grip of the Enlightenment project, starting with the critical turn towards understanding the basis of current Western social formations. Its success as a form of critique may well be due to its 'lack' of an Enlightenment answer to moral questions (Flax, 1990, page 89).

Feminists, and to a certain extent anti-racists, have cautioned, however, that many post-structuralist positions maintain an androcentric viewpoint and that others, despite their theoretical challenge to hegemonic meanings, do little to change the material conditions of marginalized groups such as women and racialized peoples (Bondi, 1990). Socialist feminists, in particular, have argued that deconstructionists are as trapped as many of their predecessors in self-referential language systems and, as a result, actually do little to change the world. For all that our notion of unnatural discourse has in common with methods of deconstruction, we too are cautious of wandering too close to the abyss of nihilism, and of dissolving political efficacy within a 'justice of multiplicities' (Lyotard, 1984; cf. Kobayashi, 1993). We are also conscious of the need to *see the problems*, (in this case embodied, landscaped), as well as to *see through* the concepts that allow their social and theoretical articulation.

Geographers are now increasingly in accord about the indivisibility of values and ideology, and about the need to recognize that social constructions usually involve the expression of dominant ideologies. The current challenge is to go beyond this position, without lapsing into essentialism or retreating to universalist arguments that recast marginalized groups within overarching structures, albeit structures that are now differentiated according to the specificity of localized and identifiable 'others' (Harvey, 1992). Our agenda involves, first, the need to analyze processes of social construction to understand how concepts associated with 'race' and gender become naturalized, both in academic discourse and in society in general, that is, how they are normalized and viewed as intrinsic to a universal order. As Jackson (1991, page 193) has recently asserted, '[w]hereas many studies take the social construction of categories like ''race'' and gender as an end point, we need now to treat the idea as a starting point and to consider what difference it makes to our analysis'. Secondly, we need to develop ways of initiating what we call 'unnatural discourse', a discourse that develops, albeit uncomfortably, along lines that challenge categories so naturalized that they may not even be viewed as problematic. It is only in recent attempts to open up to scrutiny the conceptual categories involved in feminist research and in research on racism that we have been able to expose critically the ideologies embedded in each. In the following section, we attempt to interrogate the logic of 'race' and gender, to show both the ways in which they are linked and the ways in which they have been separately conceptualized. In the process, it may be possible to clarify some of the ways in which the study of one informs the study of the other.

'Race' and gender: making connections

Although racism and sexism need to be disentangled analytically on their own terms, they cannot be understood except in light of other processes. This is especially true if we examine the late nineteenth century, by which time naturalized notions of 'race' and sex were well entrenched within the processes of capitalism/imperialism/colonialism. Racism and sexism of the late nineteenth century were neither simple nor simple minded. They were supported and sanctioned through a complex system of social mores, including religious norms. Because they were systematic, they included women and racialized peoples in such ways that these groups often became complicit in their own subjugation, submitting their own values to the standard of logical, rational, and powerful, white man. Such was more the case for women, who often recast sexist mores and norms in feminine ways, through fashion, ways of speaking and acting, in the proscription of activities and the development of complex codes of etiquette, as well as in the web of attitudes surrounding motherhood. The ideal of the nineteenth century woman was one that was portrayed as both natural and good. One of the reasons that patriarchy is such a complex and durable form of social relationship is that it contains much that women have traditionally viewed as positive; the major sources of women's happiness in 'traditional' terms have involved marriage, love, motherhood. In recent years women have not abandoned their submissiveness easily or even, in many cases, willingly since, ironically, those very feminine traits are seen as necessary to overcoming the effects of sexism. The naturalization of women's roles has thus been one of the most effective social processes – perhaps the most effective social process – ever to occur.

According to Carole Pateman (1989, Ch. 1), women are constructed historically as the disorder that opposes the order of civilized man. To admit women to the realm of the political and the public would be to disrupt the order that is necessary for civilization to advance; hence, in part, the relegation of woman to the domestic sphere, and the creation of yet another powerful dualism that separates the public from the private, reinforced by an academic discourse that traditionally privileges the public. The private realm becomes, like the nature with which it is conceptually associated, an *object* of domination and control. As an object of control, it is constructed as inferior to that associated with the cultured and the male, that which has 'risen above' the baser conditions of the natural.

Racialization occurs within this naturalizing context as the construction of subhuman others supported by a series of economic and political interests as well as cultural constructions of the exotic, the foreign and the associations that involve evil and darkness. The separate sphere of racialized peoples has occurred through the racialization of labour (Miles, 1989), housing (Smith, 1989) and culture (Jackson, 1989), to name a few. In the expression of racism, active complicity has been more constrained than with sexism. One could argue that there are aspects of mutual dependency and emotional gratification in the master/slave relationship. But this is limited. Most racialized people have been well aware of their subjugation, and it has seldom been celebrated. Passive complicity has been extensive, however, to the extent that racialized people have taken on the characteristics ascribed to them by a dominant ideology. Their material circumstances of deprivation have often provided little choice.

The fact that 'race' and sex/gender have had different histories as social constructions is reflected in the rather different ways that they have been conceptualized by scholars. For over two decades, from the late 1960s until the late 1980s, the dominant position in feminist studies was to view 'gender' as a social construction predicated upon, but separate from, the biological category of 'sex'. 'Race', on the other hand, has predominantly been viewed by anti-racist scholars as a social construction based on an arbitrary (but systematic and therefore powerful) affiliation of phenotypical characteristics with social and cultural inferiority and superiority. The former notion of gender is based on essentialist thinking; the latter notion of 'race' repudiates essentialism. Therefore, within these separate, but nonetheless fundamentally similar, ways of thinking, 'race' would not exist without racism; sex would continue to exist without sexism although gender would have a radically different meaning. Hence, a convention has emerged whereby 'race' is written in inverted commas to indicate its arbitrary connection to anatomical features and its primary political meaning. Scare quoting gives it an unnatural quality, as though 'races' could be deconstructed if only racism were sufficiently resisted. Gender, however, has had no need of such a characterization given (at least in the English language) its clear distinction from the anatomical category of sex. Moreover, in gender studies, the emphasis has shifted to a positive sense of gender relations as a means of understanding structures of power and domination, while in anti-racist studies many have eschewed the concept of 'race relations' both because the term naturalizes the notion of 'race' and because emphasis should more properly be placed upon 'racist relations' (Miles, 1993). Our understanding of these developments is schematically outlined in Table 24.1, and is elaborated below, as we attempt to show that these different conceptual histories are themselves thoroughly politicized.

Dominant conceptualizations of 'race' and gender have been important for political purposes since the popularization of women's and civil rights movements during the 1960s. The social construction of gender has allowed feminists to bypass questions of biological essentialism (despite major differ-

Table 24.1 Conceptualization of race and gender

Conceptualization	Definition	Relationship
Race and sex as biological facts	Predetermined and static physical features	Biologically determined hierarchies between groups defined by sex and race
Race and gender as social constructs	A combination of arbitrarily and subjectively chosen fixed and static physical and physiological traits that do not interact (beyond an initial interaction) with social forces and relations	Interpersonal relations between dominant and subordinate members of socially defined groups. Racism: an ideology of racial exploitation which incorporates beliefs in a particular 'race's' cultural and/or biological inferiority and uses these beliefs to justify inferior treatment for that group. Similarly, sexism as an ideology promotes the belief in the inherent superiority of the male sex, and thereby its right to dominate the female sex
'Race' and gender as political constructs	A dialectical unity of the biological and cultural which are both seen as potentially dynamic and ontologically balanced/harmonious, although subject to political conditions	A political–human status which serves to draw out both differences and commonalities between human–social groups that can affect an individual's life chances

ences among feminists on this issue) in order to pursue the common goal of overcoming patriarchal relations. For anti-racists, the argument that 'race' is socially constructed has opened the conceptual barriers behind which racist practices have gained such social power and legitimacy. Connecting the two, however, is problematic, for a number of reasons. First, to render the two logically parallel involves a choice of unacceptable outcomes. One is to provide for 'race' an essential category similar to that of sex (hair texture? skin colour?), a choice that is clearly unacceptable to most anti-racists (although increasingly being adopted by certain black separatist movements). To even pose such a suggestion, moreover, helps to expose the fundamentally flawed nature of the sex/gender distinction as it is commonly understood. Alternatively, we might consider gender as a residual category that will be transcended through the abolition of patriarchy (cf. the transcending of 'women's studies' in favour of 'gender relations'). The logical extension of such a scenario, however, is a society in which difference itself, rather than the inequalities that attend the subordinating construction of difference, is eradicated. *Both* scenarios are not only unlikely, but take us into realms of theoretical speculation that are not only unconnected to social 'reality' but also highlight the shortcomings of attempting to solve problems through logical analysis. Such attempts only reinforce the extent to which we remain trapped within an Enlightenment vision, as we draw strategically upon earlier paradigms to introduce new concepts that will not only be understood, but will have the power to effect political change.

But the sex/gender relationship has been distorted as feminists have become trapped by those Enlightenment categories. The putative sex/gender distinction is attributed to de Beauvoir (1988), but has in fact been widely revised by relativist thinking, which has served to reinforce essentialist terms. What is misunderstood by those who accept the popular sex/gender distinction is that de Beauvoir's discussion was based on the Sartrean notion of pre-reflective choice (Butler, 1987, page 131), wherein we 'become' female, drawing upon a cultural discourse in which the terms of femininity are established and modified. The 'sex' upon which gender is constructed is therefore only an abstract concept of what lived social relations are *not*. There are two major variants of this line of thought within the French tradition of feminism. The first is that of the 'psycho-analytical post-structuralists', whose work stems from Freudian analysis and Lacanian linguistic theory. This approach sees the historical construction of femininity as 'negativity', the excluded other in masculine discourse, that emerges through unconscious symbolization, rooted in the symbolic intersubjectivity between all mothers and all children:

> Because women are themselves mothered by women, they grow up with the relations, capacities and needs, and psychological definition of self-in-relationship, which commit them to mothering. Men, because they are mothered by women, do not. Women mother daughters who, when they become women, mother (Chodorow, 1992, page 168).

Here, the essential quality is not sex, but the female capacity for motherhood. A second approach, seen in the materialist interpretation of Christine Delphy (1984) among others, captures more fully the social context in which gender constitution occurs by recognizing that naturalizing discourse is simultaneously a mode of intersubjectivity and a metaphor played out within a mode of production. The materialist perspective owes much to Foucault (1980) in its rejection of a 'natural' category of sex as prior to gender, and its rejection of necessary gender relations as anything more than historical moments whose consequences are, depending on the outcome of productive and strategic forms of power, contingent (Butler, 1987, page 137). Foucault's influence is perhaps even more important, however, for the way in which it mediates between psycho-analytical and materialist (socialist) feminist thought in the recognition that:

> our subjectivity, our identity and our sexuality are intimately linked; they do not exist outside of or prior to language and representation, but are actually brought into play by discursive strategies and representational practices (Martin, 1992, page 278).

What has been missed, therefore, by those who adopt the sex/gender distinction in essential and therefore simplified terms, either biological or cultural, is the fact that in either of these variants, and certainly in the original discussion put forward by de Beauvoir, 'sex' is not an essence, but rather a 'nothing', an ideal construct set up to create the possibility of dialectical understanding of gender. Sex is not, therefore, a condition prior to gender, but the quality that emerges as an ideal feature of existence.[4]

The logic of this disconnection of sex and gender also applies in refuting the essential qualities of heritable traits such as skin colour, since terms such as 'black' and 'white' are actually rather vague signifiers dependent more upon a complex set of 'encultured characteristics' (Goldberg, 1992) than upon actual pigmentation. If we attempt to apply this debate to a discussion of 'race', however, it is immediately apparent that the logical categories according to which feminist debates have occurred cannot easily be translated. Feminism is not logically parallel to racism; gender relations are not logically parallel to 'race' relations. But gendered relations parallel racialized relations; patriarchy operates in ways similar to racism. Does it follow, therefore, that there is a logical consistency between the categories of 'race' and gender? We would argue not. Why? At the most fundamental level, it is because in mainstream feminist theory, until recently, the theoretical distinction between sex and gender has served the political purpose of allowing society to believe that it can change its sexist behaviour (because gender is a social construction) without challenging the more deeply naturalized idea that men and women are essentially different.

But there has not emerged within studies of racism a category that is conceptually equivalent to that of the sexed body. We argue that this difference is due largely to the acceptance of scientific findings that phenotypical attributes, although inherited, do not distinguish human beings genetically (United Nations Educational, Cultural and Scientific Organization [UNESCO], 1969; Gould, 1981; Committee of the Inquiry into the Rise of

Racism and Racism in Europe, 1985); there is no natural antecedent (such as skin colour, for example) to the notion of 'race'. This finding has served to help overcome racism by refuting arguments of phenotypical and genotypical inferiority. The political strategy of the civil rights and anti-racism movements has been to challenge rather than to celebrate 'difference' based on biology.

It is difficult to judge which strategy, the essentialist position of mainstream feminism or the anti-essentialist strategy of civil rights, has been more effective politically. Intellectually, both 'race' and sex are increasingly being discarded as categories of analysis, except as the (socially and culturally constructed) products of racism and sexism, although gender as a social product continues to be conceptualized very differently from 'race'. To reject biological determinism, however, is not to treat 'race' or gender as irreducibly social. Alternative conceptions of 'race' and gender are ones that recognize:

> . . . that a full understanding of the human condition demands an integration of the biological and the social in which neither is given primacy or ontological priority over the other but in which they are seen as being related in a dialectical manner, a manner which distinguishes epistemologically between levels of explanation relating to the individual and levels relating to the social without collapsing one into the other or denying the existence of either (Lewontin *et al.*, 1982, page 11).

In summary, there is no *necessary* connection between racism and sexism, either theoretically or in practice. Their different stories, their potentially different subjects, and the myriad of forms in which they emerge speak against universal generalizations; but both gender and 'race' are forms of power relations, two distinctive but interconnected modes of historical domination. The ways in which connections between them have been theoretically constructed, as well as the ways that they have been mutually constituted in actual historical circumstances, are important to the expression of social power. To limit a discussion of their connection to the plane of logical categories of analysis, however, does not provide an effective means of resolving social issues, and leaves us trapped within the very confines of Enlightenment thinking that feminists and anti-racists seek to supersede. The question that confronts feminists and anti-racists today concerns the ways in which we continue to define gender and racial differences, and the ways in which such differences are embedded in dominant ideologies. Moreover, as any number of writers have now pointed out:

> Because there are conflicts, in some cases, between feminism and anti-racist objectives, because feminism is capable of racism (just as anti-racism is capable of sexism) feminists and anti-racists sometimes have to make difficult choices about where their allegiances lie (Knowles & Mercer, 1992, page 123).

We turn now to the issue of developing political strategies for unnaturalizing dominant notions of the biological and the cultural, thus changing the ways in which 'race' and gender are used to create social identities and to construct relations of domination.

Political strategies

The social construction approach has been criticized from another angle, for moving too quickly along theoretical paths that leave political problems unresolved. According to Jackson:

> In challenging the *naturalness* of categories like 'race' and gender, we run the political danger of evacuating the very concepts around which people's struggles against oppression are being organized . . . We are searching for non-essentialist conceptions of race and gender around which it is possible to mobilize politically, but we may have to face a dilemma in which we support one argument politically while rejecting its basis intellectually (1991, page 193).

This problem has led some writers (Stasiulis, 1990) to adopt positions of 'strategic essentialism' (see Spivak, 1989; Fuss, 1989) in order to retain political efficacy utilising what Stuart Hall terms 'arbitrary closure', or representation as a kind of formative political strategy (Hall, 1992). The dilemma is genuine, and it emphasizes the painful fact that despite our attempts to merge theory with practice, it is not so easily accomplished. Nor does theory in itself, however correct, provide a reliable basis for changing the world.

It is important to resist the temptation of strategic essentialism, nonetheless, for a number of reasons. First, our theories really are not just stories; interwoven with the world, their outcomes, like the dusty ideas of the Enlightenment, have real and durable effects. We need to find ways of merging theory and action in explicit as well as implicit ways. Theories can nonetheless constitute a radical form of discourse (Cohen, 1992). Secondly, the urge to essentialize, like the urge to universalize (Harvey, 1992) can be a cover for the fact that we don't have an answer. Giving in to essentialism, however promising the short-term goals, may retard the process of finding more effective solutions (bearing in mind that, in opposition to instrumental rationalism that sees solutions as progress, solutions are never more than moments of particular coalescence). Thirdly, essentialism tends to dissolve differences in broad generic categories. In political terms, such generalization is tempting, for the perpetration of images of generic suffering can mobilize considerable political force in a world where the generic and the natural are often closely intertwined. It becomes too easy, however, to mask the differences that exist among non-existent generics or, *in order to serve a political purpose*, to erect the same kinds of social constructions that have served the political purposes of racism and sexism. There are indeed no short cuts, and to take them is to risk setting amongst us the 'Trojan Horse' of feminist/anti-racist ethnocentrism (Spelman, 1990). Finally, the notion of strategic essentialism falls too readily into the hands of those nationalists throughout the world who, in the name of cultural freedom, democracy and identity, are killing and maiming and destroying all possibility for freedom, democracy and identity.

Part of the alternative to strategic essentialism lies in a form of politics that focuses on specific conditions as temporary historical moments, and attempts to establish effective remedies to the results of sexism or racism (or other forms of oppression) according to situational terms, 'built around issues, not biogenetic categories' (Knowles & Mercer, 1992, page 111). In other words, rather than succumbing to essentialism, essentials can be upended through 'a mode of politics which engages with the details of the oppression and which is capable of ending it' (page 110). Such strategies are important. They will make a difference to people's lives if they can be effectively mounted, and if they can avoid the risk of isolating the specific while ignoring the fact that specific conditions are often expressions of larger and less readily identified or organized processes, thus masking complexity as well as the amibguity of 'effects' interpreted from a variety of points of view (Brah, 1992). It is a difficult irony of such strategies that they, too, must rely upon a generalized consensus of both the problem and the solution while at the same time they deny the power of generalizations to control their lives.

We are still, however, no further out of our dilemma, restated by the Social and Cultural Geography Study Group Committee (1991, page 20):

> A curious intellectual and strategic difficulty maybe surfaces here, though, in that on the one hand the 'naturalness' of social categories is being questioned whilst on the other hand groups are themselves drawing upon such categories as a source of self-identity and internal bonding, but this difficulty is itself very much part of the cultural–normal terrain under scrutiny and thus cannot be resolved in any once-and-for-all correct fashion, either theoretically or pragmatically.

Despite the rationalist tendencies to which we have been acculturated, we are willing to accept the proposition that not only do we not know the answer *yet*, but we probably never shall; understanding is at best a temporary insight upon an otherwise invisible and intangible tangle of events. To live in a world in which lack of understanding (as opposed to misunderstanding) is a 'normal and acceptable condition (one in which contingency and uncertainty are not only theoretical propositions but the lived experience of a post-modern nightmare) is not only hopeless but also foolishly naive, for some reactionary would most certainly come along and attempt to impose order upon those made most vulnerable in their celebration of the contingency of difference: women and people of colour. Our willingness to admit a lack of understanding, however, may work to our advantage, for, as Harding (1986, page 193) has remarked, 'The greatest resource for would-be "knowers" is our non-essential, nonnaturalizable, fragmented identities and the refusal of the delusion of a return to an "original unity".'

What we can hope, for the moment, is to become more adept at the practice of unnatural discourse, as a means of exposing the conditions under which naturalizing tendencies have worked, so that we not only come to terms with the unsettling implications of post-modernism, but also disrupt the discourses set in motion by Enlightenment thought. This disruption involves two general goals:

(1) A continuing analysis and critique of the processes of social construction, as (almost universally) processes of naturalization, in order to understand the concepts, meanings, representations, practices and political forms through which 'race' and gender are constructed as normal, that is, viewed as part of a fundamental and unchanging (or slowly changing) order that is 'second nature'. Such understanding comes about not

especially through sophisticated theorizing, but through patient and determined empirical work that investigates the details of people's lives, their taken-for-granted worlds, and that asks questions that no one had previously thought to ask.

(2) The development of ways of initiating 'unnatural discourse', using both language (as concept) and political practice in such a way that natural categories are challenged. A method of deconstruction, or disassembly, is needed to allow us to understand not just the explicit statements of belief about what is 'natural', but also those values that are so naturalized that they have not previously come under question. This project requires, in part, a recovery of the significant past, and of the mnemonic qualities that are invoked in every use of language to give it its naturalizing power. As geographers, we need to invoke ideology to uncover the terrain that is uncontested because it is deemed to be ruled by common sense.

These challenges are huge, and no pretence will be made to outline here all that is involved. Perhaps imagination is our greatest asset, since most of what is involved is not even on the academic agenda. We can begin to speculate, however, about some of the elements of the unnatural, and upon how the political strategies might be conceptualized if not put in place.

The appeal of unnatural discourses lies in their use of practices that would in the past have been designated as (if you will) witchcraft or voodoo, practices associated with the unnatural, and therefore evil because threatening, powers of women or people of colour. The unnatural, in other words, is anything that falls outside the parameters of the naturalized, safe and known world of rational man. Included within those parameters are all practices that involve the imposition of social order based on a notion of a 'natural state of man'. Such a position could exclude, for example, violence (viewed by some as a natural tendency of man) as a means of bringing about social change. More radical are possibilities that instead project social actions that are beyond the realm of current imagination, so far, perhaps, that they may even be construed as 'supernatural'.

'The unnaturalization of everyday language' is an important aspect of this project. In language is codified the normative categories through which human relations are constructed, and communication provides the only means through which the conditions of change are expressed. The creation of 'race' is one such example of an effective means of unnaturalizing deeply held convictions; if that particular linguistic tool is now losing its efficacy so much the better, for it is an indication not that the term was inadequate, but that it provoked some of the changes that were sought. Linguistic strategies rely upon the imposition of conceptual disorder and ideological confounding against the power of naming as one of the most important means of imposing order and dominion. Using language as a force of the unnatural to deconstruct the power of words to create oppression is not to destroy the power of words to have meaning, and so create lives that are truly meaningless in a nihilistic void. Rather, we need to gain control of language in order to understand the ways in which discourses are constructed as ideological traditions that gain efficacy through repetition, inscription, and representation (Goodrich, 1990, Ch. 4). Such processes will go on, but hopefully at a higher level of competence.

This project involves an encounter with the concept of 'nature' itself (Fitzsimmons, 1989), and with the contradiction involved in the modern paradox that nature is both other and internalized, an external constellation of otherness to be subdued and denigrated while at the same time drawing upon an internal bundle of resources which constitute the 'second nature' that justifies human domination. This is by no means a consistent or unambiguous relationship. It has been the basis of much philosophical debate in which the two natures are sometimes complementary, sometimes co-existent, often at odds. The debate itself is the legacy of rationalism, however, and it has been invoked to powerful ends. Even so, the logic of the natural order is close to the breaking point in today's society, for reasons that involve environmental crises as well as crises of faith in the rationality of control. It is important that in the process of such crises, those who suffer be supported and the conditions that brought crises about be unnaturalized at their foundations.

Geographers may have a particular role in this project. The idea that all social relations are spatial, and *take place* within particular physical contexts (landscapes) whose attributes do matter, is an article of faith for most, if not all, geographers. We are of the school, furthermore, which contends that the social production of space is an ideological project. As Lefebvre notes:

> We should have to study not only the history of space, but also the history of representations, along with that of their relationships – with each other, with practice, and with ideology (1991, page 42).

He asks:

> What is an ideology without a space to which it refers, a space which it describes, whose vocabulary and links it makes use of, and whose code it embodies? (1991, page 44).

Racism and sexism are spatial/ideological practices, which *embody* the codes through which spatial control is maintained. Lefebvre's monumental contribution is his insight upon the ways in which, at different times in history, dominant men have constructed space, or landscapes, in their own image.[5] To understand and resist sexism and racism, then, we need to unnaturalize the landscapes upon which gendered and racialized relations are played out.

If, as Kant claimed, the importance of the Enlightenment was the development of intellectual maturity to the point where 'man' could have the freedom to 'make *public use* of [his] reason' (Kant, 1970, page 55) geographers can further their understanding of the social effects of the Enlightenment by recognizing the extent to which its dominating concepts have encoded the *public use of space*.[6] Genevieve Lloyd's observation that 'Kant's theme of Enlightenment as maturity is associated with spatial metaphors as well as temporal ones' involving 'a public space of autonomous speech' (1984, page 67) needs to be taken literally as well as metaphorically, to recognize ways in which the corollary to Western man's public emergence includes the feminization of the private and the exclusion of the racialized as part of the larger project of establishing spatial power and domination. The unnatural discourse required to shift racist and androcentric practices must include the public designation of space.

Geographers might also make use of the concept of unnatural discourse to question their own disciplinary codes. Our naturalized discourse includes the question: 'Is it geography?', to mean 'Does it conform to the dominant order?' 'Is it in its proper place?' Geography is largely about placing, and we need to understand that process as similar to and congruent with the process of othering, as a basis for extending our geographical sphere by displacing that which has been established as falling within the sphere of academic geography. Above all, as geographers with a long history of fascination with landscapes as material representations, we are in a position to develop new ways of seeing. Seeing involves the embodiment of people and their construction as different within landscapes that are not only 'real', but constructed to justify vision and ideology. The unnaturalization of vision involves a replacement, a rearrangement of what falls naturally 'into place', a shedding of new light. It involves seeing through the 'Enlightenment'.

Notes

1 This project has been undertaken with respect to sexism by Gillian Rose in her . . . book, *Feminism and Geography* (1993).

2 Nominalism is taken here as the doctrine that only physically perceivable objects have real existence; linguistic relations with objects do not constitute 'reality'.

3 An exemplary illustration of this trend occurs in the work of Anne Buttimer (1974, 1990). Value as Buttimer expressed it in 1974 was a personal attribute, which may or may not be shared with other people. In 1990, she sees it as a social construction that has no meaning outside of its shared expression.

4 This point parallels Jaspers's (1955) famous existential principle: existence before essence.

5 There is an important debate on the use of 'space' versus 'place' versus 'landscape'. We choose not to enter that debate here, however, and for the sake of argument accept the notion of 'space' as it is used by Lefebvre, recognizing also that it loses something in the translation from French to English.

6 We do not mean to imply here an acceptance of Kant's concepts of either reason or space as *a priori* categories. What is signifiant is that Kant was part of the Enlightenment project that made public and accessible, for some, certain ideological standards for the use of reason and space.

Selected references

Bondi, L. (1990) Feminism, postmodernism and geography: space for women?, *Antipode*, 22, pp. 156–168.

Brah, A. (1992) Difference, diversity and differentiation, in: J. Donald & A. Rattinsi (Eds) *'Race', Culture and Difference*, pp. 126–148 (London and Newbury Park, CA, Sage Publications and the Open University).

Butler, J. (1987) Variations on sex and gender: Beauvoir, Wittig and Foucault, in: S.

Benhabib & D. Cornell (Eds) *Feminism as Critique*, pp. 128–142 (Minneapolis, University of Minnesota Press).

Buttimer, A. (1974) *Values in Geography* (Washington, Association of American Geographers Commission on College Geography Research Report No. 24).

Buttimer, A. (1990) Geography, humanism, and global concern, *Annals of the Association of American Geographers*, 80, pp. 1–33.

Calhoun, C. (1993) Postmodernism as pseudohistory, *Theory, Culture and Society*, 10, pp. 75–96.

Chodorow, N. (1992) The psychodynamics of the family, in: H. Crowley & S. Himmelweit (Eds) *Knowing Women: feminism and knowledge*, pp. 153–169 (Milton Keynes, Polity Press and the Open University).

Cohen, P. (1992) 'It's racism what dunnit': hidden narratives in theories of racism, in: J. Donald & A. Rattansi (Eds) *'Race', Culture and Difference*, pp. 62–103 (London and Newbury Park, CA, Sage Publications and the Open University).

Collins, P. H. (1990) *Black Feminist Thought* (London, Unwin Hyman).

Committee of Inquiry into the Rise of Fascism and Racism in Europe (Ed.) (1985) *Report on the Findings of the Inquiry*, (European Parliament).

De Beauvoir, S. (1988) *The Second Sex*, trans. H. M. Parshley (London, Pan).

Delphy, C. (1984) *Close to Home: a materialist analysis of women's oppression* (London, Hutchinson).

Domosh, M. (1991) Toward a feminist historiography of geography, *Transactions of the Institute of British Geographers*, 16, pp. 95–104.

Driver, F. (1991) Henry Morton Stanley and his critics: geography, exploration and empire, *Past and Present*, 133, pp. 134–138.

Driver, F. (1992) Geography's empire: histories of geographical knowledge, *Environment and Planning: Society and Space*, 10, pp. 23–40.

Fitzsimmons, M. (1989) The matter of nature, *Antipode*, 21, pp. 106–120.

Flax, J. (1990) *Thinking Fragments* (Berkeley, CA, University of California Press).

Foord, J. & Gregson, N. (1986) Patriarchy: towards a reconceptualization, *Antipode*, 18, pp. 186–211.

Foucault, M. (1980) *The History of Sexuality, Volume 1: An Introduction* (New York, Vintage Books).

Fuss, D. (1989) *Essentially Speaking* (London, Routledge).

Gier, J. & Walton, J. (1987) Some problems with reconceptualising patriarchy, *Antipode*, 19, pp. 54–58.

Goldberg, D. T. (1992) The semantics of race, *Ethnic and Racial Studies*, 15, pp. 543–567.

Goodrich, P. (1990) *Languages of Law: from logics of memory to nomadic masks* (London, Weidenfeld & Nicholson).

Gould, S. (1981) *The Mismeasure Measure of Man*, (New York, Norton).

Gregson, N. & Foord, J. (1987) Patriarchy: Comments on critics, *Antipode*, 19, pp. 371–375.

Hall, S. (1992) New ethnicities, in: J. Donald & A. Rattansi (Eds) *Race, Culture and Difference*, pp. 252–259 (London and Newbury Park, CA, Sage Publications and the Open University).

Harding, S. (1986) *The Science Question in Feminism* (Ithaca, NY, Cornell University Press).

Harvey, D. (1992) Class relations, social justice and the politics of difference (unpublished paper).

Hegel, G. W. F. (1952) *The Philosophy of Right*, trans. T. M. Knox (Oxford, Oxford University Press).

Jackson, P. (1989) *Maps of Meaning: an introduction to cultural geography* (London, Unwin Hyman).

Jackson, P. (1991) Repositioning social and cultural geography, in: C. Philo (Comp.) *New Words, New Worlds: reconceptualising social and cultural geography* (Lampeter, Social and Cultural Geography Study Group of the Institute of British Geographers), pp. 193–195.

Jaspers, K. (1955) *Reason and Existenz*, trans. William Earle (New York, Noonday Press).

Johnson, L. (1987) Patriarchy and feminist challenges, *Antipode*, 19, pp. 210–215.

Kant, E. (1970) An answer to the question: 'What is enlightenment?', in: H. Reiss (Ed.), trans. H. B. Nisbet, *Kant's Political Writings* (Cambridge, Cambridge University Press).

Knopp, L. & Lauria, M. (1987) Gender relations and social relations, *Antipode*, 19, pp. 48–53.

Knowles, C. & Mercer, S. (1992) Feminism and antiracism: an exploration of the political possibilities, in: J. Donald & A. Rattansi (Eds) *'Race', Culture and Difference*, pp. 104–125 (London and Newbury Park, CA, Sage Publications and the Open University).

Kobayashi, A. (1993) Multiculturalism: representing a Canadian institution, in: J. Duncan & D. Ley (Eds), *Place/Culture/Representation*, pp. 205–231 (London, Routledge).

Lefebvre, H. (1991) *The Production of Space*, trans. D. Nicholson-Smith (Oxford and Cambridge, USA, Blackwell Publishers).

Lewontin, R., Rose, S. & Kamin, L. (1982) Bourgeois ideology and the origins of biological determinism, *Race and Class*, 24, pp. 1–16.

Livingstone, D. (1992) *The Geographical Tradition: episodes in the history of a contested enterprise* (Oxford, Blackwell Publishers).

Lloyd, G. (1984) *The Man of Reason: 'male' and 'female' in Western philosophy* (London, Methuen).

Lyotard, J-F. (1984) *The Postmodern Condition: a report on knowledge*, trans. Geoff Bennington & Brian Massumi (Minneapolis, MN, University of Minnesota Press).

McDowell, L. (1986) Beyond patriarchy: a class-based explanation of women's subordination, *Antipode*, 18, pp. 311–321.

Martin, B. (1992) Feminism, criticism and Foucault, in: H. Crowley & S. Himmelweit (Eds) *Knowing Women: feminism and knowledge*, pp. 275–286 (Cambridge, Polity Press and the Open University).

Mayer, T. (1989) Consensus and invisibility: the representation of women in human geography textbooks, *The Professional Geographer*, 41, pp. 397–409.

Miles, R. (1989) *Racism* (London, Routledge).

Miles, R. (1993) *Racism After 'Race Relations'* (London, Routledge).

Monk, J. (1983) Integrating women into the geography curriculum, *Journal of Geography*, 34, pp. 11–23.

Pateman, C. (1989) *The Disorder of Women* (Cambridge, Polity Press).

Pratt, M. L. (1992) *Imperial Eyes: travel writing and transculturation* (London, Routledge).

Rose, G. (1993) *Feminism and Geography* (Cambridge, Polity Press).

Smith, S. (1989) *The Politics of 'Race' and Residence* (Cambridge, Polity Press).

Social and Cultural Study Group Committee (1991) De-limiting human geography: new social and cultural perspectives, in: C. Philo (Comp.) *New Words, New Worlds: reconceptualising social and cultural geography*, pp. 14–27 (Lampeter, Social and Cultural Study Group of the Institute of British Geographers).

Spelman, E. (1990) *Inessential woman: problems of exclusion in feminist thought* (London, Women's Press).
Spivack, G. C. (1989) In a word: interview, *Differences*, 1(2), pp. 124–156.
Stasiulis, D. K. (1990) Theorizing connections: gender, race, ethnicity and class, in: P. Li (Ed.) *Race and ethnic Relations in Canada* pp. 269–305 (Toronto, Oxford University Press).
Sydie, R. A. (1988) *Natural Women, Cultured Men* (Scarborough, Ontario, Nelson).
United Nations Educational Scientific and Cultural Organization (1969) *Four Statements on the Race Question* (Paris, UNESCO).
Williams, R. (1983) *Keywords* (London, Fontana Press).

25 Michael Brown

Ironies of Distance: An Ongoing Critique of the Geographies of AIDS

Reprinted in full from: *Environment and Planning D: Society and Space* 13, 159–83 (1995)

Geographies of bodies and spaces?

Figs 25.1–25.4 relay an equivocal geography of AIDS in British Columbia that at once reveals and erases. They are standard representations of 'the geography of AIDS'; they mimic figures found in current literature on the

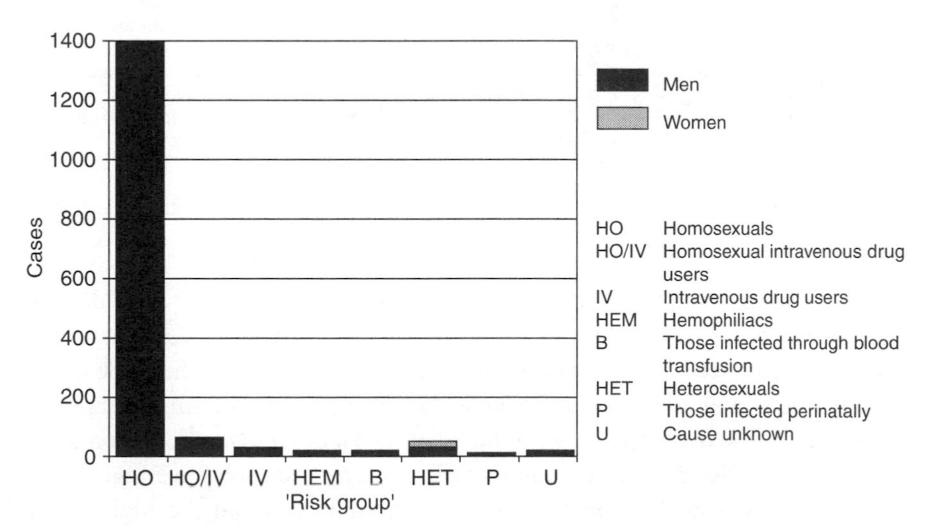

Fig. 25.1 AIDS in British Columbia by 'risk' factor, 1985–September 1993 (source: Rekart and Roy, 1993)

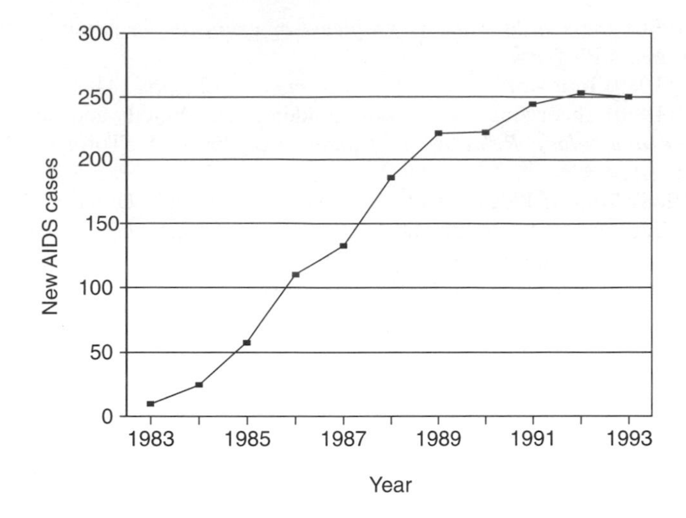

Fig. 25.2 Reported AIDS cases in British Columbia by year, 1983–September, 1993 (source: Rekart and Roy, 1993)

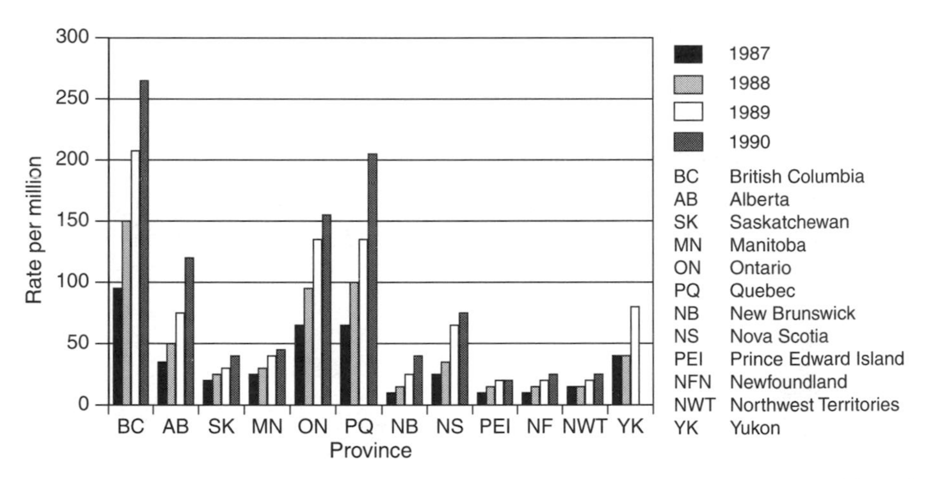

Fig. 25.3 Provincial per capita AIDS rates in Canada, 1987–1990 (source: Health and Welfare Canada, 1990)

subject, at least to the best estimate of the available data. As Fig. 25.1 alludes, it has been seen as a gay disease (see Edwards, 1992). Fig. 25.2 indicates the climbing incidence of the virus within the province through the 1980s, and suggests that rates are increasing at a decreasing rate. British Columbia's incidence of HIV is compared with the other provinces and territories on a *per capita* basis in Fig. 25.3. Although Ontario and Quebec each have higher raw numbers of HIV-positive people, British Columbia has had the highest *per capita* incidence of HIV. In other words, British Columbia (and Vancouver by implication, in light of Fig. 25.4) is the most intense location of the AIDS

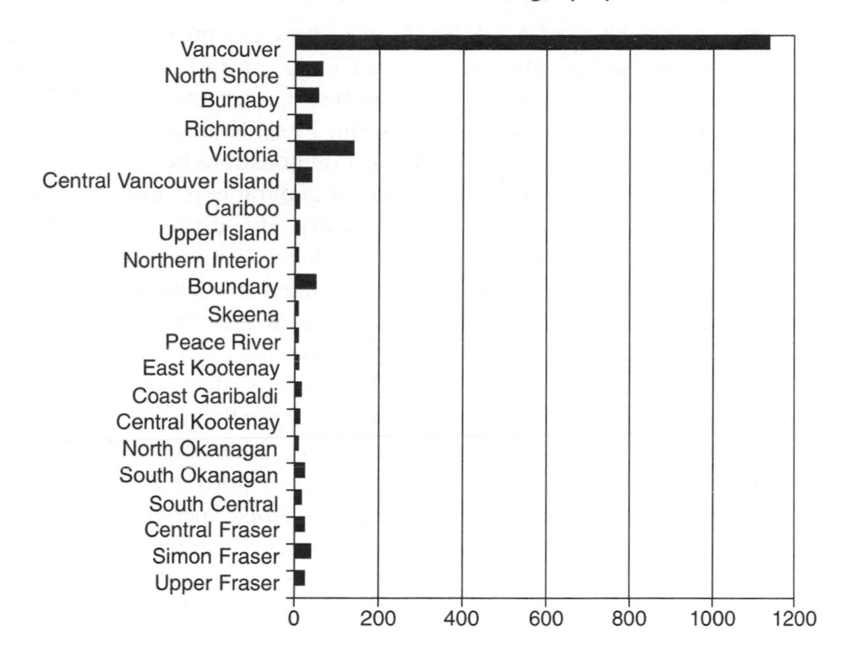

Fig. 25.4 Reported AIDS cases in British Columbia, by health unit to September, 1993 (source: Rekart and Roy, 1993)

crisis in Canada. Certainly these representations tell us much about 'the geography of AIDS' in Vancouver, Canada through the past decade. Yet this rendering of AIDS or HIV is intensely partial, incomplete, and – as I illustrate below – ultimately obfuscating. Simple yet pressing queries are prompted: Can the risk factors in Fig. 25.1 really be so mutually exclusive? How has Vancouver's gay community responded? These are some of the immediate critical silences Figs 25.1–25.4 evoke. They hint at a broader erasure.

Situated in scientific epistemology and ontology these representations upstage and exclude other ways of knowing about the places and people dealing with AIDS. Such alternatives, I insist, would portray a more social geography. Yet over the past five years geography has considered the AIDS pandemic largely through depictions like the ones illustrated above. For all its panoply of methods and perspectives, the discipline has drawn almost exclusively on its spatial science tradition to frame AIDS. Dangerous implications result. The epidemiological model legitimates a heightened focus on the virus and its travels across space. The hegemony of scientific discourse in social life is replicated and reinforced, in turn, by the discipline's singular tack. Most importantly, however, gay men have been closeted by this spatial-scientific geography of AIDS. This consequence is deeply ironic given the gains in visibility made because of their response to AIDS (Altman, 1988; Signorile, 1993).

In this paper I bring long-standing cultural critiques of AIDS science to bear on current geographies of AIDS by showing how gay men and the spaces they occupy are represented *and simultaneously erased*, in spite of the holocaust that AIDS has wrought on gay communities (Carter and Watney, 1989; Crimp, 1988; Crimp and Rolston, 1990; Epstein, 1991; Kramer, 1989; Patton, 1985; 1990; Rappoport, 1988; Wallis, 1989). I demonstrate how the erasures (by scientific representations) ironically conceal geographers' distances from gay men. I am not using the word metaphorically. By 'distance'[1] I mean a textually endorsed gap that emphasizes geographers' ignorance of gay men and their communities. This process is quietly hidden by science's exclusive focus on the virus itself. Gay men and their spaces are foregrounded unidimensionally, asocially, and only occasionally as nodal points in an epidemiological epic. These people are textually, socially distanced as bodily carriers. The viral focus reduces the already marginalized gay body to a mere vector for illness. Further, geographers have taken an Archimedean, abstracted – hence distanced – account of space in its portrayal of the AIDS crisis. Only certain places are discussed from a global perspective, while others – specifically gay communities in North America – are all but ignored.

The implication of these ironic distances is an erasure or closeting of an already stigmatized social group that has transformed places fundamentally in its response to AIDS. Ethnography, I demonstrate, can reveal such geographies in spite of the distances perpetuated by spatial science. The linchpin of my argument is that sociospatial distance sacrifices geographic knowledge; that ethnography overcomes distance. However, I must acknowledge the role distance played in my own ethnographic research of AIDS politics in Vancouver. To sharpen my critique of distance in spatial representations of AIDS, in the second half of this paper I describe the ironic ways social distance can potentially reveal geographies. In my concluding remarks, I specify the contexts and consequences of distancing as a point of critique, rather than the ever-existing span between the researcher and the researched itself.

Distant bodies

The viral focus

At first glance, it is scarcely intriguing that the AIDS pandemic has been taken up almost exclusively within medical geography, given the subdiscipline's long-standing concern over the relationship between health and physical environment (Smith, 1986, page 293). Although the spatial arrangements of health care and support are vital elements (for example, Willms et al, 1991), the spatial diffusion of AIDS has been the paradigmatic focus (Cliff and Haggett, 1989; Shannon et al, 1991; Smallman-Raynor and Cliff, 1990; Wood, 1988). Geography has entered the debates over HIV's origins and has tracked the spread of either HIV or AIDS across the globe in the past decade (Gould, 1993; Shannon, 1991; Shannon and Pyle, 1989). Similarly, predictive models anticipating future geographies have been developed (Gould, 1991a; Loyotonnen, 1991). Finally, of course, the diffusion of the virus or syndrome at various national and subnational scales has been charted, particularly in the United States (Dutt et al, 1987; 1990; Gardner et al, 1989; Gould, 1991a). Traditionally the purpose of mapping morbidity and mortality has been to shed light on the *cause* of disease, or the role certain elements of places play in the progression of the illness (Smith, 1986).

I would stress that what is being plotted, mapped, etc, across these geographies is the *virus* rather than the people dealing with it. People with HIV are not brought any closer – in terms of 'social distance' – by this brand of geography; the virus is. It is always the main character of these narratives. The virus and the syndrome are the central concerns for geographers. So, for instance, on page 17 of Shannon et al's (1991) book we are offered a full-page plate of the virus, a cross-section on page 21 (see also Shannon and Pyle, 1989, pages 4–5). Cliff and Haggett (1989, page 220) reveal HIV through electron microscopy as well. These texts, as well as Gould's (1993), offer extensive travelogues of the mobility of the virus and its operations inside the body (compare Fauci, 1991; Greene, 1993). Such discussions, placed at the front ends of texts, serve as scholarly anchors in scientific discourse, legitimating the subsequent truth claims, lending them a certain authoritative scientific power. Haraway (1991; 1992) and Patton (1990) have both stressed the legitimacy of science gained by such exploratory metaphors. Science can show us what we cannot see ourselves. It can 'boldly go where no man has gone before' [sic]. For geography, cartographies within the body can confidently and authoritatively explain (and potentially control) cartographies of the body.

Body–virus–science

Consequently the body is explained and portrayed as *the vector* of AIDS. The general scientific authority through which the body is depicted solely as a transmission node is furthered in spatial science. The fixation on 'Patient Zero', an airline attendant presumed to have been one of the earliest and most nefarious carriers of the disease, is indicative of this vampire trope, and has been incorporated into geography's explanations, albeit sceptically (for example, Shannon and Pyle, 1989, page 17; compare Shannon et al, 1991, page 118; Gould, 1993, pages 38–39).[2] Patient Zero becomes a black arrow across a map, defined narrowly by his sexual acts and his antibody status. The media sensationalism and stigma this sort of explanation fuels are not taken up. His roles in media hype and cultural mythologies of contagion are never discussed by geographers (compare Shilts, 1987;[3] Williamson, 1989; Hanson, 1991). He is important to geographers only to the extent that he spread the virus across space. Similarly, Gardner et al (1989, pages 37–39) place abstracted, smoothed HIV/bodies as plots on distance-decay graphs to describe the New York, Miami, Houston, and San Francisco metropolitan areas. Through these renderings gay men's bodies become reduced to biological hosts. Their contacts, their travels, their behaviour are important only to the extent that they enable the virus to move across space. There are no other contexts offered for these bodies. The multiple social discourses through which these social bodies are networked are reduced down to behavioural–biological attributes of the host. Thus in extended footnotes Shannon and Pyle (1989, footnotes 1, 2) cite scientific studies of gay men's sexual practices to seal their identity to particular sexual acts and promiscuities (compare Crimp, 1988; Watney, 1987).

Geographical discussion, then, reproduces the modern scientific tendency to fragment the body to such an extent that the social subject is irretrievably lost (for example, Rabinow, 1992). It is instructive to quote Patton (1990, page 55) on this process with respect to scientific discourses around AIDS in general:

> In AIDS medical science, the body becomes a screen or agar plate on which the disease is in play. The complex of symptoms, diseases, in themselves, produces repetitions. . . . Diagnostic medicine abstracts the symptoms from the body to produce a totalizing explanation with a single or primary cause, a pathology. Because the immune system, understood metaphorically, transcends the place of the body, the abstraction 'AIDS' folds back to correspond exactly to the *space* of the body. The virus is lost and, metaphorically speaking, the homosexual/prostitute/African/injecting-drug-user/hemophiliac body *becomes* AIDS (emphasis in the original).

Although her concern is over the social consequences of scientific hegemony of certain bodies, Patton notes that these outcomes are predicated on the textual erasure of social context for that authority (see also Horton, 1989). She insists that science produces knowledge by filtering out the social world. It reinscribes these forms of knowing as 'data' about (for instance) 'aggregate behaviour change' (page 53). Geographical discussions of AIDS, in this way, perpetuate a distance that is legitimated by scientific epistemologies of AIDS and HIV. That distance, Rosenthal (1989) suggests, lends a necessary objectivity to the emotionally charged accounts of the epidemic. It must be stressed, however, that as a thoroughly social distance, the span between spatial science and people living with AIDS is structured with considerable social authority. Science has become a master narrative, an ultimate arbiter, when alternative discourses conflict. Treichler (1988) claims the gay body is at once read from but at the same time does not exist in science, stressing many scientists' gaping ignorance, unease, and lack of contact with gay men generally, a point concurred with within geography most recently by Knopp (1992). Spatial science, consequently, never considers or justifies its (lack of) renderings about gay men because of this ambivalence. Rose (1993, page 77) has also called attention to the scientific power exerted over the body in geography through her critique of the masculinist, objectifying gaze towards bodies of the Other. Distance validates the objectivity of the geographer's framing of the body.

The hegemony of scientific authority over the bodies of gay men is all the more amplified in light of the long-standing historical *scientific* discourses that stigmatized the gay body clinically. The category 'homosexual' was a 19th-century concatenation of a series of behaviours and body types deemed threatening to the social order (Foucault, 1980). Bayer (1981) reminds us that it has only been in the last two decades that homosexuality was declassified as an (bodily) illness by the American Psychiatric Association. AIDS, however, has publicly reinforced this pathology. These bodies have become objects of power through scientific discourse, and the AIDS crisis has augmented that historical process (Epstein, 1991). Gay men's disease-racked bodies – emaciated, scarred by Kaposi Sarcoma lesions – have been so used as markers for suffering and death across the media, images that join with scientific renderings of contagion, contamination, threat, and guilt (Crimp, 1992a; Watney, 1988; 1989). To be scientifically represented as vectors of transmission, the gay body is available for surveillance and regulation. People living with AIDS are reduced to 'victims' in popular thought – to be both pitied and feared, the already dead (Navarre, 1988). By not showing us social subjects, only bodies, the discipline does not confront – and hence potentially augments – the social power of these scientific-spectre images. Such a distant, scientific, authoritative gaze at the body, Watney argues, is crucial to this hegemony (1989, page 68):

> AIDS is thus embodied as an exemplary and admonitory drama, relayed between the image of the miraculous authority of clinical medicine and the faces and bodies of individuals who clearly disclose the stigmata of their guilt. The principal target of this sadistically punitive gaze is the body of 'the homosexual'.

Bodies of risk

The concept of a 'risk group' (as in Fig. 25.1) further plays into this stigma. Risk groups have defined the virus through the body, especially the gay male body. Bodies only differ according to their risk capacity for transmission. Loyotonnen's (1991, page 133) figure 2 (mimicked here as Fig. 25.1) illustrates the tactic, as does table 2 in Smallman-Raynor and Cliff (1990, page 154) and figure 5 in Dutt et al (1987, page 463). Shannon et al's broad conceptual model is also framed through risk groups (1991, pages 157–164). Wood (1988) derives distinctive diffusion models based on bodies. Sexual practices are mapped over social identity. Categories tightly reinscribe the bodily boundaries where there are often fluid transgressions (Butler, 1990). For example, gay and straight couples may both engage in 'high risk' anal intercourse, but would be graphed in completely different – yet supposedly meaningful – categories in Fig. 25.1. Thus scientific discourse exercises its power through other social discourses, while cultural studies has criticized the notion of 'risk group' as it flows out of epidemiology towards the political sphere (Grover, 1988, pages 27–28):

> In the media and in political debate, the epidemiological category of *risk group* has been used to stereotype and stigmatize people already seen as outside the moral and economic parameters of 'the general population'. Jesse Helm's success in October 1987 in getting the Senate to prevent federal dollars from being spent on safe sex information for gay men – the hardest hit 'risk group' in the US, and the only group in which reported transmission of the virus has declined . . . due to safe sex education by gay men themselves – makes clear the social and political as opposed epidemiological, functions of the *risk group* concept: to isolate and condemn people rather than to contact and protect them.

Altman (1988, page 301) concurs, but more directly challenges spatial science's will to control when he traces scientific ontologies in the public sphere and the consequences of their itinerant relocations:

> This distinction between behaviour and identity, which often seems academic, is in fact vital to a rational understanding of AIDS. Because the media and the public generally do not make these distinctions, 'gay' and 'AIDS' have become conflated, so that public perception of homosexuality becomes largely indistinguishable from its perceptions of AIDS. This, in turn, has two consequences: (1) It causes unnecessary discrimination against all those who are identified as gay (including, in some cases, lesbians), and (2) it also means that people who are not perceived (and do not perceive themselves) as engaging in high risk behaviours can deny they are at risk of HIV infection.

Private bodies – public virus

Although the stigmatized relation between gay men and HIV diffusion has not eluded geography, it is hardly taken up critically. The distancing of gay men's bodies takes yet another form in the contradictions over the private sphere. Deconstructing images of people living with AIDS, Crimp (1992a, page 120) notes the irony that:

> . . . the privacy of the people portrayed is both brutally invaded and brutally maintained. Invaded, in the obvious sense that these people's difficult personal circumstances have been exploited for public spectacle, their most private thoughts and emotions exposed. But at the same time, maintained: The portrayal of these people's personal circumstances never includes an articulation of the public dimension of the crisis, the social conditions that made AIDS a crisis and continue to perpetuate it as a crisis. People with AIDS are kept safely within the boundaries of their private tragedies.

Geography too, I would argue, is highly susceptible to this critique. At least two pieces of geographic scholarship (Gardner et al, 1989; Loyotonnen, 1991) use data that have been collected through an invasion of privacy: through state policies of mandatory testing and/or reporting.[4] The authors do not reflect on the ethics of these data collections (compare Gould, 1993, pages 168–177; Last, 1990). Only Gould takes up the issue in a discussion of confidentiality. Yet he has been bothered by what he sees as an overconcern with confidentiality in surveillance data. Indeed, he offers a map of people living with AIDS in Los Angeles and challenges the reader actually to locate any of these people (1993, page 173). He states (1991a, page 31):

> It is argued that the geographic location of people with AIDS, specified by a very fine coordinate system or street address could become an identifier. I believe this to be wrong for perfectly understood ethical reasons. In all mathematical modelling, geographers have not identified anyone. . . . The loss of confidentiality is a genuine fear, but it has been taken to absurd lengths . . . so extreme and absurd that a vital component of our understanding is now in total disarray.

The concern for more precise and reliable data sets appear to trump the very real social issues confronting gay men (Patton, 1985). Gould reads the issue of confidentiality far too narrowly. Again, spatial science fails to consider the social discerning of these bodies *generally*. The issue is not so much the identification of particular persons with AIDS (though that does remain![5]), but rather the overall ignorance of their social situatedness. For instance, not only are the major risk activities for HIV infection *illegal* in many US states, simply *being* gay or lesbian – that is, the gay or lesbian body – can be inscribed as illegal and pathological in specific areas (Signorile, 1993; Wallis, 1989). As behaviour is scientifically reduced to the body, the state can outlaw both. Science is not the only discourse to conflate behaviour with social identity. The social stigma of the gay body or acts associated with it has been completely ignored by spatial science's discussion of data sets. Accordingly, the location and antibody status of the body – its raison d'être for this scientific endeavour – is knowledge more important than the social construction of the body, its social geography.

Bodies threatening the body politic

The distancing of gay men's bodies proceeds in geography in another fashion, appearing when we look across its audience. Gay men are conspicuous by their absence. Grover (1992, page 231) sums up this orientation when she writes:

> This, it seems to me, is a central dilemma in much American writing and broadcasting on AIDS: it ignores the communities with the greatest stake in AIDS as *subjects*, as viewers or readers, and uses them only as *objects* of its discourses.

It is widely acknowledged that control over the virus by geographic modelling and predictive mapping means control for people who are not infected. These have been textualized as 'the general population', which presumably gays do not constitute (Beauchamp, 1991). Gays are well aware of the proximity of HIV and AIDS in their lives, their communities. The constant stress on 'prevention' of diffusion can set up an us–them dichotomy socially founded on fear and homophobia. Gay bodies threaten the public body; 'they' are seen to threaten 'us' (again, in ways that often only science can allow us actually to see, like electron microscopy or geographic information systems). Indeed, Gould's book (1993) has a running 'we' through it. Given that he does not specifically discuss gay men or gay space, who could 'they' be, but not-'us'? Juxtapose, for instance, the geographies of AIDS with the arch conservative account of 'the gay plague' proferred by Gairdner (1992). It details the allegedly direct threats (one of which is AIDS) to the Canadian family (and by implication to Canadian society) from gays and lesbians. Situated in such a social context, the geography of AIDS, to date, has not been a geography written for gay men. They are the object – not the subject – of research. The discipline distances them by default because it does not speak to them, or their experience, all the while claiming an all-encompassing perspective by writing '*the* geography of AIDS' (Shannon et al, 1991; compare Gould, 1993). That scientific-authoritative erasure of gay men helps to explain the rise of a vehement cultural critique of scientific representations of AIDS and HIV.

Such critiques have just begun around geographic representations of AIDS. Consider, for instance, the angry, confrontational tone of Geltmaker's (1992) recent urban political geography of ACT UP Los Angeles that stresses the violence done to the gay body through institutional responses to AIDS. Newman (1990) and Watts and Okello (1990) have both criticized spatial science's extremely awkward encounters with African culture. Even within the spatial science discussion there are repeated calls for further, more detailed study of cultural practices among risk groups that facilitate transmission – but a simultaneous absence of the undertaking by any medical geographer or diffusionist to date (for example, Shannon et al, 1991). It can hardly be argued that gay men are *ignored* by the geography of AIDS. Rather, when they are considered, they are drawn as bodies that carry the disease. In this way, their social geographies are so distanced that they are erased.

To date, then, gay men with AIDS inter alia seem to be important to geography only as data points or modes by which the virus spreads across space. Geography's gaze reproduces a threatening image of the gay male body (especially ones that carry HIV); the us–them dichotomy is fuelled by the spectre of contamination. There is a social distance, then, between spatial science and gay men or space. Challengers to scientific, medical authority have been at pains to claim their lives and experiences – their knowledge – as valid, reliable ways of knowing about the AIDS crisis to which more powerful discourses must listen. This point has been most forcefully made through the rallying cry of ACT UP: 'Silence = Death', empowering lesbians and gay men to transgress social distance, as well as the persons with AIDS invective, 'Stop *looking at us*! Start *listening to us*!' (Crimp, 1992a, page 118).

Distant spaces

Straight space

Persistent debates about HIV's global origins and diffusions subsume the significance of the local responses to AIDS in places. Moreover, when local effects have been charted, the spaces are largely heterosexual ones (for example, Gould, 1993). Geographers note the decline in seropositivity rates in San Francisco or New York gay men (for example, Gould, 1991a), but say nothing about the incredible responses launched there (Fernandez, 1991; Kayal, 1993; Kuklin, 1989). If we take an inventory of the places that are discussed in the geography of AIDS, the absence of gay spaces is ironic, given the risk-group approach of spatial science. A certain heterosexism and homophobia underlies the absence to date of any geography written about an urban gay neighbourhood's experience with AIDS. That lacuna is all the more striking when one considers the history of such locales. For instance, by the late 1980s, it was estimated that one out of every two gay San Franciscans was HIV positive (Perrow and Guillen, 1990)! Thus there has been no mention of Greenwich Village in the geography of AIDS. Instead, the section of New York City we are likely to be brought to is the Bronx (Gould, 1993; Wallace and Fullilove, 1991). That heterosexual inner city is a familiar, and hence closer, terrain to geographers compared with the terra incognita of the gay ghetto, in spite of Lauria and Knopp's (1985) call for research into such places ten years ago. Indeed, as recently as 1992 gay and lesbian spaces had to remain anonymously concealed in geographic writing, reflecting structures of homophobia as much as an ethic of confidentiality (Adler and Brenner, 1992). Ignoring gay neighbourhoods while focusing on the AIDS crisis creates an ironic distance for geography, both in its concern to trace diffusion accurately, and its concern for changing facets of urban space. It denies the recent work on the intensely heterosexual structurings of the city, as well as the historic concealment or 'closeting' gays and lesbians have undergone (Valentine, 1993).

Although gay spaces are not completely ignored by the geographies of AIDS, they are heavily coded in scientific metaphors that deny their social geography. This mode of representation further distances gay space from geographers. Spaces like New York and San Francisco are cartographically described – literally – as 'epicenters' (Gould, 1993, page 124). In another passage, places exhibiting high infection rates are labelled 'incubators', signifying birth ironically in the location of so much death (Shannon and Pyle, 1989, page 13). The invasion or ground-zero metaphors surely contribute to the us–them distancing as they decontextualize what are often gay neighbourhoods or districts. As incubators on a metropolitan scale, gay men's bodies and spaces become distanced, serving as biological hosts to the natal virus.

Straight from space

The social and political limitations of AIDS geography are nowhere better illustrated than in its cartography. Armed with a will to represent the virus in an exclusive scientific focus, geographers portray HIV and AIDS like weather systems passing across a place (Shannon and Pyle, 1989, figure 5, pages 14–15). Likewise choropleth maps of the United States are deployed (Dutt et al, 1987, figures 1–2, pages 458–459; Gardner et al, 1989, figures 2–8, pages 30–36; Shannon et al, 1991, figures 6.1–6.6, pages 115–117). Distance-decay lines extend away from metropolitan areas (Gardner et al, 1989, figures 9–10, page 37; Gould, 1993, figures 9.2 and 9.3, pages 112–113). How do these representations help us to understand directly the social and cultural contexts of places which have been irrevocably, thoroughly changed by the crisis? As space is represented from a satellite, orbital (and hence scientifically augmented) position outside of where the virus is, its position reinforces distance while establishing science's authority to reveal the truth (Cosgrove, 1985).

Again, geography's lacunae speak louder than its contributions. The geography of AIDS hardly ever takes us into these places (compare Gould, 1993). Very little is actually conveyed about what these places are like. There is no 'obtrusive' or ethnographic research that could convey these knowledges, in spite of considerable work done in other disciplines (Gamson, 1991; Kotarba, 1990; Perrow and Guillen, 1990) and the merits of ethnography argued from within geography (Jackson, 1985). Even where specific locations are discussed, the orientation of the geographer is Archimedean. Place is brought forward to us, rather than us being brought to that place. For example, Gould takes us on a worldwide tour of the AIDS pandemic: from Thailand to the South Bronx. Although surely the comparative method is useful in stressing the unique experiences of places with a common problem, a rich context must be sacrificed. Furthermore, it is difficult to understand the point of such cartographies. Unlike cholera or smallpox, HIV is not spread through casual contact, and transmission modes do not always hinge on residential proximity. Maps illustrating HIV like 'a wine stain spreading across a tablecloth' (Gould, 1993, page 67) decontextualize how it is spread. Would it not make more sense to map people's sexual contacts across space? Would these not be more precise and accurate renderings of the geography of AIDS?

If geography connotatively or denotatively reduces gay men – or others – to mere bodily carriers of HIV and AIDS, it distances these people and compromises our potential to understand their struggles. Moreover, if the discipline only understands places like this scientifically – as an epicentre of a virus – it will *ironically* upstage a social process of distancing that goes on there. People and places are simply not passive conduits for the virus – even 'promiscuous gay men' and 'epicentres' like New York, San Francisco or (according to Figs 25.1–25.4) Vancouver. The travels of the virus are not the only geographies to be chronicled. In turn, these people and places are not simply webbed together by viral trails. Spatial science, then, potentially does not serve its own professed ends of controlling the virus. Lacking a cultural treatment of either gay men or their spaces, spatial science neglects the social construction of disease and identity, which are key elements to revealing the hidden geographies of AIDS. Textualizing gay men as biological carriers and their spaces as epicentres, while deflecting social attention away from them and towards 'risk groups' will not control the spread of the virus. It will only augment the erasure of gay men's histories and geographies.

Against erasure: ethnography

Moved by the often erased efforts of one very special AIDS activist, I began a year and a half long research project into the local responses to AIDS in Vancouver, Canada in 1991.[6] My research has been heavily informed by poststructuralism, and I took an ethnographic approach to the fieldwork. It culminated in the collection of 120 oral histories of people involved in the city's AIDs politics. Ethnography overcame the social distance of spatial science in several obvious ways. As a method it was self-evidently more intimate and social than unobtrusive methods, hence richer, more contextual data were collected. Experiential authority – 'You are there . . . because I was there' (Clifford, 1988, page 22) – however problematic, decisively overcame the distances written into current geographies of AIDS. Most broadly, it revealed a complex, articulated set of responses to HIV and AIDS across the state, the voluntary sector, and the family. I describe their efforts in greater detail elsewhere (Brown, 1994). Here, however, I want to draw out the implications for the erasures that spatial science perpetuates by conversely discussing the way ethnographic proximity *revealed* geographies of AIDS. I show below how gay men's bodies and their spaces act *socially and contextually* to prevent the diffusion of AIDS and HIV, emphasizing how important it is to broaden the scope of 'the geography of AIDS' beyond viral diffusion.

Gay bodies block HIV

In spite of a concern for prevention, no geographer has detailed gay men's roles in educating themselves, each other and others about safer sex. Should this not be a facet to the story of how AIDS – through gay men's bodies – passes across *or is prohibited from* – places? For example, Fig. 25.5 reports on a 1993 Canadian survey of gay and bisexual men's AIDS awareness. It demonstrates that a majority of gay men across Canada have modified their sexual practices considerably since (or because of) AIDS (Meyers et al, 1993, page 38). Fig. 25.6 is taken from an AIDS Vancouver's Man-to-Man[7] brochure. Along with Figs 25.7 and 25.8 it explicitly links the gay body to safe sex practices, visibility, and self-esteem. In these examples gay bodies are precisely *not* vectors for HIV's diffusion, yet more distanced geographies fail to convey that inhibitive role for the gay body. By failing to relate spatial science's social distance with the stigmatized context of the gay body, geographic discussion potentially perpetuates a discourse of threat and contagion.

These examples are supplemented by the fact that during my interviews, gay men were forthright in labelling themselves as gay. Many even insisted I use their names explicitly. To them, confidentiality was a means of erasure. They were emphatic about gay men who *were* involved in the fight against AIDS. They were reacting to representations such as Figs 25.1–25.4 that portray gay men simply as a vector of illness. Through ethnography a very

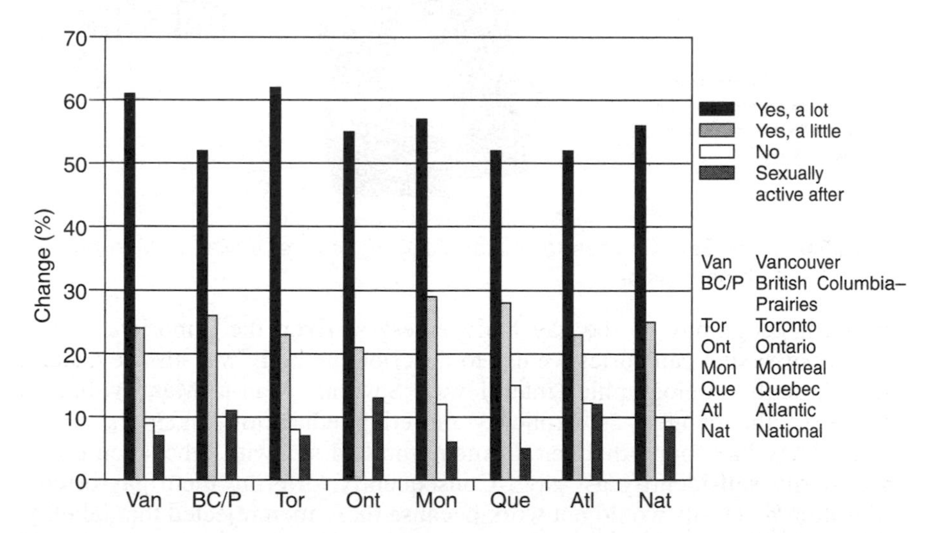

Fig. 25.5 Changes in sexual practices since AIDS for gay and bisexual men. The categories British Columbia–Prairies and Ontario exclude Vancouver and Toronto, respectively. In this survey, gay men were asked whether or not they had modified their sexual practices since AIDS (source: Meyers et al, 1993)

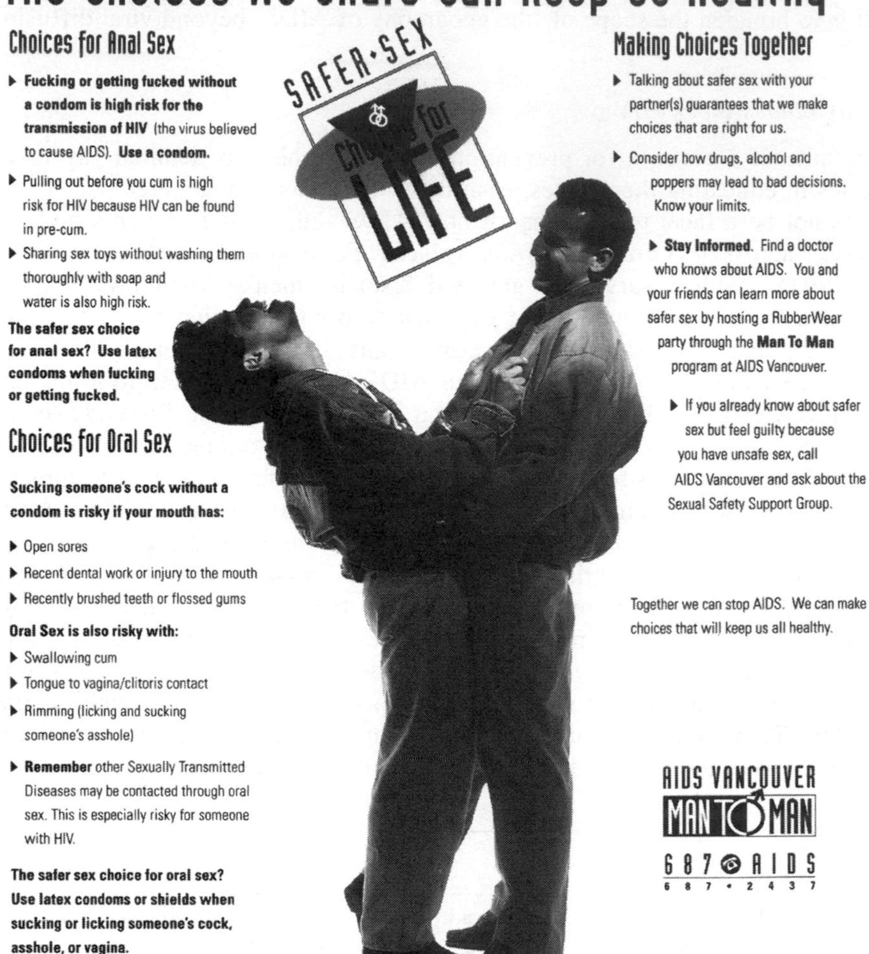

Fig. 25.6 'Safer Sex: Choices to Share': AIDS Vancouver's Man-to-Man brochure (reprinted with permission)

different geography of the gay body emerges. Even the importance of the fluidity of sexual categories we use to describe the body was also revealed to me through ethnographic interviews. Several Man-to-Man volunteers described the dangers in explicitly targeting education materials at 'gay men'. They had found that certain men who had sex with other men did not necessarily self-identify as 'gay'. Consequently, offering them gay-oriented education materials would not work because these men rejected that label, yet still engaged in high-risk sex with other men. The Vancouver ethnography revealed a group of gay men committed to overcoming the educational obstacles that the stigmatized gay body presented. Here, the gay body also blocks the virus, but in a far more complex way than spatial science could ever map.

Family Values
Gay men's relationships, no matter how they are defined, are as valid as those of any 'traditional' family.

Built on respect, caring, and honesty, our families remain strong under the toughest of circumstances. Practising safer sex *all of the time* is an important, healthy expression of our values.

We can still have great sex in the age of AIDS—with NO FEAR—whatever your HIV status. Our community supports your decision to have sex safely.

Right to Life
Gay men have the same right to life as anyone. The right to life includes protecting ourselves and others – by always using a latex condom every time we have anal sex.

Someday everyone will fully recognize our rights. Until then, we will protect and care for each other.

The Moral Majority
Gay men are part of the solution to the HIV epidemic. We can be proud to our success in fighting this disease.

Yet we're rarely acknowledged or congratulated for our achievements. Often, we're bashed, harassed, and discriminated against. At best, we're ignored.

Safe Sex *is* the norm for gay men today. *This* Moral Majority is made up of HIV-negative and HIV-positive men who express their sexuality in a healthy way.

By maintaining safe sex, we are showing the world how to successfully stop the spread of HIV. By strengthening our commitment to safe sex we continue to be part of the solution.

Fig. 25.7 Edited version of 'Family Values' brochure, San Francisco AIDS Foundation, 1992 (reprinted with permission)

Fig. 25.8 'Pride': acrylic on masonite, by Tom Laws, 1993; AIDS Vancouver's Man-to-Man promotional material, 1993 (reprinted with permission)

Gay spaces block HIV

As Fig. 25.6 illustrates, innovative, culturally sensitive educational programs have been extensively deployed in (and by) the gay community. These practices are often intensely spatial: safe sex (or 'rubberwear') parties in the home, condom and literature distributions, media campaigns, and metropolitan hotlines. Safe sex education and materials were also ubiquitous at lesbian and gay events in Vancouver like the Stonewall Festival in June, or the Pride Parade in August. As well, there are several AIDS Vancouver Information Centres in Vancouver's gay bars, bookstores, bathhouses, even restaurants. Thus, like the body, gay spaces do not merely exhibit the diffusion of HIV, but also its blockade. Figs 25.7 and 25.8 are frequent images across Vancouver's gay neighbourhoods. They are part of Man-to-Man's annual safe sex campaigns. These public images link gay bodies and space by stressing their role in *preventing* the spread of HIV. Consequently, Figs 25.6–25.8 potentially control the spread of the virus better than Figs 25.1–25.4, the representations of spatial science. Here, ethnography shows that there are significant diffusions *within* a place, aside from viral transmissions, that are part of 'the geography of AIDS'. Again, these diffusions are so important because they block the spread of HIV.

Ethnographic interviews also revealed the diffusion of information and resources that link Vancouver to other places. Because of the rapid onset of the crisis just after the emergence in the 1970s of gay subcultures and neighbourhoods, there were few historical resources to cope with AIDS (Crimp, 1992b; Kramer, 1989; Shilts, 1987). Most responses were fashioned from scratch. Consequently, information networks and exchanges quickly developed between and within locales. Places become linked together not simply by the diffusion of HIV but also by information and resources. These connections are made in various ways. Vancouver is a source of information and materials for many rural communities across British Columbia. Members of ACT UP Vancouver attended the San Francisco and Montreal International AIDS Conferences to aid local protesters (the former in violation of US immigration law). A founder of the Gay Mens Health Crisis in New York helped a group of gay men in the West End to form AIDS Vancouver in 1983 (Shilts, 1987). The relatively easy access to AZT in Canada created informal smuggling routes down to the USA. These spatial connections represent links within the international and intranational gay communities, as well as other affected communities. They demand that there is more to the geography of AIDS than the diffusion of the virus. Connections can also span and link identities and struggles as they connect places globally. John Gates (1992, page 4), a keynote speaker at the 1992 Canadian AIDS Society's general meeting, made this point poignantly:

> My proposal is this: we have over the last ten years lobbied successfully for the early release of drugs. Tonight I'm asking you to change tactics and reverse that process. I am asking that people living with AIDS and with HIV and their organizations call on governments and multinational organizations to delay the release of any new vaccine or cure for AIDS until three conditions can be met: that the drug or

> vaccine be affordable worldwide, that it be accessible worldwide, and that it be available worldwide. Without these conditions being met, we would make it clear that we would not be in favour of the release of those drugs or vaccines.[8]

Beyond diffusion as the geography of AIDS

So far I have shown how ethnographic proximity reveals other diffusions that impact significantly on spatial science's narratives. Ethnographic proximity also stresses the horrific impacts of HIV on people and their places. The scientific vision employed by geographers surely fails to capture and convey – hence it misrepresents – the pain, suffering, and struggles that have certainly altered places where AIDS has struck (compare Kuklin, 1989; Wolf, 1991). Reflect on Russo's (1990) passionate geography of New York City that pivots precisely on his incredulity at the social distancing, with which the geography of AIDS is complicit:

> Living with AIDS in this country is like living in the Twilight Zone. Living with AIDS is living through a war which is happening only for those people who are in the trenches. Every time a shell explodes you look around to discover you've lost more of your friends. But nobody else notices – it isn't happening to them. They're walking the streets as though we weren't living through a nightmare . . . [I]t's worse than wartime because during a war people are united in a shared experience. This war has not united us – it's divided us. It's separated those of us with AIDS and those of us who fight for people with AIDS from the rest of the population.[9]

These sentiments were also expressed in Vancouver. One man sketched a similar urban geography:

> The whole thing is a sort of smouldering powder keg of people who are just so angry at having *lost*. You know, you're not just talking about losing one friend. You're talking about losing dozens. Here in Vancouver, in the West End, there's nobody I could meet that doesn't know somebody who's got it or who hasn't had somebody die of it. Everybody's been touched by this thing. And it's a Goddamn nightmare that people want to run away from, but there's nowhere to run!

These narratives portray urban neighbourhoods that have been irrevocably changed because of AIDS and the responses of gay men to it. They insist that there are alternate geographies of AIDS to tell, beyond the travels of a virus, which – after over a decade – geographers have yet to discuss.

A final example of the significant implications ethnographic research has for the geographies of AIDS: while volunteering at the Vancouver display of the Canadian NAMES Project AIDS Memorial Quilt in May of 1993, I heard the name of the man quoted directly above being read out, indicating that he had died. He was quite sick when I had interviewed him the previous summer; a nurse hastily arranged our meeting because she did not think he would live much longer. His was a rich and candid interview. His work in the local response to AIDS was often quiet and behind the scenes, yet it had considerable efficacy. He was the first person to die from my roster of informants. In December of 1993, one other interviewee died; no doubt more will follow. As data points on the figures that opened this paper, these two people will always reveal a geography. Without ethnography, however, the other geographies they helped construct are lost.

Distances in ethnography

To end my critique here, however, would be dishonest because ethnography also relies on distances, ironically. In the spirit of poststructuralism's perpetual critique and political reflection (White, 1988), I wish to consider the ways that social distance functioned positively in my ethnographic research. Ethnography is not immune from distancing (Hammersley, 1992; Stacey, 1988), but erasure may not be a necessary consequence of social distance. In this section I will sketch out the ironic necessity of distance for the dialogic and polyphonic modes of ethnographic authority.[10]

Although I became a familiar face on the local AIDS scene, I was nonetheless distanced as an outsider by my academic purpose and affiliation. I did not become a formal volunteer at any of the local AIDS service organizations (ASO) for fear that I would jeopardize my standing with other local AIDS groups that I was nurturing. I did not want to be seen as being 'from' a particular ASO.[11] Distance worked for this ethnography polyphonically because it allowed me to gain access to multiple voices endemic to urban politics. It brought me closer to a variety of subjects I would have lost had I entrenched myself completely in one form of volunteering or organization. Unaligned, I was trusted. It became important for people to get their own interview to balance my time spent with others. For instance, it was very difficult to secure interviews at the Vancouver Persons With AIDS (PWA) Society (a self-help, empowerment group exclusively for HIV-positive people) because I was not a member. Members were suspicious of being coopted, used, and most especially misrepresented by the HIV-community (after all, their mandate was self-empowerment: to speak for themselves). Many did not want to be reduced or appropriated in an academic tome. Yet, once my project was explained to potential subjects by fellow PWA members, some became eager to speak with me because they wanted to make sure their stories were recorded too. They saw refusing to participate as a distancing by erasure. My distance, then, was trusted as a vehicle of balance. If a scholar from outside came 'in' and only talked to AIDS Vancouver affiliates, the PWA voice would be erased or drowned out, it was thought.[12]

Likewise, I was surprised to find how helpful people were in pointing me towards *opposite* directions and opinions than their own, across often acrimonious and bitter local divides. Had I been too closely affiliated to one group, I am not sure I would have been sent in those directions, nor would I have been so well received once I arrived there. One director of a local AIDS organization deliberately insisted that I speak with an activist who spent much of our interview berating that director for his complicity with power structures in the city. My strategic distance worked inside organizational structures in a similar way. Vancouver evinces a rapidly growing and bureaucratizing AIDS voluntary sector. Growing pains were evident in all organizations I touched, spawning much bitterness, alienation, and mistrust. Had I been, for instance, seen by volunteers at AIDS Vancouver as too closely aligned with staff, I am not sure they would have been so candid with me about their own misgivings and criticisms. For example, in the spring of 1992 a 'volunteer caucus' developed in response to the alleged neglect of AIDS Vancouver volunteers by its (increasing) paid staff. I could discuss the caucus with both volunteers and staff, yet neither could broach the subject with the other without becoming rather officious and tense. I crossed a similar tension between current and former (often very disillusioned) affiliates of local organizations. My distance allowed me to get at a wider range of local activists than a closer, more immersed strategy would have.

Dialogically, my social position vis-à-vis the interviewees gave *them* a distance on their experiences, which we shared in the interviews. My social distance gave interviewees the chance to step back and substantially reflect on what they had done. I had not anticipated this opportunity as a benefit of my research, yet repeatedly people thanked me for it. They thanked me for giving them the opportunity to sit down for an hour and simply tell someone else what they had done *comprehensively*. This sort of critical reflection is a rare opportunity because of the all-encompassing and never-ending qualities of AIDS, the constant grieving, and intensity of emotion. Its absence can be noted in the high degree of burnout and turnover in AIDS work. As one person confided in me after an interview, 'You can't have a good day around here. You can never feel good about your work, because people are dying. How can I be doing a good job if people are still dying? How can I have a good day?' It was my distance – my lack of affiliation coupled with a genuine interest in their experiences – that gave employees and volunteers the opportunity to grasp some distance of their own, alleviating some of their frustrations. It was clear to me that many people had not realized how much they had done locally, because they never had the opportunity in the middle of the crisis to stop and think about it. To do so would risk an alleged selfishness; AIDS work is self-consciously directed towards another's pain. In a very real way, then, the distance working in the ethnographic interview allowed me to bear witness as well as achieve a fair degree of contextual breadth and understanding.

Social distance also fostered a strong desire to make sure I understood local culture and context. Once I explained that I was from another place – a place that had a completely different local politics of AIDS – they were at pains to ensure that I completely understood what they had done and its context. We could share instances of homophobia and oppression in AIDS politics, I could identify with them along the lines of sexuality and grieving, but I was still an outsider, and therefore needed to be let in on the details of this place to ground my understandings of the Vancouver context. Gaps in my understanding hardly raised suspicions about my dedication or abilities as a researcher. If anything, they were accepted, anticipated, but always to be filled in through our conversations. My distance was something they could learn from as well. Many subjects were interested in 'how is this geography?' or 'citizenship' or even 'city politics' (the terms of my theoretical agenda). As I explained my own thinking about the Vancouver situation, some would sincerely try to rethink what they were doing along these templates of social theory. Others refused to succumb to those takes on their work, and insisted on their own interpretations. So now I find it inappropriate to slap my scholarly notions of citizenship and city politics *completely* over relationships that could readily be characterized as friendship, love, or faith. Distance, I would argue, allowed a conversation about the local responses to AIDS in Vancouver, about new forms and locations of local politics, and about lesbian and gay lives to emerge.

Distance reconsidered

If, as I have argued above, distance can augment geographic knowledge, it is incumbent on me to hone the critique of distance advanced at the front of this paper (Marcus, 1992), even if it is a different kind of separation (I take this point up below). The front half of this paper is radically incomplete because ethnography does not entirely mitigate the span, even if it does provide a corrective to some of its consequences. Although I remain critical of spatial science's geography of AIDS, I must insist as well that distance can work to reveal in crucial ways. It is important to know, to the best of our ability, about where HIV is on the globe, especially as it varies across permutations of HIV. Spatial-scientific Archimedean perspectives do enable us to see the globalness of the crisis, the *pan*demic. Further, it is important to track the movements of HIV across space. The AIDS crisis, as activists intone, is *not* over; spatial science can illustrate this quite convincingly. Predictive models are helpful, especially with respect to state funding and preventative measures. In these ways spatial science's investigative role cannot be denied, and Gould's (1991b) pedagogy is hardly without merit. Spatial science can play a (limited) role in AIDS education and prevention in each of these ways.

An example from one of my interviews illustrates this point. In Vancouver, science's hegemony ultimately forced a homophobic provincial cabinet with strong Christian Fundamentalist leanings to begin to fund AIDS organizations (albeit covertly) (see Brown, 1994 for a fuller discussion). Bureaucrats produced figures like Figs 25.1–25.4 that convinced an intensely homophobic government to act. A British Columbia Ministry of Health bureaucrat recalled:

> In the spring of '87 I made a presentation to Cabinet. [Peter] Dueck was Health Minister. And I talked about AIDS and how it was growing in our province. And I made some predictions as to how bad the problem could become if it continued to grow at the same rate. Now, in the last couple of years those of us in public health have been taken to task for those predictions being too dire. On the other hand, those predictions were never stated as our opinion of what would happen. It was always stated as, 'Given what we know now, if things continue at this rate, we will have this . . .'. And for some reason, Cabinet clicked. And they realized that they had to get in and do something. And I think $1.5 million was designated to it.

Precisely because of the social power of scientific discourse, there is critical *potential* of spatial-scientific geographies. Here Epstein's (1991) drive for a 'democratic science' of AIDS is instructive. He argues that poststructural critiques and cultural activism's challenges to scientific insular authority and hegemony represent an historic democratization to that discourse. Those challenges must be seen in a context of profound ambivalence. AIDS activists *need* science: it is useful and, most likely, treatments (will) come from it. Those working against scientific knowledge's problematic nature must make use of scientific claims in advancing their own positions. Scientists and activists, then, actually share a great deal: a significant degree of cultural capital, symbolic power, and both can speak the persuasive language of science. Inspecting that ironic lack of distance, Epstein (1991, page 53) muses over the future distances between scientists and gay men on the one hand and marginal groups who lack their shared commonalities:

> The special role of intellectuals and professionals therefore bears close scrutiny, particularly in light of the shifting epidemiology of the epidemic in the United States. As AIDS increasingly becomes a disease of an urban underclass lacking not just the material but the educational resources of, say, the middle class gay white community, what are the prospects for grassroots participation in the social processes of science?

Distance can be an inevitable aspect of social relations between researcher and researched and, as I have tried to show in this section, it does not necessarily function to erase. Indeed, it can work to reveal hidden geographies. But the caveats offered in this section should not be taken as a refutation of the criticisms lodged in the first portion of this paper. Rather, they insist that a more qualified critique of distance be fashioned. In the following section, I suggest a way to retain a focused critique of distance while acknowledging its inevitability throughout the research process.

Reconsidering the ironies of distance: conclusions

One might argue that I have overworked the concept of distance in this paper, that essentially I have juxtaposed two completely different and incomparable processes. The scientific distance is epistemological, founded upon a subject–object dualism endemic to the objective stance of the researcher. By contrast, ethnographers hold a more social distance from their subjects, along apertures of class, race, gender, sexuality; but also geography. The ethnographer physically enters a space; the spatial scientist does not. Ethnographic distance is therefore one that derives from the researcher's presence inside of a situation but outside of its context. If these separate definitions were accepted, the two forms of distance would be incomparable, having actually little to do with each other. I believe this potential criticism is misplaced for two reasons. Foremost, both spans hold true to Eyles's (1986) definition cited above, which is a substantial one. In both cases I am referring to the separation of scholars from research subjects and spaces; the implication of an unequal balance of power across that divide has also been demonstrated in this paper. More to the point, the distinction between philosophical stance and empirical practice is rather too clearcut in that potential criticism for these postmodern times. For the past decade, geographers have insisted that social relations are fundamentally spatial (and vice versa). Presumably this includes our *own* social relations. We never stand completely outside the worlds we study any more than we stand completely inside of them. Thus, although I agree that the forms of distance I describe are distinctive, I also insist that they are similar insofar as they refer to the relationships which are just as social as the ones between ourselves and those (places) we study. To understand 'social distance' only in the narrow or classical context of, say, redlining, presumes an impermeable divide between social theory and social practice that is out of step with the times.

How are we to reconcile these opposing vignettes on social distance? Alcoff (1991) provides a means to maintain a critical stance on spatial science's erasures, while acknowledging the inevitability of social distance in the research process. Concerned about the widespread reluctance of academics and activists to 'speak for the other', she argues that although scholars can no longer pretend to transcend their own social locations, and that certain locations are socially privileged, she reminds us that truth claims are not wholly dependent on social location or distance. As she puts it (1991, page 16):

> To say that location *bears* on meaning and truth is not the same as saying that location *determines* truth.

Concomitantly, she maintains that there are situations where *not* speaking for the other is quite dangerous and irresponsible. Besides, she insists, speaking only from a distanced location will always implicate and affect others because of the very connectedness of social life. Alcoff concludes that, if speaking for the other is inevitable, we must be vigilant in limiting its dangers. The issue of speaking for others is intensely related to the gaps and distances I have noted above: both refer to the power relations that define the relationships between researcher and researched. Her argument substantiates the critical points against spatial science developed in this paper by stressing the irresponsibility and dangers of erasing gay men and their spaces in current geographies of AIDS, while honestly acknowledging the inevitability of social distance. She suggests that we must be constantly reflexive, always considering the potential damage done by speaking for the other. Alcoff's point, to my mind, focuses the edge of a critique of distance. Social distance is inevitable in the research process. It is therefore not so much its manifestations that need to be rejected, but rather we must constantly (re)consider its damages and erasures if we are to provide critical geographies of AIDS.

To return to the figures that opened this paper, they do potentially erase gay men and their space. Nonetheless, they concisely impart the crises that AIDS has created in Vancouver and British Columbia. Further, although they can distance and erase, they potentially can also bring AIDS closer, showing us the extent of the pandemic. There can be many ways of knowing what AIDS is (Treichler, 1988; compare Sontag, 1988). We should be critically aware, therefore, that there can also be many geographies to know through this crisis, and their implications must be perpetually interrogated. In this paper I have noted a variety of distances and proximities mediating geographic research.

To privilege one venue over another exclusively, even within a progressive agenda, can feed into 'the dangers of difference'; the point of geographic inquiry becomes lost in ethical posturing and categorical affinities (Pratt and Hanson, 1994).

Though fractured between spatial science and urban ethnography, this critique is united by a consideration of the ironic roles distance plays in writing geographies of the AIDS crisis. I have argued that spatial science manufactures a distance predicated in part on the bodies of gay men: the virus is foregrounded, as is its spread across space. Gay men and the places they change as they respond to AIDS are ironically distanced by this brand of geography. Ethnography can, of course, write a corrective geography. It does not, however, eradicate distance; in fact it uses it constructively to reveal often concealed geographies. In one sense, distance erases, in another it provides perspective for critical geographical reflection. To endorse a rejection *tout court* of medical geography and spatial science of the AIDS pandemic based on their social distance alone is ill founded because distance is not essentially a good or a bad social relation. Distance has not erased gay men and their spaces from the geographies of AIDS, rather an ongoing failure to consider its implications is to blame.

Notes

1 Eyles (1986, page 435) defines *social distancing* as 'The separation of two or more distinct social groups for most activities, by either mutual desire, . . . or by the dictate of the superordinate group'.
2 Patient Zero died in March 1984.
3 Randy Shilts died in February 1994.
4 Gardner et al rely on data collected from the US military applicant-screening programs. Although they acknowledge the potential for self selection out of the sample, they nonetheless maintain its reliability. It is not clear if applicants know they are being tested for HIV antibodies. Loyotonnen's Finland data are based on mandatory reporting to the National Board of Health. Again, the virtue of the reliability of the data set is pressed, but the ethics behind collection are not considered.
5 For instance Kansas Reverend Fred Phelps and his followers have made it a point to protest about homosexuality at the funerals of people who have died of AIDS, carrying signs that read, 'God Hates Fags! ' and 'Praise God for AIDS! ' (see Bull, 1993).
6 Terry Brown died in November 1991.
7 Man-to-Man is the AIDS Vancouver department that deals with education and prevention programs specifically targeted to gay men.
8 John Gates died in 1993.
9 Vito Russo died in November 1990.
10 Clifford (1988) defines dialogic (or discursive) authority as, 'a constructive negotiation involving at least two, and usually more, conscious, politically significant subjects' (page 41). Polyphonic authority, 'represents speaking subjects in a field of multiple discourses' (page 46). Its encounter is 'particularly open to readings not specifically intended' (page 52).
11 This was because of the ever-present tensions across local AIDS organizations. As

well, there was a long waiting list for volunteer training, I would displace a more 'authentic' community volunteer; as well there was a time commitment that I was unable to meet.

12 Previously strained relations between the PWA Society and AIDS Vancouver help explain this fear. In 1987 a coalition of persons living with AIDS broke away from AIDS Vancouver, and eventually became an autonomous organization. They left because they felt the needs and concerns of PWAs were not being recognized. That tension as well as the different mandates and (until 1992) locations of the organizations subsequently placed them on rather separate trajectories.

References

Adler S, Brenner J, 1992, 'Gender and space: lesbians and gay men in the city' *International Journal of Urban and Regional Research* **16** 24–33

Alcoff L, 1991, 'The problem of speaking for others' *Cultural Critique* **20** 5–32

Altman D, 1988, 'Legitimation through disaster: AIDS and the gay movement', in *AIDS: The Burdens of History* (Eds) E Fee, D M Fox (University of California Press, Berkeley, CA) pp 301–315

Bayer R, 1981 *Homosexuality and American Psychiatry: The Politics of Diagnosis* (Basic Books, New York)

Beauchamp D E 1991, 'Morality and the health of the body politic', in *The AIDS Reader: Social, Political and Cultural Issues* Ed. N MacKenzie (Meridian, New York) pp 408–421

Brown M P, 1994, 'The work of city politics: citizenship through employment in the local response to AIDS' *Environment and Planning A* **26** 873–894

Bull C, 1993, 'Us vs. them: Fred Phelps' *The Advocate* issue 641, 2 November, pages 41–45

Butler J P, 1990 *Gender Trouble: Feminism and the Subversion of Identity* (Routledge, New York)

Carter E, Watney S (Eds), 1989 *Taking Liberties: AIDS and Cultural Politics* (Serpent's Tail, London)

Cliff A D, Haggett P, 1989, 'Spatial aspects of epidemic control' *Progress in Human Geography* **13** 315–347

Clifford J, 1988 *The Predicament of Culture: Twentieth-century Ethnography, Literature and Art* (Harvard University, Cambridge, CA)

Cosgrove D, 1985, 'Prospect, perspective and the evolution of the landscape idea' *Transactions of the Institute of British Geographers* **10** 45–62

Crimp D, 1988, 'How to have promiscuity in an epidemic', in *AIDS: Cultural Analysis/Cultural Criticism* Ed. D Crimp (MIT Press, Cambridge, MA) pp 237–271

Crimp D, 1992a, 'Portraits of people with AIDS', in *Cultural Studies*, Eds L Grossberg, C Nelson, P Treichler (Routledge, New York) pp 117–133

Crimp D, 1992b, 'AIDS demo graphics', in *A Leap in the Dark: AIDS, Art and Contemporary Culture*, Eds A Klusacek, K Morrison (Vehicule, Montreal) pp 47–57

Crimp D, Rolston A, 1990 *AIDS Demo Graphics* (Bay Press, Seattle, WA)

Dutt A K, Miller D, Dutta H M, 1990, 'Reflections on the AIDS distribution pattern in the United States of America', in *London Papers in Regional Science 21. Spatial Epidemiology* Ed. R W Thomas (Pion, London) pp 183–196

Dutt A K, Monroe C, Dutta H M Prince B, 1987, 'Geographical patterns of AIDS in the United States' *Geographical Review* **77** 456–471

Edwards T, 1992, 'The AIDS dialectics: awareness, identity, death and sexual politics', in *Modern Homosexualities: Fragments of Lesbian and Gay Experience* Ed. K Plummer (Routledge, New York) pp 151–159

Epstein S, 1991, 'Democratic science? AIDS activism and the contested construction of knowledge' *Socialist Review* **21** 35–64

Eyles J, 1986, 'Social distance', in *The Dictionary of Human Geography* 2nd edition Eds. R J Johnston, D Gregory, D M Smith (Basil Blackwell, Oxford) page 435

Fauci A, 1991, 'The human immunodeficiency virus: infectivity and mechanisms of pathogenesis', in *The AIDS Reader: Social, Political, Ethical Issues* Ed. N MacKenzie (Meridian, New York) pp 25–41

Fernandez E, 1991, 'A city responds', in *The AIDS Reader: Social, Political, Ethical Issues* Ed. N MacKenzie (Meridian, New York) pp 577–585

Foucault M, 1980 *The History of Sexuality* (Vintage, New York)

Gairdner W D, 1992 *The War Against the Family: A Parent Speaks Out on the Political, Economic, and Social Policies That Threaten Us All* (Stoddart, Toronto)

Gamson J, 1991, 'Silence, death, and the invisible enemy: AIDS activism and social movement "newness"', in *Ethnography Unbound: Power and Resistance in the Modern Metropolis* Ed. M Burawoy (University of California Press, Berkeley, CA) pp 35–57

Gardner L I, Brundage J F, Burke D S, McNeil J G, Visintine R, Miller R N, 1989, 'Spatial diffusion of the human immunodeficiency virus infection epidemic in the United States, 1985–1987' *Annals of the Association of American Geographers* **79** 25–43

Gates J, 1992, 'Global responsibility' *Vancouver PWA Newsletter* Vancouver Persons with AIDS Society, 1107 Seymour Street, Vancouver, BC V5B 5S8, pages 1–4

Geltmaker T, 1992, 'The Queer Nation Acts Up: health care, politics, and sexual diversity in Los Angeles' *Environment and Planning D: Society and Space* **10** 609–650

Gould P, 1991a, 'Modelling the geographic spread of AIDS for educational intervention', in *AIDS and the Social Sciences* Eds R Ulack, W F Skinner (University of Kentucky Press, Lexington, KY) pp 30–44

Gould P, 1991b, 'Thinking like a geographer' *Canadian Geographer* **35** 324–331

Gould P, 1993 *The Slow Plague* (Blackwell, Winchester, MA)

Greene W C, 1993, 'AIDS and the immune system' *Scientific American* **269**(3) 98–105

Grover J Z, 1988, 'AIDS: Keywords', in *AIDS: Cultural Analysis/Cultural Activism*, Ed. D Crimp (MIT Press, Cambridge, MA) pp 17–30

Grover J Z, 1992, 'AIDS, keywords and cultural work', in *Cultural Studies* Eds L Grossberg, C Nelson, P Treichler (Routledge, New York) pp 227–239

Hammersley M, 1992 *What's Wrong With Ethnography?* (Routledge, New York)

Hanson E, 1991, 'Undead', in *Inside/Out: Lesbian Theories, Gay Theories* Ed. D Fuss (Routledge, New York) pp 324–340

Haraway D J, 1991 *Simians, Cyborgs, and Women: The Reinvention of Nature* (Routledge, New York)

Haraway D J, 1992, 'The promise of monsters: a regenerative politics for inappropriate/d others', in *Cultural Studies* Eds L Grossberg, C Nelson, P Treichler (Routledge, New York) pp 295–337

Health and Wealfare Canada, 1990 *AIDS Surveillance Report* (Supply and Services Canada, Ottawa)

Horton M, 1989, 'Bugs, drugs, and placebos: the opulence of truth, or how to make a treatment decision in an epidemic', in *Taking Liberties: AIDS and Cultural Politics* Eds E Carter, S Watney (Serpents Tail, London) pp 161–182

Jackson P, 1985, 'Urban ethnography' *Progress in Human Geography* **9** 157–176

Kayal P M, 1993, *Bearing Witness: Gay Men's Health Crisis and the Politics of AIDS* (Westview Press, Boulder, CO)

Knopp L, 1992, 'Sexuality and the spatial dynamics of capitalism' *Environment and Planning D: Society and Space* **10** 651–669

Kotarba J A, 1990, 'Ethnography and AIDS: returning to the streets' *Journal of Contemporary Ethnography* special issue on ethnography and AIDS **10**(3) 259–270

Kramer L, 1989 *Reports from the Holocaust: The Makings of an AIDS Activist* (St Martin's Press, New York)

Kuklin S, 1989 *Fighting Back: What Some People Are Doing About AIDS* (G P Putnam's Sons, New York)

Last J M, 1990, 'Epidemiology and ethics', in *Ethics in Epidemiology: International Guidelines* Eds Z Barkowski, J H Bryant, J M Last, proceedings of the Council for the Organization of Medical Sciences Conference, Geneva, Switzerland, pp 14–29; care of World Health Organization, Ave. Appia, CH-1211, Geneva 27

Lauria M, Knopp L, 1985, 'Toward an analysis of the role of gay communities in the urban renaissance' *Urban Geography* **6** 152–169

Loyotonnen M, 1991, 'The spatial diffusion of the human immunodeficiency virus type I in Finland, 1982–1987' *Annals of the Association of American Geographers* **8** 127–151

Marcus G, 1992, '"More (critically) reflexive than thou": the current identity politics of representation' *Environment and Planning D: Society and Space* **10** 489–493

Meyers T, Godin G, Calzavara L, Lambert J Locker D, 1993, 'The Canadian survey of gay and bisexual men and HIV infection: men's survey', Canadian AIDS Society, #701, 100 Sparks Street, Ottawa, Ontario K1P 5B7

Navarre M, 1988, 'Fighting the victim label: PWA Coalition portfolio', in *AIDS: Cultural Analysis/Cultural Activism* Ed. D Crimp (MIT Press, Cambridge, MA) pp 147–166

Newman J L, 1990, 'On the transmission of AIDS in Africa' *Association of American Geographers* **80** 300–301

Patton C, 1985 *Sex and Germs: The Politics of AIDS* (South End, Boston, MA)

Patton C, 1990 *Inventing AIDS* (Routledge, New York)

Perrow C, Guillen M F, 1990 *The AIDS Disaster: The Failure of Organizations in New York and the Nation* (Yale University Press, New Haven, CT)

Pratt G, Hanson S, 1994, 'Geogrpahy and the construction of difference' *Gender, Place, and Culture* **1** 5–29

Rabinow P, 1992, 'Severing the ties: fragmentation and dignity in late modernity' *Knowledge and Society: The Anthropology of Science and Technology* **9** 167–187

Rappoport J, 1988 *AIDS Inc: Scandal of the Century* (Human Energy Press, San Bruno, CA)

Rekart M L, Roy J L, 1993 *AIDS Update, Quarterly Report: Third Quarter 1993* British Columbia Centre for Disease Control, 811 West 11th Avenue, Vancouver, BC V5Z 1L8

Rose G, 1993 *Feminism and Geography: The Limits of Geographical Knowledge* (University of Minnesota Press, Minneapolis, MN)

Rosenthal D B, 1989, 'Passion, compassion, and distancing in the AIDS literature' *Urban Affairs Quarterly* **25** 173–180

Russo V, 1990, 'A test of who we are as a people', in *Democracy: A Project by Group Material* Ed. B Wallis (Bay Press, Seattle, WA) pp 299–302.

Shannon G W, 1991, 'AIDS: a search for origins', in *AIDS: and the Social Sciences* Eds R Ullack, W F Skinner (University of Kentucky Press, Lexington, KY) pp 8–29

Shannon G W, Pyle G F, 1989, 'The origin and diffusion of AIDS: a view from medical geography' *Annals of the Association of American Geographers* **79** 1–24

Shannon G W, Pyle G F, Bashshur R L, 1991, *The Geography of AIDS: Origins and Course of an Epidemic* (Guilford Press, New York)

Shilts R, 1987 *And the Band Played On: Politics, People, and the AIDS Epidemic* (Penguin Books, New York)

Signorile M, 1993 *Queer in America: Sex, the Media, and the Closets of Power* (Random House, New York)

Smallman-Raynor M R, Cliff A D, 1990, 'Acquired Immunodeficiency Syndrome (AIDS): the global spread of Human Immunodeficiency Virus Type 2 (HIV-2)', in *London Papers in Regional Science 21. Spatial Epidemiology* Ed. R W Thomas (Pion, London) pp 139–182

Smith A, 1986, 'Medical geography', in *Dictionary of Human Geography* Eds R J Johnston, D Gregory, D M Smith (Basil Blackwell, Winchester, MA) page 293

Sontag S, 1988 *AIDS and Its Metaphors* (Farrar, Straus, and Giroux, New York)

Stacey J, 1988, 'Can there be a feminist ethnography?' *Women's Studies International Forum* **11** 21–27

Treichler P, 1988, 'Biomedical discourse: an epidemic of signification', in *AIDS: Cultural Analysis/Cultural Activism* Ed. D Crimp (MIT Press, Cambridge, MA) pp 31–70

Valentine G, 1993, '(Hetro)sexing space: lesbian perceptions and experiences of everyday spaces' *Environment and Planning D: Society and Space* **11** 395–413

Wallace R, Fullilove M T, 1991, 'AIDS deaths in the Bronx 1983–1988: spatiotemporal analysis from a sociogeographic perspective' *Environment and Planning A* **23** 1701–1724

Wallis M, 1989, 'Grasmsci-the-goalie: reflections in the bath on gays, the Labour party, and socialism', in *Coming on Strong: Gay Politics and Culture* Eds S Shepard, M Wallis (Unwin Hyman, London) pp 287–300

Watney S, 1987 *Policing Desire: Pornography, AIDS, and the Media* (University of Minnesota Press, Minneapolis, MN)

Watney S, 1988, 'The spectacle of AIDS', in *AIDS: Cultural Analysis/Cultural Activism* Ed. D Crimp (MIT Press, Cambridge, MA) pp 71–86

Watney S, 1989, 'The subject of AIDS', in *AIDS: Social Representations, Social Practices* Eds P Aggleton, G Hart, P Davies (Falmer, New York) pp 64–86

Watts S J, Okello R, 1990, 'Medical geography and AIDS' *Annals of the Association of American Geographers* **80** 301–303

White S, 1988, 'Poststructuralism and political reflection' *Political Theory* **16** 186–208

Whitmore G, 1988, *Someone Was Here: Profiles in the AIDS Epidemic* (New American Library, New York)

Wigwood R, 1993, 'AIDS takes highest toll of all diseases' *Vancouver Sun* 11 September, page A-1

Williamson J, 1989, 'Every virus tells a story: the meanings of HIV and AIDS', in *Taking Liberties: AIDS and Cultural Politics* Eds E Carter, S Watney (Serpents Tail, London) pp 69–80

Willms S M, Hayes M V, Hulchanski J D, 1991, 'Choice, voice, and dignity: housing issues and options for persons with HIV infections in Canada: a national study', Centre for Human Settlements, University of British Columbia, Vancouver

Wolf E, 1991, 'A week on Ward 5A', in *The AIDS Reader: Social, Political, and Ethical Issues* Ed. N MacKenzie (Meridian, New York) pp 527–533

Wood W B, 1988, 'AIDS north and south: diffusion patterns of a global epidemic and a research agenda for geographers' *Professional Geographer* **40** 266–279.

26 Michael J. Watts

Mapping Meaning, Denoting Difference, Imagining Identity: Dialectical Images and Postmodern Geographies

Excerpts from: *Geografiska Annaler* 73, 7–16 (1991)

> I first saw Elvis [Presley] live in 1954. It was at the Big D Jamboree in Dallas and first thing, he came out and spat on the stage. . . . I didn't know what to make of it. There was just no reference point in the culture to compare it.
>
> Roy Orbison

> . . . there's no such thing as 'England' any more . . . welcome to India brothers! This is the Caribbean! . . . Nigeria!. . . . There is no England, man. This is what is coming. Balsall Heath is the center of the melting pot, 'cos all I ever see when I go out is half-Arab, half-Pakistani, half-Jamaican, half-Scottish. . . . I know 'cos I am [half-Scottish, half-Irish] . . . who am I? Tell me who do I belong to? They criticise me, the good old England. Alright where do I belong? You know I was brought up with blacks, Pakistanis, Africans, Asians, everything . . . who do I belong to?
>
> White reaggae fan from Balsall Health, Birmingham, UK

> Culture is the constant process of producing meanings of and from our social experience and such meanings necessarily produce a social identity for the people involved.
>
> John Fiske

I want to root my talk in the same ground as Roy Orbison's reflections on one of rock and roll's iconic figures; namely the dislodging effect of performance, and more precisely of seeing culture as a sort of reference point, as a set of markers, as a map of meanings.[1] I do so from a particular vantage point, however, namely of trying to understand what metamorphoses capitalism is capable of, of endeavoring to grasp its seemingly infinite capacity for reconfiguration and recombination, its desperate propensity for creative destruction, and its relentless drive toward the commodification of virtually everything. In short, I am, quite immodestly, trying to examine both the radical character of capitalism and also the complex articulations of capitalism, modernity and culture understood as a field of struggle, what Fred Jameson has called the 'attempt to theorise the specific logic of the cultural production of [late capitalism]' (1989, page 33).

And what a curious, disquieting moment to assess the radical and the symbolic in the belly of capitalism! In the wake of a hideous and brutal Gulf War, fought with smart bombs against a raggedy-arsed nineteenth century Iraqi army in the name of a New World Order; amidst the shocking market religiosity and capitalist triumphalism in Eastern Europe; and during a growing economic polarization between North–South, indeed a crisis of development, in which some reaches of the Third World, Africa most palpably, seem to be sliding inexorably off the face of the world economy into a terrifying Heart of Darkness. Against this less than sanguine backcloth, cultures afford an intellectual entry point into capitalist modernity, but it should not carry the dangerous implication that the political economic character of contemporary capitalism is as it were 'resolved' (or of diminished significance). Indeed, I consider it to be a signal weakness of the post-modern sensibility, whether as an aesthetic affectation or as a critical response to grand history and metanarratives, that many of the most compelling questions of political economy – flexible accumulation, the internationalization of the money capital, new labor processes – are so often sidelined, or worse are presumed irrelevant, to an exploration of the postmodernity and the lived experience of capitalism (cf. Harvey, 1989).[2]

The canvas for my remarks is provided by cultural studies broadly defined. Indeed, I want to attempt to breach the hard edges and contours of capitalist accumulation *culturally*, by focusing on, as it were, commodity fetishism rather than the commodity circuit. This is not the place, and I am not the person, to review the complex historical trajectories of (and the important national differences between) cultural studies on both sides of the Atlantic. Rather I should like to begin with an observation by Renato Rosaldo in *Culture and Truth* which might serve as some sort of leitmotif for the field:

> Whether speaking about shopping in a supermarket, the aftermath of nuclear war, Elizabethan self-fashioning, academic communities, tripping through Las Vegas, Algerian marriage practices, or ritual among the Ndembu of Central Africa, work in cultural studies sees human worlds as constructed through historical and political processes, and not as brute timeless facts of nature (1989, page 7).

Contained within Rosaldo's vision is, in fact, a dense and complex set of claims about cultural diversity and creativity, cultural construction and symbolic contestation, and the relations between cultures, especially between mass and popular cultures. I want to problematize culture by starting, somewhat arbitrarily, with several formative essays published over the last fifteen years.

First, culture was seen to be political, and popular culture was therefore more than 'flying ducks on the wall and garden gnomes' (Hall 1979, page 234). Rather, the vernacular in its multiplicity of guises was a site of struggle, an arena which engaged dominant and powerful cultures, a place where, God forbid, socialism might be constituted. *That* is why, Hall stated with admirable frankness, popular culture matters, 'otherwise, to tell you the truth, I don't give a damn about it' (1979, page 239). Second, all of these critics sought to work actively with culture in a dialectical sense, in terms of the delicate polarities of containment and resistance, reification and utopia. Rather than capitulating to a view of mass culture as something which is simply a form of class control or alternatively an authentic form of subversion – in other words in terms of unified, coherent oppositions and voices – they relocated these polarities by suggesting that cultural creation under capitalism was divided against itself (Denning, 1990). Alongside the banalities of commercial culture lay 'elements of recognition and identification . . . to which people are responding' (Hall 1979, page 223). Jameson notes, in the same way, that works of mass culture cannot be ideological without being utopian, 'they cannot manipulate unless they offer some genuine shred of content as a fantasy bribe to the public about to be so manipulated' (1979, page 144). In this sense there is no authentic culture as such, but complex interpenetrations, dependencies, contests and negotiations, 'the continuous and necessarily uneven and unequal struggle by the dominant culture to constantly disorganise and reorganise popular culture' (Hall 1979, page 233). Third, and relatedly, culture is active, not merely marking or living out wider social contradictions but, as Paul Willis put it, working upon them 'to achieve partial resolutions, recombinations, limited transformations' (1977, page 124). Culture is, in this sense, a form of symbolic creativity, and a part of *necessary work* as Willis himself says (1990), which contains its own 'grounded aesthetic'.[3] And finally, there is a sensitivity to what Jameson called the false problem of value, the modalities by which culture is distinguished from non-culture, how cultures are themselves separated, evaluated, judged and classified. Boundaries represent, then, an exercise in cultural power which itself should be an object of scrutiny (Denning 1990, page 6).

If Hall and company assisted in the process by which the popular, and subcultures more generally, were liberated from their second class cultural citizenship, during the 1980's cultural studies has emerged as a crucible within which the central questions of class, race and gender have been broached. If an early concern of, for example, the Birmingham School was to link culture and class, it was quickly subverted (challenged?) by what one might call the new pluralism. In fact the growth and proliferation of gender, sexuality, race and ethnicity within wide-ranging cultural debates actually rested upon the meltdown of – or at least a growing scepticism toward – the class reactor. Retrieving culture was critical but was somehow less compelling if it was linked to a notion of class as a master narrative, which typically produced a revolutionary subject (the white working class male) who always showed up at the right historic moment, if sometimes temporarily delayed, but rarely duped by, the tappings of false consciousness. This grand view of class was replaced, of course, by the multiplication of sites of struggle, by a scattering of antagonisms, indeed by a fragmentation of the political itself. Blown in part by the Gallic winds which placed the idea of totality in some jeopardy, or at least in brackets, and enhanced by the complex multi-culturalism and post-colonialism of many globalized advanced capitalist societies, cultural studies became a safe harbour for all sorts of emergent discourses.

Broadly speaking, I think that Aronowitz (1991) is quite correct in his assessment that cultural studies in this expanded sense contributed to, and radically challenged, three primary domains: *epistemic* (including forms of knowledge and facticity, and the constitution and limits of conventional disciplines), *discursive* (how social identity is constructed by communities and how communities are constructed discursively), and *aesthetic* (the historic context of the aesthetics of everyday life). Defined in this way cultural studies was an intellectual and academic expression of new social movements, what has come to be labelled 'identity politics'. In sum, what began tentatively as an interrogation of culture has produced, what Cornell West (1990) calls 'the new cultural politics of difference'.

> Distinctive features of the new cultural politics of difference are to trash the monolithic and homogenous in the name of diversity, multiplicity and heterogeneity; to reject the abstract, general and universal and pluralize by high-lighting the contingent, provisional, variable . . . (1990, page 93).

If these gestures are now new, what makes them distinctive according to West is the weight and gravity lent to *representation*, and how and what constitutes *difference*.

*

I would like to now turn to difference of another sort, a concern with difference that emanates from geographical interrogations of postmodernity and globalization. Perhaps one might begin here with an observation by Henderson and Castells (1987) that the internationalization of post-war capitalism has produced a lived experience in which 'the space of flows . . . supersede[s] the space of places' (page 7).[4] Deterritorialization, to employ the language of Deleuze and Gattari (1977), speaks to the complex interpenetrations, the constellation of the local and the global in the contemporary epoch. There are many ways of talking about both the experience of, and the analytics appropriate to, the contradictory unity and disintegration of modernity. Jameson, for one, dwells on the insertion of selves into a 'multidimensional set of radically discontinuous realities', in which 'the truth of the experience no longer coincides with the place in which it takes place' (1988, pages 349 and 351); Henderson and Castells talk of 'powerless places and placeless power' (1987, page 7). I do not wish to linger on these issues which have been treated in a serious fashion elsewhere (Gregory, 1994) but only to note that one can productively situate the local–global conundrum on the wider canvas of the production of capitalist space, and the periodic mutations by which capitalism 'constructs objective conditions of space and time sufficient unto its needs and purposes of material and social reproduction' (Harvey 1990, page 419). Periods of accelerated change and reconfiguration within capitalism – the 'creative' and pitiless destruction of everything it cannot use – produces what geographers have referred to as space–time convergence, or space–time compression (Harvey 1989). The erosion, some might say demise, of the glorious age of post-war fordism, and its displacement by some form of flexible or post-fordist form of accumulation, represents in this context the most recent modulation in a recursive, wave-like pattern of space–time compressions. Discussions of flexibility, New Times, industrial divides and so on connote, in this sense, a speeding up, a hypermobility, a fast capitalism. Postmodernism as the contempoary manifestation of space–time compression represents, therefore, not so much the end of, or a rupture with, modernity – indeed it harkens back to early modernism[5] – as much as one expression of serial or multiple modernisms and modernities (Soja 1989, Pred and Watts 1992).

In what ways, then, does space–time compression speak to the question of difference, especially at a moment distinguished by a capitalist horizon which is, to an unprecedented extent, global and international? One account is provided by Ed Soja in his brilliant Borgesian narrative on the internationalization of Los Angeles, in which he argues that there is not so much a global duplication of Los Angeles as much as the appropriation and reproduction of other urban experiences within the metropolitan fabric; in short a city of simulacra (1989, page 221). Another quite different account can be traced to Trotsky, in particular his idea of combined and uneven development in which the archaic and the contemporary are amalgamated (his term) to produce startling heterogeneities across space. It is, in both of these views, the footloose and mobile character of capital which seizes hold of difference to reconfigure the local.[6] The experience of the 'large abstractions we choose to call capitalism', always arrives as Jim Scott wryly notes, 'in quite personal, concrete, localized, mediated forms' (1985, page 348). Globalization does not signal the erasure of local difference, but in a strange way its converse; it revalidates and reconstitutes place, locality and difference (Alger 1988; Swyngedouw 1989). As a counterweight to the dominant logic of flows there is the irreducible local experience which defends local interest and identity around places.

It is to this common ground of the multiplication of difference – emanating from a dialectic between cultural studies on the one side and postmodern geographies on the other – that I wish to speak. More concretely, I want to argue that mapping the spectrum of cultural forms, onto spatial, class and social identities in the context of global interconnectedness might constitute an important frontier for geographic inspection. The common ground I refer to is, it seems to me, fundamentally about identity, space and the politics of difference. And it strikes me in this regard – if I might quite arbitrarily pull out of the hat two books – that the analyses of modernity provided by Anthony Giddens in his *Consequences of Modernity* and by Ed Soja in his *Postmodern Geographies* are both singularly lacking. In Soja's marvellous description of the 'extraordinary crazy quilt [and] dazzling patchwork quilt' (1989, page 245) of Los Angeles, or in Giddens' invocation of the dissonances and heterogeneities which compose a larger structure of modernity, there is little consideration of how people define themselves, how these identities are cobbled together and contested in order to act. In the hyperspace of postmodernity, space is not irrelevant but rather has been reterritorialised; and as Gupta and Ferguson (1992) note, this forces a radical rethinking of community, solidarity, identity and cultural difference. How are places forged from spaces, who has this power to forge, how is it contested and fought over, and what is at stake in these contestations?

Living in a global economy necessarily blurs the distinction between 'here' and 'there'. More and more people live in what Said (1979, page 18) calls a 'generalized condition of homelessness'. This is necessarily so for the millions of refugees, migrants, and displaced persons. But the experience is more general. We inhabit a world of diasporic communities linked together by a transnational public culture and global commodities; not only has the old international division of labor disappeared but so has the old identity between people and places. Pre-revolutionary Iran arises in contemporary Houston; 'Pakistanis' in northern England and Pakistan simultaneously engage in the book burning of Rushdie's *Satanic Verses*; Third World simulacra appear in the First World and vice versa. Not only the displaced experience displacement. The identities within the 'periphery' are as surely unhinged as they are in the 'metropole' and for similar reasons. There is an erosion of the spatially-circumscribed homeland or community and a compensatory growth of the imaginary homeland from afar, an imaginary always shaped however by specific political-economic determinations (Gupta and Ferguson 1992).

Amidst the shards of modernist disengagement, fragmentation and difference, what types of identity can be constructed? As Clifford (1988, page 275) says, at the end of the twentieth century 'what processes rather than essences are involved in present experiences of cultural identity?' How can we approach the experience of the Korean Buddhist chemical engineer, recently arrived from three years in Argentina, who becomes a Christian greengrocer in Harlem? Or the Guyanese Indian New Yorker who served on the Howard Beach trial jury? Or the life on the border as Guillermo Gomez-Pena describes the state of being Mexican-American:

> Who are we exactly? The off-spring of the synthesis, or the victims of the fragmentation; the victims of double colonialism or the bearers of a new vision. . . . What the hell are we? De-Mexicanised Mexicans, pre-Chicanos, cholopunks, or something that still has no name (Gomez-Pena 1987, page 1).

And perhaps the border is the appropriate metaphor for the postmodern subject, the site of what Bhaba calls hybridity, a place of 'incommensurable contradictions within which people survive, are politically active and change' (1989, page 67). The borderland becomes the home of the postmodern. But what can this mean when only a Protean self can seemingly occupy this space? Proteus was, after all, divine (Sennett 1991, page 6).

I want to focus on the complexities of the process of interpellation, of how the geographer's concern with the nexus of the global and the local can be explored by 'mapping' the labile, sliding identities forged in specific yet globalized sites. Here I would yet again turn to Stuart Hall for advice:

> I use the term identity . . . precisely to try to identify that meeting point where the processes that constitute and continuously reform the subject have to act and speak in the social and cultural worlds. . . . I understand identities therefore as points of suture, points of attachment, points of temporary identification . . . [O]ne only discovers who one is because of *the identities which one has to take on in order to act* . . . always knowing that they are always representations [which] can never be adequate to the subjective processes which are temporarily invested in them. . . . I think identities is sort of . . . like a bus, you just have to get from here to there, the whole of you can never be represented in the ticket you carry but you just have to buy a ticket in order to get from here to there (Hall, 1989, no pagination, emphasis mine).

Accounting for the processes by which we acquire our bus ticket(s) – not least in a world in which the routes are many and global – is a worthy and rather important project. How exactly are individuals interpellated by multiple and often contradictory cultural and symbolic practices rooted in historically constituted communities and places? What are the processes by which a sense of self-construction is shared with others? Why and in what ways are such representations made more or less appealing, and how are they contested (Radway 1990, page 24)?

In posing these questions, two rather obvious points need to be made. First, to the extent that identity is constructed across difference and in relation to other identities, then we are necessarily brought back to the question of mass, popular and dominant cultures. In the same way that hegemony is hard work (the adage is Stuart Hall's) and always subject to negotiation, so personal and community identities are malleable and flexible (but not infinitely so) because of the failure to exhaust the meanings of popular experience from below (cf. Anderson 1990, Ryan 1989). And second, identities are complex sorts of 'holding operations', stories told by ourselves about ourselves and therefore fictions if *necessary* fictions (Hall 1989). As such they are imaginary, straddling so to speak the Real and Desire, from which it seems to me they derive their weight and effect.

My own conceit in exploring such lofty issues is to turn not to Los Angeles or to Paris but to Africa, a geographic and ideological space which Joseph Conrad believed – and I suspect many still believe – took one not to the heart of modernity but 'to the earliest beginnings of the world'. My focus on Africa reflects in part a belief that certain realities at the periphery of world capitalism can throw the phantom objectivity of that system into stark relief (cf. Taussig 1980), but more critically a sense that under 'planetary capitalism' (the term is from Gayatri Spivak) every national society creates and debates its own modernity (Paz 1990). A central question, then, concerns the trajectories that shape modernization and the modalities by which some voices 'get to play with modernity' (Appadurai 1990). It is this public culture as Appadurai calls it, a zone of cultural debate, encounter and interrogation, which I seek out in an African context. In so doing, I am following Spivak (1990, page 94) in her request that 'what goes on over here be defined in terms of what goes on over there'.

*

> To the form of the new means of production that in the beginning is still dominated by the old one (Marx), there correspond in the societal superstructure wish images in which the new is intermingled with the old in fantastic ways.
>
> Walter Benjamin

In some parts of rural West africa there is a widespread belief in powerful spirits who roam the highways in search of human prey.[7] Sometimes they assume the form of beautiful Eurasian women who lure and subsequently kill male travellers; sometimes they may appear as trucks – loud, rumbling juggernauts with blinding headlights – which terrorize villagers by endlessly circling their huts at night. Tire tracks on the ground are never found. Adeline Masquelier (1990) tells of one particular road spirit in western Niger who causes head-on collisions along the major highway. According to her informants, the spirit was finally captured by an enterprising Frenchman and his camera (with of course electronic flash), and revealed a beautiful, fair-skinned slender woman whose lower body was a rubber tyre onto which was inscribed the word DUNLOP. The image, derived from advertising in many of the Western magazines which circulate in rural Africa, froze the spirit in the act of transformation from siren to machine, preparing to take more life by driving full tilt into oncoming, unsuspecting vehicles.

What sort of images are these road spirits? And what sort of dialectical vision do they demand?

It is tempting, indeed necessary, to locate the popular vision of cultural commodities on the wider canvas of political economy. Nigeria, for example, where some of these road spirits and road narratives are widely talked about, has the worst traffic accident rate in the world, and there is a sense in which the anarchy, chaos and wealth generated by the oil boom is embalmed in the wreckage, peril and waste associated with Nigerian vehicular travel. Roads and vehicles have, moreover, other deep associations with political economy, namely with the arrival of the Europeans, with forced labor, migration for wage work and so on.[8] But this does not exhaust the meanings of roads and vehicles as cultural commodities. They speak of course in some cases to the questions of sexuality and its danger – the Eurasian DUNLOP siren – and in some cases as mnemonic devices. White (1991) has shown how in parts of eastern and central Africa, certain sorts of vehicles (fire-engines, government landrovers) and occupations were associated directly with vampirism, bloodsucking and body snatching. These cultural commodities aroused suspicions of bloodsucking not because they were associated with colonialism per se but, according to White, because they aroused anxieties about technology and work. Vehicles, machines and bloodsucking were, if you like, mneumonics for debates about the labor process and work.

These cultural commodities – the means by which space is transformed under capitalism – speak directly to what Walter Benjamin calls the fantastic intermingling of the old and the new. . . . They correspond to 'wish images', collective representations wedded to elements of urhistory which 'attempt to transcend and to illuminate the incompleteness of the social order of production' (cited in Buck-Morss 1989, page 114). Starting from what Brecht called the 'bad new things', Benjamin brings the very old into shocking conjunction with the very new (Eagleton 1990). But how can these contradictory images – the dialectical images of African modernity – speak to questions of modernist identity?

In seeking to say something about identity and modernity, I want to focus on one specific intermingling of the old and the new – a series of Muslim millenarian insurrections that occurred among Hausa speaking people in northern Nigeria during the early 1980s (Watts 1990). The desire to construct an alternative Muslim community – widely referred to as an heretical and fanatical exercise in Nigeria and elsewhere – was inspired by a selfproclaimed prophet, Alhaji Maitatsine. Schooled in the science of Qu'ranic exegesis, Maitatsine lent his public teaching a highly idiosyncratic, antimaterialist and syncretic style damning all those who read any text other than the Qu'ran, who carried money, who rode bicycles or cars, who smoked cigarettes, wore buttons or watched TV. Critical of orthodox Muslim practices – including the physical details of the performance of prayer – he vilified the influential clergy and *ulema* in northern Nigeria, a Sunni stronghold of some 50 million Muslims. Maitatsine established a community in the walled city of Kano – some 10 000 strong by 1980 – united to their jacobin, populist reading of Islam, and their condemnation of corruption, profligacy and Westernization.

To situate Maitatsine as a cultural and political movement requires two minimal moves. The first is to recognize that it was an attempt to forge new Muslim identities from *within* Islam. And second, that it must be situated within the rhythms and jagged breaks of a frenzied Nigerian capitalism during the 1970s fed by a huge influx of oil dollars.[9] As I have described elsewhere (Watts 1990), the explosive surge of growth, urban construction and commodities in circulation was mediated directly through the Nigerian state, the significance of which is central in understanding popular reactions to the oil boom. Oil money became a source of political accumulation, as large, corrupt bureaucracies distributed oil rents to favoured constituencies in a deeply divided Nigerian polity. The unimaginable corruption and venality associated with public office linked, in the popular consciousness, petroleum, money and moral degeneracy.

Against this backdrop, the followers of Maitatsine constituted an archetypical lumpen, Fanonite class, uprooted from the countryside. Often drawn into the city through qu'ranic networks as students, they were inserted as casual laborers in the interstices of the urban informal economy. Nigerian modernity – a world awash in commodities of all sorts – was experienced as exclusion, corruption and a sort of moral decline, an erosion of the Muslim moral economy.

If Maitatsine's followers occupied a particular class position within the structure of Nigerian petrolic accumulation, they brought with them to Kano a conflation of other identities. An ethnic identification as Hausa, a Muslim identity rooted in a particular exposure and interpretation of Islam through the exegesis of the qu'ran, a legacy of animist belief in the spirit world, and what I can only describe as a commoner consciousness (in Hausa the term is *talaka*) inherited from the precolonial Muslim Caliphate. In this sense, their experience of proletarianization under a corrupt and flabby form of oil based capitalism was extremely heterogeneous and fragmented. The question, then, is 'how is the feeling of community or homogeneity created within social relations that are neither communal nor homogeneous?' (Roseberry 1989, page 224).

My argument is simply that Maitatsine articulated such a feeling, a particular identity, through what one might call the local hermeneutics of Islam.[10] Rather than seeing Islam as prescriptive in simple ways, providing rigid rules and unambiguous guidance, it is a text-based religion which is made socially relevant through enunciation, performance, citation, reading and interpretation. It is this dialogic and interpretive tradition around a central text which is fundamentally enigmatic which necessarily leads to questions of how and which texts are used, by whom and with what authority (Fischer and Abedi 1990). In this sense Maitatsine was renegotiating and contesting a certain sort of Muslim tradition, building certain identities through idiosyncratic interpretations and readings. He focused solely on the qu'ran and gave a literalist reading – certain Arabic characters, for example, resembled a pictorial representation of a bicycle and hence spoke to Maitatsine's antimaterialism – in a way that cobbled together both a particular sort of Muslim and ethnic identity, and a wish-image of a Benjaminian sort (a future community which, like the Angel of History, was being propelled backwards by looking to the past). Maitatsine drew upon the global resources (and commodities) of capitalism and the Muslim diaspora to assemble, in local terms, a *bricolage* which derived its power in part from the historic tension between the Muslim community (*umma*) and state in Muslim discourse. To this degree, Maitatsine was part and parcel of wide ranging global debates within Islam over modernity.

The symbolic domain represents, then, a field of struggle and negotiation. Maitatsine's interpellation of a particular Muslim subject, amidst a clamor of other Muslim voices and interpretations, addresses at least two basic processes. The first is what Bourdieu (1987) calls the definition of legitimized culture, and hence the principles of legitimized domination. The second, is how the hegemonic vision of a dominant Muslim clergy failed to exhaust the meanings of Islam from below. Maitatsine provided an imaginary community and helped to create a sliding sense of Muslim identity which appealed to many by reinterpreting, and contesting the dominant map of Nigeria's cultural economy. That it was undertaken in the name, and on the terrain, of Islam made it in some sense subject to the same referents as the dominant Muslim discourse and at the same time extremely threatening to it. In this sense, Maitatsine did not, as Gramsci put it, break through religion to see political goals, and in the final analysis was crushed by the state precisely because of its 'heretical' claims.[11]

*

> I too . . . am a fantasist. I build imaginary countries and try to impose them on ones that exist. I, too, face the problem of history: what to retain, what to dump, how to hold onto what memory insists on relinquishing, how to deal with change.
>
> Salman Rushdie

What [can] one derive from a Maitatsine insurrection in relation to modernity and identity? Let me begin, however, with what I am not saying. I am not suggesting that we need more studies of 'subaltern resistance', more studies of the poor and powerless who, as Spivak says, are always assumed to speak with one unified voice. I have some grave misgivings about the conceptions of power in some of these studies, and in any case in the wake of an imperialist war we probably need more work on the rich and powerful. Neither am I appealing in an unproblematic way to religion as a basis for understanding identity politics. The Muslim populism of Maitatsine was deeply flawed and its horizons limited.

Difference and identity is produced and reproduced within a field of power relations rooted in interconnected spaces linked by political and economic relations. Maitatsine's appeal can only be grasped in relation to the Muslim diaspora, world commodities, and the oil based political economy of Nigeria. Second, there are fundamental relations between identity and the contesting and negotiating of difference in cultural terms. Maitatsine's community does say something about both the symbolic materials, and the symbolic creativity through which, to return to Stuart Hall, one gets from here to there. The symbolic contestation of modernity is instantiated in shifting identities rooted in enunciation, dialogue and what Paul Willis (1990) in another context calls the symbolic creativity of mismatching. Insofar as these sliding identities [are] narratives they invoke and draw sustenance from the imaginary, the wish images which recapitulate the ceaseless configuration and reconfiguration of the old and the new in a commodity society. Cultural and symbolic struggle, the contesting of meaning, are the means by which identities are given shape, and which allow us to act. And perhaps here resides a different sense of hegemony, complementary to the usual sense of the manufacture of consent via orchestration, in which:

> socially engaged selves, already split and decentred . . . [which] have to act to hold together within themselves a society that has split and turned against itself in irreconcilable and mutually inconstruable ways (Rebel 1989, page 129).

And finally, there is the question of how identity which rests on difference and splitting can produce a common ground for politics. How, as Hall (1989) notes, can we think about forms of political action which are able to understand these processes of identification rooted in difference rather than try to transcend it? And it seems to me that this project requires not a flight from class but a crying need to retheorize where class has gone to, and to rethink it – and to reintegrate it – in non-essentialist terms.

These are bold claims and I am making even bolder associations. Namely, that these questions which, I believe, strike to the heart of certain experiences of modernity for us, can be discussed (to return to Spivak's injunction) in terms of what goes on there as much as what goes on here. That in the dialectical images thrown up by African modernity, from the fantastic intermingling of the old and the new, is a possibility that 'our own expectations and understanding [be brought] to a momentary standstill' (Taussig 1990, page 224).

Notes

1 In this regard, my dependence on the work of Walter Benjamin is considerable. His rabbinical, apocalyptic and messianic vision leaves me somewhat cold, but his exploration of dialectical images and of his interrogation of capitalism through the archaeology of the commodity – his juxtaposition of the very old with the very

new – seems to me to be brilliant and original. In this light, I have also benefitted substantially from the work of Mick Taussig.

2 See Jameson (1989: 34) and his comment that postmodernism demands historical analysis and a diagnosis of the political and ideological function of the concept as it pertains to our 'imagery resolutions of real contradictions'. By the same token, much of the debate over post-fordism and flexible accumulations seems bereft of any sensitivity to the social, symbolic and cultural specificities inscribed within these particular modalities as Gregory and Turner (1992) put it.

3 'The creative element in a process whereby meanings are attributed to symbols and practices, and where symbols and practices are selected, reselected, highlighted and recomposed to resonate further appropriated and particularized meanings. . . . Grounded aesthetics are the yeast of common culture' Paul Willis, *Common Culture*, Boulder, Westview, 1990, p. 21.

4 Manuel Castells in his book *The City and the Grassroots* (University of California Press, Berkeley, 1983) refers to a 'space of variable geometry . . . [in which] Space is dissolved into flows . . . [and] cities into shadows' (p. 314).

5 'Modernism and postmodernism are not chronological eras, but political positions in the century-long struggle between art and technology. If modernism expresses utopian longing by anticipating the reconciliation of social function and aesthetic form, postmodernism acknowledges their nonidentity and keeps fantasy alive. Each position represents a partial truth; each will recur anew so long as the contradictions of commodity society are not overcome'. Susan Buck-Morss, *The Dialectics of Seeing*, Boston: MIT Press, 1989, p. 359.

6 Suzanne Berger explains that the nature of capitalism is not to 'create a homogeneous social and economic system but rather to dominate and draw profit from the diversity . . . that remain[s] in permanence', in *Discontinuity in the Politics of Industrial Society, Dualism and Discontinuity in Industrial Societies*, Cambridge: Cambridge University Press, 1980, p. 136.

7 This is drawn from the special of panel of papers on roads delivered at the African Studies Association, Baltimore, MD, USA, November 1990.

8 As Luise White (1991) notes, they also may make issues like mobility, safety and access open for popular debate and reflection.

9 Nigeria is a member of OPEC and experienced massive windfall oil revenues in 1973 and 1979.

10 'The specific problems raised by the translation of objective meaning of written language into the personal act of speaking . . . [is an act of] appropriation . . . The nature of the texts and knowledge to be drawn from them in any given historical context are shaped by the sociology or political economy of knowledge: how textual knowledge is reproduced . . . what social factors mediate access to texts, who is able to read and in what manner, who has the authority to represent . . . and how challenges to such authority are manifested' (Lambek 1990, pp. 23–24).

11 At least 5000 followers of Maitatsine were killed when the community was bombed by the Nigerian air force and attacked by ground forces.

Selected references

Alger, C. (1988): Perceiving, Analyzing and Coping with the Local-Global Nexus, *International Social Science Journal*, 117, pp. 321–340.

Anderson, P. A. (1990): Culture in Contra Flow II, *New Left Review*, 182, pp. 85–137.

Appadurai, A. (1990): Disjuncture and Difference in the Global Culture Economy, *Public Culture*, 2/2, pp. 1–24.

Aronowitz, S. (1991): An Introductory Essay on Cultural Studies, unpublished manuscript, Sociology Department, The Graduate Centre, CUNY, New York.
Bhabha, H. (1989): Location, Intervention, Imcommensurability, *Emergencies*, 1/1, pp. 63–88.
Bourdieu, P. (1987): *Distinction.* Cambridge, Harvard University Press.
Buck-Morss, S. (1989): *The Dialectics of Seeing*. Boston, MIT Press.
Clifford, J. (1988): *The predicament of culture: twentieth-century ethnography, literature and art*. Cambridge MA, Harvard University Press.
Denning, M. (1990): The End of Mass Culture. *International Labor and Working Class History*, 37, pp. 4–18.
Deleuze, G. and Gattari, F. (1977): *The Anti-Oedipus*. New York, Viking.
Eagleton, T. (1990): *The Ideology of the Aesthetic*. Oxford, Basil Blackwell.
Fischer, M. and ABEDI, M. (1990): *Debating Muslims*. Madison, University of Wisconsin Press.
Giddens, A. (1990): *The consequences of modernity*. Cambridge: Polity Press, Stanford: Stanford University Press.
Gomez-Pena, G. (1987): The Conflicts and Culture of the Borderlands, *Utne Reader*, July, p. 1.
Gregory, D. (1994): *Geographical Imaginations*. Oxford UK and Cambridge MA: Blackwell.
Gregory, D. and Turner, B. (1992): The Local Global Continuum, in Abler, R. et al., (eds): *Geography's Inner Worlds*. New Brunswick: Rutgers University Press.
Gupta, A. and Ferguson, I. (1992): Beyond Culture: Space, Identity and the Politics of Difference, *Cultural Anthropology*, 7, pp. 6–23.
Hall, S. (1979): Notes on deconstructing the popular. In R. Samuel, (ed.) (1981) *Peoples history and socialist theory*. London, Routledge, pp. 227–40.
Hall, S. (1989): Imaginary identification and Politics, transcript of a talk given at the Institute of Contemporary Arts, London.
Harvey, D. (1989): *The Connection of Postmodernity*. Oxford, Basil Blackwell.
Harvey, D. (1990): Between Space and Time, *Annals of the Association of American Geographers*, 80/3, pp. 435–447.
Henderson, J. and Castells, M. (1987): Introduction, in *Global Restructuring and Territorial Development*, London, Sage, pp. 1–17.
Jameson, F. (1979): Reification and utopia in mass culture. Social text 1, pp. 134–46.
Jameson, F. (1988): Cognitive Mapping, in Nelson, C. and Grossberg, L. (eds): *Marxism and the Interpretation of Culture*, Urbana, University of Illinois Press, pp. 347–360.
Jameson, F. (1989): Marxism and Postmodernism, *New Left Review*, 176, pp. 31–46.
Lambek, M. (1990): Certain Knowledge, Contestable Authority. *American Ethnologist*, 17/1, pp. 23–40.
Masquelier, A. (1990): Road Sirens, Dangerous trucks and Vengeful Spirits. Paper presented to the African Studies Association, Baltimore, MD.
Meyer, W. B., Gregory, D., Turner, B. L. and McDonell, P. F. (1992): The local-global continuum. In Abler, R., Marcus, M. and Olson J. (Eds), *Geography's inner worlds: pervasive themes in contemporary American geography*. New Brunswick NJ: Rutgers University Press, pp. 255–279.
Paz, O. (1990): *In Search of the Present*. New York: Harcourt.
Pred, A. (1990): *Lost Words and Lost Worlds*. Cambridge, Cambridge University Press.
Pred, A. and Watts, M. (1992): *Reworking modernity: capitalisms and symbolic discontent*. Rutgers University Press.

Radway, J. (1990): Response, *International Labor and Working Class History*, 37, pp. 19–26.
Rebel, H. (1989): Cultural hegemony and class experience, *American Ethnologist*, 16/1, pp. 117–136.
Rosaldo, R. (1989): *Culture and Truth: The Remaking of Social Analysis*. New York, Basic.
Roseberry, W. (1989): *Anthropologies and Histories*. New Brunswick, Rutgers University Press.
Ryan, M. (1989): *Politics and Culture*. Baltimore: Johns Hopkins University Press.
Said, E. (1979): Zionism form the Standpoint of its Victims, *Social Text*, pp. 7–58.
Scott, J. C. (1985): *Weapons of the weak: everyday forms of peasant resistance*. New Haven: Yale University Press.
Sennett, R. (1991): Fragments Against the Ruin. *Times Literary Supplement*, February 8th, pp. 6–7.
Soja, E. (1989): *Postmodern geographies: the reassertion of space in critical social theory*. London, Verso.
Spivak, G. (1990): Gayatri Spivak on the Politics of the Postcolonial Subject, *Socialist Review*, 20/3, pp. 81–90.
Swygedouw, E. (1989): The Heart of the Place, *Geografiska Annaler* 71 B, pp. 31–42.
Taussig, M. (1980): *The Devil and Commodity Fetishism in South America*. Durham, University of North Carolina Press.
Taussig, M. (1990): Violence and Resistance in the Americas. *Journal of Historical Sociology*, 3/3, pp. 209–224.
Watts, M. (1990): The Shock of Modernity, *The Attwood Lecture*, Clark University.
West, C. (1990): The New Cultural Politics of Difference, *October*, 53, pp. 93–109.
White, L. (1991): Cars Out of Place. *Working Papers of the Social History Workshop*, University of Minnesota.
Willis, P. (1977): *Learning to Labor*, London, Gower.
Willis, P. (1990): *Common Culture*, Boulder, Westview.

GLOSSARY

The brief definitions that follow are confined to some of the more technical terms that we use in our introductory essays. These are 'bare-bones' summaries, and this is not the place to look for subtlety: in most cases fuller definitions of these terms will be found in the latest edition of the *Dictionary of Human Geography* (Oxford, UK and Cambridge, MA: Blackwell).

Terms in *italics* refer to other entries in the Glossary.

agency In human geography 'agency' usually refers to the capabilities of human beings, either individually or collectively, to act in the world.

androcentrism An approach that privileges the experiences, interests, values and actions of men and marginalises or ignores those of women (see also *masculinism*).

anthropocentrism An approach that privileges the experiences, interests, values and actions of human beings and marginalises or ignores the non-human world: the term is frequently used in politico-ecological critiques of the human degradation of the physical environment (see also *biocentrism*).

areal differentiation The study of the spatial variation (or 'differentiation') of human and physical phenomena on the surface of the earth; the term has often been used as a synonym for traditional regional geography or *chorology*, but the development of a critical human geography has seen a renewed interest in the particularities of place, in processes of uneven development, and in the connections between localization and globalization.

Berkeley School A school of cultural geography associated with Carl Sauer (1889–1975) at the University of California, Berkeley, and characterised by an emphasis on fieldwork (frequently among non-Western societies) and historical reconstructions of the evolution of landscapes as a product of changing interactions between culture and ecology.

biocentric/biocentrism An approach that treats the world in terms of all living things; unlike *androcentrism*, 'biocentrism' insists on a respect for the biosphere as a whole.

chorology A descriptive study of the different parts of the world: chorology is the oldest (classical) tradition of Western geographical inquiry, but the modern case for geography as a 'chorographic science' concerned primarily with *areal differentiation* was argued most forcefully by Richard Hartshorne in *The nature of geography* (1939).

contingent relations Relations among phenomena that exist because of particular external circumstances: the connections between them are thus circumstantial or 'conjunctural', in contrast to the *necessary relations* that are the special concern of analytical approaches informed by *realism*.

cultural landscape As formulated by the geographer Carl Sauer (see *Berkeley School*) the 'cultural landscape' was the product of the collective *agency* of culture working over time on the 'natural landscape', but the rise of the *new cultural geography* has entailed a series of more complex redefinitions that pay more attention to the intersections between power, practice and representation in the production of landscapes.

deconstruction A method developed by the French philosopher Jacques Derrida (1930–) that seeks to destabilize the truth claims made by any text; it involves identifying key oppositions at work within a text (e.g. culture/nature) and then showing that these fail to provide a secure foundation for the argument because they are used in a contradictory manner. The text thus fails on its own terms; its unity is ruptured and its closures prised open. So, for example, in *Leviathan* the political philosopher Thomas Hobbes argued that political stability depends on 'plain language' and that *metaphor* ought to be erased from the political lexicon because its sliding meanings allowed for differences in interpretation and hence political dissent: and yet the very title of Hobbes' book turned on the metaphor of a monster ('Leviathan') to characterise the state.

deep ecology An environmental philosophy predicated on *biocentrism*, which holds that all parts of nature should be accorded a deep moral respect and that human beings must take direct action in defense of the earth as a whole.

discourse In everyday language, 'discourse' usually means simply 'speech' or 'language' and, by extension, 'text'; but in the late twentieth century it has also come to have a more specialised, academic meaning derived from *post-structuralism*. 'Discourse' in this second sense implies a mobile web or network of concepts, statements and practices that is intimately involved in the production of particular knowledges.

ecofeminism A political and intellectual movement organised around the claim that the oppression of women and the degradation of nature are structurally connected: both are the result of capitalist *patriarchy* so that the liberation of the one will not occur without the liberation of the other.

empirical–analytic science A term derived from critical theory to identify forms of knowledge that are primarily concerned with explanations of an object-world, in contrast to the so-called 'historical-hermeneutic sciences' which are concerned primarily with interpretations of a subject-world (see *hermeneutics*).

empiricism An approach that privileges empirical observations over theoretical statements and which, in its most extreme form, assumes that the facts 'speak for themselves': that they are independent of concepts and theories which are thus supposed to be parasitic upon them. Critics argue that in privileging observations in this way empiricism ignores the concepts and technologies that make its empirical observations possible.

emplotment The process of telling stories such that they describe and make intelligible in a single, cohesive account the circumstances, objectives, interactions and unintended results of the actors involved.

Enlightenment project A tradition of learning that traces its origins to the Enlightenment of the eighteenth century and which places a special emphasis on the political, intellectual and moral virtues that flow from the public exercise of Reason (rather than faith, superstition or revelation); the tradition was not confined to Europe and it was far from homogenous, but in speaking of it as a 'project' the intention is to emphasise its continuation by later generations of philosophers and other thinkers.

environmental determinism A belief that human activities are controlled in decisive ways (rather than merely 'influenced') by the physical environment.

epistemology/epistemological The study of knowledge in its most general form, through which rules are established to identify what is to count as 'true'.

essentialism The idea that objects, social relations or categories of thought are defined by basic, unchanging and sometimes universal characteristics: for example, the claim that the essence of human beings is the ability to use language.

ethnocentrism The assumption – often taken-for-granted rather than explicitly formulated – that one's own way of seeing the world and doing things is 'natural', 'normal' or even universal.

ethnography A predominantly qualitative set of field methods based on first-hand observation of a place and its inhabitants that typically focuses on the representation and interpretation of the webs of meanings within which those people's daily lives are led.

exceptionalism The belief that geography and history are methodologically distinct from other fields of enquiry because they are peculiarly concerned with the study of particular places and unique events; the basic idea was derived from the German philosopher Immanuel Kant, but it was introduced into modern geography during debates over *spatial science*.

flexible accumulation A term derived from the Regulation School of political economy, 'flexible accumulation' describes a late twentieth-century configuration of computer-based production technologies, labour practices and inter-firm linkages directed towards greater flexibility of operation; many critics suggest that such an economic regime is closely connected to social polarization, neo-conservatism and particular versions of postmodern culture (see *postmodernism*).

Fordism A term derived from the mass-production assembly-line system developed by Henry Ford in his Detroit car factory in the early twentieth century, but since applied to the more general system of economic practices and social relations that came to dominate Western Europe and North America from the 1950s to c. 1975; it was oriented towards the mass production and mass consumption of standardised commodities, and it was connected to a compromise between capital and labour negotiated through the welfare state.

genetic methodology Any method which seeks to explain a situation with reference to its historical origins and subsequent developments.

grand theory A term first used by the American sociologist C. Wright Mills to describe (and criticise) attempts to explain the social order by elaborating and manipulating highly abstract concepts with little or no attention to concrete particularities or variations in time and space.

hermeneutics The study of interpretation and the clarification of meaning.

historical materialism The series of propositions that lies at the core of Marxism. While those ideas have undergone considerable development and debate since Marx formulated them, they usually spiral around the importance of the 'material foundation' for social action (hence the central role of political economy in the analysis of contemporary capitalism), the succession of different *modes of production* in human history (hence the importance of historical depth and specificity), and the political possibility of 'making history' (and in particular transcending capitalism) through class struggle.

historicism Most generally, 'historicism' refers to a critical tradition that insists on the importance of historical context to the interpretation of all kinds of texts: thus *historical materialism, post-colonialism* and *post-structuralism* might all be described as 'historicist' on this reading. But 'historicism' is often used in a much more restrictive (and, nowadays, usually negative) sense to describe intellectual traditions that assume human history to have an inner logic, overall design or direction.

historico-geographical materialism The ongoing development of *historical materialism* to place questions of geography alongside questions of history at the centre of its politico-intellectual agenda; this has involved, among others, the critical analysis of productions of space and nature, place and landscape.

human ecology The projection of concepts and methods from ecology to the study of the relations between human beings and their physical environments.

humanism In its most general form 'humanism' usually refers to a belief system that emphasizes the importance of secular human values and choices arrived at through the exercise of an autonomous reason; but it also has a more precise, almost technical meaning that implies a belief in a common humanity and subjectivity – in essential features that are supposed to define all human beings and which thus provide the central terms for all explanations of human action.

humanistic geography An approach to human geography underwritten by the philosophy of *humanism* and distinguished by the central and active role given to 'the human subject': in contrast to *spatial science,* for example, with its special emphasis on spatial patterns and spatial structures, humanistic geography emphasizes the intentions and meanings embedded in human actions and the creative capacities of human agency as these are inscribed within places and landscapes.

hybridity A term used most frequently in *postcolonialism* to describe the complexity of identities in the modern world – never fixed, always ambiguous; never essential, always conjunctural – and to draw particular attention to the ways in which colonialism and imperialism continue to be involved in processes of globalization that reinforce the interpenetration of cultures.

iconography The symbolic analysis of visual images that takes into account the cultural context of their production in time and space.

idealism In its most general form, 'idealism' usually refers to the belief that reality is constituted by the realm of ideas; but in human geography the term has come to have a much narrower meaning, where it refers to supposedly a-theoretical approaches that attempt to understand human action by putting oneself in the place of the original actors and rethinking their thoughts in the light of their knowledges, beliefs and motivations.

identity The constellation of attributes that is associated with a particular subject position – e.g. a black, heterosexual, middle-class woman – and which is the product of various competing *discourses*.

identity politics Forms of political thought and action that are organised around the construction of *identity*, where 'identity' is seen as a complex and often contradictory social construction shaped by class, 'race', gender, sexuality and other markers. Geographers have become particularly interested in the ways in which place and space enter into the construction of identities and subjectivities.

ideology At its most general, 'ideology' usually means a system of meaning and belief, but in a more particular sense it directs attention to the two-way traffic between ideas and the world and hence implies that particular relations of power are inscribed in – and often legitimised through – the articulation of those meanings and beliefs.

idiographic Concepts and approaches directed towards the elucidation of the unique and the particular (cf. *nomothetic*).

imaginative geographies A term derived from the writings of Edward Said to describe not only representations of other places – of peoples, landscapes and cultures – but also, more significantly, the ways in which such representations project the desires, fantasies and preconceptions of their authors and the grids of power between them and their subjects.

intertextuality The process by which meaning is produced through the interaction between texts – through allusions, cross-references, commentaries, etc – rather than through the interaction between texts and the world.

language game The implicit rules and background conditions that animate the practices and conventions necessary for words to take on specific meanings.

lifeworld The familiar meanings and routine practices that make up and mark the sites of ordinary, everyday life.

locality studies A theoretical approach and a research project that originated in Britain in the early 1980s, 'locality studies' drew on both human geography and sociology to explore the intrinsically two-way connections between large-scale processes of economic and social restructuring and localised practices, political struggles and outcomes.

logical positivism A twentieth-century development of the philosophy of *positivism* that claimed that only two kinds of statements were scientifically meaningful: (a) analytical statements, like the propositions of mathematics, which are true by definition (hence 'logical'); and (b) 'synthetic' statements, whose truth has to be established by empirical verification.

logocentrism A term used by the French philosopher Jacques Derrida to describe and criticise the belief in a completely ordered world whose inner secret or logic – the ultimate 'word' or logos – provides the core, foundation and meaning for all thought, language and experience. Derrida argued that this belief, a search for the philosopher's stone or the holy grail, runs throughout the mainstream Western philosophical tradition (see also *deconstruction*; *post-structuralism*).

masculinism/masculinist Approaches and arguments that represent the world of men as the norm; much of this work at least tacitly claims to be comprehensive and inclusive, but feminist critics have shown how it reproduces the privileges of masculine subject-positions in its topics, its concepts, its research practices and its pedagogy (see also *phallocentrism*).

metanarrative A synthesizing framework constructed around a central organizing principle; metanarratives typically describe the world from a single point of view, which is treated as either (a) a privileged claim to truth, which marginalises other points of view, or (b) a universal claim to truth, which erases competing perspectives altogether.

metaphor A rhetorical trope in language in which the characteristics of one, usually familiar object (the primary referent) are transferred to those of another, usually less familar object (the secondary referent).

mode of production A key concept in *historical materialism*, the 'mode of production' describes the different ways in which human societies organize their productive activity, the extraction of surplus value and the reproduction of social life. Each mode of production has its own economic foundation – organized around the labour process and the division of labour – and a distinctive matrix of social, cultural and political relationships, and the whole complex structure is both inscribed within and dependent upon the production of characteristic landscapes and geographies.

modernism A twentieth-century movement in the arts and literature that sought to both criticise and radicalize the cultures of modernity. It was not a homogenous movement, but its main themes and emphases were: (a) the constructedness of all representations, which challenged the idea of art as a direct reflection of the world; (b) simultaneity and juxtaposition, which undercut the power of historical narratives; (c) paradox, ambiguity and uncertainty, which refused the position of a single and coherent point of view; (d) the demise of the centred subject, which disrupted the fiction of the sovereign individual. After the Second World War a different, 'high modernism' emerged, focusing on a universalism – a formal purity and a stark functionalism – that was criticised by the emergence of *postmodernism* in the 1980s.

morphological laws Within *spatial science*, a systematic connection between one spatial pattern and another: if pattern A, then pattern B.

narrative The telling of a story in such a way that heterogeneous elements are integrated and linked together by the very structure of the story itself.

necessary (relations) Relations that must hold because of the very nature of things they connect: e.g. 'teachers' necessarily imply 'students' (cf. *contingent relations*). Approaches informed by *realism* are particularly concerned with identifying necessary relations that constitute 'structures'.

neoclassical economics The mainstream school of economic theory in the UK and North America, which presumes that when rational individuals with perfect information enter the market they will, by pursuing their own freely determined objectives, realize both individual and social optimality.

new cultural geography An approach to human geography that emerged in Britain and North America in the 1980s and which, in contrast to the emphases of the *Berkeley School*, makes explicit use of social, cultural (and, increasingly, psychoanalytical) theory to understand places and landscapes; it has a particular interest in the relations between power, identity and representation and, like the rise of 'cultural studies' more generally, the new cultural geography has been drawn into *post-colonialism*, *post-modernism* and *post-structuralism*.

nomothetic Concepts and approaches directed towards the explanation of the general and the universal (cf. *idiographic*).

objectivity Conventionally an absolutely true representation of the world, uninfluenced by subject-position or location, values or beliefs, prejudices or biases.

ontology A theory which claims to describe what the world is like – in a fundamental, foundational sense – for authentic knowledge of it to be possible (cf. *epistemology*).

patriarchy/patriarchical A social arrangement in which men dominate women, often through their power over key institutions and resources; it is intimately connected to ideologies of *masculinism* and *phallocentrism*.

phallocentric/phallocentrism A diagnosis of *masculinism* that is derived from psychoanalytic theory and hence rooted in the movement of desire within a masculine 'libidinal economy'. Phallocentrism operates by constructing and fixing meanings through a series of binary oppositions that are both hierarchized and sexualized: thus in the couplet Culture-Nature the first term would be privileged (as 'active, masculine') while the second term would be marginalized (as 'passive, feminine').

phenomenology A philosophy founded on the importance of reflecting on the ways in which the world is made intelligible and meaningful. In its original and most radical form, its objective was to establish a philosophy that would recover the fundamental structures that are so deeply taken-for-granted in our attempts to obtain knowledge of the world that they amount to an unproblematic 'natural attitude'; but as phenomenology was transferred into the humanities and the social sciences, so the project focused less on the rigorous and objective disclosure of 'fundamental structures' and much more on the subjective construction of the *lifeworld*.

positivism A philosophy of science founded on *empiricism* but going far beyond its focus on empirical observations as the leading edge of scientific enquiry. Positivism claims to underwrite 'the' scientific method – thereby disavowing other philosophies of science like *realism* – by seeking to generalise about its observations of particular events: empirical testing of hypotheses under carefully specified and controlled conditions is supposed to yield formal scientific laws whose progressive unification will eventually produce a single system of knowledge and truth.

possibilism A doctrine that was formulated to temper the excesses of *environmental determinism*, 'possibilism' claims that the physical environment provides a number of different opportunities (or 'possibilities') for human responses – a bounded space of discretion – and that there is thus considerable scope for individual or group choice.

post-colonialism A critical movement that explores the impact of colonial power on both colonizing and colonized peoples in the past and, of strategic importance, the continued reproduction of colonial relations, practices and *imaginative geographies* in the supposedly 'post-colonial' present; postcolonialism has developed most rapidly in the arts and humanities, where particular attention has been paid to the agency of colonised peoples and their place as historical subjects.

post-Marxism A variety of politico-intellectual approaches that, while critical of *historical materialism* and seeking to go beyond its special focus on class and capitalism, none the less retain important thematic continuities with it: most of all, perhaps, with *Western Marxism*, since post-Marxism has a particular interest in cultural and political questions. Its ideas have informed (and been informed by) both *identity politics* and *radical democracy*.

postmodernism A critical movement that challenges the high *modernism* that dominated Western culture in the decades following the Second World War; postmodernism challenges the closures and certainties of conventional *metanarrative* and celebrates instead heterogeneity and difference: it places considerable emphasis on polyvocality ('multiple voices'), on working between different and often dissonant theoretical traditions, and on experimentation with different forms of representation.

post-structuralism A retrospective label, usually applied by American and British writers, to the work of a divergent group of post-Second World War French philosophers and thinkers that includes Jacques Derrida, Michel Foucault, Julia Kristeva and Jacques Lacan. Post-structuralism radicalizes the fascination with language that characterized structuralism, but insists that the play of language is much less stable: hence Derrida's development of *deconstruction* as a way of calling into question the closures and certainties of texts, and Foucault's disclosure of the grids of power that are at work in *discourse*. Post-structuralism also radicalizes the structuralist critique of humanism and the human subject by drawing attention to the complex and often contradictory constitution of subjectivities (the 'decentring of the subject') and, most obviously through Lacan, the analytical importance of the unconscious.

power The ability to achieve particular ends either directly or indirectly.

production of nature Within *historical materialism*, the 'production of nature' refers to the multifacted transformation of 'first nature' through the operations of capitalism; but recent work on the boundaries of human geography, political ecology and cultural studies challenges the clinical separations between 'nature', 'culture' and 'technology' that remain embedded within this view and pays particular attention to the discursive production of nature.

quantitative and theoretical revolution At once a period and a movement that convulsed geography in Britain, Sweden, North America and Australasia in the late 1950s and 60s; often referred to as simply 'the quantitative revolution', since the introduction of statistical techniques was its most obvious hallmark, the movement also had important theoretical aspirations in its attempt to reconfigure geography as a *spatial science*.

radical democracy A politico-intellectual project most closely associated with the political philosophers Ernesto Laclau and Chantal Mouffe which seeks to challenge conventional models of society as a *totality* and to further the development of new, radical social movements that create a space for a genuinely pluralist politics. Closely associated with *post-Marxism*, radical democracy is based on a recognition of the complexity of social positions and struggles and on the need for strategic alliances in order to widen the traditional – liberal and limiting – conceptualisation of 'the political'.

rational choice theory A theory of individual decision-making which postulates that agents will always make choices in such a way as to attain the most from the least.

realism In human geography, 'realism' usually refers to a philosophy of science that is sharply critical of *empiricism* and *positivism*. It places a premium on conceptual abstraction in order to identify the causal powers contained within structures; those powers are then only concretely realized under contingent conditions.

recursive A continuing mutual process of interaction between two entities such that each is involved in constituting the other: for example, the relation between 'speech' and 'language' is recursive since each is implicated in the other.

reflexivity Self-awareness and self-questioning: the bending back of an idea so that it calls into question its own foundation and legitimation.

relativism The denial of any single philosophical, intellectual or ethical system to have universal validity and an assertion that truth, knowledge and justice are relative to factors that are historically and culturally contingent.

semiology, semiotics The science of the production and meaning of signs, which are the basis of all languages.

signifier/signified The signifier is the sound, image or written mark used to convey meaning (for example, the three letters 'DOG') while the signified is the meaning that is connoted by the signifier (in this case the animal figured by those three letters).

solipsistic/solipsism An extreme form of skepticism that claims that an individual can only be certain of his or her own existence and not about anything else.

spatiality 'Socially produced space'; the term is used in human geography to draw attention to the complex ways in which social life literally 'takes place' and in which social relations and subjectivities are constituted over space. In this view, 'space' is neither the passive stage on which the drama of history unfolds nor the plate on which the positivities of human action and meaning are inscribed: instead it is intimately involved in the making of biographies and histories.

spatial science A school of geography whose diagnostic moment was the *quantitative and theoretical revolution* of the 1960s and 70s and which regards spatial analysis, the modelling of spatial systems and the geometric conceptualisation of spatial structure as central to the discipline.

structural functionalism An approach in sociology and some of the other social sciences that explains the structure of any given society (usually conceptualised as a 'social system') by the functional requirement of system maintenance and survival.

structural Marxism A school of *Western Marxism* closely identified with the French philosopher Louis Althusser. Structural Marxism is fiercely critical of both *empiricism* and *humanism*: it treats capitalism as a totality made up of political, economic and ideological levels, structured in complex ways that do not immediately reduce to an economic determinism, and it pays particular attention to the ways in which human subjects are socially constituted (rather than simply 'given').

structuration (theory) A social theory developed by the British sociologist Anthony Giddens (1938–) and which examines the *recursive* relationship between capable and knowledgeable human agents and the wider social structures and systems in which they are implicated.

subjectivism The idea that the most important vantage point in making judgements is one's own experiences and values.

text *Hermeneutics* treats all cultural practices as intrinsically meaningful texts that can be interpreted: from artefacts to words to landscapes.

time–space compression A term coined by the geographer David Harvey (1939–) to describe the ways in which transportation and communications technologies have increased the speed with which capital, commodities, people and information move around the world, making the world seem (selectively) 'smaller', and so restructuring our experiences and representations of space and time.

totality/totalisation The view that the world is an ordered whole ('totality'), the various components of which are held together in a structured configuration by superordinate forces.

Western Marxism A heterogeneous school of *historical materialism* that developed in Western Europe and North America, not only outside the Soviet Union and the Communist bloc but also in critical opposition to its usually stark and often brutal focus on the economy. Western Marxism paid particular attention to questions of philosophy and to the development of political and cultural theory, and developed dialogues with both *humanism* and *structuralism*; but it has also been criticised for its *ethnocentrism*: its assumption of 'the West' as a privileged vantage point.

INDEX